Omics Approaches for Tomato Yield and Quality Trait Improvement

Muhammad Waseem
Channapatna S. Prakash
Sajid Fiaz • Zhengguo Li
Editors

Omics Approaches for Tomato Yield and Quality Trait Improvement

Springer

Editors
Muhammad Waseem
College of Tropical Agriculture
and Forestry
Hainan University
Haikou, Hainan, China

Sajid Fiaz
Institute of Molecular Biology
and Biotechnology
The University of Lahore
Lahore, Pakistan

Channapatna S. Prakash
College of Arts and Sciences
Tuskegee University
Tuskegee, AL, USA

Zhengguo Li
School of Life Sciences
Chongqing University
Shapingba, Chongqing, China

ISBN 978-981-96-3889-5 ISBN 978-981-96-3890-1 (eBook)
https://doi.org/10.1007/978-981-96-3890-1

This Springer imprint is published by the registered company Springer Nature Singapore Pte Ltd.
The registered company address is: 152 Beach Road, #21-01/04 Gateway East, Singapore 189721, Singapore

Preface

The global need for resilient and high-quality crops has never been more pressing. As climate change, resource constraints, and evolving consumer demands shape agricultural landscapes, the humble tomato has become a focal point for scientific advancement in crop improvement. Tomatoes are a cornerstone in diets worldwide, valued for their versatility, nutritional benefits, and economic significance. However, as with many crops, the tomato faces challenges, including susceptibility to environmental stresses, the need for improved yield, and demands for enhanced nutritional qualities. To address these challenges, scientific breakthroughs in omics technologies—genomics, transcriptomics, proteomics, and metabolomics—have unlocked new avenues for targeted, efficient crop enhancement.

This book, *Omics Approaches for Tomato Yield and Quality Trait Improvement*, is a compilation of state-of-the-art research and practical applications that leverage omics technologies to advance tomato breeding and production. In assembling these chapters, our goal was to bridge the latest molecular research with real-world agricultural needs. Each chapter explores a critical aspect of tomato improvement, from enhancing resilience to environmental stressors to optimizing yield, quality, and shelf life. By employing techniques such as molecular breeding, marker-assisted selection (MAS), and advanced genome editing, we have the tools to address these objectives with precision and sustainability.

The chapters in this book provide a comprehensive overview of omics applications tailored to the diverse needs of modern tomato production. Beginning with a foundational exploration of omics in agriculture, we delve into specific strategies for combating abiotic stresses, essential for climate-proofing tomato production. The book then explores nutrigenomics and metabolomics, detailing approaches to enrich tomato health benefits and extend post-harvest quality. With a focus on genomic insights and precision editing through CRISPR/Cas9, this book examines the latest tools that hold promise for specific trait improvements, from stress tolerance to enhanced flavor and nutrition. Other chapters address the integration of omics data into breeding programs, such as through marker-assisted selection and genomic selection (GS). These chapters offer insights into sustainable, eco-friendly tomato production that align with the evolving needs of both consumers and the environment. As we look to the future, this book also examines emerging technologies and the potential challenges on the horizon for tomato breeders and researchers alike.

This book is a reflection of the collective efforts of numerous researchers, students, and agricultural innovators dedicated to advancing tomato crop science. We are deeply grateful to all who contributed to this work and to those whose insights and encouragement made this publication possible. It is our hope that this book will serve as a resource for scientists, students, and agricultural practitioners, inspiring further exploration into the promising field of omics-based crop improvement. In these pages, we glimpse the future of sustainable and resilient agriculture, where science empowers us to cultivate crops that can thrive under changing conditions, delivering not only yield but quality and health benefits. We are confident that the knowledge shared here will contribute to impactful advancements in tomato breeding, benefiting growers, consumers, and the planet alike.

Haikou, Hainan, China — Muhammad Waseem
Tuskegee, AL, USA — Channapatna S. Prakash
Lahore, Pakistan — Sajid Fiaz
Shapingba, Chongqing, China — Zhengguo Li

Contents

Editors and Contributors

About the Editors

Muhammad Waseem is Associate Professor/Researcher at College of Tropical Agriculture and Forestry, Hainan University, Haikou, Hainan, China. He received a doctoral degree (PhD, specialization in Plant Molecular Biology) from the School of Life Sciences, Chongqing University, Chongqing, PR China. He also has four years of postdoctoral research experience. His interests include transcriptional regulation of flowering time regulation, fruit ripening and development, and abiotic stresses such as salinity and drought. In addition, he has hands-on experience in next-generation sequencing data analysis, transcriptomics, and data visualization. Currently, Dr. Waseem is actively investigating flowering time regulation in *Brassica* crops. He has been awarded with a National Science Foundation Project, Research Fund for International Young Sciencetist (RFIS). In brief, he has published more than 56 research articles, 5 review articles, and 9 book chapters in various high-impact factor publications including *New Phytologist*, *International Journal of Macromolecules*, *Planta*, *GM Crops & Food*, *Environmental and Experimental Botany*, *Physiologia Plantarum*, and *Frontiers in Plant Science*. He has recently edited a book titled "ncRNAs: Mediated Regulation: The Unarguable Biological Function to Environmental Constrains" with Springer.

Channapatna S. Prakash is Dean of the College of Arts and Sciences at Tuskegee University, USA, and a Professor of crop genetics. Dr. Prakash's research expertise is on genetic improvement research on food crops of importance to developing countries. His lab was among the first to develop transgenic sweet potato and peanut plants and conduct pioneering genomic studies on the peanut. Dr. Prakash has been a global leader in enhancing the societal awareness of crop and food biotechnology issues, and was recognized for his outstanding work on agricultural biotechnology outreach with 2015 Borlaug CAST Communication Award, by the Council of Agricultural Science and Technology, which credited him as "arguably done more than anyone else in academia or industry to promote agricultural technologies that can help feed the world's growing population." He also serves as the Editor-in-Chief of the journal *GM Crops & Food*.

Sajid Fiaz, PhD is an Assistant Professor working at the Institute of Molecular Biology and Biotechnology, the University of Lahore, Pakistan. Dr. Fiaz received his PhD in Crop Genetics and Breeding from Chinese Academy of Agricultural Sciences. Dr. Fiaz's research interests include marker-assisted selection for agronomic traits, mutation breeding for biotic and abiotic stress resistance, Quantitative Trait Loci (QTL) mapping, genome-wide association analysis, and genome editing for yield and quality traits in cereals. He has been awarded with a Start-Up Research Grant Project from Higher Education Commission of Pakistan worth 1.0 million PKR. He has published more than 160 research, 10 review articles, and 13 book chapters with prestigious scientific journals like *International Journal of Molecular Sciences*, *Frontiers in Plant Sciences*, *Plant Physiology and Biochemistry*, and *Saudi Journal of Biological Sciences*. He is working as academic editor for *PLOS ONE*, guest associate editor for *Frontiers in Plant Sciences*, review editor for *Frontiers in Genetics*, and editorial board member for *GM Crops & Food*. He has recently edited books, *Principle and Practices of OMICS and Genome Editing for Crop Improvement*, *Legumes Biofortification*, *Sustainable Agriculture in the Era of the OMICs Revolution*, and *OMICs-Based Techniques for Global Food Security*, with international publishers. Currently, his lab group is working to identify genetic factors controlling drought stress tolerance in exotic and local landraces of rice.

Zhengguo Li is a highly respected professor at Chongqing University, with an academic background that includes studies at Southwest University (1984–1997) and the Institut Polytechnique de Toulouse (1997–2002). Since 2002, he has been a faculty member at Chongqing University. His research interests are broad, covering plant hormone signal transduction mechanisms, the regulation of fruit and vegetable development and senescence, genetic engineering for plant stress resistance, and food biotechnology. Dr. Li has held leadership positions, including serving as Vice Chairman for several professional organizations such as the Chinese Horticultural Society, Chongqing Plant Society, Chongqing Horticulture Society, and the Chongqing Nutrition Society. He is also the Deputy Director of the Food and Environment Branch of the Chongqing Analytical Testing Society. Over the years, Dr. Li has published more than 120 research papers, with over 100 of them appearing in prestigious SCI journals such as *Plant Cell*, *Plant Journal*, *New Phytologist*, and *Plant Biotechnology Journal*. He has also authored 7 books, secured 10 national invention patents, and made significant contributions to research and education. Dr. Li has mentored numerous students, including 10 postdoctoral researchers, over 30 PhD students, and more than 60 master's students, solidifying his influence on the development of future scholars in the field of plant science.

Contributors

Banzeer Ahsan Abbasi Department of Botany, Rawalpindi Women University, Satellite Town, Rawalpindi, Pakistan

Aqsa Abid Department of Biochemistry and Biotechnology, The Women University Multan, Multan, Pakistan

Muhammad Sabtain Abid Department of Biosciences, COMSATS University Islamabad, Sahiwal Campus, Sahiwal, Pakistan

Shabbir Ahmad Faculty of Food and Home Sciences, Muhammad Nawaz Shareef University of Agriculture, Multan, Pakistan

Shumaila Ahmad Department of Plant Sciences, Faculty of Biological Sciences, Quaid-i-Azam University, Islamabad, Pakistan

Yumna Ahmad Department of Plant Sciences, Faculty of Biological Sciences, Quaid-i-Azam University, Islamabad, Pakistan

Mazhar Ali Department of Biotechnology, COMSATS University Islamabad, Vehari Campus, Vehari, Pakistan

Muhammad Amjad Ali Department of Plant Pathology, University of Agriculture, Faisalabad, Pakistan

Abhishek Anand The Neotia University, Sarisha, West Bengal, India

Saima Mir Arain Nuclear Institute of Agriculture (NIA), Tando Jam, Pakistan

Sadaf Arbab Department of Botany, Bacha Khan University, Charsadda, Khyber Pakhtunkhwa, Pakistan

Areeba Aslam Faculty of Food and Home Sciences, Muhammad Nawaz Shareef University of Agriculture, Multan, Pakistan

Aliza Batool Faculty of Food and Home Sciences, Muhammad Nawaz Shareef University of Agriculture, Multan, Pakistan

Aliza Erum Faculty of Food and Home Sciences, Muhammad Nawaz Shareef University of Agriculture, Multan, Pakistan

Muhammad Faheem Nuclear Institute of Agriculture (NIA), Tando Jam, Pakistan

Ayesha Faiz Centre of Agricultural Biochemistry and Biotechnology (CABB), University of Agriculture, Faisalabad, Pakistan

Abu Bakr Umer Farooq Department of Biotechnology, COMSATS University Islamabad, Vehari Campus, Vehari, Pakistan

Umar Farooq Faculty of Food and Home Sciences, Muhammad Nawaz Shareef University of Agriculture, Multan, Pakistan

Amin Fathi Department of Agronomy, Ayatollah Amoli Branch, Islamic Azad University, Amol, Iran

Kinza Fatima Integrative Omics and Molecular Modeling Laboratory, Department of Bioinformatics and Biotechnology, Government College University Faisalabad (GCUF), Faisalabad, Pakistan

College of Natural & Agricultural Sciences, University of California, Riverside, USA

Nida Firdous Faculty of Food and Home Sciences, Muhammad Nawaz Shareef University of Agriculture, Multan, Pakistan

Zubaria Haakim Department of Plant Sciences, Faculty of Biological Sciences, Quaid-i-Azam University, Islamabad, Pakistan

Shuvadeep Halder The Neotia University, Sarisha, West Bengal, India

Ehtisham Hussain Centre of Agricultural Biochemistry and Biotechnology (CABB), University of Agriculture, Faisalabad, Pakistan

Shumaila Ijaz Department of Burn and Plastic Surgery, Shenzhen Institute of Translational Medicine, The First Affiliated Hospital of Shenzhen University, School of Biomedical Engineering, Shenzhen University Medical School, Shenzhen University, Shenzhen, China

Madiha Imtiaz Department of Botany, Bacha Khan University, Charsadda, Khyber Pakhtunkhwa, Pakistan

Safeena Inam Functional Genomics and Bioinformatics Labs, National Institute for Genomics and Advanced Biotechnology (NIGAB), National Agricultural Research Centre, Islamabad, Pakistan

Javed Iqbal Department of Botany, Bacha Khan University, Charsadda, Khyber Pakhtunkhwa, Pakistan

Rashid Iqbal Department of Agronomy, Faculty of Agriculture and Environment, The Islamia University of Bahawalpur, Bahawalpur, Pakistan

Department of Life Sciences, Western Caspian University, Baku, Azerbaijan

Zafar Iqbal Punjab Food Authority, Lahore, Pakistan

Samra Irum Functional Genomics and Bioinformatics Labs, National Institute for Genomics and Advanced Biotechnology (NIGAB), National Agricultural Research Centre, Islamabad, Pakistan

Monika Jain The Neotia University, Sarisha, West Bengal, India

Fareena Jamil Faculty of Food and Home Sciences, Muhammad Nawaz Shareef University of Agriculture, Multan, Pakistan

Hafiz Muhammad Rashad Javeed Department of Environmental Sciences, COMSATS University Islamabad, Vehari Campus, Vehari, Pakistan

Faiz Ahmad Joyia Centre of Agricultural Biochemistry and Biotechnology (CABB), University of Agriculture, Faisalabad, Pakistan

Md Kamaruzzaman The Neotia University, Sarisha, West Bengal, India

Sobia Kanwal Department of Biology and Environmental Sciences, Allama Iqbal Open University, Islamabad, Pakistan

Asma Khalil Integrative Omics and Molecular Modeling Laboratory, Department of Bioinformatics and Biotechnology, Government College University Faisalabad (GCUF), Faisalabad, Pakistan

Salman Khalil Faculty of Food and Home Sciences, Muhammad Nawaz Shareef University of Agriculture, Multan, Pakistan

Muhammad Zain Farid Khan Functional Genomics and Bioinformatics Labs, National Institute for Genomics and Advanced Biotechnology (NIGAB), National Agricultural Research Centre, Islamabad, Pakistan

Muhammad Ramzan Khan Functional Genomics and Bioinformatics Labs, National Institute for Genomics and Advanced Biotechnology (NIGAB), National Agricultural Research Centre, Islamabad, Pakistan

Muhammad Sarwar Khan Centre of Agricultural Biochemistry and Biotechnology (CABB), University of Agriculture, Faisalabad, Pakistan

Shoaib Khan Department of Biology and Environmental Sciences, Allama Iqbal Open University, Islamabad, Pakistan

Department of Chemistry, Abbottabad University of Science and Technology (AUST), Abbottabad, Pakistan

Urooj Khan Department of Biotechnology, COMSATS University Islamabad, Vehari Campus, Vehari, Pakistan

Khalil Ahmad Laghari Nuclear Institute of Agriculture (NIA), Tando Jam, Pakistan

Tariq Mahmood Department of Plant Sciences, Faculty of Biological Sciences, Quaid-i-Azam University, Islamabad, Pakistan

Muhammad Majeed Department of Botany, University of Gujrat, Gujrat, Punjab, Pakistan

Shehzad Mehmood Department of Biotechnology, COMSATS University Islamabad, Vehari Campus, Vehari, Pakistan

Lubna Anjum Minhas Department of Plant Sciences, Faculty of Biological Sciences, Quaid-i-Azam University, Islamabad, Pakistan

Muhammad Mubeen Department of Biotechnology, COMSATS University Islamabad, Vehari Campus, Vehari, Pakistan

Amna Muhammad Functional Genomics and Bioinformatics Labs, National Institute for Genomics and Advanced Biotechnology (NIGAB), National Agricultural Research Centre, Islamabad, Pakistan

Abdul Samad Mumtaz Department of Plant Sciences, Faculty of Biological Sciences, Quaid-i-Azam University, Islamabad, Pakistan

Ghulam Murtaza School of Agriculture, Yunnan University, Kunming, Yunnan, China

Ghulam Mustafa Centre of Agricultural Biochemistry and Biotechnology (CABB), University of Agriculture, Faisalabad, Pakistan

Sarita Pandey The Neotia University, Sarisha, West Bengal, India

Aqsa Parvaiz Department of Biochemistry and Biotechnology, The Women University Multan, Multan, Pakistan

Kousar Parveen Centre of Agricultural Biochemistry and Biotechnology (CABB), University of Agriculture, Faisalabad, Pakistan

Sobia Parveen Department of Biotechnology, COMSATS University Islamabad, Vehari Campus, Vehari, Pakistan

Maria Qubtia Department of Biotechnology, COMSATS University Islamabad, Vehari Campus, Vehari, Pakistan

Muhammad Asif Raheem Centre of Agricultural Biochemistry and Biotechnology (CABB), University of Agriculture, Faisalabad, Pakistan

Vidadala Rajendra M.S. Swaminathan School of Agriculture, Centurion University, R. Sitapur, Odisha, India

Iqrar Ahmad Rana Centre of Agricultural Biochemistry and Biotechnology (CABB), University of Agriculture, Faisalabad, Pakistan

Nazia Rehman Functional Genomics and Bioinformatics Labs, National Institute for Genomics and Advanced Biotechnology (NIGAB), National Agricultural Research Centre, Islamabad, Pakistan

Muhammad Sadaqat Integrative Omics and Molecular Modeling Laboratory, Department of Bioinformatics and Biotechnology, Government College University Faisalabad (GCUF), Faisalabad, Pakistan

UMR CNRS 6553 Ecosystèmes, Biodiversité, Evolution (ECOBIO), Université de Rennes 1, Rennes, France

Muhammad Saeed Centre of Agricultural Biochemistry and Biotechnology (CABB), University of Agriculture, Faisalabad, Pakistan

Muhammad Sajjad COMSATS Institute of Information Technology, Islamabad, Pakistan

Ammara Saleem Institute of Botany, University of the Punjab, Lahore, Pakistan

Mazhar Saleem Department of Biotechnology, COMSATS University Islamabad, Vehari Campus, Vehari, Pakistan

Shobana Sampath Department of Biotechnology, Vel Tech Rangarajan Dr. Sagunthala R&D Institute of Science and Technology, Chennai, India

Hamna Saqib Faculty of Food and Home Sciences, Muhmmad Nawaz Shareef University of Agriculture, Multan, Pakistan

Amna Sayal Faculty of Food and Home Sciences, Muhmmad Nawaz Shareef University of Agriculture, Multan, Pakistan

Aneela Shadab Department of Biochemistry and Biotechnology, The Women University Multan, Multan, Pakistan

Centre of Agricultural Biochemistry and Biotechnology (CABB), University of Agriculture, Faisalabad, Pakistan

Tayyaba Shahzadi Department of Chemistry, Government College for Women, University Sialkot, Sialkot, Pakistan

Mahboob Ali Sial Nuclear Institute of Agriculture (NIA), Tando Jam, Pakistan

Muhammad Sibt-e-Abbas Faculty of Food and Home Sciences, Muhammad Nawaz Shareef University of Agriculture, Multan, Pakistan

Tahira Siddique Islamabad Food Authority, Islamabad, Pakistan

Mishal Subhan Department of Biochemistry and Biotechnology, The Women University Multan, Multan, Pakistan

Muhammad Tahir Department of Biotechnology, COMSATS University Islamabad, Vehari Campus, Vehari, Pakistan

Muhammad Tahir ul Qamar Integrative Omics and Molecular Modeling Laboratory, Department of Bioinformatics and Biotechnology, Government College University Faisalabad (GCUF), Faisalabad, Pakistan

Ghulam Ullah Nuclear Institute of Agriculture (NIA), Tando Jam, Pakistan

Zahid Ullah Department of Botany, Bacha Khan University, Charsadda, Khyber Pakhtunkhwa, Pakistan

Zakir Ullah Department of Plant Sciences, Faculty of Biological Sciences, Quaid-i-Azam University, Islamabad, Pakistan

Muhammad Usman Faculty of Food and Home Sciences, Muhammad Nawaz Shareef University of Agriculture, Multan, Pakistan

Department of Plant Pathology, University of Agriculture, Faisalabad, Pakistan

Muhammad Uzair Functional Genomics and Bioinformatics Labs, National Institute for Genomics and Advanced Biotechnology (NIGAB), National Agricultural Research Centre, Islamabad, Pakistan

Rooma Waqar Department of Plant Sciences, Faculty of Biological Sciences, Quaid-i-Azam University, Islamabad, Pakistan

Tabassum Yaseen Department of Botany, Bacha Khan University, Charsadda, Khyber Pakhtunkhwa, Pakistan

Saba Zafar Department of Biochemistry and Biotechnology, The Women University Multan, Multan, Pakistan

Kainat Zahra Department of Biochemistry and Biotechnology, The Women University Multan, Multan, Pakistan

Farishta Zarshan Department of Botany, Bacha Khan University, Charsadda, Khyber Pakhtunkhwa, Pakistan

"Omics" Applications for Tomato Production

1

Aliza Batool, Umar Farooq, Nida Firdous, Hamna Saqib, Zafar Iqbal, Tahira Siddique, and Amna Sayal

Abstract

In agricultural techniques, sustainability by design seeks a balance between environmental direction and productivity. Integrating "Omics" technologies such as genomics, proteomics, transcriptomics, and metabolomics presents a groundbreaking method for attaining environmentally friendly cultivation in tomato production. By applying these advanced techniques, scientists can understand the physiological, biochemical, and genetic processes that promote tomato development, stress reaction, and effective nutrient utilization. This research decreases the need for chemical inputs by allowing the production of tomato varieties that are high-yielding and resistant to pests, illnesses, and climate variability. Applications such as "Omics" help precision agriculture techniques by maximizing the use of resources including water, fertilizers, and energy. Modern science and environmentally friendly farming methods can work together to improve tomato production's environmental impact while maintaining the quality and security of the food supply. This chapter explores the "Omics" technologies that are strategically implemented in the breeding and alteration of genetic material of plants to enhance the quality and quantity, study the DNA sequencing to change the structure of tomatoes functionally and chemically, and also help increase the production processes, highlighting their critical significance in pro-

A. Batool · U. Farooq · N. Firdous (✉) · H. Saqib · A. Sayal
Faculty of Food and Home Sciences, Muhmmad Nawaz Shareef University of Agriculture, Multan, Pakistan
e-mail: nida.firdous@mnsuam.eud.pk

Z. Iqbal
Punjab Food Authority, Lahore, Pakistan

T. Siddique
Islamabad Food Authority, Islamabad, Pakistan

M. Waseem et al. (eds.), *Omics Approaches for Tomato Yield and Quality Trait Improvement*, https://doi.org/10.1007/978-981-96-3890-1_1

moting sustainable agricultural ecosystems and creating environmentally friendly surroundings.

Keywords

Abiotic stress · DNA sequencing · Physiological · QTLs · CRISPR

1.1 Introduction

In this chapter, Omics technologies and their effective utilization in the field of food technology for enhancing and sustaining agricultural production are discussed. The main focus of this chapter is the enhancement of the production of tomatoes with the use of Omics technologies.

1.2 Overview of Sustainable Agriculture

By 2055, there will be more people on the planet than the current 7.8 billion. Food supply faces significant challenges as a result of this fast population growth. On the one hand, more grains are required to supply the essential calories for humans. However, as living standards rise, people's diets are shifting to include more cattle and dairy products on average, particularly in developing nations. Therefore, to close the gap between food production and demand, agricultural yields must be increased. In the meantime, food nutrition is becoming increasingly important to support modern, industrialized lives (Groen et al. 2020). Another significant issue is the volatility of food production brought on by climate change. The Earth's temperature has increased by more than one degree since 1880, and in recent decades, the rate of global warming has accelerated. Extreme climate change, such as high temperatures, droughts, and floods, has grown more common (Kwon et al. 2020). Future crops will need to adjust to this novel and predictably changing environment. Since insects and plant diseases are predicted to be impacted by climate change, crop types resistant to biotic stressors are also required. More crucially, we require a food production system that can meet long-term development needs as well as societal demands at the same time. Grower has been mostly reliant on heavy inputs of pesticides and nitrogen since the 1960s' Green Revolution. Pollution of the environment results from this, which is not sustainable in the long term (Fernie and Yan 2019). To enable sustainable agriculture, a new breeding program is therefore desperately needed. This program should include new techniques for creating crop varieties and varieties with high potential, high yield stability, and superior grain quality and nutrition. However, it should also take environmental protection into account by using less water, fertilizers, and chemicals (Li and Yan 2020).

1.3 Significance of Tomatoes

Tomatoes (*Solanum lycopersicum*) are an excellent source of minerals, iron, potassium, lycopene, folate, and vitamin C. Tomatoes include phenolic components such as flavonoids, hydroxycinnamic acid, chlorogenic acid, homovanillic acid, and ferulic acid in addition to lycopene and vitamin C. With their nutritional worth intact, tomatoes can be eaten raw or cooked and play a significant role in a balanced diet. More than 80% of tomatoes that are farmed commercially are used to make processed foods like ketchup, soup, and juice. Because of their high antioxidant content, tomatoes and dishes made with them are known to offer several health benefits. Tomato farming is a major industry, with predicted global production of 182 million tons in 2018 and 186 million tons in 2020. It is well recognized that factors affecting plant growth and metabolism, such as the availability of water, can affect yield. Temperature, salinity, pollutants, and water availability are the main growth-limiting variables for crops. Many aspects of tomato growing can be controlled with greenhouse systems, including humidity, temperature, and light intensity. The amount of water available has an impact on tomato farming, which is increasingly using plant fertigation in conjunction with a drip irrigation system. These systems are useful for applying regulated irrigation as well as for consistently and dependably watering tomatoes. These systems help with both the consistent and dependable watering of tomatoes as well as the application of a regulated amount of fertilizers at specific intervals during the growth stage (Collins et al. 2022).

1.4 Omics Technologies

Even though we are confronting difficulties, there is a lot of amazing potential as well, particularly given the rapidly advancing field of omics technologies (Luo et al. 2020). For thousands of species, high-quality reference genomes are becoming accessible; some species have multiple reference genomes. The identification of core- and pan-genomes is made possible by genome-wide resequencing of many kinds. This deepens our comprehension of crop improvement and domestication (Li and Yan 2020). Numerous layers, including but not limited to genomes, epigenomes, transcriptomes, epi-transcriptomes, and proteomes, are producing omics data. Numerous omics data that were previously generated from tissues with a single genotype are now being extended across both micro- and macro-scales (Xu et al. 2019). Additionally, omics data are being produced at a reasonable cost. It is crucial to remember that a large number of matching informatics tools are being created quickly to extract biological significance from omics data (Groen et al. 2020).

1.5 Genomics

Genomic techniques focus on analyzing interindividual differences at the germline and somatic levels by sequencing the genome of interest. The development of DNA microarray technology; first-generation Sanger sequencing; second-generation massively parallel sequencing, also known as next-generation sequencing (NGS); and eventually, third-generation long reads sequencing (TGS) have made it possible to sequence an entire genome or exome with enough depth to characterize the mutational landscape of a given sample (Dai and Shen 2022).

1.6 Transcriptomics

Unlike the genome, the transcriptome is dynamic and composed of a wide range of players. Changes brought about by the stage of cell development, internal and/or external stimuli, and the signals' measurement duration all affect it. Although mRNA (circRNA), long noncoding RNAs (incRNAs), and microRNAs are commonly referred to as "transcriptomes," identifying and quantifying the RNA molecules being transcribed from a particular genome at a given time is the aim of transcriptomics techniques (Dai and Shen 2022).

1.7 Proteomics

Proteomics examines the information flow via protein signaling to determine the functional significance of every expressed protein in a cell, tissue, or organism. Since proteins are responsible for the majority of biological functions, it is critical to accurately assess changes in the proteome when cells undergo state transitions, such as during the development of cancer (Dai and Shen 2022).

1.8 Metabolomics

Metabolites, or low molecular weight biomolecules (<1500 Da) involved in the endogenous metabolism of cells, are metabolic intermediaries that control proteins, provide energy, and function as signaling molecules in complex biological systems. It has been shown that metabolites play a significant role in multilayer omics as biological modulators that support cellular homeostasis. The term "metabolome" refers to the entirety of metabolites within a cell, which includes all biomolecules except for the metals, transcriptome, proteome, and genome (Dai and Shen 2022) (Fig. 1.1).

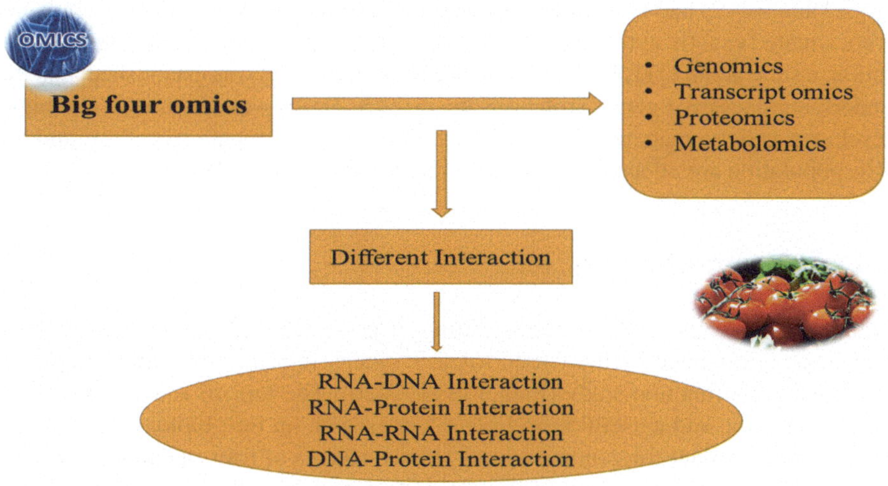

Fig. 1.1 Advancement and different technologies in Omics

1.9 Application of Genomics in Tomato Production

Plants frequently experience biotic and abiotic stress reactions due to fluctuating climate conditions. In many crop plants, the main factors influencing crop production and nutritional quality are abiotic stressors. Thanks to developments in high-throughput techniques and genome sequencing, scientists may now utilize genome editing tools to functionally characterize a large number of genes that could be improved for crops (Salava et al. 2021).

1.10 Genetic Mapping

A highly saturated genetic linkage map of a tomato (*Solanum lycopersicum* L.) recombinant inbred line (RIL) population was created by using genotyping-by-sequencing (GBS). The RIL population was generated by mapping the cultivar NC EBR-1 with the wild tomato *S. pimpinellifolium* L. accession LA2093 (Rowan et al. 2015). A setup was generated to convert single nucleotide polymorphism (SNP) input into genomic bins, allowing for the recognition of candidate genes and fine-tuning of quantitative trait loci (QTL). To create genomic bins, the pipeline uses a hidden Markov model technique to determine recombination breakpoints. It is used in the SNPbinner Phyton script (Ichihashi et al. 2016). The overall length of the newly generated high-resolution genetic map was 1.2 times more than that of the previous map that was produced using restriction fragment length polymorphism (RFLP) and polymerase chain reaction. The map was used to boost and ensure two fruit quality indicators that had previously been mentioned for the RIL population: fruit weight (FW) and fruit lycopene content (LYC). Two well-described FW QTL (fw2.2 and fw3.2) had QTL intervals that were slashed by two to ten times, and they were situated exactly at the known causative genes that underlie them. At high resolution, a substantial QTL for LYC concentration (Lyc 12.1) was established, and the underlying causative gene carotene isomerase (SlZISO) was identified. The high-resolution genetic map, the user-friendly SNPbinner genotyping workflow, and the RIL population are all available to the public (Gonda et al. 2019).

1.11 Genome Editing

Plants are currently using genome editing, a groundbreaking molecular biology technique symbolized by the CRISPR/Cas9, extensively for agricultural improvement and gene function study (Maher et al. 2020). The tomato is a key source of human nutrition and a useful model plant for research on fruit biology, making it a highly relevant plant for genome editing. Many aspects of fruit ripening need to be reexamined, according to the reevaluation of tomato genes crucial for fruit ripening by CRISPR/Cas9-mediated targeted mutagenesis (Murovec et al. 2018). Additionally, genome editing has been used in tomato breeding to increase fruit quality and productivity, enhance resilience to stress, hasten the domestication of wild tomatoes, and most recently, change tomato cultivars for urban cultivation. Furthermore, genome editing is constantly developing, and several innovative methods, like prime editing, a recent advancement in precise genome editing, have been used in plants (Xia et al. 2021).

1.12 Breeding Program

Genomic selection (GS) is a predictive method aimed at reducing the generation gap and increasing the rate of genetic gain per unit of time by the use of genome-wide markers in breeding operations (Hao et al. 2019). It has been emerged as a valuable method for improving complex aspects regulated by numerous genes with small effect gene regulation GS enables the prediction of a candidate genotype breeding value for selection (Esposito et al. 2019). We cover important challenges associated with GS and its uses in the plant setting, with a particular focus on tomato breeding. It is important to carefully consider the genomic restrictions and crucial factors influencing the prediction accuracy, such as the number of markers, statistical model, phenotyping and trait complexity, training population size, and composition. Additionally, covered is a comparison of GS techniques that makes it easier to choose superior genotypes of tomatoes for breeding initiatives. It has already been demonstrated that GS may be used in tomato breeding (Gao et al. 2019). This study showed how the GS schemes are beginning to take shape, how computer science can aid in future selection strategies, and how GS can boost the rate of gain in backcross and descendent systems, in addition to elite line selection. A fresh and innovative breeding framework is beginning to take shape to enhance tomato development methods (Cappetta et al. 2020) (Fig. 1.2).

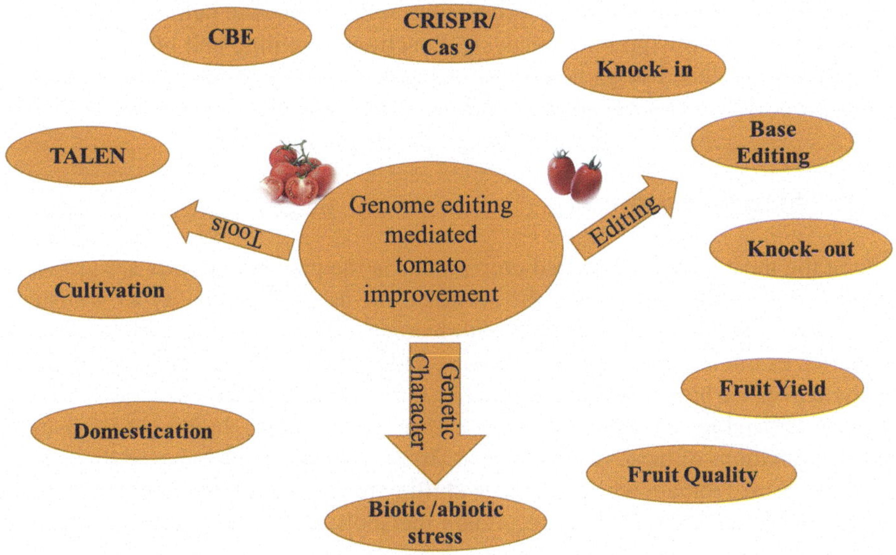

Fig. 1.2 Genome editing–mediated tomato improvement

1.13 Role of Transcriptomics in Tomato Production

Many biotic and abiotic stresses negatively impact plant development and productivity all around the world. Understanding tolerance mechanisms cannot be gained by examining individual genes since these stresses happen frequently, often in combination with one another, and because many genes are involved in these systems. The identification of transcription factors (TFs) in regulatory networks has been made feasible by high-throughput genomic data (Ashrafi-Dehkordi et al. 2018).

1.14 Gene Expression Analysis

Transcriptomic studies frequently employ gene expression data analysis to comprehend the roles and relationships of individual molecules within a cell . Transcriptomic studies employ gene expression data analysis extensively to comprehend molecular activities and interactions within cells. Differential co-expression analysis identifies gene modules whose co-expression patterns differ between situations to study illnesses and phenotypic differences (Zhu et al. 2018). They examine the optimal methods for analyzing gene expression data, taking into account comparisons and microarray and RNA-seq data, and analyzing co-expression, co-expression networks, differential networking, and differential connectedness. They draw attention to challenges in RNA-seq data analysis by employing techniques designed for microarrays (Gong et al. 2018). Throughout the study, they go over the instruments required for gene expression analysis. Furthermore, we clarified the analysis of scRNA-seq data by integrating preprocessing, scRNA-seq, and helpful tools unique to scRNA-seq in co-expression analysis. Functional profiling and biological interpretation are included to gain insights. Last, they offer analyst recommendations and research difficulties and obstacles that need to be solved (Chowdhury et al. 2019).

1.15 Stress Responses

One of the most widely cultivated crop plants on the planet, tomatoes are grown in greenhouses and fields all over the world. Cultivated crops' ability to withstand stress can be improved by breeding with recognized wild species qualities. In this work, we examined how the transcriptome and primary and secondary metabolites in the leaves of both wild and cultivated tomatoes responded to various abiotic stressors, including low nitrogen, cold or warm temperatures, high light intensities, and combinations of these (De Buck et al. 2020). The wild type behaved differently from the farmed tomato when it came to shifting temperature conditions. A nitrogen deficit specifically triggered the secondary metabolism in both species, causing the largest reactions; the cultivated tomato was more severely impacted. Their study supports the idea of adding advantageous secondary metabolites to green waste products from horticultural production, which would not ordinarily be composted until after fruit harvest (Wenger and Stern 2019). More specifically, the cultivated tomato showed a substantial induction in the mono-caffeoylquinic acid group in response to nitrogen constraint. Additionally, by contrasting how produced and wild tomatoes respond to stress, it is possible to find new breeding targets for enhanced stress tolerance (Reimer et al. 2021).

1.16 Improving Yield and Quality

Due to worldwide overpopulation, improving plant quality traits has become imperative. For example, one of the primary goals of modern breeding projects is to produce crop species with improved nutrients and chemicals that promote health. However, rather than breeding for crop quality, which primarily depends on the presence and accumulation of highly valuable metabolites in the plant's edible parts, breeders have historically focused on traits like yield or pest resistance due to the complexity of the plant metabolome and the challenges associated with accurately phenotyping it. The remarkably diverse plant metabolome may now be efficiently uncovered by using high-throughput transcriptomics, metabolomics, and genomics platforms, which are examples of modern technical advancements. Moreover, they allow us to determine the relationship between genotype and metabolite composition and unravel the genetic architecture of complex biochemical pathways, such as the accumulation of secondary metabolites in plants, many of which are critical components of human meals. In this study, they emphasize the significance of combining transcriptomic, metabolomic, and genomic methods to select crop varieties with better quality attributes and nutritional value (Pott et al. 2021).

1.17 Proteomics for Enhanced Tomato Production

The analysis of global gene products in diverse tissues and cell physiological states is made possible by proteomics. Proteomics has grown to be one of the main fields for studying functional genomics as a result of advances in protein mapping and genomic sequencing. Developing crops that can tolerate harsh environments requires an understanding of plant stress response. In plant resistance and response to abiotic stress, proteomics plays an important role. Proteomics has many efficient metabolic pathways that can help treat these conditions. Many techniques like phosphoproteomics can help modify and regulate the metabolic system. Tomato is used as the model crop in the molecular study and *Solanaceae* genetic due to its abundance of genetic resources and can easily diploidize genetic transformation. Many more studies and research are required to introduce effective methods related to tomatoes and in the proteomics field (Sant'Ana and Lefsrud 2018).

1.18 Protein Profiling

By modulating Nicotinamide adenine dinucleotide (phosphate) NAD(P) metabolism, it focused on tomato fruit growth and found that the pyridine nucleotides are regulators of mitochondrial and central metabolism. The working hypothesis was based on the regulation of the pyridine nucleotide metabolism by the transcript and protein profiling during fruit development in tomatoes. NAD(P) has strong effects on fruit growth phases, redox homeostasis, and energy supply. Further studies on the molecular mechanisms of nicotine metabolism and enzyme activities concerning Nicotinamide adenine dinucleotide NAD-related pathways in fruit growth are necessary. On the other hand, a better understanding of NAD metabolism in agriculture could mean higher yields and the production of better fruits in terms of quality. Several research works have proven that ripening is a consequence of metabolic processes, enhancing the sweetness and nutritional value of the fruit through soluble sugar and amino acid accumulation from proteolysis and starch breakdown. They opine that further study is necessary to understand the molecular and biochemical mechanism by which the pyridine nucleotides regulate fruit growth (Decros et al. 2019).

1.19 Posttranslational Modification in Tomato Production

Redox proteome studies have unraveled tomato fruit ripening as a posttranslational oxidative modification process. Fruit ripening is a complex process involving oscillations in the levels of reactive oxygen species. Thirteen redox-sensitive proteins, including those related to cell wall disintegration and ethylene generation during tomato fruit ripening, were identified by the redox proteomic approach. The study noted that with increased protein oxidation during the process of ripening, this can be regulated by ROS through changes in the redox states of enzymes. The mechanism of action behind the regulation of fruit ripening via protein redox alterations is brought to light by studies (Wang et al. 2021).

1.20 Disease Resistance

Brassinosteroid (BR) as a modulator of tomato (*Solanum lycopersicum*) fruit infected by *Alternaria solani*: changes in proteome, gene expression, and enzyme activity associated with the disease resistance and pathogenesis: It was shown that BR pretreatment delayed the onset of disease and enhanced the activity of enzymes involved in disease resistance. Solani will likely infect more in brassinazole-treated fruits, an inhibitor of BR production. All genes in the quinone and phenolic compound biosynthetic pathways, those upstream and downstream in the pathways of systematic acquired resistance, may be upregulated by BR. Proteome analysis showed that BR upregulated the expression of two proteins and downregulated the expression of 13 other proteins. Results imply that some SAR genes downstream, like phenylalanine ammonia-lyase, are shared in tomato fruits with BR-mediated disease resistance, while some others can be involved in distinct pathways (Ding et al. 2021).

1.21 Metabolomics for Sustainable Cultivation of Tomatoes

Tomatoes (*Solanum lycopoersicum* L.) are the second most important fruit and vegetable crop in the *Solanaceae* family, behind potatoes (*Solanum tuberosum* L.). It is grown for processed goods and fresh fruit. Among the many components of tomatoes that are good for you are vitamins, phenolic compounds, and carotenoids. Tomatoes are standard fare these days when it comes to researching the evolution of fleshy fruits, even beyond their culinary and commercial significance. Tomatoes are climacteric fruits, meaning that as they mature, their metabolism changes significantly. They provide a summary of what we now know about tomato fruit metabolism. First, they describe the hormonal and genetic regulation of fruits, emphasizing the metabolism of sugar, organic acids, and amino acids. Further highlighting the connections between primary and secondary metabolic pathways is the significance of volatiles, flavonoids, and pigments for tomato fruit quality. Last, since tomato plants are susceptible to a variety of abiotic stressors, they provide a brief overview of how unfavorable environmental factors affect tomato fruit metabolism and quality (Quinet et al. 2019).

1.22 Metabolic Pathways

Two essential components of crop improvement and medication creation are the identification and comprehension of metabolic pathways. Ontology and gene annotation serve as the foundation for the common method for their detection (Campos-Lozada et al. 2022). A further approach is a correlation-based network analysis, which arranges metabolites into networks. The tomato pericarp's metabolite correlation networks were mapped as subgraphs onto metabolites of known tomato pathways and random sets of metabolites (Toubiana et al. 2019).

1.23 Nutritional Quality

The tomato, a popular vegetable crop, can significantly reduce dietary deficits in people. Tomatoes have abundant micronutrients and other bioactive substances (such as minerals, vitamins, and carotenoids) that are important or advantageous for human health. There is also discussion about possible applications for tomato processing industry by-products, such as peel and seed extract, which are especially high in oils and proteins (Botinestean et al. 2015). Recently, CRISPR/Cas-mediated gene-editing technology has significantly improved the nutritional value of agricultural goods, particularly tomatoes. Studies on transcriptomics and metabolomics provide more insight into the molecular control of tomato nutrients. At a sizable tomato processing facility, tomato pomace, a by-product produced during tomato processing, was gathered. The by-product had a high moisture content (66.58 g.100 g^{-1} wet basis) and was primarily made up of tomato skin (61.5%). Dietary fiber had the highest content of all the nutrients, followed by fat and proteins (50.74, 20.91, and 14.14 g/100 g d.w., respectively) (Vinha et al. 2014). The in vitro antioxidation capacity of the pomace is considerable, particularly when assessed using the TEAC assay (224.81 μmol Trolox equivalent 100 g^{-1} d.w.). This must be because of the high concentration of lycopene (446.9 μg.g^{-1} d.w.) that remains in the by-product after processing. Using the skin fraction as the source material instead of the entire pomace separated the waste into seed and skin fractions by sedimentation, increasing the lycopene yield by 55%. There is a lot of promise in using this by-product as an ingredient source (Silva et al. 2019).

1.24 Pesticide Reduction

Vegetables and fruits are essential to a balanced diet. Fruits and vegetables provide the body with a variety of nutrients, including water, sugars, fats, proteins, fiber, vitamins, minerals, pigments, and antioxidants. The majority of fruits and vegetables have a favorable texture, color, flavor, and simplicity of use in addition to being readily available in a broad range and tasting excellent practically all year round (Miranda et al. 2022). Agricultural food contaminated with pesticides may be harmful to customers' health, so decontamination techniques must be successful

(Botinestean et al. 2015). Frontline extension workers and local specialists taught smallholder farmers how to use integrated pest management (IPM) on outdoor tomatoes, or *Solanum lycopersicum*. To reduce pesticide residue, among the technological treatments available are 4% vinegar solution and 3% salt solution with and without sonication. Tomato samples infected with pesticides were subjected to each treatment for 10 minutes as part of the experiment. To investigate the decrease in pesticide residue, tomato juices, tomato puree, and sun-dried tomatoes were also investigated (Ahmad et al. 2024).

Reducing the use of pesticides has become a common goal among European countries and a major issue in public policy due to the negative impact that pesticides have on the environment and human health. Diverse conclusions have been drawn from scientific studies assessing how well home remedies reduce pesticide residues (Hartman et al. 2018). Certain treatments have been shown in certain studies to be efficient in reducing surface residue, which may reduce the chance of being exposed to hazardous substances. For example, washing fruit in water alone or with vinegar has demonstrated potential in lowering pesticide residues. It is believed that some pesticide residues can be broken down and removed from the surface of fruits and vegetables by vinegar's acidic properties (Jacquet et al. 2022).

1.25 Integration of Omics Technologies

The development of omics technologies, including transcriptomics, proteomics, metabolomics, and genomics, has started to make highly comprehensive molecular medicine possible. Each of these technologies has played a part in the advancement of medicine that is starting to be applied in clinical settings (Çetin 2022). However, the biological intricacy of the majority of human diseases is too complex for any one technique to fully represent. Combining different technologies has become a popular way to present a more complete picture of biology and illness (Yang et al. 2021).

1.26 Predicting Modeling

In recent decades, some "omics" techniques have been shown to be effective technologies for plant systems. Next-generation sequencing (NGS) has opened the door for a new wave of mics research, including transcriptomics, proteomics, and genomics. In crop science, however, phenomics, genomics, and metabolomics have been extensively studied (Méabed et al. 2018). High-throughput techniques coupled with multi-omics methods have contributed immensely to the understanding of growth, senescence, yield, and responses to biotic and abiotic stresses in many crops. The application of omics methods has been made on several crops, such as rice (*Oryza sativa* L.), cotton (*Gossypium hirsutum* L.), millet (*Setaria italica* L.), wheat (*Triticum aestivum* L.), and *Medicago truncatula*. Such incorporation of functional genomics with other omics shall go a long way in underscoring the relationships existing between crop genomes and their phenotypic outputs under certain physiological and environmental conditions (Ahmad et al. 2024).

1.27 Challenges and Future Perspective

Genomics, transcriptomics, proteomics, and metabolomics are the terms used for omics technologies to decipher an organism's genome sequence, gene expression profiles, protein content, and metabolites. During the past two decades, there has been a huge technological development, and analytical techniques involved in omics investigations were made easier, quicker, more accurate, and inexpensive as never before. Through omics research in crop biotechnology, important agronomic features such as grain yield and genetic sequences can be procured. Genomic sequence data can be coupled with functional analysis data to provide information on accurate gene annotation, functional markers, QTL maps, etc. These results may be helpful in the construction of better agricultural types. Multidimensional omics investigations provide an enhanced understanding of the factors that direct the phenotypic features, metabolic processes, and functions of biomolecules involved in various biosynthetic pathways in crop plants (Shalini et al. 2018).

1.28 Conclusion

The "omic" technologies have brought a new frontier in the adoption of sustainable agriculture in tomato growing. These advanced methodologies, such as transcriptomics, genomics, proteomics, and metabolomics, provide an overview of the intricate molecular processes involved in the control of formation, growth, and stress responses in tomatoes. Knowing the genetic and molecular underpinnings of these processes will thus enable the development of tomato varieties that can better resist biotic and abiotic challenges such as pests, diseases, and climatic variability. Farming may become more environmentally friendly and sustainable because of its resilience—i.e., with reduced chemical fertilizers and pesticides. Moreover, "omics" technology makes the execution of precision farming easier. Knowledge about molecular levels of plant reactions and requirements regarding the tomato plant can be helpful to farmers in using resources such as energy, nutrients, and water more effectively.

Metabolomics, for example, provides information on the nutritional profile of tomatoes to guide the application of fertilizers so that it coincides with the requirements of a specific plant. On the other hand, transcriptomics can point out genes responsible for drought resistance and hence provide directions for designing cultivars needing less water. This generally improves resource efficiency and reduces environmental impacts from tomato cultivation. "omics" advantages also apply outside the fence of the industries. With improved varieties having better shelf life and nutritional profile, both food security and quality are enhanced. Technologies of this nature also afford stable yields against environmental changes by establishing and mitigating the effects of climate change on tomato production. In sum, there are considerable possibilities for bettering agricultural sustainability by applying "omics" technology in tomato cultivation. Using "omics" technology, high-yielding, resource-efficient, resilient tomato varieties are developed by fusing cutting-edge science with practical agricultural techniques. Such a holistic approach will help combat the twin challenges of increased food production and protection of environmental health. If we are to further develop and increase the use of "omics" applications in agriculture, policy support at the forefront should go in line with farmer education and cooperative research. The use of "omics" in tomato production is thus the first step toward developing a sustainable agricultural future that shall feed the growing world while saving the environment.

References

Ahmad MOH, Abdel-Tawab SA, Ahmed LAR, Roby MH (2024) Quality control of tomato products and its influenced by some technological treatments on reducing pesticide residues. Labyrinth: Fayoum J Sci Interdiscip Stud:21–34

Ashrafi-Dehkordi E, Alemzadeh A, Tanaka N, Razi H (2018) Meta-analysis of transcriptomic responses to biotic and abiotic stress in tomato. PeerJ 6:e4631

Bhatkar NS, Shirkole SS, Mujumdar AS, Thorat BN (2021) Drying of tomatoes and tomato processing waste: a critical review of the quality aspects. Dry Technol 39(11):1720–1744

Botinestean C, Gruia AT, Jianu I (2015) Utilization of seeds from tomato processing wastes as raw material for oil production. J Mater Cycles Waste Manag 17:118–124

Campos-Lozada G, Pérez-Marroquín XA, Callejas-Quijada G, Campos-Montiel RG, Morales-Peñaloza A, León-López A, Aguirre-Álvarez G (2022) The effect of high-intensity ultrasound and natural oils on the extraction and antioxidant activity of lycopene from tomato (Solanum lycopersicum) waste. Antioxidants 11(7):1404

Cappetta E, Andolfo G, Di Matteo A, Barone A, Frusciante L, Ercolano MR (2020) Accelerating tomato breeding by exploiting genomic selection approaches. Plants 9(9):1236

Çetin N (2022) Comparative assessment of energy analysis, drying kinetics, and biochemical composition of tomato waste under different drying conditions. Sci Hortic 305:111405

Chowdhury HA, Bhattacharyya DK, Kalita JK (2019) (Differential) co-expression analysis of gene expression: a survey of best practices. IEEE/ACM Trans Comput Biol Bioinform 17(4):1154–1173

Collins EJ, Bowyer C, Tsouza A, Chopra M (2022) Tomatoes: an extensive review of the associated health impacts of tomatoes and factors that can affect their cultivation. Biology 11(2):239

Dai X, Shen L (2022) Advances and trends in omics technology development. Front Med 9:911861

De Buck V, Polanska M, Van Impe J (2020) Modelling biowaste biorefineries: a review. Front Sustain Food Syst 4:11

Decros G, Beauvoit B, Colombié S, Cabasson C, Bernillon S, Arrivault S, Guenther M, Belouah I, Prigent S, Baldet P, Gibon Y (2019) Regulation of pyridine nucleotide metabolism during tomato fruit development through transcript and protein profiling. Front Plant Sci 10:1201

Ding Y, Sheng J, Cheng F (2021) Assessment of the role of brassinosteroid in regulating the disease resistance of postharvest tomato fruit by proteomic analysis. J Food Process Preserv 45(9):e15708

Esposito S, Carputo D, Cardi T, Tripodi P (2019) Applications and trends of machine learning in genomics and phenomics for next-generation breeding. Plants 9(1):34

Fernie AR, Yan J (2019) De novo domestication: an alternative route toward new crops for the future. Mol Plant 12(5):615–631

Gao L, Gonda I, Sun H, Ma Q, Bao K, Tieman DM, Burzynski-Chang EA, Fish TL, Stromberg KA, Sacks GL, Thannhauser TW (2019) The tomato pan-genome uncovers new genes and a rare allele regulating fruit flavor. Nat Genet 51(6):1044–1051

Gonda I, Ashrafi H, Lyon DA, Strickler SR, Hulse-Kemp AM, Ma Q, Sun H, Stoffel K, Powell AF, Futrell S, Thannhauser TW (2019) Sequencing-based bin map construction of a tomato mapping population, facilitating high-resolution quantitative trait loci detection. Plant Genome 12(1):180010

Gong W, Kwak IY, Pota P, Koyano-Nakagawa N, Garry DJ (2018) DrImpute: imputing dropout events in single-cell RNA sequencing data. BMC Bioinf 19:1–10

Groen SC, Ćalić I, Joly-Lopez Z, Platts AE, Choi JY, Natividad M, Dorph K, Mauck WM 3rd, Bracken B, Cabral CLU, Kumar A, Torres RO, Satija R, Vergara G, Henry A, Franks SJ, Purugganan MD (2020) The strength and pattern of natural selection on gene expression in rice. Nature 578(7796):572–576

Hao Y, Wang H, Yang X, Zhang H, He C, Li D, Li H, Wang G, Wang J, Fu J (2019) Genomic prediction using existing historical data contributing to selection in biparental populations: a study of kernel oil in maize. Plant Genome 12

Hartman K, van der Heijden MG, Wittwer RA, Banerjee S, Walser JC, Schlaeppi K (2018) Cropping practices manipulate abundance patterns of root and soil microbiome members paving the way to smart farming. Microbiome 6:1–14

Ichihashi Y, Zamir D, Headland L, Fulop D, Chitwood DH, Maloof JN, Ofner I, Sinha NR, Covington MF, West D, Ranjan A (2016) A new advanced backcross tomato population enables high-resolution leaf QTL mapping and gene identification. G3: Genes, Genomes, Genet 6:3169–3184

Jacquet F, Jeuffroy MH, Jouan J, Le Cadre E, Litrico I, Malausa T, Reboud X, Huyghe C (2022) Pesticide-free agriculture as a new paradigm for research. Agron Sustain Dev 42(1):8

Kwon CT, Heo J, Lemmon ZH, Capua Y, Hutton SF, Van Eck J, Park SJ, Lippman ZB (2020) Rapid customization of Solanaceae fruit crops for urban agriculture. Nat Biotechnol 38(2):182–188

Li Q, Yan J (2020) Sustainable agriculture in the era of omics: knowledge-driven crop breeding. Genome Biol 21:1–5

Luo C, Fernie AR, Yan J (2020) Single-cell genomics and epigenomics: technologies and applications in plants. Trends Plant Sci 25(10):1030–1040

Maher MF, Nasti RA, Vollbrecht M, Starker CG, Clark MD, Voytas DF (2020) Plant gene editing through de novo induction of meristems. Nat Biotechnol 38(1):84–89

Méabed EMH, Abou-Sreea AIB, Roby MHH (2018) Chemical analysis and germicidal effectiveness of the aqueous extract of Cymbopogon citratus Stapf. Parasitol Res 117:1745–1755

Miranda NGM, Gomes FDS, Cabral LMC, de Jesus MSC (2022) Thermal concentration effects of tomato pulp and carotenoids extraction from the residue. Sci Agric 80

Murovec J, Guček K, Bohanec B, Avbelj M, Jerala R (2018) DNA-free genome editing of Brassica oleracea and B. rapa protoplasts using CRISPR-Cas9 ribonucleoprotein complexes. Front Plant Sci 9:1594

Pott DM, Durán-Soria S, Osorio S, Vallarino JG (2021) Combining metabolomic and transcriptomic approaches to assess and improve crop quality traits. CABI Agric Biosci 2(1):1

Quinet M, Angosto T, Yuste-Lisbona FJ, Blanchard-Gros R, Bigot S, Martinez JP, Lutts S (2019) Tomato fruit development and metabolism. Front Plant Sci 10:475784

Reimer JJ, Thiele B, Biermann RT, Junker-Frohn LV, Wiese-Klinkenberg A, Usadel B, Wormit A (2021) Tomato leaves under stress: a comparison of stress response to mild abiotic stress between a cultivated and a wild tomato species. Plant Mol Biol 107(3):177–206

Rowan BA, Patel V, Weigel D, Schneeberger K (2015) Rapid and inexpensive whole-genome genotyping-by-sequencing for crossover localization and fine-scale genetic mapping. G3: Genes, Genomes, Genet 5(3):385–398

Salava H, Thula S, Mohan V, Kumar R, Maghuly F (2021) Application of genome editing in tomato breeding: mechanisms, advances, and prospects. Int J Mol Sci 22(2):682

Sant'Ana DVP, Lefsrud M (2018) Tomato proteomics: tomato as a model for crop proteomics. Sci Hortic 239:224–233

Shalini S, Singla A, Goyal M, Kaur V, Kumar P (2018) Omics in agriculture: applications, challenges and future perspectives. In: Crop improvement for sustainability. Daya Publishing House, pp 343–360

Silva YPA, Borba BC, Pereira VA, Reis MG, Caliari M, Brooks MSL, Ferreira TA (2019) Characterization of tomato processing by-product for use as a potential functional food ingredient: nutritional composition, antioxidant activity, and bioactive compounds. Int J Food Sci Nutr 70(2):150–160

Souza da Costa B, García MO, Muro GS, Motilva MJ (2023) A comparative evaluation of the phenol and lycopene content of tomato by-products subjected to different drying methods, vol 179. LWT- Food Sci Technol, p 114644

Toubiana D, Puzis R, Wen L, Sikron N, Kurmanbayeva A, Soltabayeva A, del Mar Rubio Wilhelmi M, Sade N, Fait A, Sagi M, Blumwald E (2019) Combined network analysis and machine learning allow the prediction of metabolic pathways from tomato metabolomics data. Commun Biol 2(1):214

Van Dam S, Vosa U, van der Graaf A, Franke L, de Magalhaes JP (2018) Gene co-expression analysis for functional classification and gene-disease predictions. Brief Bioinform 19(4):575–592

Vats S, Bansal R, Rana N, Kumawat S, Bhatt V, Jadhav P, Kale V, Sathe A, Sonah H, Jugdaohsingh R, Sharma TR (2022) The unexplored nutritive potential of tomatoes to combat global malnutrition. Crit Rev Food Sci Nutr 62(4):1003–1034

Wang P, Li X, Wang Y, Wang W, Tian S, Qin G (2021) Redox proteomic analysis reveals the involvement of oxidative post-translational modification in tomato fruit ripening. Postharvest Biol Technol 178:111556

Wenger J, Stern T (2019) Reflection on the research on and implementation of biorefinery systems–a systematic literature review with a focus on feedstock. Biofuels Bioprod Biorefin 13(5):1347–1364

Xia X, Cheng X, Li R, Yao J, Li Z, Cheng Y (2021) Advances in the application of genome editing in tomato and recent development of genome editing technology. Theor Appl Genet 134(9):2727–2747

Xu J, Chen G, Hermanson PJ, Xu Q, Sun C, Chen W, Kan Q, Li M, Crisp PA, Yan J, Li L, Springer NM, Li Q (2019) Population-level analysis reveals the widespread occurrence and phenotypic consequence of DNA methylation variation not tagged by genetic variation in maize. Genome Biol 20:243

Yang Y, Saand MA, Huang L, Abdelaal WB, Zhang J, Wu Y, Li J, Sirohi MH, Wang F (2021) Applications of multi-omics technologies for crop improvement. Front Plant Sci 12:563953

Zhu L, Lei J, Devlin B, Roeder K (2018) A unified statistical framework for single cell and bulk RNA sequencing data. Ann Appl Stat 12(1):609

Molecular Strategies for Confronting Abiotic Stress Tolerance in Tomatoes

2

Rooma Waqar, Javed Iqbal, Lubna Anjum Minhas, Banzeer Ahsan Abbasi, Shumaila Ijaz, Zakir Ullah, Shobana Sampath, Rashid Iqbal, Ghulam Murtaza, Abdul Samad Mumtaz, Muhammad Majeed, Tayyaba Shahzadi, and Zahid Ullah

Abstract

Tomato (*Solanum lycopersicum*) is one of the most widely cultivated and economically important crops globally. However, its productivity and quality are severely hampered by different abiotic stresses, including drought, lack of essential nutrients, salinity, and extreme temperatures. These stresses trigger a cascade of molecular and physiological changes that adversely affect the growth of the

Rooma Waqar and Javed Iqbal contributed equally with all other contributors.

R. Waqar · L. A. Minhas · Z. Ullah · A. S. Mumtaz
Department of Plant Sciences, Faculty of Biological Sciences, Quaid-i-Azam University, Islamabad, Pakistan
e-mail: roomawaqar@bs.qau.edu.pk; lubnaminhas@bs.qau.edu.pk; asmumtaz@qau.edu.pk

J. Iqbal (✉) · Z. Ullah
Department of Botany, Bacha Khan University, Charsadda, Khyber Pakhtunkhwa, Pakistan

B. A. Abbasi
Department of Botany, Rawalpindi Women University, Satellite Town, Rawalpindi, Pakistan

S. Ijaz
Department of Burn and Plastic Surgery, Shenzhen Institute of Translational Medicine, The First Affiliated Hospital of Shenzhen University, School of Biomedical Engineering, Shenzhen University Medical School, Shenzhen University, Shenzhen, China

S. Sampath
Department of Biotechnology, Vel Tech Rangarajan Dr. Sagunthala R&D Institute of Science and Technology, Chennai, India

M. Waseem et al. (eds.), *Omics Approaches for Tomato Yield and Quality Trait Improvement*, https://doi.org/10.1007/978-981-96-3890-1_2

plant and yield. Recent advances in molecular biology and biotechnology offer promising strategies to enhance tomato tolerance to these abiotic stresses. Research on abiotic stresses is limited because defensive feedback mechanisms to abiotic factors require the regulation of changes that activate numerous genes and pathways, which are not yet thoroughly explored. The involvement of these genes highlights the complexity of plant stress response and adaptation. An integrative approach can begin by targeting the multigenic nature of plant responses to abiotic stress, which involves genomic sequences, protein organisation, transcripts, collections of transcriptomes specific to stress and cellular conditions, and screenings of mutants. Therefore, this chapter focusses on the metabolomics, proteomics, and genomics of tomatoes under abiotic stress.

Keywords

Proteomics · Abiotic stress · Metabolomics · Genomics · Tomato omics

2.1 Introduction

Solanum lycopersicum L., also known as tomato, is among the most important fruits that are classified nutritionally as vegetables. The plant contains phenolic compounds (flavonoids) (Kumar et al. 2021), carotenoids (lycopene and carotene), phytosterols (β-sitosterol, campesterol, and stigmasterol), and glycoalkaloids (tomatine) (Ali et al. 2020; Chaudhary et al. 2018). Fernandes et al. (2021) found a high amount of monounsaturated and polyunsaturated fatty acids in tomatoes, including arachidic, oleic, palmitic, linolenic, and stearic acids. Furthermore, tomatoes are high in lycopene, a pigment that shields both the fruit and the consumer from UV radiation. Lycopene also acts as an antioxidant, enhancing immunological responses, decreasing cholesterol, and promoting cardiovascular health. Lycopene is also effective in treating periodontitis and gingivitis. The presence of lutein and

R. Iqbal
Department of Agronomy, Faculty of Agriculture and Environment, The Islamia University of Bahawalpur, Bahawalpur, Pakistan

Department of Life Sciences, Western Caspian University, Baku, Azerbaijan
e-mail: rashid.iqbal@iub.edu.pk

G. Murtaza
School of Agriculture, Yunnan University, Kunming, Yunnan, China

M. Majeed
Department of Botany, University of Gujrat, Gujrat, Punjab, Pakistan
e-mail: m.majeed@uog.edu.pk

T. Shahzadi
Department of Chemistry, Government College for Women, University Sialkot, Sialkot, Pakistan
e-mail: tayyaba.shehzadi@gcwus.edu.pk

zeaxanthin in tomatoes protects the eyes and lowers the risk of degeneration of the macula. Given their multiple health benefits, tomatoes are in high demand, but they are vulnerable to unfavourable growing conditions. To increase yield, it is critical to investigate the effects of various stress factors, such as salinity, low temperatures, and drought, as well as incorporate natural resistance mechanisms into the crop.

Abiotic stress is a major global danger to crop yields, cutting average production rates by more than 50% for many agricultural plants and restricting their genetic potential. The integration of advancements in current technology has led to the establishment of several multidimensional omics disciplines, including transcriptomics, metabolomics, proteomics, and genomics, which investigate the molecular components of living creatures (Mathivanan 2021; Ahmad et al. 2022). Advances in genomic research have enabled the identification of multiple gene families and mechanisms that improve abiotic stress tolerance, resulting in higher crop yields. These genes may encode protein kinases such as calcium-dependent kinases, which activate cis-acting elements and transcription factors (TFs) that regulate stress response pathways (Khoso et al. 2022).

Additionally, investigating plant–environment interactions offers a holistic view of molecules across tissue, cellular, and organismal scales (Ahmad et al. 2022). Systems biology, an emerging field of life science, combines multiple omics approaches to provide valuable biological information, enabling visualisation and explanation of how plants deal with abiotic stress integration of diverse knowledge (Debnath et al. 2011).

This chapter covers recent advances in proteomics and effective approaches for studying resources linked to abiotic stress tolerance in tomatoes. It focusses on discussing tomato genomes, metabolomics, and proteomics (Fig. 2.1).

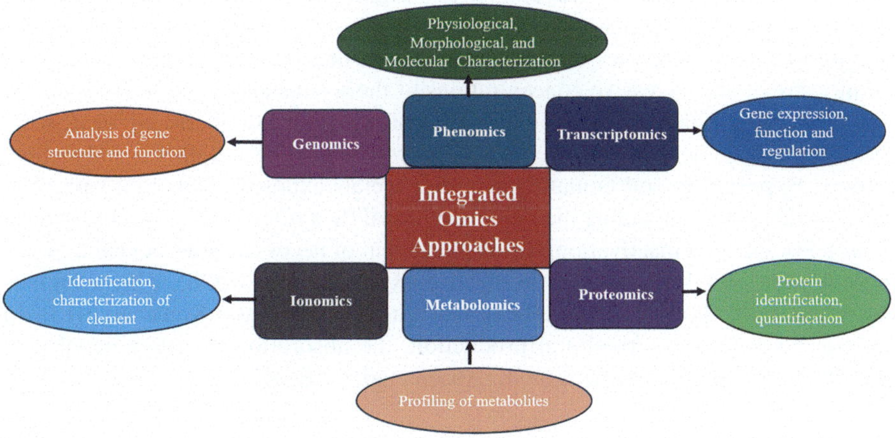

Fig. 2.1 Various omics disciplines are being utilised both individually and in an integrated fashion in plant science

2.2 Primary Abiotic Stressors in Tomatoes

Abiotic stress encompasses a broad spectrum of circumstances, including changes in temperature, osmotic pressure, waterlogging, salt, drought, bruising, ozone, harmful substances, UV radiation, and bright light. These stressors are dangerous conditions that set off a chain reaction of morphological, molecular, physiological, and biochemical alterations that impair general metabolism and cause cellular damage. These environmental stressors function as ecological barriers, impairing plant growth and agricultural productivity by upsetting cellular homeostasis (Kosová et al. 2011). It is uncommon to identify a single abiotic stressor because of the complexity and variety of its causes. As a result, in reaction to diverse stressors, plants display a broad range of stress-responsive components and pathways. One important abiotic component controlling cellular homeostasis is temperature. Heat speeds up the rates at which soil water evaporates and leaves transpire, generating water stress in field settings and quickening the dynamics of biomolecules, potentially leading to protein misfolding (Kosová et al. 2018). Cold, on the other hand, slows the kinetics of biomolecules and enzymes, which reduces the fluidity of cell membranes. In the soil, snow can condense into ice crystals that freeze the water inside.

The most important abiotic element affecting plant evolution is water availability, which might have a major impact on more than half of all arable areas by 2050 (Mustafa et al. 2019). Drought upsets the water balance in the soil, which hinders osmotic adjustment and results in a decrease in cytosol osmolarity because of the build-up of osmolytes and hydrophilic proteins (LEA proteins). Moreover, it induces stomatal closure in leaves, upsetting the equilibrium between carbon absorption and electron transport pathways, increasing thermal energy dissipation and photoinhibition (Kosová et al. 2018; Ahmad et al. 2024a). Plant roots ferment because of anaerobiosis brought on by flooding stress. Higher levels of organic acids in the cytoplasm during fermentation cause it to become more acidic, which inhibits the activity of several enzymes. Osmolytes and ions like Na^+ accumulate because of salinity stress, and excessive concentrations of these substances can be hazardous.

The global problem of heavy-metal soil pollution is made worse by weathering rocks through geochemical processes, acid rain, volcanic eruptions, and human activity. Exposure to soil contaminated with metals exposes plants to a variety of detrimental effects, including the binding of metal (heavy) ions to sulfhydryl groups in proteins, enzyme inactivation, the displacement of necessary ions detaching from binding sites, and the generation of reactive oxygen species (ROS), which can oxidatively damage protein, nucleic acid and lipid. (Hossain and Komatsu 2013). In addition to reducing agricultural production, the absorption of heavy metal puts both flora and wildlife at serious risk (Ahsan et al. 2009) (Fig. 2.2).

2.3 Molecular Mechanisms of Abiotic Stress Responses

Abiotic stress responses in tomatoes involve complex networks of signalling pathways, gene expression changes, and metabolic adjustments. Key components include:

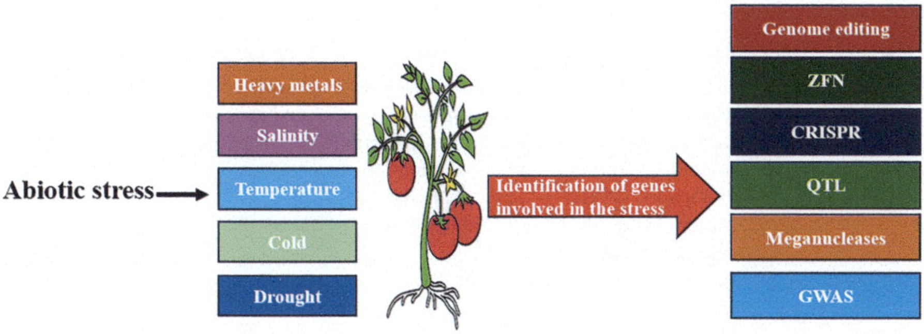

Fig. 2.2 Impact of different stresses on tomato plants: responses and pathways

2.3.1 Genomics of Abiotic Stress

With the latest sequencing of the tomato and potato genomes, the field of solanaceous crop genomics is developing at an exciting pace. Tomato (*Solanum lycopersicum* L.), which is native to South America, is among the most widely cultivated vegetable crops in the world. It features interesting developmental characteristics like compound leaves, juicy fruits, a quick generation period, and simple diploid genetics. Tomato is a great species for both applied and basic plant research because of all these features together (Gerszberg and Hnatuszko-Konka 2017). Many genetic and genomic resources existed before the tomato genome sequencing project got underway. To generate a genetic map of tomatoes, researchers used isozyme markers and morphological markers (Wang et al. 2022). Following that, they determined which 12 linkage groups were associated with the chromosomes that were visible under cytology (Ranjan et al. 2012). Breeders were able to identify quantitative trait loci (QTLs), thanks to this extensive molecular linkage map, which improved our knowledge of the genetic foundation for a variety of quantitative traits. A wealth of information about tomato genetic and genomic resources can be found on the Solanaceae Genomics Network website. The creation of new kinds has been made considerably easier by recent developments in genomics. Molecular markers have made it much easier to characterise genetic diversity in the germplasm pool of nearly any crop species (Dheer et al. 2020; Ahmad et al. 2024b). Compared to morphological or biochemical markers, these DNA polymorphism–based markers have a number of benefits, such as straightforward tests, repeatability, user-friendliness, high availability, stability in the face of changing environmental factors, and representation across whole genomes (Cambiaso et al. 2019). Numerous applications in genetics, plant breeding, and genomics, especially with tomatoes, have been made possible by DNA markers. Many omics disciplines that concentrate on the molecular components of cellular biology have emerged as a result of technological improvements (Chaudhary et al. 2019). The genomic method offers a thorough viewpoint for the structural and functional examination of genes and the discovery of genetic variations that can be used to help tomatoes resist abiotic stress. Table 2.1 presents the genomics to enhance abiotic stress tolerance in tomatoes.

Table 2.1 Using genomics to enhance abiotic stress tolerance in tomatoes

Gene	Role	Expression	Outcome	References
BEL1-like gene	In plants, transcription factors from the three-amino-loop-extension (TALE) protein superfamily regulate a wide range of biological processes	Exhibited diverse tissue-specific expression patterns and responded to cold, heat, and drought stressors	Plant growth and abiotic stress response	He et al. (2022)
SlMBP8, SlAGO4A	Drought and salt stress tolerance	Overexpression	Drought and salt stress resilience	Yin et al. (2017)
SlAIM1	Tolerance to salt and oxidative stresses	Overexpression of SlAIM1 enhances tolerance to salt and oxidative stress, whereas silencing SlAIM1 increases sensitivity to these abiotic stimuli	Resilience to abiotic stress	Wai et al. (2020) and AbuQamar et al. (2009)
PpSnRK1α	Increased metabolism of reactive oxygen species through the upregulation of antioxidase gene expression and antioxidant enzyme activity	Overexpression	Resistance to salinity	Jia et al. (2021)
SlMPK7i, SlMAPK3	Tolerance to chilling, drought, and Cadmium stresses	Overexpression	Ability to withstand chilling and drought stresses	Muhammad et al. (2019)
SlbZIP1	Tolerance to drought and salt stresses	Expression	Ability to endure salt and drought stresses	Chaudhary et al. (2018)
SlMBP8	Increased resilience to drought and salt stress	Gene silencing	Enhanced tolerance to drought and salt stresses	Yin et al. (2017)
SlMAPK3i	Reducing the build-up of reactive oxygen species and enhancing the gene expression linked to ethylene signalling	Overexpression	Tolerance to saline stress	Shu et al. (2022)

(continued)

Table 2.1 (continued)

Gene	Role	Expression	Outcome	References
SlLBD40	An inhibitor of drought tolerance, it was involved in Jasmonic acid signalling	CRISPR/Cas9-mediated targeted mutagenesis (gene knockout)	Resistance to drought	Liu et al. (2020a)
SlGRAS10	Enhanced the expression of peroxidase, catalase, and superoxide dismutase to mitigate the impact of reactive oxygen species	Reduction through RNA interference (RNAi)	Tolerance to environmental stresses	Habib et al. (2021)
SlDEAD23, SlDEAD35	Biotic and abiotic stress responses	Overexpression	Improved resistance to salt and cold stresses	Pandey et al. (2019)
Solyc03g020030	Proteinase inhibitor-II	Downregulation of genes	Heat tolerance	Balyan et al. (2020)
SlBZR1D	Enhanced salt tolerance through upregulation of multiple stress-related genes	Overexpression and increased expression	Tolerance to saline conditions	Zhu et al. (2018b)
SlNL33	Ascorbate accumulation	Reduced expression	Stress tolerance	Ye et al. (2019)
MdSWEET17	Response to drought stress and the regulation of fructose	Present in tomatoes	Drought stress	Lu Jing et al. (2019)
SiDHN	Overexpression of the dehydrin gene from *Saussurea involucrata*	Overexpression	Tolerance to drought and cold	Guo et al. (2019b)

2.3.2 Tomato Genome Sequencing

Abiotic stressors have a detrimental effect on tomato crop production in open-field farming. This leads to low fruit quality and output, which, in turn, impacts market prices and falls short of demand. Farmers frequently grow tomato crops in greenhouses in order to reduce the effects of abiotic stressors, particularly temperature and flooding, on open fields (Zhu et al. 2018c). Greenhouse farming, on the other hand, is costlier and causes the soil to quickly accumulate NO_3, PO_4, and salts, which degrade the soil and contaminate both ground and surface water. Enhancing tomato cultivars' resistance to abiotic stress is hence sustainable and profitable (Szymanski et al. 2017; Ali et al. 2022). Extreme water and temperature regimes, soil nutrient imbalances, elemental toxicity, and excessive salinity are examples of stress (abiotic) conditions that present significant obstacles to tomato development. Abiotic pressures get more complex in agricultural situations where several stressors occur simultaneously. The creation of robust, high-productivity varieties with enhanced tolerance to a range of abiotic challenges is imperative in order to meet the world's food needs (Kausar et al. 2022). Plant responses differ in the field when various elements are present at the same time, even though some components of stress have been addressed under controlled settings (Krishna et al. 2019). Conventional breeding has made considerable improvements in several features over the past 10 years, including yield components, quality-related variables, and resistance to both abiotic and biotic stress. Conventional cultivars, however, are vulnerable to various stressors in various environments. Using more intensive and integrated methods from various disciplines gives a better strategy for enhancing stress tolerance in present crops, given the environmental interactions and genetic complexity (Chaudhary et al. 2019).

2.4 Discovering and Validating Genes Linked to Tolerance to Abiotic Stress

The use of genetic engineering is one method for improving modern tomato plants. The identification of numerous gene families that affect plant responses to abiotic stresses has been made possible by advancements in functional genomics, which has increased yield (Gerszberg and Hnatuszko-Konka 2017). Unlike traditional selective breeding, genetic modification (GM) technology enables effective and faster development of tomato plants resilient to abiotic stresses, boosting food production (Gerszberg and Hnatuszko-Konka 2017). According to Yang et al. (2015), overexpression of the genes SlSnRK22.1, which regulates abscisic acid signalling, and SlSnRK2.2, which transduces the osmotic stress signal, increased osmotic stress sensitivity. According to Yu et al. (2016) and Ijaz et al. (2023), the SlMPK7 gene enhanced tolerance to low temperatures and regulated ROS homeostasis by activating cellular antioxidant mechanisms. According to Liu et al. (2015), overexpression of the ShDHN/*S habrochaites* genes resulted in the accumulation of proteins that have detergent and chaperone-like qualities, enhancing plants' ability to withstand salinity, cold, and drought. Li et al. (2015) found that SpWRKY1, a transcription factor involved in the transcriptional control of genes related to stress, boosted abiotic tolerance when overexpressed. Meng et al. (2015) found that the anthocyanin biosynthesis pathway's structural genes were elevated when the LeAN2 gene was overexpressed, which resulted in an accumulation of anthocyanin and enhanced heat resistance. Tomato plants have been discovered to be resistant to chilling when they possess the AtGRX gene, which is responsible for preserving cellular redox equilibrium (Hu et al. 2015). Similarly, Hu et al. (2016) found that overexpression of the MdSOS2L1 gene enhanced salt tolerance. This gene is implicated in signal transduction proteins that influence ion-driven transport processes. The SlSAM1 gene in tomatoes was investigated by Gong et al. (2014). This gene is implicated in polyamine and ethylene production, as well as alkali tolerance. It facilitates the conversion of ATP and L-methionine into S-adenosylmethionine. Metwali et al. (2015) silenced the vis-1 gene to increase tomato fruit quality under stress (heat) using small interfering RNA technology. To generate tomato crops with varying degrees of stress (abiotic) tolerance, researchers have used a variety of strategies across the globe (Naeem et al. 2022). It was demonstrated by Habib et al. (2021) that GRAS transcription factors (TFs) are engaged in various biological activities. The role of the SlGRAS10 gene in salt and drought stress was identified; plants (dwarf) with shorter internodes, smaller leaves, and higher flavonoid accumulation also showed tolerance to heat. SlAREB2 and SlAREB1 components are involved in grown tomatoes. It has been demonstrated that overexpressing SlAREB1 and SlAREB2 increases drought and salinity tolerance in the roots (Orellana et al. 2010). Researchers discovered that proteins involved in transcription control, oxidative stress, and late embryogenesis-abundant protein are encoded by genes increased in SlAREB1-overexpressing lines. These genes may respond to harmful microorganisms and are probably involved in abiotic stress responses. Using phylogenetic analysis, gene structure, and motif composition, Wang et al. (2022) categorised 66 putative Golden 2 (G2)–like genes in tomatoes into five categories (I–V). These G2-like genes, which included four sets of tandemly duplicated SlGlk genes, were unevenly distributed over all 12 chromosomes. Research on the SlGlks' promoter areas' cis-regulatory elements (CREs) identified several CREs linked to stress and hormones. SlGlks were expressed in reaction to various abiotic stresses, as demonstrated by RNA-seq. SlCPD, SlDWARF, and BIN2 gene expression was significantly upregulated during drought stress in SLB3-silencing seedlings, whereas the expression of TCH4-related genes was reduced. These results suggest that SLB3 gene silencing decreases tomato plants' drought tolerance through BR signalling transduction (Wang et al. 2020b). Wai et al. (2020) discovered 19 PDI genes spread over eight of the 12 tomato chromosomes. Three paralogous gene pairs were revealed to have segmental duplications. Most PDI genes exhibited varied expression across several fruit organs, based on expression profiling. Furthermore, high temperatures, salt, and abscisic acid (ABA) treatments robustly activated most PDI genes, whereas freezing and challenges including water deficiencies activated a small number of PDI genes. The PDI gene family members, including SlPDI1-1, SlPDI1-4, SlPDI1-3, SlPDI2-1, SlPDI5-1, and SlPDI4-1, exhibit dominant expression in response to ABA treatment and abiotic stressors, indicating their potential involvement in the regulation of tomato stress (abiotic) responses.

2.5 Genome Editing Systems Applied in Tomatoes

Tomatoes have been a primary target for various genome editing systems, encompassing the three primary genome editing technologies (TALEN, ZFN, and CRISPR/Cas systems), as well as the cytidine base editor (Koçberber and Dönmez 2007), a type of DNA base editor (Hilioti et al. 2016; Shimatani et al. 2017). However, unlike the widely utilised CRISPR/Cas systems, ZFN and TALEN achieve sequence specificity by combining the DNA cleavage domain of the FokI endonuclease, with zinc fingers, enabling site-specific cleavage (Xu et al. 2019). These techniques are less commonly utilised in tomatoes.

2.6 Widely Used and Continuously Optimised CRISPR/Cas Systems

The third generation of genome editing tools, CRISPR/Cas systems, first appeared in 2012 and gained popularity fast because they were more straightforward and user-friendly than TALEN and ZFN (Zhang et al. 2018). In order to attain sequence specificity in genome editing, CRISPR/Cas depends on RNA–DNA binding, which was first discovered to be an efficient acquired immunity mechanism in bacteria against viral infections (Zhang et al. 2018). Based on the different RNA combinations and Cas proteins, these systems are divided into five major categories; type II includes CRISPR/Cas9, which is the most studied (Makarova et al. 2015). CRISPR/Cas9 has been used extensively in the tomato genome, and it is now the main genome editing method for determining the function of individual genes and precision plant breeding in this species (Wang et al. 2019a; Xu et al. 2019). Furthermore, a number of plant species, including cotton, maize, and most recently, tomato and tobacco, have gradually benefited from the application of CRISPR/Cpf1 (Cas12a) (Vu et al. 2020; Wang et al. 2018). The potential of CRISPR/Cpf1 in tomato genome editing was demonstrated by the notable increase in homology-directed repair-based genome editing efficiency in tomatoes when combined with geminiviral multireplicons, as opposed to a Cas9-based single-replicon system (Vu et al. 2020).

Despite the widespread application of CISPR/Cas9 in many different animals, a great deal of technological advancements (Table 2.2) have been made to maximise plant genome editing using CRISPR/Cas9 including on tomatoes. These advancements enable mutations in particular cell types, organs, and tissues; enhance editing efficiency; reduce the possibility of off-target consequences; and accelereate the development of edited genomes. For instance, the likelihood of acquiring significant deletions in the mutant tomato plants produced by CRISPR/Cas9 rises when multiplex sgRNAs are used instead of a single sgRNA (Santillán Martínez et al. 2020). Researchers have reportedly achieved fruit-specific gene editing in tomatoes using a spatio-temporally regulated CRISPR/Cas9 toolkit, where Cas9 expression is controlled by a fruit-specific promoter derived from the phosphoenolpyruvate carboxylase 2 gene. This advancement has significantly expedited studies in fruit biology (Feder et al. 2020).

2.7 Increasingly Used Cytidine Base Editor

DNA base editors are CRISPR/Cas9-derived instruments that can precisely substitute one base for another (A-to-G or C-to-T) in genomic DNA without creating double-strand breaks (DSBs). Examples of these editors are cytidine base editing

Table 2.2 Summary of key technological advancements aimed at enhancing CRISPR/Cas9-mediated genome editing in plants, with a focus on tomatoes

Technological improvement	Applied in plants	References
Improving the codon usage of the Cas9 protein through a plant-specific codon-optimised Cas9 gene	*Arabidopsis*, maize, and rice	Castel et al. (2019) and Ma et al. (2015)
Choosing sgRNAs with high efficiency	Rice and *Arabidopsis*	Castel et al. (2019)
Choosing various promoters for driving the Cas9 protein	Maize and *Arabidopsis*	Castel et al. (2019) and Feng et al. (2018)
Choosing a robust terminator following Cas9	*Arabidopsis*	Castel et al. (2019)
Employing multiple sgRNAs simultaneously instead of a single sgRNA	Alfalfa, tomato, and rice	Santillán Martínez et al. (2020) and Wolabu et al. (2020)
Improvement of T-DNA structure	*Arabidopsis*	Castel et al. (2019)
Thermal treatment	*Arabidopsis*	LeBlanc et al. (2018)
Refinement of T-DNA structure	*Arabidopsis*	Castel et al. (2019)
Inhibiting RNA silencing	*Arabidopsis*	Mao et al. (2018)
Utilising chemically modified DNA as a donor template	Rice	Lu et al. (2020)
Editing plant genomes using nanoparticles through a straightforward foliar spray-coated with CRISPR/Cas9 constructs	Wheat	Doyle et al. (2019)
Utilising shortened sgRNA	*Arabidopsis*	Osakabe et al. (2016)
Utilising Cas9 variants such as xCas9 and Cas9-NG	Rice	Zhong et al. (2019)
Initiating gene-edited meristems from scratch to bypass the need for time-consuming tissue culture	*Nicotiana benthamiana*	Maher et al. (2020)
Incorporation of a cassette harbouring the GRF4–GIF1 chimera into the CRISPR/Cas9 construct	Citrus and wheat	Debernardi et al. (2020)
Genome editing in plants facilitated by nanoparticles through a straightforward application via foliar spray, where nanoparticles are coated with CRISPR/Cas9 constructs	Wheat	Doyle et al. (2019)
Genome editing in plants facilitated by nanoparticles through a straightforward foliar spray-coated with CRISPR/Cas9 constructs	Tobacco	Chen et al. (2018)
Incorporating an Flowering locus T FT expression element into the CRISPR/Cas9 construct	Tobacco	Liu et al. (2019)
Editing plant genomes using nanoparticles through a straightforward foliar spray-coated with CRISPR/Cas9 constructs	Wheat	Doyle et al. (2019)
Elimination of transgenes induced by drug treatment	Rice	Lu et al. (2017)
Delivering CRISPR/Cas9 RNPs into plant cells using biolistics	Rice, *Arabidopsis*, and tobacco	Luo et al. (2015)
Self-elimination of transgenes mediated by a programmed suicide gene	Rice	Yubing et al. (2019)
Using a promoter specific to fibres or fruits to control Cas9 expression	Tomato	Feder et al. (2020)

enzymes (CBEs) and adenine base editing enzymes (ABEs) (Nishida et al. 2016). A uracil glycosylase inhibitor and a cytidine deaminase combined with a Cas9 nickase (nCas9) make up the CBE genome editing system, which enables C-to-T base conversion in genomic DNA (Eid et al. 2018). With a base editing efficiency ranging from 26.2% to 53.8%, CBE was first used on tomatoes in 2017. It effectively edited the two tomato hormone signalling genes, ETR1 and DELLA (Shimatani et al. 2017). The biosynthesis route for branched-chain amino acids involves the acetolactate synthase (ALS) gene (McCourt and Duggleby 2006). Previous study reported that a mutation at the proline-186 residue in tomato ALS1 can give resistance to chlorsulfuron. Chlorsulfuron-resistant tomato plants were successfully created through CBE-mediated mutations in the tomato ALS1 gene (Veillet et al. 2019).

2.8 Genome Editing Facilitates the Re-evaluation of Essential Genes in Tomato Fruits

Prior to the advancement of genome editing technologies like CRISPR/Cas9, RNA-interference gene silencing mediated by RNA interference (RNAi) and the analysis of spontaneous mutants were commonly used to assess tomato gene function (Wang et al. 2019c). But with the development of several genome editing technologies, most notably CRISPR/Cas9, tomato mutations may now be produced quickly. Through targeted mutagenesis, this has made it easier to functionally validate a large number of tomato genes related to development, ripening, yield and quality, and stress responses (Xu et al. 2019).

2.9 Reassessing the Role of Transcription Factors in Fruit Ripening

The phytohormone ethylene is essential to the ripening of tomatoes, a traditional climacteric fruit (Liu et al. 2020b). Giovannoni (2007) reports that many spontaneous mutations in tomatoes have notable characteristics lacking in ripening. Transcription factor–encoding genes were found to underlie the three main ripening-deficient mutants: colourless non-ripening (Cnr), non-ripening (nor), and ripening inhibitor (rin) (Manning et al. 2006). It has long been known that rin, nor, and Cnr are master regulators of tomato fruit ripening. They regulate the transcriptional activation of genes associated with downstream ripening, such as cell wall–modifying enzymes, ethylene biosynthesis genes, and other transcription factors (Seymour et al. 2013). Other tomato transcription factors AP2a, FUL2, FUL1, TAG1, MADS1, and HB1 have also been studied using RNA interference (RNAi), elucidating their functions in fruit ripening (Chen et al. 2020).

Six transcriptional regulators RIN, AP2a, CNR, NOR, FUL2, and FUL1 have had their roles in tomato fruit ripening re-evaluated recently, thanks to CRISPR/Cas9-mediated targeted mutagenesis (Table 2.2). Fascinatingly, ripening phenotypes incongruous with spontaneous mutants or RNAi plants have been reported in mutations induced by CRISPR/Cas9 in the master regulators nor, rin, and Cnr (Wang et al. 2019a). For example, CRISPR/Cas9-mediated rin-knockout mutations showed partial ripening and moderate red colouration, whereas the rin mutant traditionally displayed near-complete ripening inhibition. These findings suggest that rin is essential for full ripening progression but not necessary for ripening initiation (Li et al. 2020). Similar to the spontaneous nor mutants, which displayed severe non-ripening, the CRISPR/Cas9-generated nor mutants only displayed partial non-ripening phenotypes, suggesting a complicated role for nor in regulating fruit ripening (Gao et al. 2020). Three other transcription factors FUL1, AP2a, and FUL2 have shown consistent ripening phenotypes between RNAi plants and CRISPR/Cas9 mutants in contrast to rin, nor, and Cnr (Wang et al. 2019c). According to Karlova et al. (2011), AP2a acts as a negative regulator of ripening initiation for instance, AP2a-RNAi plants and AP2a acts as a negative regulator of the initiation of ripening. This plant commence fruit ripening earlier but do not go full ripening. Comparing FUL1/FUL2 double mutants and FUL1 and FUL2 mutants to wild-type plants, they also exhibit poor fruit ripening, indicating their shared functions in ripening control (Wang et al. 2019b). Overall, these results highlight the intricacy and redundancy of the transcriptional regulatory network governing tomato fruit ripening and imply that spontaneous mutants like nor, rin, and Cnr may not correctly represent loss-of-function mutations. In subsequent research, CRISPR/Cas9-mediated targeted mutagenesis provides a potent tool to explore the functions of additional tomato transcription factors crucial for fruit ripening.

2.10 Re-evaluating the Role of Cell Wall–Modifying Enzymes in Fruit Ripening

A crucial part of the ripening process for fleshy fruits is softening, which greatly enhances flavour development and general palatability (Klee and Giovannoni 2011). Transgenic techniques have been used for the past 30 years to target tomato genes that encode cell wall–related enzymes, such as expansin, xyloglucan endo-transglycosylase, polygalacturonase (PG), pectin methylesterase, galactanase (TBG), and pectate lyase (PL), to better understand their functions in controlling fruit softening. Except for the reduction of pectate lyase (PL), which markedly decreased fruit softening, these transgenic efforts, however, frequently produced modest results or no significant influence on the softening of tomato fruit (Uluisik et al. 2016; Yang et al. 2017). Recently, CRISPR/Cas9-mediated targeted mutagenesis was used to re-evaluate the roles of three important encoding pectin-degrading enzyme genes PL, TBG4, and PG2a in tomato fruit softening (Table 2.2). The CRISPR/Cas9-targeted PL mutants exhibited stable ripening phenotypes in contrast to PL-RNAi plants, leading to notably firmer fruits. This highlights the critical function of PL in the softening of tomato fruit (Wang et al. 2019a). On the other hand, prior research revealed that fruit firmness was not significantly affected by the silencing of TBG4 and PG2a (Smith et al. 2002). Remarkably, CRISPR/Cas9-induced mutations in TBG4 and PG2a did not change the hardness of the fruit

(Wang et al. 2019a). Remarkably, these mutations did impact tomato fruit ripening in ways that were not previously noted in their corresponding silencing investigations (Smith et al. 2002). These factors included fruit colour and carotenoid production. These findings highlight CRISPR/Cas9-mediated targeted mutagenesis as a powerful tool for precisely elucidating the roles of specific genes in tomato fruit ripening, shedding new light on the complexities of fruit softening regulation beyond traditional transgenic approaches.

2.11 Other Aspects of Fruit Ripening Still Need to Be Reassessed

Besides cell wall–modifying enzymes and transcription factors, enzymes, ethylene biosynthesis ethylene receptors, microRNAs (miRNAs), and epigenetic regulators are significant factors influencing fruit ripening (Wang et al. 2020a). However, RNA interference (RNAi) techniques have been the main tool used to study several of these tomato fruit ripening factors. Thus, it is imperative to use genome editing tools to re-evaluate these characteristics. Our comprehension of the regulatory networks controlling fruit ripening is expected to be further enhanced by this method, especially regarding the crucial function of ethylene in coordinating important regulatory pathways during climacteric fruit ripening.

2.12 Applications of Genome Editing in Tomato Improvement

Genome editing has greatly advanced tomato genetic enhancement efforts in addition to basic research (Kwon et al. 2020). The utilisation of genome editing techniques, specifically CRISPR/Cas9, has been beneficial in tomato breeding as it has accelerated the domestication of wild tomato varieties, improved fruit yield and quality, and customised tomato cultivars for urban agriculture through the facilitation of precise knockout mutations.

2.13 Enhancing the Yield and Quality of Tomato Fruits

Fruit setting rate, flowering speed, fruit size at harvest, and other characteristics are the main determinants of tomato production, a crucial agronomic attribute in crop development. Important tomato genes linked to yield have been found through genetic research (Zsögön et al. 2017). A potential precision breeding technique, CRISPR/Cas9-driven promoter mutagenesis, is especially useful for generating differences in quantitative parameters for agricultural development. Genome editing has also proven crucial in improving tomato fruit quality, as customer demands for higher fruit quality continue to rise. According to Bai and Lindhout (2007), fruit quality includes characteristics including colour, shape, size, sweetness, scent, acidity nutrient content, and shelf life. Although red tomatoes are widely available, other coloured tomatoes are also in demand. In Asia, for example, pink tomatoes are favoured over red ones (Zhu et al. 2018a). Pink tomatoes from a variety of red cultivars have been effectively created by CRISPR/Cas9-mediated disruption of tomato (MYB12), a crucial regulator in the biosynthesis of flavonoids (Yang et al. 2019). Similar to this, orange and yellow tomatoes have been produced by mutations in genes related to the carotenoid biosynthesis pathway, such as phytoene synthase 1 (PSY1) (Dahan-Meir et al. 2018). This kind of fruit colour modification, though, might affect the amounts of nutrients like flavonoids and carotenoids. Anthocyanin content, which contributes to the health-promoting features of purple tomatoes, is highly appreciated in these tomatoes, as demonstrated by the recently established ‘Sun Black’ line. Purple tomatoes have been effectively created by placing a strong promoter upstream of the native anthocyanin biosynthetic gene SlANT1 using CRISPR/Cas9 or TALEN, exhibiting effective targeted gene insertion in plant breeding (Čermák et al. 2015).

Tomato fruits are an important part of our daily diet since they are full of nutrients including lycopene, vitamins, and minerals (Zhu et al. 2018c). The lycopene content in tomato fruits has increased significantly (by around 5.1 times) as a result of multiplex editing of five genes—SGR1, LCY-B1, LCY-E, and Blc—using CRISPR/Cas9 (Li et al. 2018). These genes are implicated metabolic pathways of carotenoids. Similarly, major increases in total carotenoids and lycopene levels have resulted from CBE-mediated nucleotide changes in genes involved in carotenoid accumulation, such as SlDDB1, SlDET1, and SlCYC-B (Hunziker et al. 2020). Gamma-aminobutyric acid (GABA), which promotes health and has several functional advantages, is also present in tomato fruits GABA content in tomato fruits was significantly increased (1.34–3.50 times) as a result of CRISPR/Cas9-mediated multiplex editing that targeted four GABA metabolism-related genes, including GABA-TP2, CAT9, GABA-TP3, and SSADH (Li et al. 2018).

Owing to its superior flavour and numerous industrial uses, parthenocarpy—fruits without seeds—is a desirable characteristic in horticulture crops (Ueta et al. 2017). Auxin is a phytohormone that controls this characteristic. Tomato fruits without seeds have been produced using CRISPR/Cas9-mediated mutations in genes like SlIAA9 and transcription factors like SlARF7 and SlARF5 (Ueta et al. 2017). Significantly, under heat stress, mutations in the MADS-box gene SlAGL6 have also produced parthenocarpic fruits with increased yield. Klap et al. (2017) suggested SlAGL6 as a promising gene for breeding seedless fruits, though more research is needed to determine its precise regulatory role in parthenocarpy.

Another important factor in fruit quality is its shelf life, especially when considering the needs of the consumer. Tomato fruits now have a longer shelf life, thanks to CRISPR/Cas9-induced mutations in the tomato alcohol dehydrogenase 1 (ALC) gene, which were found in the spontaneous mutant alcobaca (alc) (Yu et al. 2017). Similar to this, fruit firmness and shelf life have increased due to mutations in the pectate lyase gene (PL) without affecting other fruit qualities (Yu et al. 2017). ALC and PL are valuable genetic resources for improving tomatoes, as evidenced by the enhanced resistance to plant diseases shown by PL-RNAi fruits and alc spontaneous mutants (Yang et al. 2017). One important objective of tomato breeding is to increase plants' tolerance to biotic and abiotic stressors (Tieman et al. 2017). *Oidium neolycopersici*, the fungus that causes powdery mildew, is a global problem that affects tomatoes cultivated in fields and greenhouses. To create tomato plants resistant to powdery mildew, CRISPR/Cas9-mediated mutations in susceptibility (S) genes, like MLO and PMR4, have been used (Zaidi et al. 2018).

Within a comparatively short period of time, transgene-free resistant tomato cultivars were developed as a result of the mutation of SlMlo1, which greatly decreased sensitivity to powdery mildew (Nekrasov et al. 2017). According to Santillán Martínez et al. (2020), SlPMR4 mutations have also increased tomato resistance to *O. neolycopersici*. Another major hazard to tomatoes during growth and after harvest is grey mould, which is brought on by the necrotrophic fungus *Botrytis cinerea* (Williamson et al. 2007). Recent research has demonstrated that mutations in acetylenase-encoding genes, such as ACET1b and ACET1a, can boost resistance to this pathogen, even though genome editing for improving tomato resistance to *B. cinerea* is still understudied (Jeon et al. 2020). Through RNA interference (RNAi) or overexpression investigations, several possible susceptibility genes to *B. cinerea* in tomatoes have been found. These genes could be candidates for genome editing to create tomato cultivars that are resistant to the disease (Cai et al. 2020).

Tomato resistance to a variety of diseases, such as bacteria, viruses, and oomycetes, has also been strengthened by genome editing. The use of genome editing to genetically increase tomato resistance to abiotic stressors is still comparatively unexplored. A key abiotic stress that affects tomato farming is drought. Recently, tomatoes have shown dramatically enhanced drought tolerance due to mutations in the transcription factor gene SlLBD40 (Liu et al. 2020a). Moreover, increased resistance to the sulfonylurea herbicide chlorsulfuron has been seen as a result of CRISPR/Cas9-mediated mutations in the branched-chain amino acid synthesis gene ALS1 (Veillet et al. 2019). In conclusion, genome editing represents a powerful tool for enhancing multiple facets of tomato production, from yield and quality to stress resistance. Its application continues to advance, offering unprecedented opportunities for precision breeding and sustainable agricultural practices in tomato cultivation (Fig. 2.3).

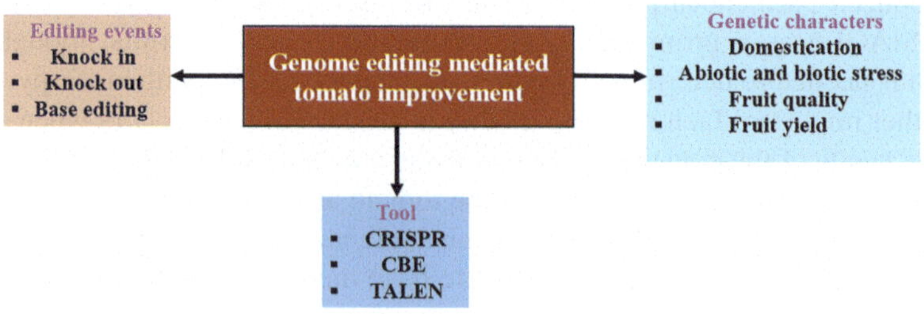

Fig. 2.3 A chart depicting the uses of genome editing for improving tomatoes

2.14 Tomato Proteomics

Proteome analysis was used to examine how aluminium affected the development of the cotyledon in tomatoes, and the results showed that 49 proteins accumulated differently when exposed to aluminium stress (Zhou et al. 2009). Another study identified 40 differently expressed proteins by 2D-gel electrophoresis after using proteomics to examine tomato seedlings exposed to salt and silicon stress. In particular, in root libraries, silicon supplementation increased the expression of 24 proteins, while salt stress caused the opposite effect (Muneer and Jeong 2015). Tomato leaves were shown to express 52 distinct proteins, some of which were unique, in response to waterlogging stress. Proteome analysis was used to examine how aluminium affected the development of the cotyledon in tomatoes, and the results showed that 49 proteins accumulated differently when exposed to aluminium stress (Zhou et al. 2009). Proteomics was employed in another investigation. These proteins are involved in energy metabolism, disease resistance, stress defence systems, photosynthesis, and protein production, among other functions (Ahsan et al. 2007; Ijaz et al. 2024). Furthermore, hot water treatment was looked into as a way to avoid protein denaturation and activate antioxidant chemicals, which could help tomato fruits develop a tolerance to chilling harm (Salazar-Salas et al. 2017). Additionally, proteomic investigations have investigated the tolerance to temperature, dryness, and stressors such as NaCl and $NaHCO_3$. For example, employing MALDI-TOF/TOF MS and 2D-gel electrophoresis, 67 differentially expressed proteins were found in tomato seedlings under high-temperature stress. A proteomic study has recently brought attention to alterations that occur when different tomato types are stressed by low temperatures (Sang et al. 2017). Even with the abundance of proteomic research on tomato abiotic stress responses, there are still major gaps in data interpretation and analysis. Public databases with increased protein annotations are becoming better, though, and this will help find more regulatory candidate proteins and eventually help create crops that can withstand stress better in terms of both yield and quality. Table 2.3 lists the various proteins involved in stress tolerance in tomatoes and their respective functions.

2.15 The Protein Profile of Tomato Fruit

Since the dawn of human civilisation, the *Solanaceae* family of flowering plants has been crucial for human sustenance. Proteomics research has significantly enhanced our knowledge of the organisation and role of *Solanaceae* species. Because of their economic significance, much of the proteomics research within the *Solanaceae* family has focussed on tomatoes, potatoes, and tobacco. Extracting stable protein mixture derived from plant tissues is extremely challenging due to the high levels of proteases and other enzymes, as well as secondary chemicals in plant cells that often hinder protein isolation and subsequent analysis (Ghosh and Xu 2014). Despite these obstacles, the fruit proteome of tomatoes has been extensively studied. Ripening is a crucial factor in determining fruit quality, encompassing five

Table 2.3 Proteins aimed at enhancing tolerance to abiotic stress in tomatoes

Protein	Function	Results	References
LeHSP21.5	Enhances resilience against tunicamycin-induced endoplasmic recticulum (ER) stress	Increased tolerance	Zhao et al. (2007)
MdVHA-B	Enhances tolerance to drought stress	Enhanced resilience	Hu et al. (2016)
Histidine kinase	Mitigates abiotic stresses	Regulation of the balance of sodium ions within cells	SinghMohan and BhallaPrem (2022)
SlbZIP1	Decreases stress mediated by ABA (abscisic acid)	Decreased sodium levels and water scarcity	Zhu et al. (2018c)
Golden 2–like	Promotes plant growth and mitigates abiotic stress	Decreased tolerance to drought stress due to reduced levels of superoxide dismutase (SOD) and peroxidase enzymes	Wang et al. (2022)
RING zinc finger	Promotes plant growth while alleviating abiotic stresses	Alleviated stress conditions for the plant	Han et al. (2022)
2,3-butanediol	Responds to drought and cold stress	Mitigated adverse impacts of abiotic stresses	Park and Choe (2019)
E42 and LA3120	Promote plant growth while alleviating stressors	Enhanced plant growth in conditions of water stress	Fernandes et al. (2021)
Dehydrins (DHNs)	Improve resilience to salinity and drought stress	Challenges posed by salinity and drought	Muñoz-Mayor et al. (2012)
LEA proteins	Prevent membrane leakage and stabilise membranes and proteins	Regulation of water equilibrium and ion retention	Olvera-Carrillo et al. (2011)
SpPKE1	Interacts with an F-box protein associated with drought resistance	Drought tolerance	Li et al. (2018)
P5CS protein	Participates in proline production and is activated during salt stress to enhance drought resilience	Tolerance to drought and salinity	Karthikeyan et al. (2011) and Ogden et al. (2020)
COR15 protein	Folds and attaches to chloroplast membranes to bolster leaf cell resilience during freezing	Cold tolerance	Thalhammer and Hincha (2014) and Zhang et al. (2021)
Systemin peptide	Excludes sodium, protease inhibition, and antioxidant activity	Reduced attractiveness to herbivores and strong tolerance to high salt stress	Cirillo et al. (2022)

developmental stages: green, breaker, turner, light-red, and red (Ghatak et al. 2017). These stages are studied to identify the proteins associated with ripening, regulated proteins, and cultivar-specific proteins. The evaluation process employs various high- and low-throughput methods, starting with a gel-based approach (2DE) to separate total fruit proteins at all stages and identify differentially altered areas. By combining quantitative reverse transcription analyses with proteomic data, researchers can discover additional genes and proteins, such as oxalyl-CoA decarboxylase and glutathione S-transferase (GST)/peroxidase (Kok et al. 2008). A study of the cherry tomato pericarp proteome showed a rise in protein levels associated with protein synthesis and amino acid degradation during cell division. During cell growth, proteins associated with photosynthesis were temporarily expressed, while proteins related to glucose metabolism increased during fruit development (Faurobert et al. 2007).

2.16 Importance of Protein Isoforms and Post-Translational Modifications (PTMs)

PTMs regulate protein stability, function, and cellular localisation. Notable changes in the post-translational modifications of certain proteins are frequently connected with plants' ability to withstand abiotic stress (Wu et al. 2016). Furthermore, the varied configurations of 20 different amino acids—the alterations including the addition of various chemical groups such as phosphates, carbohydrates, proteins, and lipids—are critical in determining protein structure. However, the resulting protein array varies according to cell type, function, shape, tissue localisation, environment, and, most critically, the cell's developmental stage. Furthermore, proteins exist in several isoforms that undergo processes such as metabolism, degradation, splicing, and frequent interactions to form specialised complexes. Rocco et al. (2006) looked at the proteome profiles of two tomato cultivars (local and commercial elite ecotypes) as they ripened. These cultivars shared proteins involved in redox control, stress response, and energy production, such as multiple thioredoxin peroxidase 1, HSPs, adenylate kinase 2, GST isoforms, and meta caspase 1 and cytosolic malate dehydrogenase. Many reactions include numerous protein isoforms. Protein changes in later metabolic processes, such as isoprene and carotene production, frequently necessitate close coordination. For example, fruit peel and pulp have unique isoform expression patterns, such as GGPS2 and GGPS1 (Szymanski et al. 2017). Plants' molecular responses to stress (abiotic) are complex, with the transcriptional activity of stress-related genes being the most important factor. While many PTMs have been found, fully analysing all potential protein changes remains difficult. Phenotyping plants for crop enhancement measures and evaluates plant responses to stress (abiotic) at the PTM level (Wu et al. 2016; Ullah et al. 2023a, b). This strategy improves our understanding of agricultural plant stress adaptation processes and the acquisition of stress tolerance by identifying and quantifying PTMs on a broad scale.

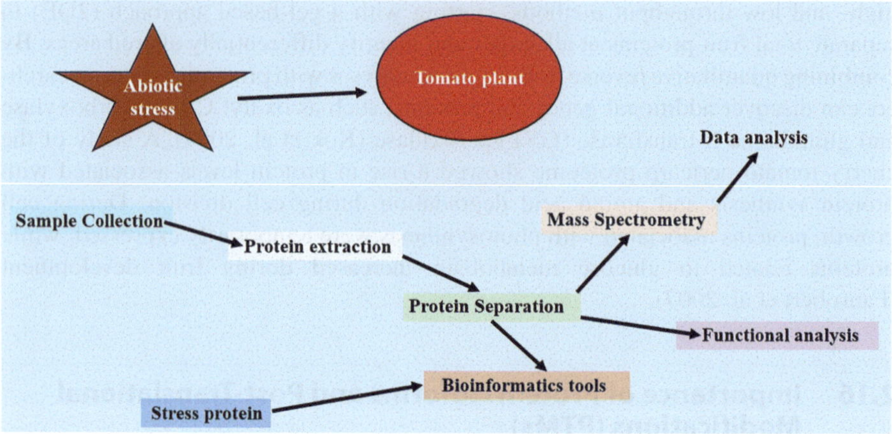

Fig. 2.4 Sketch illustrating the proteomics of abiotic stress in tomatoes

2.17 Proteomic Analysis of Protein Functions in Biology

As mentioned earlier, a proteome refers to the entire complement of proteins expressed by an organism's genome, encompassing post-translational modifications. It is dynamic and heterogeneous, varying across different conditions. The study of proteomics primarily focus on the protein quantification, localization, and identification, and the analysis of post-translational modifications (PTMs), functional and structural characteristics, and protein-protein interactions. Proteomics not only unveils the complexity of life but also emphasises the relevance of cellular viability and how it responds to various environmental or climatic situations (Szymanski et al. 2017). Proteomic research provides a thorough understanding of the mechanisms behind both normal and abnormal cellular functioning (Beynon 2005). Despite decades of molecular and physiological study into biological systems, systems biology methodologies continue to uncover previously unknown or unanticipated links between cell metabolism and activities (Chakravarti et al. 2010) (Fig. 2.4).

2.18 Tomato Metabolomics

A promising method known as metabolomics offers a biochemical snapshot of an organism's phenotype. This technique allows for the systematic identification and quantification of low-molecular-weight compounds that are crucial for understanding nutritional and toxicological traits. Understanding the complete biological complexity necessitates researching the wide range of primary and secondary metabolites that are present, even while the knowledge of genes, proteins, and transcriptomes offers insights. The roles of metabolites in plants under drought and high salinity stress have been studied extensively, using methods like nuclear magnetic resonance (NMR) (Schripsema 2010; Ullah et al. 2023c) and gas chromatography–mass spectrometry (Lee et al. 2012). Studies on different stress responses in tomatoes have been conducted; however, abiotic stress responses have received less attention. Five cherry tomato cultivars under water stress showed lower concentrations of phenolic compounds and shikimate (Sánchez-Rodríguez et al. 2011). Another study looked at how plant organs responded metabolically to low oxygen stress brought on by storing fruits and vegetables in tight spaces. Low oxygen exposure of cultured tomato cells resulted in changed metabolic profiles with increased lactate and sugar alcohol levels, as well as an accumulation of glycolysis intermediates (Ampofo-Asiama et al. 2014; Haider et al. 2021a). Nunes-Nesi identified tomato leaf metabolite quantitative trait loci (mQTL), highlighting 76 negative and 42 positive mQTL that are important in controlling the primary nitrogen and carbon metabolism of leaves, especially during stressful situations (Nunes-Nesi et al. 2019). Studies on tomato metabolism have greatly improved our knowledge of the metabolite networks and pathways connected to different economic features. By using metabolomics to investigate abiotic stress, it is possible to clarify the molecular processes that underlie stress reactions and create tomato cultivars that are more resilient and produce more.

2.19 Applications of Metabolomics in Plant Stress Responses

Plant stress refers to any changes in growth conditions that disrupt metabolic equilibrium and require metabolic pathway adaptation via acclimatisation. Metabolites are spread throughout tissues and intra-cellular regions. Plants are constantly challenged by various environmental conditions, including abiotic and biotic stimuli, which affect their metabolite composition. These alterations impede overall plant growth, development, and productivity, posing a threat to food security, particularly as the world's population grows. Metabolomics throws light on plant metabolism during development and in response to diverse stresses by identifying compounds such as stress metabolism derivatives, stress signal transduction molecules, and chemicals involved in the plant's acclimatising response. It aids in a thorough understanding of plant stress biology. Metabolites are crucial for plant development, and their molecular structures and abundances are generally well-conserved throughout the plant kingdom. In contrast, secondary metabolites and their control are highly sensitive to environmental variations, such as light, salinity, UV radiation, temperature, nutritional deficiencies, and oxidative stress (Feng et al. 2021), as well as interactions with other plant species (Szakiel et al. 2011; Haider et al. 2021b). These stressors influence the expression of genes involved in their manufacture (Sharma 2018). Under stress, plants tend to produce a variety of secondary metabolites to cope with negative impacts, allowing for metabolic modifications and homeostasis restoration. Thus, knowing the fluctuation of plant secondary metabolites in reaction to stress is critical. Metabolites that contribute to abiotic stress tolerance in tomatoes are given in Table 2.4.

Table 2.4 Metabolites contributing to abiotic stress tolerance in tomatoes

Gene	Function	Results	References
Melatonin	Protection of membranes and protein from damage	Ability to resist biotic stress	Martinez et al. (2018)
Solyc04g014600 (universal stress protein)	Analysis of protein composition in phloem and its exudates	Ability to tolerate drought	Irfan et al. (2021)
SpUSP, an annexin-interacting universal stress protein	Reduction of oxidative stress by inhibiting ROS production and activating various stress-responsive genes, leading to the accumulation of osmo-protective solutes	Ability to tolerate drought	Loukehaich et al. (2012)
C2H2-type zinc finger protein	Enhancement of overall growth and development of plant tissue	Reactions to abiotic stress	Ming et al. (2020)
SlMAPK3	Increased seedling growth and germination rate, and additionally, the transgenic plants exhibited higher chlorophyll content and increased root biomass accumulation under Cd^{2+} stress	Ability to withstand Cd stress in addition to drought tolerance	Muhammad et al. (2019)
GOBLE	Determination of the boundaries of leaflets in compound leaves	Reduction of excessive transpiration	Berger et al. (2009)
SlNAC1	Enhancement of viral replication and cold tolerance via interactions with replication accessory proteins	Ability to withstand phytopathogens and cold temperatures	Ma et al. (2013) and Selth et al. (2005)
SlNAM2	Morphogenesis of flower boundaries	Prevention of excessive water loss	Hendelman et al. (2013)
SlNAC4	Ripening of fruit and accumulation of carotenoids	Response to oxidative stress	Zhu et al. (2014)

2.20 Tomato Ionomics

Ionomics studies how species, organs, variety, and environment affect the accumulation of metals, metalloids, non-metals, and essential and non-essential elements in plants. In this field, high-throughput methods such as inductively coupled plasma (ICP) atomic emission spectrometry and inductively coupled plasma mass spectrometry (ICP-MS) are essential (Salt et al. 2008). Silicon (Si) has been the subject of extensive ionomics research, especially in dicots and the *Solanaceae* family, whose tissues normally contain less than 0.5% silicon. By postponing wilting and lowering reactive oxygen species (ROS) in situations of heavy metal stress, silicon improves drought tolerance (Ahmad et al. 2019). Exogenous silicon decreased sodium (Na) and chloride (Cl) concentrations in roots, stems, and leaves without influencing translocation to shoots, according to studies examining the impact of silicon on tomatoes' ability to withstand salt (Li et al. 2015). According to research conducted in hydroponic systems, silicon helps tomatoes grow under salinity stress by enhancing photosynthesis, chlorophyll concentration, and leaf water content (Haghighi and Pessarakli 2013). Moreover, a study conducted on different tomato cultivars demonstrated that silicon supplementation had a major impact on fruits' absorption of arsenic (Marmiroli et al. 2014; Haider et al. 2022a). Extensive research is still needed to fully understand the roles of several elements in various biochemical forms in plants.

2.21 Advancements in Phenomics for Enhancing Abiotic Stress Tolerance in Tomatoes

High-throughput investigation of phenotypic variations that result from intricate interactions between phenotypes, genotypes, and the environment is the focus of phenomics. Research on individual genomes and phenomes (which include all phenotypes) in big populations or individuals complement each other (Ichihashi and Sinha 2014). When using high-throughput sequencing technology to identify alleles, plants with tolerant phenotypes are useful genetic resources. While phenotypic characterisation has advanced more slowly than genotyping efficiencies due to breakthroughs in sequencing, especially for quantitative traits linked to stress tolerance, this has not always been the case (White et al. 2012). However, advances in phenotyping techniques have made it possible to precisely identify traits with the aid of sophisticated imaging systems, sensors, automation, and computational tools. As a result, phenomics has become a high-throughput method that can evaluate hundreds of phenotypic parameters across thousands of genotypes at the same time (Tardieu et al. 2017). Numerous phenomics platforms have been created to study physiological characteristics in tomato plants during drought, such as the scan analyser 3D (Laxman et al. 2018). Given the intricacy of gathering phenomics data, cooperation among scientists with various specialties is crucial. Furthermore, gathering phenomics data takes a lot of time and resources; therefore, integrating technology developments to cut costs and increase throughput is required.

2.22 Integration of Omics Technologies

Profound datasets that can reveal new genetic and chemical components controlling a variety of physiological processes have been produced by recent advances in omics methods (genomics, proteomics, transcriptomics, metabolomics, phenomics, and ionomics) (Cohen et al. 2017). But no single omics technique is sufficient to fully comprehend the complexity of plant stress responses (Fig. 2.2). Instead, many omics approaches must be integrated. Interpreting response mechanisms is highly challenging when high-throughput data from several omics methods are analysed. While there are many software tools available for basic omics data analysis, understanding the complex biochemical and physiological reactions under stress settings requires collaborative approaches. Integrating various tools, techniques, and approaches appears crucial for developing climate-smart crop varieties, as evidenced by recent breakthroughs in omics approaches (proteomics, genomics, transcriptomics, ionomics, and so on) (Chaudhary et al. 2019; Haider et al. 2022b). Plant molecular scientists still face enormous challenges in figuring out the specific molecular roles of genes. Genome-wide association studies (GWAS) finds regulatory loci impacting several variables at once and is adaptable to a variety of germplasms (Sonah et al. 2015). In soybeans, candidate genes have been successfully discovered by combining QTL and GWAS techniques, even though QTL regions are usually extensive and contain many genes. Furthermore, combining gene expression profiles with RNA sequencing offers vital information about the roles of unidentified genes (Chen et al. 2016). Thus, combining transcriptome profiling with QTL and GWAS offers the potential for finding candidate genes with differential expression (Guo et al. 2019a). For example, a study using differentially expressed genes, integrated QTLs, and expression Quantitative Trait Loci (eQTLs) in tomatoes identified potential genes influencing tolerance to water stress (Albert et al. 2017). The effective adoption of computational approaches by plant breeders, despite the huge and multi-gene QTL areas, is contingent upon the availability of user-friendly interfaces, online lessons, manuals, and interactive features. As shown in Table 2.5, numerous easily navigable databases have been created for tomatoes to incorporate omics-scale data from various methods.

2.23 Challenges in Improving Tomato Tolerance to Abiotic Stresses

Improving the resilience of tomato varieties to abiotic stresses requires addressing several key challenges and implementing strategic recommendations.

2.23.1 Genetic Complexity and Multifactorial Stress Conditions

Tomatoes possess a complex genome with numerous stress-responsive genes. Developing varieties that can simultaneously tolerate multiple abiotic stresses, such as high temperatures and drought, is more challenging than addressing single stress factors.

2.23.2 Unpredictable Environmental Conditions

Climate change results in unpredictable weather patterns, complicating the breeding of tomatoes that can thrive under a wide range of stress conditions.

Table 2.5 Online databases created for integrated omics research in tomatoes

Database	Description	References
Tomato Functional Genomic Database (TFGD)	Database of microarray expression, analysis of metabolite profiles, and RNA-Seq data	Fei et al. (2010)
Tomato Genomic Resources Database (TGRD)	Interactive exploration of tomato genes, microRNAs, simple sequence repeats (SSRs), and key quantitative trait loci (QTLs)	Suresh et al. (2014)
Tomatomics	Complete mRNA sequences, expression profiles, gene architectures, and functional annotations of genes	Kudo et al. (2017)
Tomatoma (Micro-Tom Database)	Available resources including metabolite information, phenotype details, and TILLING data	Shikata et al. (2016)
MiBASE	A database containing Micro-Tom expressed sequence tags (ESTs) and tomato Unigene EST sequencing, EST annotations, SSRs, single nucleotide polymorphisms (SNPs), metabolic pathways, gene ontology, gene expression data, and sequence similarities	Yano et al. (2006)
KaFtom	A database containing full-length complementary DNA(cDNA) clones of Micro-Tom and libraries of full-length cDNAs for EST sequencing	Aoki et al. (2010)
Kazusa Tomato Genomic Database (KaTomics DB)	A database containing DNA markers, annotations of SNPs, and genome sequences	Shirasawa and Hirakawa (2013)
CoxPathDB	Predicting the function of tomato genes based on the results of analyses of functional enrichment in genes that are co-expressed	Nekrasov et al. (2017)
Metabolome Tomato Database (MoToDB)	Liquid chromatography–mass spectrometry	Moco et al. (2006)
Sol Genomics Network	Exploring the tomato genome, identifying sequence similarities, and retrieving annotations	Shimatani et al. (2017)

2.23.3 Gene–Environment Interactions and Yield Considerations

The effectiveness of stress tolerant genes varies with environmental conditions, posing challenges in predicting the outcome of genetic modifications. Enhancing stress tolerance can sometimes lead to trade-offs such as reduced yield or fruit quality.

2.23.4 Regulatory and Consumer Acceptance

The regulatory landscape and consumer perceptions of genetically modified organisms (GMOs) influence the adoption of new stress-tolerant tomato varieties.

2.23.5 Limited Genetic Diversity

Many commercial tomato varieties have a narrow genetic base, limiting the traits available for breeding stress tolerance.

2.24 Future Recommendations

To advance the development of stress-tolerant tomatoes, the following future recommendations in molecular strategies are proposed:

2.24.1 Integration of Cutting-Edge Genomic Technologies

Continue leveraging next-generation sequencing (NGS) for comprehensive genome analysis to identify new stress-responsive genes and regulatory elements.

2.24.2 Precision Breeding and Genome Editing

Utilise advanced CRISPR/Cas9 technologies and emerging genome editing tools for precise modification of stress-responsive genes.

2.24.3 Exploiting Natural Variation and Wild Relatives

Explore wild tomato species and traditional landraces to identify stress-tolerant traits and use marker-assisted introgression to transfer these traits into elite cultivars.

2.24.4 Multi-omics Integration and Systems Biology

Integrate genomics, transcriptomics, proteomics, metabolomics, and epigenomics to understand stress response pathways comprehensively and identify targets for improving stress tolerance.

2.24.5 Epigenetic Modifications and Synthetic Biology

Apply epigenome editing tools to modify gene expression under stress conditions and design synthetic metabolic pathways to enhance stress tolerance.

2.24.6 Enhancing Physiological Traits

Optimise root system architecture and modify stomatal regulation to improve water and nutrient uptake efficiency under stress.

2.24.7 High-Throughput Phenotyping and Data Analytics

Deploy advanced phenotyping technologies and utilise artificial intelligence for real-time monitoring and analysis of stress responses in large populations.

2.24.8 Collaborative Research and Translational Trials

Strengthen international collaborations and engage farmers in participatory breeding programmes to ensure the adaptation and effectiveness of stress-tolerant varieties under field conditions.

By adopting these strategies and recommendations, researchers can accelerate the development of tomato varieties resilient to abiotic stresses, contributing to sustainable agriculture and global food security amidst changing environmental conditions.

2.25 Conclusion

Tomato (*Solanum lycopersicum*) is an important crop with global economic and nutritional benefits. However, abiotic factors such as drought, nutritional shortages, salt, and severe temperatures have a significant impact on productivity and quality. These stresses cause a complex network of molecular and physiological reactions, which can significantly reduce plant growth and output. Understanding the molecular processes that drive abiotic stress responses in tomatoes is critical for creating effective stress tolerance measures. Recent advancements in molecular biology and biotechnology have provided vital insights into these pathways, resulting in new techniques to mitigate the negative impacts of abiotic stressors. CRISPR/Cas9 gene editing technologies have emerged as potent tools for precisely modifying stress-related genes, allowing the production of tomato cultivars that are more resistant to abiotic stress. Genetic engineering techniques have also been used to transfer stress-responsive genes from other species, which improves tomatoes' stress resilience. Furthermore, marker-assisted breeding has facilitated the identification and inclusion of favourable features into new tomato cultivars, hastening the production of stress-tolerant types. These molecular techniques show significant promise for boosting tomato tolerance to abiotic stressors and increasing yield and quality. However, their successful application necessitates a thorough understanding of the intricate interconnections of genetic, environmental, and management factors. Long-term advances in tomato stress tolerance will require a combination of modern molecular tools, conventional breeding strategies, and sustainable agricultural practices. Finally, the combination of cutting-edge molecular technology and classic breeding procedures provides a solid foundation for improving tomato resistance to abiotic stressors. Continued research and innovation in this field will be vital to ensuring long-term tomato production in the face of rising environmental problems.

References

AbuQamar S, Luo H, Laluk K, Mickelbart MV, Mengiste T (2009) Crosstalk between biotic and abiotic stress responses in tomato is mediated by the AIM1 transcription factor. Plant J 58(2):347–360

Ahmad P, Tripathi DK, Deshmukh R, Pratap Singh V, Corpas FJ (2019) Revisiting the role of ROS and RNS in plants under changing environment. Environ Exp Bot 161:1–3

Ahmad Y, Haider S, Iqbal J, Abbasi BA, Yaseen T, Mahmood T (2022) The mechanisms of genome editing technologies in crop plants. In: Principles and practices of Omics and genome editing for crop improvement. Springer International Publishing, Cham, pp 295–313

Ahmad Y, Haider S, Iqbal J, Naseer S, Attia KA, Mohammed AA, Mahmood T (2024a) In-silico analysis and transformation of OsMYB48 transcription factor driven by CaMV35S promoter in model plant–Nicotiana tabacum L. conferring abiotic stress tolerance. GM Crops Food 15(1):130–149

Ahmad Y, Haakim Z, Iqbal J, Abbasi BA, Mahmood T, Kazi M (2024b) Technological innovations for abiotic stress resistance in horticultural crops. In: Omics-based techniques for global food security. Wiley, pp 233–244

Ahsan N, Lee DG, Lee SH, Kang KY, Bahk JD, Choi MS, Lee IJ, Renaut J, Lee BH (2007) A comparative proteomic analysis of tomato leaves in response to waterlogging stress. Physiol Plant 131(4):555–570

Ahsan N, Renaut J, Komatsu S (2009) Recent developments in the application of proteomics to the analysis of plant responses to heavy metals. Proteomics 9(10):2602–2621

Albert E, Sauvage C, Bouchet J-P, Bitton F, Beukers M, Carretero Y, Causse M (2017) Integration of qtl, eqtl and allele specific expression to unravel genotype by watering regime interaction in cultivated tomato. np

Ali MY, Sina AAI, Khandker SS, Neesa L, Tanvir E, Kabir A, Khalil MI, Gan SH (2020) Nutritional composition and bioactive compounds in tomatoes and their impact on human health and disease: a review. Foods 10(1):45

Ali I, Salah KBH, Sher H, Ali H, Ullah Z, Ali A, Mahmood T (2022) Drought stress enhances the efficiency of floral dip method of Agrobacterium-mediated transformation in Arabidopsis thaliana. Braz J Biol 84:e259326

Ampofo-Asiama J, Baiye V, Hertog M, Waelkens E, Geeraerd A, Nicolai B (2014) The metabolic response of cultured tomato cells to low oxygen stress. Plant Biol 16(3):594–606

Aoki K, Yano K, Suzuki A, Kawamura S, Sakurai N, Suda K, Kurabayashi A, Suzuki T, Tsugane T, Watanabe M (2010) Large-scale analysis of full-length cDNAs from the tomato (Solanum lycopersicum) cultivar Micro-Tom, a reference system for the Solanaceae genomics. BMC Genomics 11:1–16

Bai Y, Lindhout P (2007) Domestication and breeding of tomatoes: what have we gained and what can we gain in the future? Ann Bot 100(5):1085–1094

Balyan S, Rao S, Jha S, Bansal C, Das JR, Mathur S (2020) Characterization of novel regulators for heat stress tolerance in tomato from Indian sub-continent. Plant Biotechnol J 18(10):2118–2132

Berger Y, Harpaz-Saad S, Brand A, Melnik H, Sirding N, Alvarez JP, Zinder M, Samach A, Eshed Y, Ori N (2009) The NAC-domain transcription factor GOBLET specifies leaflet boundaries in compound tomato leaves. Development 136:823–832

Beynon RJ (2005) The dynamics of the proteome: strategies for measuring protein turnover on a proteome-wide scale. Brief Funct Genomics 3(4):382–390

Cai J, Chen T, Wang Y, Qin G, Tian S (2020) SlREM1 triggers cell death by activating an oxidative burst and other regulators. Plant Physiol 183(2):717–732

Cambiaso V, Pratta GR, da Costa JHP, Zorzoli R, Francis DM, Rodríguez GR (2019) Whole genome re-sequencing analysis of two tomato genotypes for polymorphism insight in cloned genes and a genetic map construction. Sci Hortic 247:58–66

Castel B, Tomlinson L, Locci F, Yang Y, Jones JD (2019) Optimization of T-DNA architecture for Cas9-mediated mutagenesis in Arabidopsis. PLoS One 14(1):e0204778

Čermák T, Baltes NJ, Čegan R, Zhang Y, Voytas DF (2015) High-frequency, precise modification of the tomato genome. Genome Biol 16:1–15

Chakravarti B, Mallik B, Chakravarti DN (2010) Proteomics and systems biology: application in drug discovery and development. In: Systems biology in drug discovery and development: methods and protocols. Humana Press, pp 3–28

Chaudhary P, Sharma A, Singh B, Nagpal AK (2018) Bioactivities of phytochemicals present in tomato. J Food Sci Technol 55:2833–2849

Chaudhary J, Khatri P, Singla P, Kumawat S, Kumari A, Vikram A, Jindal SK, Kardile H, Kumar R, Sonah H (2019) Advances in omics approaches for abiotic stress tolerance in tomato. Biology 8(4):90

Chen W, Yao Q, Patil GB, Agarwal G, Deshmukh RK, Lin L, Wang B, Wang Y, Prince SJ, Song L (2016) Identification and comparative analysis of differential gene expression in soybean leaf tissue under drought and flooding stress revealed by RNA-Seq. Front Plant Sci 7:209431

Chen L, Li W, Katin-Grazzini L, Ding J, Gu X, Li Y, Gu T, Wang R, Lin X, Deng Z (2018) A method for the production and expedient screening of CRISPR/Cas9-mediated non-transgenic mutant plants. Hortic Res 5:13

Chen T, Qin G, Tian S (2020) Regulatory network of fruit ripening: current understanding and future challenges. New Phytol 228(4):1219–1226

Cirillo V, Molisso D, Aprile AM, Maggio A, Rao R (2022) Systemin peptide application improves tomato salt stress tolerance and reveals common adaptation mechanisms to biotic and abiotic stress in plants. Environ Exp Bot 199:104865

Cohen H, Szymanski J, Aharoni A (2017) Assimilation of 'omics' strategies to study the cuticle layer and suberin lamellae in plants. J Exp Bot 68(19):5389–5400

Dahan-Meir T, Filler-Hayut S, Melamed-Bessudo C, Bocobza S, Czosnek H, Aharoni A, Levy AA (2018) Efficient in planta gene targeting in tomato using geminiviral replicons and the CRISPR/Cas9 system. Plant J 95(1):5–16

Debernardi JM, Tricoli DM, Ercoli MF, Hayta S, Ronald P, Palatnik JF, Dubcovsky J (2020) A GRF–GIF chimeric protein improves the regeneration efficiency of transgenic plants. Nat Biotechnol 38(11):1274–1279

Debnath M, Pandey M, Bisen P (2011) An omics approach to understand the plant abiotic stress. Omics J Integr Biol 15(11):739–762

Dheer P, Rautela I, Sharma V, Dhiman M, Sharma A, Sharma N, Sharma MD (2020) Evolution in crop improvement approaches and future prospects of molecular markers to CRISPR/Cas9 system. Gene 753:144795

Doyle C, Higginbottom K, Swift T, Winfield M, Bellas C, Benito-Alifonso D, Whitney H (2019) A simple method for spray-on gene editing in planta. BioRxiv 805036

Eid A, Alshareef S, Mahfouz MM (2018) CRISPR base editors: genome editing without double-stranded breaks. Biochem J 475(11):1955–1964

Faurobert M, Mihr C, Bertin N, Pawlowski T, Negroni L, Sommerer N, Causse M (2007) Major proteome variations associated with cherry tomato pericarp development and ripening. Plant Physiol 143(3):1327–1346

Feder A, Jensen S, Wang A, Courtney L, Middleton L, Van Eck J, Liu Y, Giovannoni JJ (2020) Tomato fruit as a model for tissue-specific gene silencing in crop plants. Hortic Res 7:142

Fei Z, Joung J-G, Tang X, Zheng Y, Huang M, Lee JM, McQuinn R, Tieman DM, Alba R, Klee HJ (2010) Tomato Functional Genomics Database: a comprehensive resource and analysis package for tomato functional genomics. Nucleic Acids Res 39(suppl_1):D1156–D1163

Feng C, Su H, Bai H, Wang R, Liu Y, Guo X, Liu C, Zhang J, Yuan J, Birchler JA (2018) High-efficiency genome editing using a dmc1 promoter-controlled CRISPR/Cas9 system in maize. Plant Biotechnol J 16(11):1848–1857

Feng Z, Ji S, Ping J, Cui D (2021) Recent advances in metabolomics for studying heavy metal stress in plants. TrAC Trends Anal Chem 143:116402

Fernandes I, Leça JM, Aguiar R, Fernandes T, Marques JC, Cordeiro N (2021) Influence of crop system fruit quality, carotenoids, fatty acids and phenolic compounds in cherry tomatoes. Agric Res 10:56–65

Gao Y, Wei W, Fan Z, Zhao X, Zhang Y, Jing Y, Zhu B, Zhu H, Shan W, Chen J (2020) Re-evaluation of the nor mutation and the role of the NAC-NOR transcription factor in tomato fruit ripening. J Exp Bot 71(12):3560–3574

Gerszberg A, Hnatuszko-Konka K (2017) Tomato tolerance to abiotic stress: a review of most often engineered target sequences. Plant Growth Regul 83:175–198

Ghatak A, Chaturvedi P, Weckwerth W (2017) Cereal crop proteomics: systemic analysis of crop drought stress responses towards marker-assisted selection breeding. Front Plant Sci 8:239189

Ghosh D, Xu J (2014) Abiotic stress responses in plant roots: a proteomics perspective. Front Plant Sci 5:66977

Giovannoni JJ (2007) Fruit ripening mutants yield insights into ripening control. Curr Opin Plant Biol 10(3):283–289

Gong B, Li X, VandenLangenberg KM, Wen D, Sun S, Wei M, Li Y, Yang F, Shi Q, Wang X (2014) Overexpression of S-adenosyl-l-methionine synthetase increased tomato tolerance to alkali stress through polyamine metabolism. Plant Biotechnol J 12(6):694–708

Guo T, Yang J, Li D, Sun K, Luo L, Xiao W, Wang J, Liu Y, Wang S, Wang H (2019a) Integrating GWAS, QTL, mapping and RNA-seq to identify candidate genes for seed vigor in rice (Oryza sativa L.). Mol Breed 39:1–16

Guo X, Zhang L, Wang X, Zhang M, Xi Y, Wang A, Zhu J (2019b) Overexpression of Saussurea involucrata dehydrin gene SiDHN promotes cold and drought tolerance in transgenic tomato plants. PLoS One 14(11):e0225090

Habib S, Lwin YY, Li N (2021) Down-regulation of SlGRAS10 in tomato confers abiotic stress tolerance. Genes 12(5):623

Haghighi M, Pessarakli M (2013) Influence of silicon and nano-silicon on salinity tolerance of cherry tomatoes (Solanum lycopersicum L.) at early growth stage. Sci Hortic 161:111–117

Haider S, Iqbal J, Naseer S, Yaseen T, Shaukat M, Bibi H, Mahmood T (2021a) Molecular mechanisms of plant tolerance to heat stress: current landscape and future perspectives. Plant Cell Rep 40:1–25

Haider S, Iqbal J, Shaukat M, Naseer S, Mahmood T (2021b) The epigenetic chromatin-based regulation of somatic heat stress memory in plants. Plant Gene 27:100318

Haider S, Raza A, Iqbal J, Shaukat M, Mahmood T (2022a) Analyzing the regulatory role of heat shock transcription factors in plant heat stress tolerance: a brief appraisal. Mol Biol Rep 49(6):5771–5785

Haider S, Iqbal J, Naseer S, Shaukat M, Abbasi BA, Yaseen T, Mahmood T (2022b) Unfolding molecular switches in plant heat stress resistance: a comprehensive review. Plant Cell Rep 41(3):775–798

Han G, Qiao Z, Li Y, Yang Z, Wang C, Zhang Y, Liu L, Wang B (2022) RING zinc finger proteins in plant abiotic stress tolerance. Front Plant Sci 13:877011

He Y, Yang T, Yan S, Niu S, Zhang Y (2022) Identification and characterization of the BEL1-like genes reveal their potential roles in plant growth and abiotic stress response in tomato. Int J Biol Macromol 200:193–205

Hendelman A, Stav R, Zemach H, Arazi T (2013) The tomato NAC transcription factor SlNAM2 is involved in flower-boundary morphogenesis. J Exp Bot 64(18):5497–5507

Hilioti Z, Ganopoulos I, Ajith S, Bossis I, Tsaftaris A (2016) A novel arrangement of zinc finger nuclease system for in vivo targeted genome engineering: the tomato LEC1-LIKE4 gene case. Plant Cell Rep 35:2241–2255

Hossain Z, Komatsu S (2013) Contribution of proteomic studies towards understanding plant heavy metal stress response. Front Plant Sci 3:41398

Hu Y, Wu Q, Sprague SA, Park J, Oh M, Rajashekar C, Koiwa H, Nakata PA, Cheng N, Hirschi KD (2015) Tomato expressing Arabidopsis glutaredoxin gene AtGRXS17 confers tolerance to chilling stress via modulating cold responsive components. Hortic Res 2:15051

Hu DG, Ma QJ, Sun CH, Sun MH, You CX, Hao YJ (2016) Overexpression of MdSOS2L1, a CIPK protein kinase, increases the antioxidant metabolites to enhance salt tolerance in apple and tomato. Physiol Plant 156(2):201–214

Hunziker J, Nishida K, Kondo A, Kishimoto S, Ariizumi T, Ezura H (2020) Multiple gene substitution by Target-AID base-editing technology in tomato. Sci Rep 10(1):20471

Ichihashi Y, Sinha NR (2014) From genome to phenome and back in tomato. Curr Opin Plant Biol 18:9–15

Ijaz S, Iqbal J, Abbasi BA, Tufail A, Ullah Z, Sharifi-Rad J, Uddin S (2023) Biofortification: lessons from the past and strategies for future food security. In: Legumes biofortification. Springer International Publishing, Cham, pp 521–545

Ijaz S, Iqbal J, Abbasi BA, Yaseen T, Rehman S, Kazi M, Mahmood T (2024) Role of Omics-based technologies in plant sciences. In: OMICs-based techniques for global food security. Wiley, pp 45–66

Irfan M, Kumar P, Ahmad I, Datta A (2021) Unraveling the role of tomato Bcl-2-associated athanogene (BAG) proteins during abiotic stress response and fruit ripening. Sci Rep 11(1):21734

Jeon JE, Kim J-G, Fischer CR, Mehta N, Dufour-Schroif C, Wemmer K, Mudgett MB, Sattely E (2020) A pathogen-responsive gene cluster for highly modified fatty acids in tomato. Cell 180(1):176–187. e119

Jia C, Zhao S, Bao T, Zhao P, Peng K, Guo Q, Gao X, Qin J (2021) Tomato BZR/BES transcription factor SlBZR1 positively regulates BR signaling and salt stress tolerance in tomato and *Arabidopsis*. Plant Sci 302:110719

Jing LU, Sun MH, Hui KANG, Liu YJ, Hao YJ, You CX (2019) MdSWEET17, a sugar transporter in apple, enhances drought tolerance in tomato. J Integr Agric 18(9):2041–2051

Karlova R, Rosin FM, Busscher-Lange J, Parapunova V, Do PT, Fernie AR, Fraser PD, Baxter C, Angenent GC, de Maagd RA (2011) Transcriptome and metabolite profiling show that APETALA2a is a major regulator of tomato fruit ripening. Plant Cell 23(3):923–941

Karthikeyan A, Pandian SK, Ramesh M (2011) Transgenic indica rice cv. ADT 43 expressing a Δ 1-pyrroline-5-carboxylate synthetase (P5CS) gene from Vigna aconitifolia demonstrates salt tolerance. Plant Cell Tissue Organ Cult (PCTOC) 107:383–395

Kausar R, Wang X, Komatsu S (2022) Crop proteomics under abiotic stress: from data to insights. Plan Theory 11(21):2877

Khoso MA, Hussain A, Ritonga FN, Ali Q, Channa MM, Alshegaihi RM, Meng Q, Ali M, Zaman W, Brohi RD (2022) WRKY transcription factors (TFs): molecular switches to regulate drought, temperature, and salinity stresses in plants. Front Plant Sci 13:1039329

Klap C, Yeshayahou E, Bolger AM, Arazi T, Gupta SK, Shabtai S, Usadel B, Salts Y, Barg R (2017) Tomato facultative parthenocarpy results from Sl AGAMOUS-LIKE 6 loss of function. Plant Biotechnol J 15(5):634–647

Klee HJ, Giovannoni JJ (2011) Genetics and control of tomato fruit ripening and quality attributes. Annu Rev Genet 45:41–59

Koçberber N, Dönmez G (2007) Chromium (VI) bioaccumulation capacities of adapted mixed cultures isolated from industrial saline wastewaters. Bioresour Technol 98(11):2178–2183

Kok E, Lehesranta S, Van Dijk J, Helsdingen J, Dijksma W, Van Hoef A, Koistinen K, Karenlampi S, Kuiper H, Keijer J (2008) Changes in gene and protein expression during tomato ripening—consequences for the safety assessment of new crop plant varieties. Food Sci Technol Int 14(6):503–518

Kosová K, Vítámvás P, Prášil IT, Renaut J (2011) Plant proteome changes under abiotic stress—contribution of proteomics studies to understanding plant stress response. J Proteome 74(8):1301–1322

Kosová K, Vítámvás P, Urban MO, Prášil IT, Renaut J (2018) Plant abiotic stress proteomics: the major factors determining alterations in cellular proteome. Front Plant Sci 9:325823

Krishna R, Karkute SG, Ansari WA, Jaiswal DK, Verma JP, Singh M (2019) Transgenic tomatoes for abiotic stress tolerance: status and way ahead. 3 Biotech 9:1–14

Kudo T, Kobayashi M, Terashima S, Katayama M, Ozaki S, Kanno M, Saito M, Yokoyama K, Ohyanagi H, Aoki K (2017) TOMATOMICS: a web database for integrated omics information in tomato. Plant Cell Physiol 58(1):e8–e8

Kumar M, Tomar M, Bhuyan DJ, Punia S, Grasso S, Sa AGA, Carciofi BAM, Arrutia F, Changan S, Singh S (2021) Tomato (*Solanum lycopersicum* L.) seed: A review on bioactives and biomedical activities. Biomed Pharmacother 142:112018

Kwon C-T, Heo J, Lemmon ZH, Capua Y, Hutton SF, Van Eck J, Park SJ, Lippman ZB (2020) Rapid customization of Solanaceae fruit crops for urban agriculture. Nat Biotechnol 38(2):182–188

Laxman R, Hemamalini P, Bhatt R, Sadashiva A (2018) Non-invasive quantification of tomato (Solanum lycopersicum L.) plant biomass through digital imaging using phenomics platform. Indian J Plant Physiol 23:369–375

LeBlanc C, Zhang F, Mendez J, Lozano Y, Chatpar K, Irish VF, Jacob Y (2018) Increased efficiency of targeted mutagenesis by CRISPR/Cas9 in plants using heat stress. Plant J 93(2):377–386

Lee YJ, Perdian DC, Song Z, Yeung ES, Nikolau BJ (2012) Use of mass spectrometry for imaging metabolites in plants. Plant J 70(1):81–95

Li J-B, Luan Y-S, Liu Z (2015) SpWRKY1 mediates resistance to Phytophthora infestans and tolerance to salt and drought stress by modulating reactive oxygen species homeostasis and expression of defense-related genes in tomato. Plant Cell Tissue Organ Cult (PCTOC) 123:67–81

Li J, Wang Y, Wei J, Pan Y, Su C, Zhang X (2018) A tomato proline-, lysine-, and glutamic-rich type gene SpPKE1 positively regulates drought stress tolerance. Biochem Biophys Res Commun 499(4):777–782

Li S, Zhu B, Pirrello J, Xu C, Zhang B, Bouzayen M, Chen K, Grierson D (2020) Roles of RIN and ethylene in tomato fruit ripening and ripening-associated traits. New Phytol 226(2):460–475

Liu H, Yu C, Li H, Ouyang B, Wang T, Zhang J, Wang X, Ye Z (2015) Overexpression of ShDHN, a dehydrin gene from Solanum habrochaites enhances tolerance to multiple abiotic stresses in tomato. Plant Sci 231:198–211

Liu Y, Zeng J, Yuan C, Guo Y, Yu H, Li Y, Huang C (2019) Cas9-PF, an early flowering and visual selection marker system, enhances the frequency of editing event occurrence and expedites the isolation of genome-edited and transgene-free plants. Plant Biotechnol J 17(7):1191

Liu L, Zhang J, Xu J, Li Y, Guo L, Wang Z, Zhang X, Zhao B, Guo Y-D, Zhang N (2020a) CRISPR/Cas9 targeted mutagenesis of SlLBD40, a lateral organ boundaries domain transcription factor, enhances drought tolerance in tomato. Plant Sci 301:110683

Liu Y, Tang M, Liu M, Su D, Chen J, Gao Y, Bouzayen M, Li Z (2020b) The molecular regulation of ethylene in fruit ripening. Small Methods 4(8):1900485

Loukehaich R, Wang T, Ouyang B, Ziaf K, Li H, Zhang J, Lu Y, Ye Z (2012) SpUSP, an annexin-interacting universal stress protein, enhances drought tolerance in tomato. J Exp Bot 63(15):5593–5606

Lu HP, Liu SM, Xu SL, Chen WY, Zhou X, Tan YY, Huang JZ, Shu QY (2017) CRISPR-S: an active interference element for a rapid and inexpensive selection of genome-edited, transgene-free rice plants. Plant Biotechnol J 15(11):1371

Lu Y, Tian Y, Shen R, Yao Q, Wang M, Chen M, Dong J, Zhang T, Li F, Lei M (2020) Targeted, efficient sequence insertion and replacement in rice. Nat Biotechnol 38(12):1402–1407

Luo S, Li J, Stoddard TJ, Baltes NJ, Demorest ZL, Clasen BM, Coffman A, Retterath A, Mathis L, Voytas DF (2015) Non-transgenic plant genome editing using purified sequence-specific nucleases. Mol Plant 8(9):1425–1427

Ma NN, Zuo YQ, Liang XQ, Yin B, Wang GD, Meng QW (2013) The multiple stress-responsive transcription factor SlNAC1 improves the chilling tolerance of tomato. Physiol Plant 149(4):474–486

Ma X, Zhang Q, Zhu Q, Liu W, Chen Y, Qiu R, Wang B, Yang Z, Li H, Lin Y (2015) A robust CRISPR/Cas9 system for convenient, high-efficiency multiplex genome editing in monocot and dicot plants. Mol Plant 8(8):1274–1284

Maher MF, Nasti RA, Vollbrecht M, Starker CG, Clark MD, Voytas DF (2020) Plant gene editing through de novo induction of meristems. Nat Biotechnol 38(1):84–89

Makarova KS, Wolf YI, Alkhnbashi OS, Costa F, Shah SA, Saunders SJ, Barrangou R, Brouns SJ, Charpentier E, Haft DH (2015) An updated evolutionary classification of CRISPR–Cas systems. Nat Rev Microbiol 13(11):722–736

Manning K, Tör M, Poole M, Hong Y, Thompson AJ, King GJ, Giovannoni JJ, Seymour GB (2006) A naturally occurring epigenetic mutation in a gene encoding an SBP-box transcription factor inhibits tomato fruit ripening. Nat Genet 38(8):948–952

Mao Y, Yang X, Zhou Y, Zhang Z, Botella JR, Zhu J-K (2018) Manipulating plant RNA-silencing pathways to improve the gene editing efficiency of CRISPR/Cas9 systems. Genome Biol 19:1–15

Marmiroli M, Pigoni V, Savo-Sardaro ML, Marmiroli N (2014) The effect of silicon on the uptake and translocation of arsenic in tomato (*Solanum lycopersicum* L.). Environ Exp Bot 99:9–17

Martinez V, Nieves-Cordones M, Lopez-Delacalle M, Rodenas R, Mestre TC, Garcia-Sanchez F, Rubio F, Nortes PA, Mittler R, Rivero RM (2018) Tolerance to stress combination in tomato plants: new insights in the protective role of melatonin. Molecules 23(3):535

Mathivanan S (2021) Abiotic stress-induced molecular and physiological changes and adaptive mechanisms in plants. In: Abiotic stress in plants, vol 315. IntechOpen

McCourt J, Duggleby R (2006) Acetohydroxyacid synthase and its role in the biosynthetic pathway for branched-chain amino acids. Amino Acids 31:173–210

Meng X, Wang J-R, Wang G-D, Liang X-Q, Li X-D, Meng Q-W (2015) An R2R3-MYB gene, LeAN2, positively regulated the thermo-tolerance in transgenic tomato. J Plant Physiol 175:1–8

Metwali EM, Soliman HI, Fuller MP, Almaghrabi OA (2015) Improving fruit quality in tomato (Lycopersicum esculentum Mill) under heat stress by silencing the vis 1 gene using small interfering RNA technology. Plant Cell Tissue Organ Cult (PCTOC) 121:153–166

Ming N, Ma N, Jiao B, Lv W, Meng Q (2020) Genome wide identification of C2H2-type zinc finger proteins of tomato and expression analysis under different abiotic stresses. Plant Mol Biol Report 38:75–94

Moco S, Bino RJ, Vorst O, Verhoeven HA, De Groot J, Van Beek TA, Vervoort J, De Vos CR (2006) A liquid chromatography-mass spectrometry-based metabolome database for tomato. Plant Physiol 141(4):1205–1218

Muhammad T, Zhang J, Ma Y, Li Y, Zhang F, Zhang Y, Liang Y (2019) Overexpression of a mitogen-activated protein kinase SlMAPK3 positively regulates tomato tolerance to cadmium and drought stress. Molecules 24(3):556

Muneer S, Jeong BR (2015) Proteomic analysis of salt-stress responsive proteins in roots of tomato (Lycopersicon esculentum L.) plants towards silicon efficiency. Plant Growth Regul 77:133–146

Muñoz-Mayor A, Pineda B, Garcia-Abellán JO, Antón T, Garcia-Sogo B, Sanchez-Bel P, Flores FB, Atarés A, Angosto T, Pintor-Toro JA (2012) Overexpression of dehydrin tas14 gene improves the osmotic stress imposed by drought and salinity in tomato. J Plant Physiol 169(5):459–468

Mustafa G, Akhtar MS, Abdullah R (2019) Global concern for salinity on various agro-ecosystems. In: Salt stress, microbes, and plant interactions: causes and solution, vol 1. Springer Singapore, pp 1–19

Naeem M, Shahzad K, Saqib S, Shahzad A, Nasrullah, Younas M, Afridi MI (2022) The Solanum melongena COP1LIKE manipulates fruit ripening and flowering time in tomato (Solanum lycopersicum). Plant Growth Regul:1–14

Nekrasov V, Wang C, Win J, Lanz C, Weigel D, Kamoun S (2017) Rapid generation of a transgene-free powdery mildew resistant tomato by genome deletion. Sci Rep 7(1):482

Nishida K, Arazoe T, Yachie N, Banno S, Kakimoto M, Tabata M, Mochizuki M, Miyabe A, Araki M, Hara KY (2016) Targeted nucleotide editing using hybrid prokaryotic and vertebrate adaptive immune systems. Science 353(6305):aaf8729

Nunes-Nesi A, Alseekh S, de Oliveira Silva FM, Omranian N, Lichtenstein G, Mirnezhad M, González RRR, Sabio y Garcia J, Conte M, Leiss KA (2019) Identification and characterization of metabolite quantitative trait loci in tomato leaves and comparison with those reported for fruits and seeds. Metabolomics 15:1–13

Ogden AJ, Bhatt JJ, Brewer HM, Kintigh J, Kariuki SM, Rudrabhatla S, Adkins JN, Curtis WR (2020) Phloem exudate protein profiles during drought and recovery reveal abiotic stress responses in tomato vasculature. Int J Mol Sci 21(12):4461

Olvera-Carrillo Y, Luis Reyes J, Covarrubias AA (2011) Late embryogenesis abundant proteins: versatile players in the plant adaptation to water limiting environments. Plant Signal Behav 6(4):586–589

Orellana S, Yanez M, Espinoza A, Verdugo I, Gonzalez E, Ruiz-Lara S, Casaretto JA (2010) The transcription factor SlAREB1 confers drought, salt stress tolerance and regulates biotic and abiotic stress-related genes in tomato. Plant Cell Environ 33(12):2191–2208

Osakabe Y, Watanabe T, Sugano SS, Ueta R, Ishihara R, Shinozaki K, Osakabe K (2016) Optimization of CRISPR/Cas9 genome editing to modify abiotic stress responses in plants. Sci Rep 6(1):26685

Pandey S, Muthamilarasan M, Sharma N, Chaudhry V, Dulani P, Shweta S, Jha S, Mathur S, Prasad M (2019) Characterization of DEAD-box family of RNA helicases in tomato provides insights into their roles in biotic and abiotic stresses. Environ Exp Bot 158:107–116

Park J, Choe S (2019) DNA-free genome editing with preassembled CRISPR/Cas9 ribonucleoproteins in plants. Transgenic Res 28(Suppl 2):61–64

Ranjan A, Ichihashi Y, Sinha NR (2012) The tomato genome: implications for plant breeding, genomics and evolution. Genome Biol 13:1–8

Rocco M, d'Ambrosio C, Arena S, Faurobert M, Scaloni A, Marra M (2006) Proteomic analysis of tomato fruits from two ecotypes during ripening. Proteomics 6(13):3781–3791

Salazar-Salas NY, Valenzuela-Ponce L, Vega-Garcia MO, Pineda-Hidalgo KV, Vega-Alvarez M, Chavez-Ontiveros J, Delgado-Vargas F, Lopez-Valenzuela JA (2017) Protein changes associated with chilling tolerance in tomato fruit with hot water pre-treatment. Postharvest Biol Technol 134:22–30

Salt DE, Baxter I, Lahner B (2008) Ionomics and the study of the plant ionome. Annu Rev Plant Biol 59:709–733

Sánchez-Rodríguez E, Moreno DA, Ferreres F, del Mar Rubio-Wilhelmi M, Ruiz JM (2011) Differential responses of five cherry tomato varieties to water stress: changes on phenolic metabolites and related enzymes. Phytochemistry 72(8):723–729

Sang Q, Shan X, An Y, Shu S, Sun J, Guo S (2017) Proteomic analysis reveals the positive effect of exogenous spermidine in tomato seedlings' response to high-temperature stress. Front Plant Sci 8:237444

Santillán Martínez MI, Bracuto V, Koseoglou E, Appiano M, Jacobsen E, Visser RG, Wolters A-MA, Bai Y (2020) CRISPR/Cas9-targeted mutagenesis of the tomato susceptibility gene PMR4 for resistance against powdery mildew. BMC Plant Biol 20:1–13

Schripsema J (2010) Application of NMR in plant metabolomics: techniques, problems and prospects. Phytochem Anal Int J Plant Chem Biochem Tech 21(1):14–21

Selth LA, Dogra SC, Rasheed MS, Healy H, Randles JW, Rezaian MA (2005) A NAC domain protein interacts with tomato leaf curl virus replication accessory protein and enhances viral replication. Plant Cell 17(1):311–325

Seymour GB, Østergaard L, Chapman NH, Knapp S, Martin C (2013) Fruit development and ripening. Annu Rev Plant Biol 64:219–241

Sharma A (2018) Plant metabolites and regulation under environmental stress. Elsevier, pp 407–414

Shikata M, Hoshikawa K, Ariizumi T, Fukuda N, Yamazaki Y, Ezura H (2016) TOMATOMA update: phenotypic and metabolite information in the micro-tom mutant resource. Plant Cell Physiol 57(1):e11–e11

Shimatani Z, Kashojiya S, Takayama M, Terada R, Arazoe T, Ishii H, Teramura H, Yamamoto T, Komatsu H, Miura K (2017) Targeted base editing in rice and tomato using a CRISPR-Cas9 cytidine deaminase fusion. Nat Biotechnol 35(5):441–443

Shirasawa K, Hirakawa H (2013) DNA marker applications to molecular genetics and genomics in tomato. Breed Sci 63(1):21–30

Shu P, Li Y, Li Z, Sheng J, Shen L (2022) SlMAPK3 enhances tolerance to salt stress in tomato plants by scavenging ROS accumulation and up-regulating the expression of ethylene signaling related genes. Environ Exp Bot 193:104698

SinghMohan B, BhallaPrem L (2022) Biological parts for engineering abiotic stress tolerance in plants. BioDesign Res 2022:9819314

Smith DL, Abbott JA, Gross KC (2002) Down-regulation of tomato β-galactosidase 4 results in decreased fruit softening. Plant Physiol 129(4):1755–1762

Sonah H, O'Donoughue L, Cober E, Rajcan I, Belzile F (2015) Identification of loci governing eight agronomic traits using a GBS-GWAS approach and validation by QTL mapping in soya bean. Plant Biotechnol J 13(2):211–221

Suresh BV, Roy R, Sahu K, Misra G, Chattopadhyay D (2014) Tomato genomic resources database: an integrated repository of useful tomato genomic information for basic and applied research. PLoS One 9(1):e86387

Szakiel A, Pączkowski C, Henry M (2011) Influence of environmental biotic factors on the content of saponins in plants. Phytochem Rev 10:493–502

Szymanski J, Levin Y, Savidor A, Breitel D, Chappell-Maor L, Heinig U, Töpfer N, Aharoni A (2017) Label-free deep shotgun proteomics reveals protein dynamics during tomato fruit tissues development. Plant J 90(2):396–417

Tardieu F, Cabrera-Bosquet L, Pridmore T, Bennett M (2017) Plant phenomics, from sensors to knowledge. Curr Biol 27(15):R770–R783

Thalhammer A, Hincha DK (2014) A mechanistic model of COR15 protein function in plant freezing tolerance: integration of structural and functional characteristics. Plant Signal Behav 9(12):e977722

Tieman D, Zhu G, Resende MF Jr, Lin T, Nguyen C, Bies D, Rambla JL, Beltran KSO, Taylor M, Zhang B (2017) A chemical genetic roadmap to improved tomato flavor. Science 355(6323):391–394

Ueta R, Abe C, Watanabe T, Sugano SS, Ishihara R, Ezura H, Osakabe Y, Osakabe K (2017) Rapid breeding of parthenocarpic tomato plants using CRISPR/Cas9. Sci Rep 7(1):507

Ullah Z, Iqbal J, Abbasi BA, Ijaz S, Sharifi-Rad J, Yaseen T, Mahmood T (2023a) How genome editing can be helpful in the biofortification of legumes. In: Legumes biofortification. Springer International Publishing, Cham, pp 207–232

Ullah Z, Iqbal J, Abbasi BA, Akhtar W, Kanwal S, Ali I, Mahmood T (2023b) Assessment of GUS expression induced by anti-sense Os PPO gene promoter and antioxidant enzymatic assays in response to drought and heavy metal stress in transgenic Arabidopsis thaliana. Sustainability 15(17):12783

Ullah R, Ullah Z, Iqbal J, Chalgham W, Ahmad A (2023c) Aspartic acid-based nano-copper induces resilience in Zea mays to applied lead stress via conserving photosynthetic pigments and triggering the antioxidant biosystem. Sustainability 15(16):12186

Uluisik S, Chapman NH, Smith R, Poole M, Adams G, Gillis RB, Besong TM, Sheldon J, Stiegelmeyer S, Perez L (2016) Genetic improvement of tomato by targeted control of fruit softening. Nat Biotechnol 34(9):950–952

Veillet F, Perrot L, Chauvin L, Kermarrec M-P, Guyon-Debast A, Chauvin J-E, Nogué F, Mazier M (2019) Transgene-free genome editing in tomato and potato plants using agrobacterium-mediated delivery of a CRISPR/Cas9 cytidine base editor. Int J Mol Sci 20(2):402

Vu TV, Sivankalyani V, Kim EJ, Doan DTH, Tran MT, Kim J, Sung YW, Park M, Kang YJ, Kim JY (2020) Highly efficient homology-directed repair using CRISPR/Cpf1-geminiviral replicon in tomato. Plant Biotechnol J 18(10):2133–2143

Wai AH, Naing AH, Lee D-J, Kim CK, Chung M-Y (2020) Molecular genetic approaches for enhancing stress tolerance and fruit quality of tomato. Plant Biotechnol Rep 14:515–537

Wang M, Mao Y, Lu Y, Wang Z, Tao X, Zhu JK (2018) Multiplex gene editing in rice with simplified CRISPR-Cpf1 and CRISPR-Cas9 systems. J Integr Plant Biol 60(8):626–631

Wang D, Samsulrizal NH, Yan C, Allcock NS, Craigon J, Blanco-Ulate B, Ortega-Salazar I, Marcus SE, Bagheri HM, Perez Fons L (2019a) Characterization of CRISPR mutants targeting genes modulating pectin degradation in ripening tomato. Plant Physiol 179(2):544–557

Wang R, Tavano ECDR, Lammers M, Martinelli AP, Angenent GC, de Maagd RA (2019b) Re-evaluation of transcription factor function in tomato fruit development and ripening with CRISPR/Cas9-mutagenesis. Sci Rep 9(1):1696

Wang T, Zhang H, Zhu H (2019c) CRISPR technology is revolutionizing the improvement of tomato and other fruit crops. Hortic Res 6:77

Wang W-Q, Allan AC, Yin X-R (2020a) Small RNAs with a big impact on horticultural traits. Crit Rev Plant Sci 39(1):30–43

Wang Z-Y, Bao Y-F, Tong P, Wu T-R, Xu D, He M-X, Yue W, Liu Q-F, Yang H-H, Jiang J-B (2020b) Silencing the SLB3 transcription factor gene decreases drought stress tolerance in tomato. J Integr Agric 19(11):2699–2708

Wang Z-Y, Zhao S, Liu J-F, Zhao H-Y, Sun X-Y, Wu T-R, Pei T, Wang Y, Liu Q-F, Yang H-H (2022) Genome-wide identification of Tomato Golden 2-Like transcription factors and abiotic stress related members screening. BMC Plant Biol 22(1):82

White JW, Andrade-Sanchez P, Gore MA, Bronson KF, Coffelt TA, Conley MM, Feldmann KA, French AN, Heun JT, Hunsaker DJ (2012) Field-based phenomics for plant genetics research. Field Crop Res 133:101–112

Williamson B, Tudzynski B, Tudzynski P, Van Kan JA (2007) Botrytis cinerea: the cause of grey mould disease. Mol Plant Pathol 8(5):561–580

Wolabu TW, Cong L, Park J-J, Bao Q, Chen M, Sun J, Xu B, Ge Y, Chai M, Liu Z (2020) Development of a highly efficient multiplex genome editing system in outcrossing tetraploid alfalfa (Medicago sativa). Front Plant Sci 11:1063

Wu X, Gong F, Cao D, Hu X, Wang W (2016) Advances in crop proteomics: PTMs of proteins under abiotic stress. Proteomics 16(5):847–865

Xu J, Hua K, Lang Z (2019) Genome editing for horticultural crop improvement. Hortic Res 6:113

Yang Y, Tang N, Xian Z, Li Z (2015) Two SnRK2 protein kinases genes play a negative regulatory role in the osmotic stress response in tomato. Plant Cell Tissue Organ Cult (PCTOC) 122:421–434

Yang L, Huang W, Xiong F, Xian Z, Su D, Ren M, Li Z (2017) Silencing of Sl PL, which encodes a pectate lyase in tomato, confers enhanced fruit firmness, prolonged shelf-life and reduced susceptibility to grey mould. Plant Biotechnol J 15(12):1544–1555

Yang T, Deng L, Zhao W, Zhang R, Jiang H, Ye Z, Li C-B, Li C (2019) Rapid breeding of pink-fruited tomato hybrids using the CRISPR/Cas9 system. J Genet Genomics = Yi chuan xue bao 46(10):505–508

Yano K, Watanabe M, Yamamoto N, Tsugane T, Aoki K, Sakurai N, Shibata D (2006) MiBASE: a database of a miniature tomato cultivar Micro-Tom. Plant Biotechnol 23(2):195–198

Ye J, Liu G, Chen W, Zhang F, Li H, Ye Z, Zhang Y (2019) Knockdown of SlNL33 accumulates ascorbate, enhances disease and oxidative stress tolerance in tomato (Solanum lycopersicum). Plant Growth Regul 89:49–58

Yin W, Hu Z, Cui B, Guo X, Hu J, Zhu Z, Chen G (2017) Suppression of the MADS-box gene SlMBP8 accelerates fruit ripening of tomato (*Solanum lycopersicum*). Plant Physiol Biochem 118:235–244

Yu L, Yan J, Yang Y, He L, Zhu W (2016) Enhanced tolerance to chilling stress in tomato by over-expression of a mitogen-activated protein kinase, SlMPK7. Plant Mol Biol Report 34:76–88

Yu Q-H, Wang B, Li N, Tang Y, Yang S, Yang T, Xu J, Guo C, Yan P, Wang Q (2017) CRISPR/Cas9-induced targeted mutagenesis and gene replacement to generate long-shelf life tomato lines. Sci Rep 7(1):11874

Yubing H, Min Z, Lihao W, Junhua W, Qiaoyan W, Rongchen W, Yunde Z (2019) Improvements of TKC technology accelerate isolation of transgene-free CRISPR/Cas9-edited rice plants. Rice Sci 26(2):109–117

Zaidi SS-E-A, Mukhtar MS, Mansoor S (2018) Genome editing: targeting susceptibility genes for plant disease resistance. Trends Biotechnol 36(9):898–906

Zhang Y, Massel K, Godwin ID, Gao C (2018) Applications and potential of genome editing in crop improvement. Genome Biol 19(1):210

Zhang L, Guo X, Zhang Z, Wang A, Zhu J (2021) Cold-regulated gene LeCOR413PM2 confers cold stress tolerance in tomato plants. Gene 764:145097

Zhao C, Shono M, Sun A, Yi S, Li M, Liu J (2007) Constitutive expression of an endoplasmic reticulum small heat shock protein alleviates endoplasmic reticulum stress in transgenic tomato. J Plant Physiol 164(7):835–841

Zhong Z, Sretenovic S, Ren Q, Yang L, Bao Y, Qi C, Yuan M, He Y, Liu S, Liu X (2019) Improving plant genome editing with high-fidelity xCas9 and non-canonical PAM-targeting Cas9-NG. Mol Plant 12(7):1027–1036

Zhou S, Sauvé R, Thannhauser TW (2009) Proteome changes induced by aluminium stress in tomato roots. J Exp Bot 60(6):1849–1857

Zhu M, Chen G, Zhang J, Zhang Y, Xie Q, Zhao Z, Pan Y, Hu Z (2014) The abiotic stress-responsive NAC-type transcription factor SlNAC4 regulates salt and drought tolerance and stress-related genes in tomato (Solanum lycopersicum). Plant Cell Rep 33:1851–1863

Zhu G, Wang S, Huang Z, Zhang S, Liao Q, Zhang C, Lin T, Qin M, Peng M, Yang C (2018a) Rewiring of the fruit metabolome in tomato breeding. Cell 172(1):249–261. e212

Zhu H, Ai H, Cao L, Sui R, Ye H, Du D, Sun J, Yao J, Chen K, Chen L (2018b) Transcriptome analysis providing novel insights for Cd-resistant tall fescue responses to Cd stress. Ecotoxicol Environ Saf 160:349–356

Zhu M, Meng X, Cai J, Li G, Dong T, Li Z (2018c) Basic leucine zipper transcription factor SlbZIP1 mediates salt and drought stress tolerance in tomato. BMC Plant Biol 18:1–14

Zsögön A, Cermak T, Voytas D, Peres LEP (2017) Genome editing as a tool to achieve the crop ideotype and de novo domestication of wild relatives: case study in tomato. Plant Sci 256:120–130

Confronting Abiotic Stresses: Molecular Strategies for Improving Tomato Stress Tolerance

3

Saima Mir Arain, Muhammad Sajjad, Muhammad Faheem, Ghulam Ullah, Khalil Ahmad Laghari, and Mahboob Ali Sial

Abstract

Tomato ranks among the crops of higher economic importance worldwide, with its yield experiencing a reduction of high magnitude due to abiotic stresses such as drought, heat, cold, and salinity. Developing tomato varieties with tolerance to stresses underpins food safety and sustainable agriculture. This chapter examines advanced molecular approaches in breeding to enhance abiotic stress tolerance in tomatoes. It starts with the importance of tomatoes, then the negative impacts of abiotic stresses, followed by an exploration of tomato plant stress-response mechanisms: signalling pathways, and essential regulatory genes. This highlights the role of genetic engineering, transgenic, and CRISPR technologies in enhancing tolerance. The molecular details of each specific strategy for drought, heat, cold, and salinity tolerance are presented. The comprehensive discussion herein outlines the roadmap for researchers and breeders to capitalize on the emerging key points and technologies needed to increase tomato stress tolerance while drawing attention to future research directions.

Keywords

Agriculture · CRISPR/Cas · Drought · Heat · Salinity · Tomato · Transgenic

S. M. Arain (✉) · M. Faheem · G. Ullah · K. A. Laghari · M. A. Sial
Nuclear Institute of Agriculture (NIA), Tando Jam, Pakistan

M. Sajjad
COMSATS Institute of Information Technology, Islamabad, Pakistan

M. Waseem et al. (eds.), *Omics Approaches for Tomato Yield and Quality Trait Improvement*, https://doi.org/10.1007/978-981-96-3890-1_3

3.1 Introduction

Tomato (*Solanum lycopersicum* L.) belongs to the family *Solanaceae*, also referred to as nightshade, and genus, sub-family, and tribe of *Lycopersicon*, *Solanoideae*, and *Solaneae*, respectively. The family includes other vital crops like potatoes, aubergine eggplants, chilli, bell peppers, and tobacco. Originating from South America's Andean region, tomato is the second most important crop after potato, with 182.3 million tons produced annually on 4.85 million hectares. Asia leads in production at 61.1%, followed by Europe (13.5%), America (13.4%), and Africa (11.8%) (FAOSTAT, 2019). Yields vary widely, from over 508 tons per hectare in the Netherlands to less than 1.5 tons in Somalia, with a global average of 376 tons per hectare. Southern Europe is a key region, with Turkey, Egypt, Italy, Spain, and Morocco being major producers. Tomato consumption is highest in India, North Africa, China, the Middle East, Brazil, and the United States, ranging from 61.9 to 198.9 kg per capita (Quinet et al. 2019; Panno et al. 2021).

3.2 Generic Position of Tomatoes

The system of naming plants with a genus and species, initiated by Linnaeus in 1753, simplified earlier lengthy Latin descriptions. Philip Miller, an English botanist, initially used the generic name *Lycopersicon*, meaning "wolf peach," for tomatoes in 1731 but also recognized *Solanum*, including eggplants and potatoes. Linnaeus classified tomatoes under *Solanum* in his renowned book *Species Plantarum*. Despite Miller's continued use of *Lycopersicon*, he later included potatoes in this genus based on fruit characteristics. Over time, various botanists, including Dunal, Bentham, Hooker, and others have debated the classification, with some supporting *Lycopersicon* based on another morphology. However, modern consensus, backed by morphological and molecular analyses, places tomatoes firmly in the genus *Solanum* (Smith 2001; Knapp and Peralta 2016).

3.3 Origin and a Brief History

Christopher Columbus introduced tomatoes to Europe in 1493, followed by Spanish conquistadors likely bringing them to Europe in the early 1500s. In the early sixteenth century, tomatoes were known as "*mala auraea*" or golden apple and described as "*poma amoris fructu rubri*," meaning apple of love with red fruit. Tomatoes became a staple in Spain in the 1540s and gained popularity in other European countries by 1548. Italian botanist Pietro Andrea Matthiolus first recorded them in Europe in 1554 as "*pomi d'oro*" (golden apple), later called "*pomme d'amour*" (love apple) by the French. The wild ancestral tomato originated in western South America, from Chile through the Andes to Peru and Ecuador, with the greatest diversity of wild species found along the Peruvian coast. Small, yellow tomatoes migrated north to Mexico, where the Aztecs cultivated them. Tomatoes arrived in England in 1597, where they contributed to the manufacturing of ketchup and the development of tomato varieties (Gould 1974; Waheed et al. 2020).

In 1753, Linnaeus classified tomatoes under *Solanum* as *S. lycopersicum L.*, while Phillip Miller later established the genus *Lycopersicon* in 1754, naming the tomato *Solanum esculentum Mill.* In 1863, the "Trophy" variety became the first large, red, smooth-skinned tomato marketed in the United States, leading to further breeding for marketable traits. By the late 1700s, tomatoes became popular as food in North America, thanks to Italian and French immigrants. Hybridization produced orange and red varieties, which became widespread by the mid-1800s. Larger, sweeter tomatoes were cultivated, boosting their economic importance (Pua and Davey 2007).

This chapter explores cutting-edge molecular breeding methods to increase tomato stress tolerance. It explores tomato plant regulatory genes, stress-response mechanisms, and the importance of tomatoes, as well as the negative effects of abiotic stresses. It stresses the need for enhancing tolerance through genetic engineering, transgenic, and CRISPR technologies. Particular methods for improving resistance to heat, cold, drought, and salinity are described. This talk outlines future directions for study and offers breeders and researchers a road map for using cutting-edge technology and important discoveries to increase tomato stress tolerance.

3.4 Overview of Tomato Crop Importance and the Impact of Abiotic Stresses

Tomatoes are globally recognized as valuable crops with a crucial role in human nutrition, owing to their versatility in various forms, from raw to processed (Chaudhary et al. 2019). As the world's third most vital vegetable, tomatoes are integral to diets worldwide, attracting extensive research attention due to their potential health benefits (Dorais et al. 2008; Slimestad and Verheul 2009). The levels of beneficial components differ according to factors such as cultivar and growing environment. Tomatoes, being high in substances that promote health, are readily integrated into balanced meals (Martí et al. 2016) and are commonly used in processed products (Li et al. 2018b). While consumers look for functional foods more and more to prevent diseases, classics like fruits and vegetables are still quite successful in this regard (Viuda-Martos et al. 2014; Dhandevi and Jeewon 2015).

Tomatoes are important nutritionally because of their profusion of phenolic compounds, carotenoids, and vitamins (Raiola et al. 2014; Liu and Howell 2016; Martí et al. 2016; Li et al. 2018b). Among the physiological effects of the bio-active components are anti-allergenic, vasodilatory, antibacterial, anti-inflammatory, cardioprotective, antioxidant, and anti-thrombotic effects (Raiola et al. 2014). Furthermore, a major dietary source of the carotenoid lycopene in the human diet is tomatoes (Viuda-Martos et al. 2014). Tomatoes' taste, smell, and texture are all improved by the interaction between carotenoids and polyphenolic chemicals (Raiola et al. 2014; Tohge and Fernie 2015; Martí et al. 2016). Other natural antioxidants, for instance, vitamins E and C (Martí et al. 2016), and known metabolites like ascorbic acid, malate, hexoses, citrate, and sucrose are also abundant in tomatoes (Li et al. 2018b).

Besides their nutritional and economic importance, tomatoes have been used as a model plant for studying the development of other fleshy fruits (Kim et al. 2018; Li et al. 2018b). The whole sequencing of the tomato genome complemented by molecular markers and physical and genetic mapping has yielded a valuable genomic resource (Tomato Genome Consortium 2012; Suresh et al. 2014; Zhou et al. 2019). There are also substantial numbers of recognized single gene mutants, wild tomato species, TILLING populations, recombinant inbred lines (RILs), and genetic editing tools available (Martín-Pizarro and Posé 2018; Rothan et al. 2019).

The growing conditions of the tomato plant and the subsequent biosynthesis of metabolites are strongly impacted by the conditions provided during the growth cycle (Diouf et al. 2018). However, the challenges faced worldwide include water scarcity, soil salinization, and other abiotic stresses (Fahad et al. 2017; Gharbi et al. 2017; Zhou et al. 2019). Tomato farming is facing more and more challenges from limiting variables like salinity and drought, especially in areas with a Mediterranean climate like sections of southern Europe and North and South America. These difficulties make tomato-growers less competitive and increase the desertion of the rural sector, which compromises the integrity of the ecosystem.

While tomatoes are adaptable to various climatic regions worldwide due to their tropical origins, their yield potential is primarily limited by environmental stress factors. Recently, there has been an intensive study on the molecular pathways related to environmental stress tolerance, focussing on mechanisms to tolerate individual stressors (Chaudhary et al. 2019). Abiotic stress encompasses various challenges such as chilling, high temperature, osmotic shock, drought, salinity, waterlogging, exposure to ozone, UV-B radiation, excessive light, wounding, and toxic ions (Rehem et al. 2012; Bashir et al. 2019). Unfortunately, abiotic stresses are complex and influenced by various factors, including genetics and the environment, which hinder breeding strategies for crop improvement (da Silva and Costa de Oliveira 2014).

3.5 Abiotic Stresses

Abiotic stressors like drought, soil salinity, and extreme temperatures significantly impact crop production, a challenge exacerbated by climate change and population growth (Fig. 3.1). With the global population projected to reach 8.5 billion by 2030 and 9.7 billion by 2050, meeting food demand faces increasing environmental pressures (Swetha et al. 2024). Abiotic stress can reduce yields by over 60%, particularly affecting regions already experiencing extreme climate conditions (Cramer et al. 2011; Yadav et al. 2020). Recent studies suggest that climate extremes account for up to half of the global crop yield variability, with temperature-related factors playing a more significant role than precipitation under regular irrigation (Vogel et al. 2019).

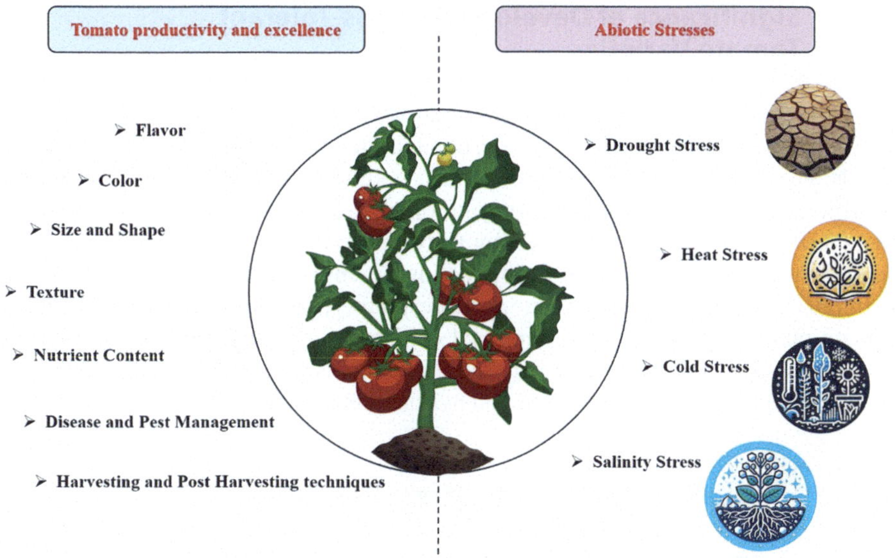

Fig. 3.1 A graphical representation of abiotic stresses and quality traits in tomatoes

The optimal growth and marketable quality of crops depend on specific conditions, defining the most suitable locations for cultivation (Wang et al. 2023b). As previously mentioned, plants endure various stresses, all of which affect global crop yields. The subsequent section will delve into the importance of developing stress-tolerant plants, particularly discussing the tomato plant.

Many stress-responsive genes working together with stress signal pathways give rise to plant tolerance or vulnerability to abiotic stresses. Improving tomato and other crops' stress tolerance requires an understanding of these processes. Although breeding and cultural methods have their limits in enhancing tomato tolerance to challenges including cold, salinity, and drought, genetic engineering shows promise by overexpressing genes that encode antioxidant enzymes, ion transporters, and osmoprotectants (Pandey et al. 2011).

Modern agriculture must accept creative ideas to feed the expanding world population as conventional farming techniques reach their limits. As a well-known instrument among several at hand, genetic engineering improves crop plants by adding alien genes and analysing gene networks for stress tolerance. Functional genomics advances highlight important gene families and mechanisms improving abiotic stress adaptation and increasing yields (Kissoudis et al. 2015). While conventional tomato breeding focusses on characteristics for processing or fresh consumption, the growing threat of abiotic stressors forces a change in emphasis to stress resistance. But genetic distance makes it difficult to transfer stress tolerance features from wild relatives to cultivated kinds, which is why contemporary biotechnological methods like transgenesis are used to get around crossing barriers (Böndel et al. 2018; Liu et al. 2018a).

3.6 Significance of Developing Stress-Tolerant Tomato Varieties

Stress-tolerant tomato varieties are essential for ensuring both sustainable agriculture and global food security in the face of environmental swings. As climate change accelerates and population pressures mount, traditional agricultural methods prove insufficient, necessitating innovative solutions. These resilient tomato cultivars fortify yields and mitigate crop losses by thriving in adverse conditions. Leveraging functional genomics and genetic engineering, scientists can precisely identify and manipulate crucial genes within stress-response pathways, thereby enhancing tomato plant resilience. These improved cultivars not only ensure consistent harvests but also contribute to robust agricultural systems capable of meeting the needs of a growing population with minimal environmental impact.

3.7 Unravelling the Tomato Stress Response

3.7.1 Overview of Plant Stress-Response Mechanisms

Plants adjust protein expression patterns in response to adverse conditions. Stress signals are initially perceived at the plasma membrane, where most receptor proteins are situated. Abiotic stress disrupts the balance between reactive oxygen species (ROS) and their scavengers, primarily produced in mitochondria and chloroplasts, adversely affecting cellular environments (Osakabe et al. 2013; Khan et al. 2023). ROS serve as signalling molecules, regulating gene and protein expression across cellular compartments (Mittler et al. 2022). Analysing the predicted sub-cellular localization of stress-regulated gene products revealed distinct responses across organelles. Mitochondria displayed prominent changes during temperature stress, with an abundance of upregulated genes. Conversely, plastid-localized proteins exhibited a higher proportion of downregulated genes, particularly under cold and salinity stress. Golgi-localized proteins, mainly downregulated under heat stress (HS), indicate a unique interaction between heat and Golgi functions. Drought and salinity stress induce similar cytosolic and extracellular responses but differ significantly in mitochondrial and plastid protein downregulation. Overall, specific abiotic stresses exhibit distinct molecular signatures (Kopecká et al. 2023).

Stress in plants is triggered by external factors that disrupt optimal living conditions, necessitating metabolic adaptability for survival, a phenomenon known as plant tolerance. This adaptability allows plants to avoid, tolerate, or recover from unfavourable environments. Adaptation to abiotic stresses is a complex process involving multifaceted responses and intricate gene interplay and molecular pathway crosstalk. These responses are elucidated through advancements in the understanding of plant abiotic responses. Abiotic stress typically progresses through stages of alarm, resistance, exhaustion, and, if conditions permit, regeneration

(Rehem et al. 2012; da Silva and Costa de Oliveira 2014). The regeneration stage, proposed previously, enables partial or full recovery of the plant's physiological function if the stress is alleviated before irreversible damage occurs (Lichtenthaler 1988). Cell wall receptors detect stress signals, which start signal transduction events involving different intermediary stress genes like those in the calcium-dependent or mitogen-activated protein kinase (MAPK) cascades. Plant stress tolerance is increased by these genes' activation of transcription factors (TFs) and cis-acting elements, which control the expression of stress-response genes. Based on their roles, stress-induced genes fall into three categories: transcription factors, regulatory proteins with unclear functions, and structural or enzymatic proteins (Yamada et al. 2009).

3.8 Stress Signalling Pathways and Key Regulatory Genes

3.8.1 Abscisic Acid Pathway

Abscisic acid (ABA) is well recognized as a stress hormone; its importance for plant growth and development even in non-stressful scenarios is well established, and it plays a primary role in reducing abiotic stress in plants (Liu et al. 2022; Singh et al. 2022). By upregulating hormone-responsive transcription factors, kinases, and phosphatases, the plant hormone ABA is essential in helping plants adapt to different abiotic stressors (Tuteja 2007; Sah et al. 2016). Multiple stress signals that these secondary signalling molecules transduce control the expression of genes that are sensitive to stress (Pandey et al. 2017). Furthermore, ABA has been reported to control the expression of many genes linked to plant dehydration tolerance (Trebichalský et al. 2017). Quantitative characteristics linked to drought, cold, and salinity have shown increases in ABA levels; comparable events have been reported in several crop species (Guóth et al. 2009; Thameur et al. 2011).

ABA controls the *Solanaceae* plants' response to various abiotic stressors in most part (Nivedithadevi et al. 2012). When faced with a drought, tomato plants close their stomata to conserve water and activate genes that confer drought tolerance. Similar ABA regulation of the ion channels in eggplants, tomatoes, and pepper maintains ion balance and prevents sodium ions from accumulating under salt stress (Ruggiero et al. 2019; Meng et al. 2020; Alkhatib et al. 2021).

Elevated ABA levels under cold stress boost the low temperature (LT) tolerance of potatoes and pepper plants and also activate genes that are sensitive to cold (Danieli et al. 2023). Moreover, ABA balances the effects of ROS and therefore helps *Solanaceae* plants manage oxidative stress by enhancing the antioxidant defence system. These adaptive responses emphasize how important ABA is for *Solanaceae* plants to be able to survive abiotic stress and for offering opportunities to boost crop yield even in difficult conditions (Back et al. 2016).

3.9 Reactive Oxygen Species Signalling

Raised ROS levels can cause genetically programmed cell death even though they are essential for plant development and act as essential secondary messengers in cell metabolism. Reactive oxygen species generation in several cell organelles can be increased by adverse environmental circumstances, hence causing oxidative stress (Tripathi et al. 2017; Banerjee et al. 2019). Each of the ROS hydrogen peroxide (H_2O_2), hydroxyl radical (OH−), singlet oxygen ($1O_2$), and superoxide radical (O_2−) has a different half-life and may damage different parts of cells (Waszczak et al. 2018). Highly reactive catalytically, ROS, OH–, and $1O_2$, in particular, may start harmful oxidative cascades when produced in large quantities under stress (Lushchak 2014).

When under stress, ROS play two functions depending on their cellular concentration. Oxidative stress can arise from increased ROS levels damaging proteins, lipids, nucleic acids, and cofactors by oxidation (John et al. 2010; Nath et al. 2016; Martinez et al. 2018). To counteract this, plants have developed detoxification mechanisms to maintain cellular homeostasis (Kushwaha et al. 2019). Conversely, at reduced or moderate concentrations, ROS serve as crucial components of signalling pathways, triggering the upregulation of anti-oxidative pathways to enhance tolerance against oxidative damage (Singh et al. 2017a, b).

ROS generated during stress are detected by ROS sensors, which use redox processes and the mitogen-activated protein kinase (MAPK) pathway to send the signal to the nucleus. This activation leads to stress-responsive transcription factors binding to their respective cis-acting elements (CREs), initiating transcriptional reprogramming and protein synthesis. NADPH oxidases (RBOH) sense the stress and become activated by calcium and phosphorylation, producing ROS. ROS can be detected by receptors, redox-sensitive compounds like transcription factors, or inhibiting phosphatases (Foyer and Noctor 2016; Gilroy et al. 2016). It is noteworthy to mention here that low levels of ROS function as signalling molecules to regulate the growth, development, and defence against both biotic and abiotic stressors (Turkan 2017).

3.10 Calcium Signalling

Changes in cytosolic calcium levels triggered by external stimuli are termed calcium waves, which are detected by calcium sensors. These sensors are proteins that undergo conformational and functional changes upon binding to Ca^{2+}. Consequently, calcium signalling acts as the mediator of various stimuli and serves as the initial trigger for subsequent responses. Numerous calcium sensors have been implicated in both abiotic and biotic stress responses, positioning them as central components in plant reactions to simultaneous abiotic and biotic stresses (Ranty et al. 2016). Growing research also suggests that MAPK cascades are important controllers of how a number of calcium sensors have been recognized as important participants in plant stress responses; some have been demonstrated to engage in both biotic and abiotic stress signalling pathways. As instances of AtCML37, AtCML42, AtCML9, and OsCPK12. The dual functions of jasmonic acid (JA) signalling in abiotic and biotic stress responses are regulated by AtCML37 and AtCML42 (Ahmad et al. 2016). By triggering defence-related genes, JA treatment in plants has been shown to provide protection against a variety of conditions including cold, freezing, salt, drought, necrotrophic diseases, insect attacks, and mechanical damage (Kazan 2015). The activation of protective genes by JA upon infection by necrotrophic diseases, insect attacks, and wounding events highlights calcium sensors as possible centres for both abiotic and biotic stress responses in plants (Yan and Xie 2015). Furthermore, linked to ABA sensitivity are several calcium sensors that respond to stress. For example, AtCML9 controls ABA sensitivity and engages in both kinds of stress, as much as AtCML37 and AtCML42. In a similar vein, OsCPK12 plays a double function in biotic and abiotic stressors, regulating defensive mechanisms against blast fungus negatively and salt stress positively by lowering ROS accumulation. This dual function is explained by the way it affects ABA signalling, an important process in both kinds of stress (Magnan et al. 2008; Asano et al. 2012).

3.11 Mitogen-Activated Protein Kinase Cascade

The MAPKs are intra-cellular signal transducers that, by a series of events involving MAPKs, MAPK kinases (MAPKKs), and MAPK kinase kinases (MAPKKKs), transform extracellular stimuli into cellular reactions. When MAPKKKs are stimulated, they activate MAPKKs, which phosphorylate and activate MAPKs in turn. Moving into the nucleus, phosphorylated MAPKs control gene expression and affect plant growth and development. Stress activates MAPK signalling pathways. NaWIPK and NaSIPK MAPKs, for example, are highly activated in *Nicotiana attenuata* in reaction to wounding, which may affect the plant's stress response by modifying gene expression (Lee et al. 2001; Heinrich et al. 2011; Wei et al. 2022). Horticultural plants respond to abiotic stresses such as UV radiation, ozone, salt, drought, and high temperatures by MAPK cascades (Yanagawa et al. 2016; Ji et al. 2017).

The foundation of the MAPK cascade is increasing data suggests that the signal transduction system underlying plant stress resistance operates as a complex network structure. The MAPK cascade efficiently amplifies and conveys various stress signals to downstream response components through a sequence of phosphorylation and dephosphorylation processes, thereby starting a series of stress reactions. Major participants in gene transcription regulation and necessary for signal transmission, MAPK cascades have been applied to enhance tolerance to abiotic stress (Nehanjali Parmar et al. 2017). Responses to heat and oxidative stresses mediated by brassinosteroids (BRs) are associated with tomato SlMPK1 and SlMPK2 MAPK genes. SlMPK1 acts as a tomato plant's negative thermotolerance regulator. Transgenic tomato plants with SlMPK1 silencing exhibit better tolerance to high temperatures; overexpression of SlMPK1 lowers tolerance and raises H_2O_2 and malondialdehyde (MDA) levels (Nie et al. 2013; Ding et al. 2018). Growing research also suggests that MAPK cascades are important controllers of how horticultural plants react to salt stress. Furthermore, the same MAPK cascade genes can be simultaneously activated by many stressors. The precise functions of MAPK genes and the mechanisms by which MAPK cascades control plant stress resistance are not well-understood despite the fact that numerous studies on horticultural plants have demonstrated the involvement of MAPK cascades in various biological processes in response to both abiotic and biotic stresses (He et al. 2020).

3.12 Physiological and Molecular Response to Abiotic Stress

Abiotic stressors induce significant physiological changes in horticultural crops, categorized as adaptive and non-adaptive responses. Non-adaptive reactions reflect damages caused by adversity, such as osmotic impacts from freezing, drought, and salinity leading to the accumulation of osmoprotectants. ROS production, as mentioned in the preceding section, is a common stress-induced reaction across abiotic stress conditions. Heat and salt stress affect ion transport and compartmentalization, while drought and salinity induce physiological water deficit and reduce stomatal openings, limiting CO_2 transport and carbon metabolism. High temperatures disrupt protein structure and membrane fluidity, prompting structural adaptations in plants like cacti. Limited research has explored the physiological responses of plants to various stimuli (Pandey et al. 2015; Zhang et al. 2022a).

Essential biological macromolecules, like transcription factors and genes, play a key role in coordinating translational and post-translational responses to several abiotic stressors. Involved in sensing environmental cues, synthesizing crucial enzymes and hormones, and balancing growth with stress response, these diverse molecules play multifaceted roles. The first line of defence for horticultural crops against abiotic stressors is environmental cue detection. Stress sensors can modify signal pathways to quickly initiate appropriate reactions. As a result, plants can translate abiotic stress inputs into functional cellular signals. According to recent studies on the sensing mechanisms of abiotic stress signals, histidine kinases (HKs) and calcium ion (Ca^{2+}) sensors are essential for identifying stress signals in a variety of horticultural crops (Balfagón et al. 2022; Xu et al. 2022). A schematic representation is illustrated in Fig. 3.2.

3.13 Molecular Strategies for Improving Tomato Stress Tolerance

3.13.1 Genetic Engineering and Overexpression Approaches

Gene overexpression and RNA interference (RNAi) are examples of genetic engineering techniques that allow for the direct introduction of a desired gene into an organism or the suppression of an unwanted gene by the use of recombinant DNA methods (Fischhoff et al. 1987; Kramer and Redenbaugh 1994). Genetic

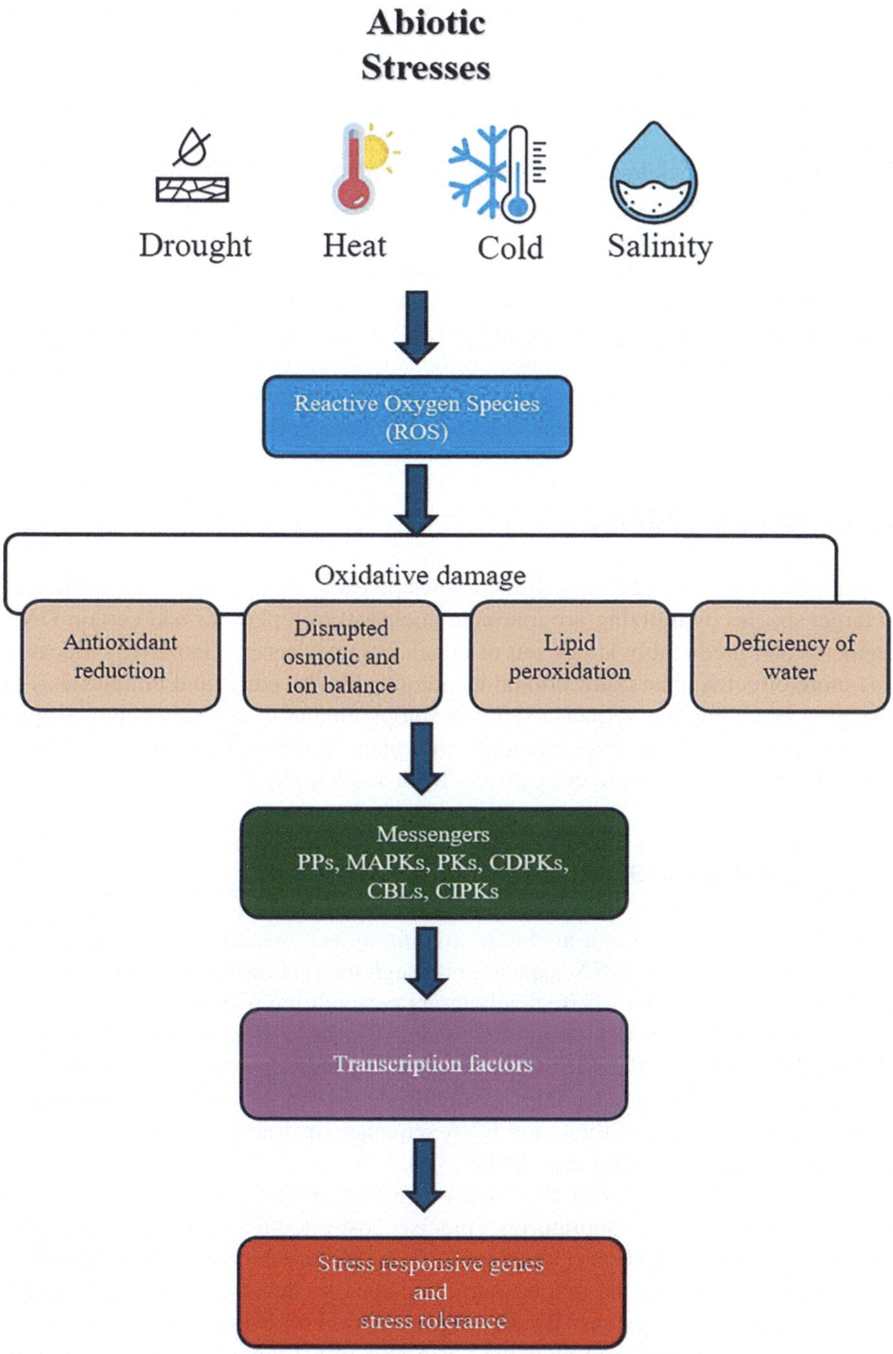

Fig. 3.2 Plant signalling response to abiotic stresses: ROS-modulated protein kinases (PKs), calcium-dependent protein kinases (CDPKs), mitogen-activated protein kinases (MAPKs), protein phosphatases (PPs), calcineurin-B-like proteins (CBLs), CBL-interacting protein kinases (CIPKs)

engineering enables the transfer of DNA sequences between many creatures and plant species, in contrast to conventional breeding techniques that are restricted to transferring genes within the same species. Researchers are given strong instruments by this adaptability to improve agronomic properties accurately and effectively.

Unlike gene knockdown or elimination techniques like RNAi and CRISPR/Cas9, overexpressing genes in a wild-type genetic background provides an easy mechanism to investigate alternative biological pathways. Tomato improvement has long included gene overexpression. For this aim, several promoters have been improved, including constitutive ones like CMV and others like the fruit-specific promoter 2A11 or the inducible promoter Lip9 (Luo et al. 2008; Wang et al. 2013). This approach has been widely used in the functional investigation of regulatory genes and in improving features associated with environmental stressors and tomato fruit quality. An overview of some studies is depicted in Table 3.1.

3.14 Genome Editing

As an innovative tool, genome editing allows precise, site-specific genetic changes in target species by utilizing programmed nucleases to replace or add certain DNA segments and predictably knock out or knock in target genes (Gao 2015). Because it is more effective than conventional techniques like breeding and mutagenesis in changing target genome sequences, this development has completely changed agricultural practices, crop improvement, and plant functional genomics research (Osakabe et al. 2010; Sikora et al. 2011; Arora and Narula 2017).

3.15 CRISPR-Cas9 Techniques

ZFNs, TALENs, and homing nucleases are among genome editing techniques that bind and identify certain DNA sequences through interactions between proteins and DNA. Therefore, in order to effectively target gene editing with these technologies, bespoke tailored nucleases must be designed, which is expensive, and time-consuming. Contrarily, CRISPR technology just needs a combination comprising a Cas-associated protein and a guide RNA (gRNA), in which the gRNA, via complementary base pairing with the target DNA sequence, instructs the Cas9 nuclease to cleave the target region (Gaj et al. 2016).

In 2012, the discovery of CRISPR/Cas9 revolutionized crop improvement and genome editing, offering laboratories a precise, cost-effective, and easily accessible tool for genetic modifications (Jinek et al. 2012). Modified tomato leaf architecture was reported to produce the first heritable site-specific mutation in tomatoes utilizing the CRISPR/Cas system (Brooks et al. 2014). Following that, tomato stress tolerance has been improved by using CRISPR/Cas9 (Lv et al. 2024; Saikia et al. 2024; Wang et al. 2024) (Table 3.1).

Table 3.1 Overview of studies related to gene overexpression that are associated with abiotic stress tolerance in tomatoes

Stress	Gene	Role	Reference
Drought	*BGC7, BGC11 and SlMYB13*	Through the enhancement of the scavenging process of reactive oxygen species and the elevation of the abscisic acid content, the tomatoes are improved	Cao et al. (2024)
	AT14A	AT14A regulates the expression of protein phosphatase 2C 51 (PP2C 51) and abscisic acid–insensitive 5 (ABI5), participating in ABA pathways to enhance drought tolerance	Xin et al. (2023)
	SlHK2	Suppressing the expression of the SlHK2 cytokinin receptor gene using RNA interference (RNAi) can enhance plant resilience to drought, heat, and combination stress	Mushtaq et al. (2022)
	SlGATA17	Through the control of the phenylpropanoid biosynthesis pathway, the SlGATA17 gene affects tomato plant drought tolerance	Zhao et al. (2021)
	SlMBP22	Plants overexpressing the SlMBP22 gene accumulated higher levels of chlorophyll, starch, and soluble sugars, which enhanced their tolerance to drought stress	Borg et al. (2020)
Heat	*SlHyPRP1*	CRISPR-Cas9-based precise domain editing of the SlHyPRP1 gene generated multi-stress-tolerant having moderate heat tolerance	Tran et al. (2023)
	SlMYB41	Two individual SlMYB41-RNA interference (RNAi) lines and one CRISPR/Cas9-mediated myb41 mutant were obtained to investigate the function of SlMYB41 in tomato thermotolerance that downregulated heat-related genes	Wang et al. (2023a)
	SlBBX17	Overexpression of SlBBX17 impacted the expression of genes involved in the GA biosynthesis pathway, photosynthesis, heat stress, ROS, and other cellular functions	Xu et al. (2022)
	CPK28	The ascorbate peroxidase (APX) activity has decreased, while the levels of ROS have increased	Hu et al. (2021)
	SlSNAT	The SlSNAT overexpression line demonstrated protected ribulose bisphosphate carboxylase oxygenase proteins and elevated responsiveness of heat transcription factors and heat shock proteins under heat stress	Wang et al. (2020a)

(continued)

Table 3.1 (continued)

Stress	Gene	Role	Reference
Cold	*SlNAC3*	Modulating SlNAC3 through upregulation or downregulation resulted in either decreased or increased cold resistance	Wang et al. (2024)
	SlREC2	SlREC2 encodes a tetratricopeptide repeat protein that positively regulates tomato cold tolerance via decreased abscisic acid (ABA) levels, photoprotection, and the expression of C-repeat binding factor (CBF)–pathway genes	Zhang et al. (2023)
	SlCYP90B3	The overexpression of SlCYP90B3 enhanced the ability to withstand low temperatures during storage by controlling the expression of SlCBF1 and the antioxidant enzyme system	Hu et al. (2022)
	LeCOR413PM	Overexpression of the LeCOR413PM2 gene reduced cell membrane damage, ROS accumulation, and photoinhibition of PSII while maintaining high antioxidant enzyme activity and osmotic regulator content	Zhang et al. (2021)
	SlHY5	Overexpression of SlHY5 increased tomato plant anthocyanin production and decreased oxidative damage, lipid peroxidation, and membrane injuries from cold stress	Han et al. (2020)
Salinity	*MdATL16*	Ectopic expression of MdATL16 in transgenic tomato enhanced salt-stress tolerance, and this was related to decreased build-up of hydrogen peroxide (H_2O_2) and superoxide radicals ($O_2^{\bullet -}$) and less electrolyte leakage and malondialdehyde (MDA)	Yuan et al. (2024)
	SlGGP-LIKE	Overexpressing SlGGP-LIKE can increase tomatoes' ability to withstand salt stress by stimulating the production of ascorbic acid (AsA)	Yang et al. (2023)
	SlABIG1	Tomato salt tolerance was greatly increased by SlABIG1, which also decreased ROS, MDA, and Na^+ build-up and increased chlorophyll content, photosynthetic ability, root dry weight, and proline levels	Ding et al. (2022)
	LeNHX2 and SlSOS2	Under NaCl cultivation conditions, transgenic tomato plants overexpressing SlSOS2, LeNHX2, or both exhibited better growth and had higher biomass	Maach et al. (2021)
	LeNHX4	Overexpression of LeNHX4 in tomato plants enhanced salt tolerance and boosted fruit production and quality under both normal and saline stress conditions	Maach et al. (2020)

3.16 Drought Stress Tolerance

In meteorology, drought is defined as an extended period without significant rainfall. Plants experience drought stress when soil water availability is low, and frequent water loss occurs due to evaporation and transpiration. Agricultural drought is a lack of sufficient moisture for plant growth and development. Drought is a highly damaging environmental stress, significantly affecting plant production, growth, biomass, and yield (Farooq et al. 2012; Munné-Bosch and Villadangos 2023). Exposure to water or osmotic stress triggers the production of harmful reactive oxygen species, causing oxidative damage and enzyme inactivation. To minimize this damage, cells produce antioxidants that neutralize these reactive species (Atkinson et al. 2011; Vurukonda et al. 2016; Nawaz and Wang 2020). Twenty per cent of agricultural land worldwide is irrigated, and of that, only 14.51% is well irrigated; the bulk is fed by rain. This prevalence of rain-fed agriculture makes drought a common occurrence, significantly impacting genetically predetermined yields (Krishna et al. 2022). Drought disrupts the physiology of tomato plants by affecting plant–water relations and the generation of ROS. While many wild species possess the tolerant genes for drought, their utilization is challenging due to high genetic distances and barriers during embryo development, both pre- and post-transcriptionally (Pessoa et al. 2023).

3.17 Specific Molecular Strategies for Drought Tolerance

Water scarcity shapes both the morphology as well as the physiology of plants, prompting urgent breeding for drought-tolerant crops amidst increasing and prolonged droughts from climate change. Molecular pathways activate genes, promoting osmolyte accumulation, antioxidant system enhancement, reduced transpiration, inhibited shoot growth, decreased tillering, and hormonal regulation. While ABA is prominent in drought conditions, inducing stress-related gene expression, ABA-independent regulatory systems also govern drought-responsive gene expression (Joshi et al. 2016). There is still little known genetically about tomato drought tolerance. While several quantitative trait loci (QTLs) have been found, the exact identification of those linked to drought tolerance is complicated by the fact that these areas frequently include several genes. Previous studies reported few drought-tolerant QTLs that are important for seed germination during drought stress and have been linked to increased yield under drought conditions (Foolad et al. 2003; Gur and Zamir 2004). Recently, high-throughput analysis has revealed thousands of genes affected by drought stress, yet limited research has delved into the significance of upregulated or downregulated genes for enhancing tomato drought tolerance (Candar-Cakir et al. 2016; Iovieno et al. 2016; Liu et al. 2018b). Wild tomato species like *Solanum pennellii* possess elite genes conferring drought tolerance, contrasting with modern cultivars, which are often sensitive to drought due to their narrow genetic basis (Bolger et al. 2014). A model was proposed outlining the drought tolerance mechanisms in *S. pennellii* (Egea et al. 2018). Gene expressions in drought-tolerant *S. pennellii* lines and a drought-sensitive *S. lycopersicum* cultivar ("M82") were examined. They found 400 genes responsive to drought that are involved in transcription, cell wall architecture, wax production, and plant height. The adaptive reactions of tomato plants to drought were highlighted by the way drought stress impacted signalling pathways associated with energy metabolism and oxidative damage (Gong et al. 2010). Genome-wide studies have identified genetic variations in fruit quality components under drought stress. The authors identified 41 putative candidate genes associated with drought stress response in tomatoes, distributed among three constitutive QTLs and 15 interactive or specific QTLs. These genes are related to various functions including protein protection, water and solute transport, sugar metabolism, and hormonal signalling (Albert et al. 2016).

3.18 Drought-Responsive Genes and Regulatory Networks

Drought stress induces endogenous ABA levels in plants, altering gene expression in cultivated tomato leaves (Vishwakarma et al. 2017). While many cultivated tomato varieties are sensitive to drought, wild relatives like *S. pennellii* exhibit high drought tolerance (Harrak et al. 2001). ABA-sensitive genes were explored in *S. pennellii* that are identical to those in *S. lycopersicum* (Kahn et al. 1993). However, only a subset of these genes responded to exogenous ABA treatment in *S. pennellii*. Additionally, transcription factors like abscisic acid responsive element binding protein (AREB) and dehydration-responsive element-binding (DREB) TFs play crucial roles in ABA-mediated stress responses, with genes like SlAREB1 and LeDREB3 implicated in drought tolerance mechanisms. By controlling genes associated with stress, SlAREB1 overexpression in tomatoes increased tolerance to salt and drought. Comparably, overexpression of LeDREB3 improved tomato plant stress tolerance (Islam and Wang 2009; Orellana et al. 2010). Moreover, ERF proteins including CBF genes and TERF2/LeERF2 regulate downstream stress-related genes in tomato stress-response pathways (Zhang et al. 2004, 2009; Zhang and Huang 2010). In tomato oxidative and abiotic stress responses is LeDREB2, another DREB gene (Guo and Wang 2011). Additionally, SlDREB negatively regulates tomato plant architecture while enhancing drought tolerance, affecting GA biosynthesis genes and endogenous GA levels (Li et al. 2012).

3.19 Drought-Tolerant Tomato Varieties and Their Characteristics

Transgenic technology has been utilized to meet global tomato demands and uphold tomato fruit quality. Its aim is to sustain predetermined genetic yield and quality standards in tomatoes. In 2003, researchers introduced the AtCBF1 gene, regulated by an ABA/stress-inducible promoter, into tomato plants to explore their yield potential and survival under drought stress. Compared to the wild type's 32.0 g, transgenic tomato lines (AC1–3) produced far more after 4 weeks without irrigation: 102.2 g, 115.0 g, and 101.0 g per plant, respectively. Furthermore, the natural type produced no surviving plants; the transgenic lines showed remarkable survival rates: AC1 (25 out of 30 plants), AC2 (28 out of 30 plants), and AC3 (29 out of 30 plants). During drought stress, the CBF1 transcription factor is essential in controlling many molecular pathways, which promotes plant growth and development (Ratcliffe and Riechmann 2002). The tomato variety Yaxin 87–5 was transformed with the HARDY gene obtained from *Christolea crassifolia*. After 16 days of drought stress, the transgenic lines produced 38.65–44.65% more than the natural type and showed better growth (Guo et al. 2017a, b). Transgenic tomatoes were developed using an aquaporin gene (MdPIP1;3) from *Malus domestica*. The yield potential of these transgenic tomatoes in artificially created drought conditions was assessed. Comparing the transgenic tomatoes to wild-type tomatoes, the former produced heavier and bigger fruits after 20 days without irrigation (drought stress). Under a drought, transgenic tomato yield increased by about 1.2%. Moreover, whereas 65% of wild-type tomato stomata remained open throughout the drought, 77% of transgenic tomato stomata closed. This implies that by making water transportation in the fruits easier during the formative stages, the aquaporin contributed to the management of drought stress. One useful approach to deal with drought stress in tomatoes could be to increase their water consumption efficiency (Wang et al. 2017). After introducing *A. thaliana* guard-cell hexokinase (AtHXK1) into tomato (cv. MP-1), another study group discovered that, under drought stress, tomato guard cells' water usage efficiency increased by 10–20% (Kelly et al. 2017).

3.20 Heat Stress Tolerance

Tomato crops are severely stressed by high temperatures; even little increases result in large losses in worldwide production (Ayenan et al. 2019). The vegetative and reproductive stages of tomato tolerance to heat stress differ, which influences pollen germination and viability (Zinn et al. 2010; Nievola et al. 2017). By creating dehydrated embryos and pollen grains, among other adaptations, plants have evolved to withstand heat stress (Raja et al. 2019). Managing agricultural productivity requires an understanding of plant reactions across reproductive stages (Zinn et al. 2010). Growing and producing at high temperatures require heat stress resistance (Janni et al. 2020). It irreversibly harms plants, impairing growth, yield, and other physiological functions and hence lowering plant productivity (Zhao et al. 2020). Increased HS results in the synthesis of ROS, which seriously damages cells and may even kill them (Frank et al. 2009). Rising ROS levels affect programmed cell death (PCD) and cause transcriptional alterations (Kotak et al. 2007). To adjust to HS, ROS engage with pathways including hormone signals, kinase cascades, and Ca^{2+} signalling (Suzuki and Katano 2018). Because of protein denaturation, heat shock proteins (HSPs) including chaperones are also induced by HS (Scharf et al. 2012).

3.21 Molecular Approaches Particular to Heat Tolerance

Protein denaturation is widely regarded as the primary trigger for the HS-induced synthesis of several heat shock proteins (HSPs), mainly chaperones (Scharf et al. 2012). Numerous heat stress sensors have been found in different plants in recent research (Liu and Howell 2016). ABF/bZIP, CBF/DREB, Ca^{2+}-CBL-CIPK, and MAPK pathways are only a few of the molecular signalling pathways that heat stress can affect and which are essential to plant growth and adaptation to significant abiotic challenges.

Thermotolerance-related molecular pathways have been discovered in *Arabidopsis*, tomatoes, and other species (Qu et al. 2013; Ohama et al. 2017). Heat shock factors (HSFs) are transcription factors (TFs) associated with heat stress. While most eukaryotes possess one to three HSFs, plants have a higher number, exceeding twenty, which are classified into categories A, B, and C. Class A HSFs function as transcriptional activators. It was demonstrated that in *Arabidopsis*, class B HSFs, specifically HSFB1 and HSFB2b, act as transcriptional repressors. They inhibit the expression of heat-induced HSFs (HSFA7a, HSFA2, HSFB2b, and HSFB1) as well as other genes responsible for producing heat shock proteins (Ikeda et al. 2011). The promoter of the dehydration-responsive element-binding protein 2A (DREB2A) contains a heat shock element that was discovered previously to be a cis-acting element involved in DREB2A's response to heat stress. Specific mutants that do not have HSFA1a, HSFA1b, and HSFA1d were shown to have reduced expressions of DREB2A in response to heat stress. As major controllers of the plant's HS response (HSR), these proteins are necessary for gene expression during normal growth (Yoshida et al. 2011). Through full-length cDNA over-expression by FOX hunting identified HSFA1d which regulates HS-responsive gene expression, thus enhancing heat tolerance. In addition, a specific region of HSFA1d prevents its transactivation activity in response to heat shock; removal of this region results in considerable thermal stability. Controlling HS-related variables, including DREB2, thus requires research that advances our understanding of the temperature-controlled mechanism of HSFAs (Higashi et al. 2013; Ohama et al. 2016).

3.22 Heat-Responsive Genes and Regulatory Networks

Plant HSR genetic foundations and the discovery of genes, transcription factors, and QTLs have advanced dramatically in the past few years. Next-generation breeding methods enable the production of plants with HS-responsive genes, confirmed on advanced phenotyping platforms, therefore improving our understanding of HSR genes and the development of heat-tolerant cultivars. Physio-morphological acclimatization, acquired thermotolerance, biological circadian clock synchronization, and immune response control are some of the plant responses to high-temperature stress (Janni et al. 2020). Tobacco thermotolerance is mediated by about 21 heat stress transcription factors (HSFs). Among them, HSFA1 stands out as the main heat-shock-response transcriptional regulator. The heat shock response in tomatoes is synergistically increased by its connection with HSFA2 (Scharf et al. 1998; Chan-Schaminet et al. 2009). Crucially, for adjusting to increased ambient temperatures, the tomato PIF4 gene shows transitory expression and temperature-dependent binding to auxin biosynthetic gene promoters. Flowering time is also regulated by MADS-box genes such as Flowering Locus M (FLM)/Mads Affecting Flowering (MAF) 1–5 in reaction to temperature variations and short vegetative phase (SVP), together with PIF4 (Lee et al. 2007; Sun et al. 2012).

Overseeing protein quality, heat shock proteins are regulated by HSFs (Scharf et al. 2012). The HSR depends critically on these important chaperone proteins. Small HSPs (sHSPs) and sub-family proteins including HSP70, HSP60, HSP90, and HSP100 are examples of the many forms of heat shock proteins (Qu et al. 2013). Analysing tomato heat stress expression levels of TFs and HSPs will help understand the molecular processes underlying the mutant's response to high temperatures.

3.23 Heat-Tolerant Tomato Varieties and Their Characteristics

Researchers did not detect the MT-sHSP in wild-type plants under optimal or high temperatures (Nautiyal et al. 2005). However, transgenic tomato plants carrying the MT-sHSP gene exhibited thermotolerance under high-temperature stress. This observation aligns with a similar positive correlation reported previously (Mahesh et al. 2013).

In another work, gene transfer is shown to improve, in laboratory or field settings, heat, UV-B, and sunscald stress tolerance in transformed tomato (*Lycopersicon esculentum* cv. Zhongshu No. 5) plants. Transgenic tomatoes with overexpression of cAPX showed far higher resilience to UV-B stress and heat (40 °C) than wild-type plants. For example, electrolyte leakage was 93% in wild-type leaf discs after 13 h at 40 °C but 24% and 52% in transgenic lines A9 and A16. Comparably, wild-type fruits brown 95% after 5 days of UV-B (2.5 mW cm^{-2}) exposure, but transgenic fruits brown 33% and 37% (Wang et al. 2006).

Tomato BZR1, also known as Tomato BRASSINAZOLE RESISTANT 1, plays a crucial role in the response to brassinosteroid (BR) and is involved in plant development as well as heat stress resistance. Apoplastic H_2O_2 serves as a signal in brassinosteroid (BR)-mediated stress tolerance, initiating the activation of ROS-scavenging enzymes and thereby enhancing abiotic stress tolerance (Zhou et al. 2014). Two tomato genotypes, CLN-1621 and H88-78-1, showed higher stress tolerance traits and upregulated the expression of heat shock proteins and stress-responsive genes under high-temperature conditions compared to the susceptible genotype Punjab Chhuhara. Class III HSP, H HSF B-2a, HSF B-3, SF A-4a, HSF30, HSF24, and CLN-1621 were found upregulated following the exposure to high-temperature stress (Karkute et al. 2021).

3.24 Cold Stress Tolerance

One of the most detrimental environmental challenges higher plants face, in addition to heat stress, is low temperature or cold stress. Depending on the surrounding temperature, cold stress is classified as freezing stress (<0 °C) or chilling stress (<20 °C). Not only does it impact the plant's growth and development, but it also severely limits the plant's geographic range (Guo et al. 2017a, b; Shi et al. 2018). It has come to light that plants found in temperate zones can withstand cold stress and even become acclimatized to it. Hence, cold stress is one of the most important abiotic stresses since it significantly reduces agricultural yields. The degree of stress and length of exposure determine how plants respond to cold stress. The cold primarily affects the plant when it is at the seedling stage. The emergence of the surface lesion, discolouration, desiccation, induced senescence, decreased shelf life, and ethylene production are typical indications (Solanke and Sharma 2008).

3.25 Molecular Response to Cold Stress

Cold acclimatization affects gene expression significantly (Hannah et al. 2005). Low temperatures trigger signalling processes to induce cold-responsive genes, enhancing tolerance (Denesik 2007). During acclimatization, free Ca^{2+} levels rise, affecting gene regulation (Eckardt 2009). Ca^{2+} signalling is crucial in the cold induction of CBF pathways (Jenks and Wood 2009). ROS act as second messengers in stress responses, with Ca^{2+} influx triggering ROS production (Rao et al. 2006). Glutathione regulates H_2O_2 during cold acclimatization (Kocsy et al. 2001). MicroRNAs (miRNAs) repress gene induction, with many cold-regulated miRNAs identified (Ghildiyal and Zamore 2009).

3.26 Genetic Regulation in Response to Cold Stress

Understanding the molecular mechanisms of cold acclimatization requires studying transcriptional changes in plants under low-temperature (LT) stress. The ICE-CBF-COR pathway is a well-documented response to LT stress, where ICE activates downstream genes encoding osmoregulatory substances. ICE1, an MYC-type basic helix-loop-helix TF, initiates cold acclimatization by inducing CBF3 expression, which is regulated by DELLAs via JAZs (Zarka et al. 2003; Zhen and Ungerer 2008; Wang et al. 2018). CBFs, belonging to the AP2/ERF TF family, activate COR gene expression by binding to the CRT cis-element in their promoters. Besides LT stress, CBFs are involved in other abiotic stresses. Some cold-responsive genes, like EgDREB1 and FtbHLH2, contain ABA and dehydration-responsive elements in their promoters, indicating their responsiveness to cold signalling (Azzeme et al. 2017; Borba et al. 2018; Londo et al. 2018; Yao et al. 2018).

Many transcription factors (TFs) have been isolated to regulate cold signalling and stress, for example, MYB15, ICE1/2, CBF3/2/1, CAMTA3, COR15a/b, and

ZAT12. The most critical TF families related to the LT stress of *Pyrus ussuriensis* are MYB, DREB, NAC, AP2/ERF, WRKY, and bHLH. Brassinosteroids (BRs) are critical plant growth hormones that have been documented to protect plants from abiotic stresses, including cold. Brassinazole-resistance 1 (BZR1) and CESTA (CES) are the two well-known TFs of BRs that directly regulate CBF (C-repeat binding factor) (Barrero-Gil and Salinas 2017; Wang et al. 2017).

3.27 Cold-Tolerant Tomato Varieties and Their Characteristics

Low temperatures are usually considered aggressive environmental conditions since they act by retarding growth and development, lead to lower yield, and can even be lethal to plants (Fedoroff et al. 2010). Low temperatures mainly affect plants through changes in carbon and nitrogen metabolism, associated with increasing the production of ROS in plants (Watanabe et al. 2008). Transgenic approaches have been influential in the alleviation of this challenge. It has been reported, through recent research, that the expression of the *Ammopiptanthus nanus* AnGolS1 gene, acting as a galactinol synthase, resulted in increased cold tolerance in tomato plants. Galactinol, in principle, has been involved in the mechanism regulating JA biosynthesis, which helps improve cold tolerance in tomatoes by the mechanism of the SlSPI5-SlLOX5-SlCPase-SlERF4-7/SlAOS2 model (Liu et al. 2024). Another mechanism of ALA improving cold tolerance in tomato plants is by its regulation. Overexpression of SlGSTU43 was shown to enhance the ability of a plant to scavenge ROS, hence increasing resistance to cold stress. The experimental results show that SlMYB88 and SlMYB4 can bind to the SlGSTU43 promoter, thus suggesting the gene SlGSTU43 to be the most useful inclusion in the repository of cold-responsive genes (Zhang et al. 2024). This research group has also undertaken preliminary studies on how 5-aminolevulinic acid improves tomato cold tolerance through multi-omics analysis and gene expression studies (Zhang et al. 2022b). To impede cold stress tolerance in tomatoes, the highly efficient genome editing technique or tool CRISPR/Cas9 has also been adopted. In this regard, Li and researchers studied the C-repeat binding factor 1 (CBF1) gene in tomatoes by the CRISPR/Cas9 technique. CBF1-mutant tomato plants showed a higher rate of electrolyte leakage and more considerable stress under cold conditions, which is proof of the possibility of genome editing (Li et al. 2018a).

3.28 Salinity Stress Tolerance

Salinity has impacted over one-third of irrigated regions already, and projections suggest that salinization will impact more than half of the world's agricultural land by 2050. This phenomenon greatly limits the usage of the land and drastically lowers crop yields (Van Zelm et al. 2020). High salt ion (Na^+) and chloride ion (Cl^-) concentrations in the soil solution induce primary strains, such as osmotic and ion imbalances, and secondary stresses, like oxidative and metabolic disorders. Na^+

accumulation elevates soil osmotic pressure, reduces water potential, and diminishes root water uptake, thus reducing water availability. Additionally, disrupted Na^+/potassium ion (K^+) ratios hinder the activity of essential K^+-dependent enzymes in cells. Chloride disrupts ion balance by interfering with the uptake or metabolism of other vital ions in plants (Bazihizina et al. 2019; Van Zelm et al. 2020; Zhao et al. 2020). Tomatoes originate from western South America, where their wild counterparts have evolved to thrive in highly saline coastal areas. However, through domestication, cultivated varieties have lost their ability to withstand salt stress (Pailles et al. 2020).

3.29 Specific Molecular Strategies for Salinity Tolerance

Salt stress triggers a cascade of effects, including increased production of ROS, decreased photosynthetic electron transport, and reduced stomatal conductance, ultimately leading to plant damage and demise. ROS, such as hydrogen peroxide and superoxide radicals, induce oxidative harm to proteins, nucleic acids, and lipids, disrupting plant metabolism. Plants activate both enzymatic and non-enzymatic antioxidant systems to mitigate ROS-induced oxidative damage (Singh et al. 2020). The stress of salt first elicits a series of responses, with the rapid increase of cytoplasmic Ca^{2+} acting as an early signal in the sensing pathway of salt. Among the potential Ca^{2+}-dependent proteins, CBL-interacting protein kinases (CIPK) as well as calcium-dependent protein kinases (CDPKs), together with calcineurin B-like proteins (CBLs), decipher the Ca^{2+} signal. The well-studied Ca^{2+}-dependent salt overly sensitive (SOS) pathway, controlled by SOS2 and SOS3, governs ionic homeostasis and salt tolerance. Tomato SOS2 is expressed in roots and acts with SOS3 to confer salt tolerance with the tomato enhancer of ENA1 and ENH1 for Na^+ sequestration and K^+ retention. Tomato SOS2 fine-tunes the V-ATPase and NHX transporters that maintain activity and salt tolerance. Tomato SOS2-CBL10 is also involved in regulating the Na/Ca^{2+} ratio and salt-protected damage of a cell. 14-3-3 proteins regulate SOS2 activity, while the generation of ROS and MAPK cascades mediate salt-stress signalling. Membrane transporters like HKT1;2 and HAK20 play crucial roles in Na^+/K^+ homeostasis, influencing salt tolerance in tomatoes (Li et al. 2013; Yang and Guo 2018; Romero-Aranda et al. 2020; Wang et al. 2020b; Zhao et al. 2020; Cho et al. 2021).

3.30 Salinity-Responsive Genes and Regulatory Networks

Plant gene expression is altered by salinity stress, as by other abiotic stimuli. Though research mostly concentrates on changes in transcription factors, knowledge of gene expression resulting from salt stress is still limited (Devkar et al. 2020). For plants under salinity stress, modifications in the expressions of genes involved in protein synthesis, transport activities, cell wall building, production of volatiles, secondary metabolites, etc., are recommended. According to studies, tomato leaves accumulated proline and Na^+ but not roots when the P5CS gene was overexpressed. In roots, higher NHX1 and NHX3 expression was associated with lower Na^+ levels than in leaves. While HKT1;1 gene was in charge of Na^+ accumulation in the roots, expression of HKT1;2 in roots was linked with Na^+ levels in the leaves and stems. In response to salinity stress, HKT1;2 expression increased in stems and roots in the cultivar Arbasson, correlating with increased stress. Na^+ accumulation in leaves was linked to low HKT1;2 expression (Almeida et al. 2014a, b). Salinity stress was also found to alter nitrogen uptake genes. Reduced expression of NRT1.1 and NRT1.2 expression but increased levels of AMT1.1 and AMT1.2 in Manitoba were detected. Conversely, *S. pennellii* showed Gs1 downregulation (Abouelsaad et al. 2016). In one of the works, the salt-stressed tomato plants presented with the upregulation of heat shock transcription factor HSF30 (Hoffmann et al. 2021). Other genes affected by salt stress include SlERF5, a member of the ERF family involved in ethylene and abscisic acid signalling (Pan et al. 2012); SlGSTT2, SlDHAR5, SlGSTU23, SlGSTL3, SlGSTU26, and SlGSTZ2, which play roles in primary metabolism, plant growth regulation, anthocyanin absorption, and detoxification (Csiszár et al. 2014); LeHAK5, whose expression decreases with increased Na^+ concentrations (Bacha et al. 2015); and SlARF8A, SlARF1, SlARF19, SlARF4, and SlARF24, upregulated under salt-stress conditions (Bouzroud et al. 2018). For a detailed overview of the genes involved, readers are suggested to review the work of Guo et al. (2022) and Roşca et al. (2023).

3.31 Salinity-Tolerant Tomato Varieties and Their Characteristics

Transgenic lines with enhanced salt tolerance have been engineered by targeting various metabolic pathways, including antioxidants, osmoregulation, ion transport, and stress proteins, as well as universal stress proteins, transcription factors, and signal transduction components. Improvement in tomatoes for salt tolerance through the genetic transformation of genes participating in oxidative stress and ion homeostasis has been achieved. RING finger E3 ligase Ring is a positive regulator in the process of salt-stress signalling through the maintenance of Na^+ and K^+ homeostasis to reduce the H_2O_2 level. Still, at the same time, it also acts positively in lipid peroxidation and in inducing the expression of genes related to stress (Qi et al. 2016). Overexpression of caffeic acid O-methyltransferase 1 (COMT1), which is a gene involved in the melatonin synthesis pathway, leads to improved salt tolerance by equilibrating Na^+/K^+ levels, activating the SOS pathway, resulting in better antioxidant capacity, and upregulating genes associated with stress-response pathways (Liu et al. 2019; Sun et al. 2020). The small SALT TOLERANCE ENHANCER1 (STE1) protein further enhances ABA-dependent salt-responsive pathways through the binding of ABA receptors and positive regulators of ABA signalling while also influencing Na^+ and K^+ homeostasis and ROS scavenging (Meng et al. 2020). Genome editing in tomatoes has been a significant advancement in enhancing their tolerance to abiotic stress. For instance, genes that improve salinity stress tolerance have been targeted with CRISPR/Cas9 in tomato lines. Such is the case, for example, an edited gene that gave salt-stress-tolerant tomato plants is associated with it. This approach has promising outcomes for meeting environmental challenges towards the sustenance of agriculture (Tran et al. 2021, 2023).

3.32 Combining Molecular Strategies for Multiple Stress Tolerance

The integration of molecular techniques has emerged as a potential strategy for dealing with the complex issues provided by numerous environmental pressures on crop productivity. This includes mixing diverse genetic features to promote tolerance to several stimuli concurrently. Using advanced biotechnological methods, the researchers are attempting to solve the shortcomings of traditional methods. Employing omics (single/combined/multi-omics) technology, they include integrated, comprehensive, sustainable, and repeatable methods for crop development programmes. Understanding and analysing the overall plant stress-responsive web networks and processes requires a thorough science that integrates several branches at a single biological platform and overarches the data of complete components existing within the cell. Next, together with other developing omics technologies, newly identified stress-responsive genes may be mapped utilizing branches like genomics, transcriptomics, and proteomics (Joshi et al. 2023) (Fig. 3.3).

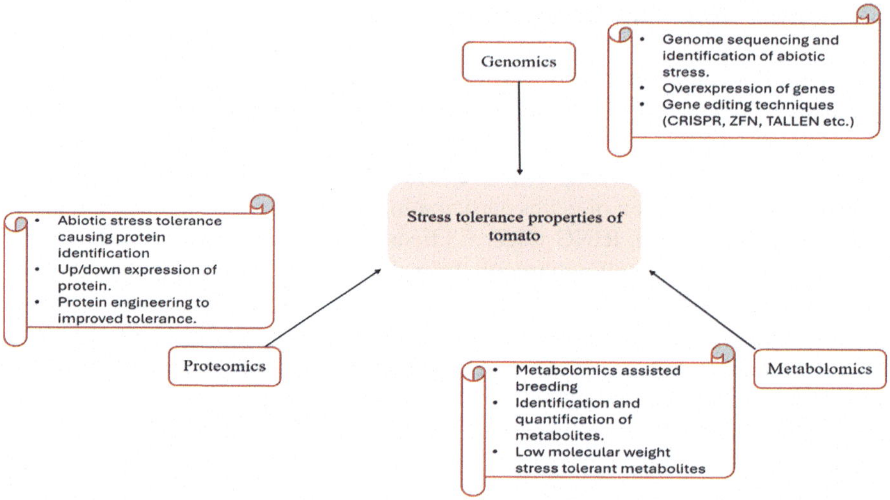

Fig. 3.3 A flowchart of genomics, proteomics, and metabolomics to study stress tolerance properties of tomatoes

3.33 Plant Biotechnology: Transformation Techniques

A new potential field has evolved with the quick development of transformation methods in plant biotechnology, which makes it easier to transfer protein-coding and non-coding genes, especially microRNA (miRNA) genes both inside and across various crossable and non-crossable plant species. These approaches, transgenesis, cisgenesis, and intragenesis, use agrobacterium-mediated T-DNA engineering to transport and express genes of interest; miRNA genes are a focal point in each (Wani et al. 2016; Basso et al. 2019).

1. *Transgenesis.*

   Transgenesis gets across impassable barriers by transferring one or more miRNA genes between non-crossable plant species (Christey 2020). Incorporating miRNA genes into previously inaccessible species enhances nutritional value, stress tolerance, and other beneficial traits.
2. *Cisgenesis.*

   Cisgenesis is the transmission of miRNA genes across naturally crossing plant species (Schouten 2022). The technique chooses a recipient plant that is compatible with the donor; superfluous genetic modifications are avoided, and traditional breeding saves time (Mujjassim et al. 2019).
3. *Intragenesis.*

   The miRNA genes spread within the same species by intragenesis. The method may be used to successfully construct and include customized gene cassettes into the DNA of the intended plant. Within intragenesis, additional promoter and terminator sequences guarantee appropriate gene expression and protect against overexpression-related problems (Rommens et al. 2007; Basso et al. 2019).

3.34 RNA-Based Method

Researchers are using artificial miRNAs (amiRNAs) increasingly to address the issue of non-specific gene silencing that results from siRNA-based methods (Zhang et al. 2018). Plant cell mature amiRNA accumulation is selectively and very effectively reduced by target gene sequence integration into a stem-loop backbone to generate a precursor that processes into amiRNAs in vivo (Van Vu and Do 2017). Another strategy to control plant biological processes is target mimicking, which is divided into endogenous (ETMs) and artificial (e.g. STTM and miRNA SPONGES). Circular and long non-coding RNAs are examples of ETMs; synthetic constructs are used in STTM and SPONGES to change miRNA levels and provide accurate gene control (Reichel et al. 2015; Thomson and Dinger 2016). According to research on *A. thaliana*, the environmental factors and developmental phases affect how effective amiRNA-mediated silencing occurs (von Born et al. 2018). Advanced instruments for stress tolerance and environmentally friendly, natural agricultural methods are provided for the development of amiRNAs and crop improvement through target mimicking methods. Advanced omics technologies are part of the current tools that offer insightful information on molecular stress reactions and help design stress-tolerant crops to solve issues with sustainable agriculture and global food security.

3.35 Targeting Gene Modulation via Genome Editing

With the current developments in genetics, new crop types like tomatoes are being produced completely differently. Molecular markers based on DNA polymorphisms have greatly increased the genetic diversity in agricultural germplasm. Molecular markers differ from morphological or biochemical markers in a few ways: they are easy to experiment with, reproducible, convenient, highly available, stable, and reflect the genome-wide distribution. Sequence-specific editing is revolutionizing the field of plant genome modification as it enables exact changes without adding foreign DNA (Shanmugavadivel et al. 2019). Important techniques including CRISPR/Cas9, zinc finger nucleases, and transcription activator-like effector nucleases modify genes by means of cellular repair processes. CRISPR/Cas9 enables efficient modification of miRNA genes, thereby controlling interactions and miRNA expression (Perez Rojo et al. 2018; Tiwari et al. 2021). Smaller in size, miRNA genes have more limited target options and regulatory networks. CRISPR/Cas9 has been successfully applied to enhance traits such as drought tolerance in rice (Chung et al. 2020).

3.36 Tomato Varieties with Multiple Stress Tolerance

Recently, Saikia and colleagues developed *SlHyPRP1* and *SlDEA1* genes by CRISPR/Cas9 genome editing. These genes are possible negative regulators of *S. lycopersicum L.'s* response to biotic stimuli (bacterial leaf spot and bacterial wilt) and abiotic stresses (drought and salt) (Saikia et al. 2024). In another study, as mentioned in Table 3.1, Tran and colleagues generated the *SlHyPRP1* gene by CRISPR-Cas9-based precise editing for multi-stress tolerance in tomatoes (Tran et al. 2023). In 2020, Saikia and coworkers identified genes *SlHyPRP1* and *DEA1* as regulators of multiple stress-responsive genes of tomato (Saikia et al. 2020). Joint stresses, drought and cold, were investigated, and it was reported that the interaction between leaf physiology and gene expressions involved in ROS, photosynthetic electron transport, phytohormone, signal transduction, and carbohydrate transport in tomatoes under individual and combined stress conditions resulted in abiotic stress-induced disorder in carbohydrate metabolism (Zhou et al. 2019).

3.37 Summary of Key Points

- Being an important crop worldwide, tomatoes suffer significant yield losses due to abiotic stresses of drought, temperature, and salinity.
- Genetic engineering, transgenic, and CRISPR technologies in the development of stress-tolerant varieties are significant and crucial for food security and sustainable agriculture.
- Various stress responses in tomatoes include many complex pathways, such as ROS signalling, calcium signalling, and MAPK cascades, while individual genes increase drought, heat, cold, and salinity tolerance.
- The objective above could be achieved by harmonizing molecular and novel breeding strategies, omics technologies for deploying stress-resilient tomato varieties.
- These innovations are envisaged to develop climate-resilient tomato cultivars that will save food production as a step towards global food security.

3.38 Overview of Emerging Strategies and Future Perspective for the Improvement of Stress Tolerance in Tomatoes

Stress tolerance is among the leading fields of study in agricultural biotechnology, developing strategies and future research initiatives to increase tomato stress tolerance. One of the most widely grown and consumed foods worldwide, tomatoes are very seriously exposed to many abiotic conditions, like heat, drought, salt, and cold. A lot of new approaches are being targeted by researchers to reduce these problems to ensure sustained output. Genetic engineering and genome editing represent some of the most essential biotechnological advances. One such applied technique, which facilitates prudent genetic changes in tomatoes, is known as CRISPR-Cas9. These are stress-responsive genes from outside infections, which can significantly bolster a tomato plant's resistance to extremes. Recent developments in omics technologies, including genomics, proteomics, transcriptomics, and metabolomics, will provide a complete insight into the molecular basis of stress responses. Their recent technologies have made it possible to identify relevant regulatory genes and metabolic pathways related to stress tolerance. This paves the way for focussed breeding and also biomimetic methods. Scientists can breed such tomato varieties that have a higher resistance against abiotic stresses by incorporating the stress resistance mechanisms, which are natural and common in other plants. Numerous desert plants, for instance, can withstand a lot of harsh conditions because of their unique traits. Breeders can try and insert the same characteristics and study these responses to develop tomatoes that withstand intense heat and drought. On the contrary, the modern techniques of breeding, like genomic selection and marker-assisted selection, have greatly helped in fast-tracking the development of stress-tolerant tomato cultivars. These methods add precision to the process of plant selection using molecular markers linked with some traits. Excitement regarding artificial intelligence (AI) is another technological principle affecting fundamental changes in tomato stress tolerance. Machine learning models and AI algorithms are exploring large datasets from omics research. In this way, AI-based assessments can predict the patterns and possible responses of large numbers of tomato cultivars under various stress levels. For example, machine learning algorithms can be developed to choose the most drought-tolerant cultivars based on gene expression patterns. One real-time area of research using technology based on artificial intelligence that finds application is tomato crop monitoring and management. Leading these frontiers of biotechnology are technologies that can enhance the tolerance of tomatoes to stress through the incorporation of new breeding methods, ecologically friendly farming methods, and a deep comprehension of the interaction between plants and the environment. Integration of existing technologies with increased new approaches will arm researchers and producers with a wide range of options to come up with tomato varieties that are strong and resilient to not only survive but also conquer under harsh conditions. This would ensure a reliable supply of this critical crop, mainly contributing to food security worldwide at times of environmental stress and climate change.

References

Abouelsaad I, Weihrauch D, Renault S (2016) Effects of salt stress on the expression of key genes related to nitrogen assimilation and transport in the roots of the cultivated tomato and its wild salt-tolerant relative. Sci Hortic 211:70–78

Ahmad P, Rasool S, Gul A, Sheikh SA, Akram NA, Ashraf M, Kazi A, Gucel S (2016) Jasmonates: multifunctional roles in stress tolerance. Front Plant Sci 7:813

Albert E, Segura V, Gricourt J, Bonnefoi J, Derivot L, Causse M (2016) Association mapping reveals the genetic architecture of tomato response to water deficit: focus on major fruit quality traits. J Exp Bot 67(22):6413–6430

Alkhatib R, Abdo N, Mheidat M (2021) Photosynthetic and ultrastructural properties of eggplant (Solanum melongena) under salinity stress. Horticulturae 7(7):181

Almeida P, de Boer G-J, de Boer AH (2014a) Differences in shoot Na+ accumulation between two tomato species are due to differences in ion affinity of HKT1; 2. J Plant Physiol 171(6):438–447

Almeida P, Feron R, de Boer G-J, de Boer AH (2014b) Role of Na+, K+, Cl−, proline and sucrose concentrations in determining salinity tolerance and their correlation with the expression of multiple genes in tomato. AoB Plants 6:plu039

Arora L, Narula A (2017) Gene editing and crop improvement using CRISPR-Cas9 system. Front Plant Sci 8:1932–1932

Asano T, Hayashi N, Kobayashi M, Aoki N, Miyao A, Mitsuhara I, Ichikawa H, Komatsu S, Hirochika H, Kikuchi S (2012) A rice calcium-dependent protein kinase OsCPK12 oppositely modulates salt-stress tolerance and blast disease resistance. Plant J 69(1):26–36

Atkinson NJ, Dew TP, Orfila C, Urwin PE (2011) Influence of combined biotic and abiotic stress on nutritional quality parameters in tomato (Solanum lycopersicum). J Agric Food Chem 59(17):9673–9682

Ayenan MAT, Danquah A, Hanson P, Ampomah-Dwamena C, Sodedji FAK, Asante IK, Danquah EY (2019) Accelerating breeding for heat tolerance in tomato (Solanum lycopersicum L.): an integrated approach. Agronomy 9(11):720

Azzeme AM, Abdullah SNA, Abd Aziz M, Wahab PEM (2017) Oil palm drought inducible DREB1 induced expression of DRE/CRT-and non-DRE/CRT-containing genes in lowland transgenic tomato under cold and PEG treatments. Plant Physiol Biochem 112:129–151

Bacha H, Ródenas R, López-Gómez E, García-Legaz MF, Nieves-Cordones M, Rivero RM, Martínez V, Botella MÁ, Rubio F (2015) High Ca2+ reverts the repression of high-affinity K+ uptake produced by Na+ in Solanum lycopersycum L.(var. microtom) plants. J Plant Physiol 180:72–79

Back K, Tan DX, Reiter RJ (2016) Melatonin biosynthesis in plants: multiple pathways catalyze tryptophan to melatonin in the cytoplasm or chloroplasts. J Pineal Res 61(4):426–437

Balfagón D, Zandalinas SI, dos Reis T, de Oliveira C, Santa-Catarina, Gómez-Cadenas A (2022) Reduction of heat stress pressure and activation of photosystem II repairing system are crucial for citrus tolerance to multiple abiotic stress combination. Physiol Plant 174(6):e13809

Banerjee A, Tripathi DK, Roychoudhury A (2019) The karrikin 'calisthenics': can compounds derived from smoke help in stress tolerance? Physiol Plant 165(2):290–302

Barrero-Gil J, Salinas J (2017) CBFs at the crossroads of plant hormone signaling in cold stress response. Mol Plant 10(4):542–544

Bashir K, Matsui A, Rasheed S, Seki M (2019) Recent advances in the characterization of plant transcriptomes in response to drought, salinity, heat, and cold stress. F1000Research 8

Basso MF, Ferreira PCG, Kobayashi AK, Harmon FG, Nepomuceno AL, Molinari HBC, Grossi-de-Sa MF (2019) Micro RNA s and new biotechnological tools for its modulation and improving stress tolerance in plants. Plant Biotechnol J 17(8):1482–1500

Bazihizina N, Colmer TD, Cuin TA, Mancuso S, Shabala S (2019) Friend or foe? Chloride patterning in halophytes. Trends Plant Sci 24(2):142–151

Bolger A, Scossa F, Bolger ME, Lanz C, Maumus F, Tohge T, Quesneville H, Alseekh S, Sørensen I, Lichtenstein G (2014) The genome of the stress-tolerant wild tomato species Solanum pennellii. Nat Genet 46(9):1034–1038

Böndel KB, Nosenko T, Stephan W (2018) Signatures of natural selection in abiotic stress-responsive genes of Solanum chilense. R Soc Open Sci 5(1):171198

Borba AR, Serra TS, Górska A, Gouveia P, Cordeiro AM, Reyna-Llorens I, Kneřová J, Barros PM, Abreu IA, Oliveira MM (2018) Synergistic binding of bHLH transcription factors to the promoter of the maize NADP-ME gene used in C4 photosynthesis is based on an ancient code found in the ancestral C3 state. Mol Biol Evol 35(7):1690–1705

Borg M, Jacob Y, Susaki D, LeBlanc C, Buendía D, Axelsson E, Kawashima T, Voigt P, Boavida L, Becker J (2020) Targeted reprogramming of H3K27me3 resets epigenetic memory in plant paternal chromatin. Nat Cell Biol 22(6):621–629

Bouzroud S, Gouiaa S, Hu N, Bernadac A, Mila I, Bendaou N, Smouni A, Bouzayen M, Zouine M (2018) Auxin response factors (ARFs) are potential mediators of auxin action in tomato response to biotic and abiotic stress (Solanum lycopersicum). PLoS One 13(2):e0193517

Brooks C, Nekrasov V, Lippman ZB, Van Eck J (2014) Efficient gene editing in tomato in the first generation using the clustered regularly interspaced short palindromic repeats/CRISPR-associated9 system. Plant Physiol 166(3):1292–1297

Candar-Cakir B, Arican E, Zhang B (2016) Small RNA and degradome deep sequencing reveals drought-and tissue-specific micrornas and their important roles in drought-sensitive and drought-tolerant tomato genotypes. Plant Biotechnol J 14(8):1727–1746

Cao P, Yang J, Xia L, Zhang Z, Wu Z, Hao Y, Liu P, Wang C, Li C, Yang J (2024) Two gene clusters and their positive regulator SlMYB13 that have undergone domestication-associated negative selection control phenolamide accumulation and drought tolerance in tomato. Mol Plant 17(4):579–597

Chan-Schaminet KY, Baniwal SK, Bublak D, Nover L, Scharf K-D (2009) Specific interaction between tomato HsfA1 and HsfA2 creates hetero-oligomeric superactivator complexes for synergistic activation of heat stress gene expression. J Biol Chem 284(31):20848–20857

Chaudhary J, Alisha A, Bhatt V, Chandanshive S, Kumar N, Mir Z, Kumar A, Yadav SK, Shivaraj S, Sonah H (2019) Mutation breeding in tomato: advances, applicability and challenges. Plan Theory 8(5):128

Cho JH, Sim S-C, Kim K-N (2021) Calcium sensor SlCBL4 associates with SlCIPK24 protein kinase and mediates salt tolerance in Solanum lycopersicum. Plan Theory 10(10):2173

Christey MC (2020) Transgenic crop plants using Agrobacterium rhizogenes-mediated transformation. Hairy Roots, CRC Press, pp 100–112

Chung PJ, Chung H, Oh N, Choi J, Bang SW, Jung SE, Jung H, Shim JS, Kim J-K (2020) Efficiency of recombinant CRISPR/rCas9-mediated miRNA gene editing in rice. Int J Mol Sci 21(24):9606

Cramer GR, Urano K, Delrot S, Pezzotti M, Shinozaki K (2011) Effects of abiotic stress on plants: a systems biology perspective. BMC Plant Biol 11:1–14

Csiszár J, Horváth E, Váry Z, Gallé Á, Bela K, Brunner S, Tari I (2014) Glutathione transferase supergene family in tomato: salt stress-regulated expression of representative genes from distinct GST classes in plants primed with salicylic acid. Plant Physiol Biochem 78:15–26

da Silva GJ, Costa de Oliveira A (2014) Genes acting on transcriptional control during abiotic stress responses. Adv Agric 2014:1

Danieli R, Assouline S, Salam BB, Vrobel O, Teper-Bamnolker P, Belausov E, Granot D, Tarkowski P, Eshel D (2023) Chilling induces sugar and ABA accumulation that antagonistically signals for symplastic connection of dormant potato buds. Plant Cell Environ 46(7):2097–2111

Denesik TJ (2007) Quantitative expression analysis of four low-temperature-tolerance-associated genes during cold acclimation in wheat (Triticum aestivum L.). University of Saskatchewan

Devkar V, Thirumalaikumar VP, Xue GP, Vallarino JG, Turečková V, Strnad M, Fernie AR, Hoefgen R, Mueller-Roeber B, Balazadeh S (2020) Multifaceted regulatory function of tomato SlTAF1 in the response to salinity stress. New Phytol 225(4):1681–1698

Dhandevi P, Jeewon R (2015) Fruit and vegetable intake: benefits and progress of nutrition education interventions-narrative review article. Iran J Public Health 44(10):1309

Ding H, He J, Wu Y, Wu X, Ge C, Wang Y, Zhong S, Peiter E, Liang J, Xu W (2018) The tomato mitogen-activated protein kinase SlMPK1 is as a negative regulator of the high-temperature stress response. Plant Physiol 177(2):633–651

Ding F, Qiang X, Jia Z, Li L, Hu J, Yin M, Xia S, Chen B, Qi J, Li Q (2022) Knockout of a novel salt responsive gene SlABIG1 enhance salinity tolerance in tomato. Environ Exp Bot 200:104903

Diouf IA, Derivot L, Bitton F, Pascual L, Causse M (2018) Water deficit and salinity stress reveal many specific QTL for plant growth and fruit quality traits in tomato. Front Plant Sci 9:279

Dorais M, Ehret DL, Papadopoulos AP (2008) Tomato (Solanum lycopersicum) health components: from the seed to the consumer. Phytochem Rev 7:231–250

Eckardt NA (2009) CAMTA proteins: a direct link between calcium signals and cold acclimation? Plant Cell 21(3):697–697

Egea I, Albaladejo I, Meco V, Morales B, Sevilla A, Bolarin MC, Flores FB (2018) The drought-tolerant Solanum pennellii regulates leaf water loss and induces genes involved in amino acid and ethylene/jasmonate metabolism under dehydration. Sci Rep 8(1):2791

Fahad S, Bajwa AA, Nazir U, Anjum SA, Farooq A, Zohaib A, Sadia S, Nasim W, Adkins S, Saud S (2017) Crop production under drought and heat stress: plant responses and management options. Front Plant Sci 8:1147

FAOSTAT, F (2019) Food and Agriculture Organization of the United Nations-Statistic Division. https://www.fao.org/faostat/en/#data, QC

Farooq M, Hussain M, Wahid A, Siddique K (2012) Drought stress in plants: an overview. In: Plant responses to drought stress: from morphological to molecular features, pp 1–33

Fedoroff NV, Battisti DS, Beachy RN, Cooper PJ, Fischhoff DA, Hodges C, Knauf VC, Lobell D, Mazur BJ, Molden D (2010) Radically rethinking agriculture for the 21st century. Science 327(5967):833–834

Fischhoff DA, Bowdish KS, Perlak FJ, Marrone PG, McCormick SM, Niedermeyer JG, Dean DA, Kusano-Kretzmer K, Mayer EJ, Rochester DE, Rogers SG, Fraley RT (1987) Insect tolerant transgenic tomato plants. Nat Biotechnol 5(8):807–813

Foolad MR, Zhang L, Subbiah P (2003) Genetics of drought tolerance during seed germination in tomato: inheritance and QTL mapping. Genome 46(4):536–545

Foyer CH, Noctor G (2016) Stress-triggered redox signalling: what's in pROSpect? Plant Cell Environ 39(5):951–964

Frank G, Pressman E, Ophir R, Althan L, Shaked R, Freedman M, Shen S, Firon N (2009) Transcriptional profiling of maturing tomato (Solanum lycopersicum L.) microspores reveals the involvement of heat shock proteins, ROS scavengers, hormones, and sugars in the heat stress response. J Exp Bot 60(13):3891–3908

Gaj T, Sirk SJ, Shui S-L, Liu J (2016) Genome-editing technologies: principles and applications. Cold Spring Harb Perspect Biol 8(12):a023754

Gao C (2015) Genome editing in crops: from bench to field. Natl Sci Rev 2(1):13–15

Gharbi E, Martínez J-P, Benahmed H, Lepoint G, Vanpee B, Quinet M, Lutts S (2017) Inhibition of ethylene synthesis reduces salt-tolerance in tomato wild relative species Solanum chilense. J Plant Physiol 210:24–37

Ghildiyal M, Zamore PD (2009) Small silencing RNAs: an expanding universe. Nat Rev Genet 10(2):94–108

Gilroy S, Białasek M, Suzuki N, Górecka M, Devireddy AR, Karpiński S, Mittler R (2016) ROS, calcium, and electric signals: key mediators of rapid systemic signaling in plants. Plant Physiol 171(3):1606–1615

Gong P, Zhang J, Li H, Yang C, Zhang C, Zhang X, Khurram Z, Zhang Y, Wang T, Fei Z (2010) Transcriptional profiles of drought-responsive genes in modulating transcription signal transduction, and biochemical pathways in tomato. J Exp Bot 61(13):3563–3575

Gould WA (1974) Tomato production, processing and quality evaluation

Guo J, Wang M-H (2011) Expression profiling of the DREB2 type gene from tomato (Solanum lycopersicum L.) under various abiotic stresses. Hortic Environ Biotechnol 52:105–111

Guo X, Zhang L, Zhu J, Liu H, Wang A (2017a) Cloning and characterization of SiDHN, a novel dehydrin gene from Saussurea involucrata Kar. et Kir. that enhances cold and drought tolerance in tobacco. Plant Sci 256:160–169

Guo X, Zhang L, Zhu J, Wang A, Liu H (2017b) Christolea crassifolia HARDY gene enhances drought stress tolerance in transgenic tomato plants. Plant Cell Tissue Organ Cult (PCTOC) 129:469–481

Guo M, Wang X-S, Guo H-D, Bai S-Y, Khan A, Wang X-M, Gao Y-M, Li J-S (2022) Tomato salt tolerance mechanisms and their potential applications for fighting salinity: a review. Front Plant Sci 13:949541

Guóth A, Tari I, Gallé Á, Csiszár J, Pécsváradi A, Cseuz L, Erdei L (2009) Comparison of the drought stress responses of tolerant and sensitive wheat cultivars during grain filling: changes in flag leaf photosynthetic activity, ABA levels, and grain yield. J Plant Growth Regul 28:167–176

Gur A, Zamir D (2004) Unused natural variation can lift yield barriers in plant breeding. PLoS Biol 2(10):e245

Han N, Fan S, Zhang T, Sun H, Zhu Y, Gong H, Guo J (2020) SlHY5 is a necessary regulator of the cold acclimation response in tomato. Plant Growth Regul 91:1–12

Hannah MA, Heyer AG, Hincha DK (2005) A global survey of gene regulation during cold acclimation in Arabidopsis thaliana. PLoS Genet 1(2):e26

Harrak H, Azelmat S, Baker EN, Tabaeizadeh Z (2001) Isolation and characterization of a gene encoding a drought-induced cysteine protease in tomato (Lycopersicon esculentum). Genome 44(3):368–374

He X, Wang C, Wang H, Li L, Wang C (2020) The function of MAPK cascades in response to various stresses in horticultural plants. Front Plant Sci 11:952

Heinrich M, Baldwin IT, Wu J (2011) Two mitogen-activated protein kinase kinases, MKK1 and MEK2, are involved in wounding-and specialist lepidopteran herbivore Manduca sexta-induced responses in Nicotiana attenuata. J Exp Bot 62(12):4355–4365

Higashi Y, Ohama N, Ishikawa T, Katori T, Shimura A, Kusakabe K, Yamaguchi-Shinozaki K, Ishida J, Tanaka M, Seki M (2013) HsfA1d, a protein identified via FOX hunting using Thellungiella salsuginea cDNAs improves heat tolerance by regulating heat-stress-responsive gene expression. Mol Plant 6(2):411–422

Hoffmann J, Berni R, Sutera FM, Gutsch A, Hausman J-F, Saffie-Siebert S, Guerriero G (2021) The effects of salinity on the anatomy and gene expression patterns in leaflets of tomato cv. micro-tom. Genes 12(8):1165

Hu Z, Li J, Ding S, Cheng F, Li X, Jiang Y, Yu J, Foyer CH, Shi K (2021) The protein kinase CPK28 phosphorylates ascorbate peroxidase and enhances thermotolerance in tomato. Plant Physiol 186(2):1302–1317

Hu S, Wang T, Shao Z, Meng F, Chen H, Wang Q, Zheng J, Liu L (2022) Brassinosteroid biosynthetic gene SlCYP90B3 alleviates chilling injury of tomato (Solanum lycopersicum) fruits during cold storage. Antioxidants 11(1):115

Ikeda M, Mitsuda N, Ohme-Takagi M (2011) Arabidopsis HsfB1 and HsfB2b act as repressors of the expression of heat-inducible Hsfs but positively regulate the acquired thermotolerance. Plant Physiol 157(3):1243–1254

Iovieno P, Punzo P, Guida G, Mistretta C, Van Oosten MJ, Nurcato R, Bostan H, Colantuono C, Costa A, Bagnaresi P (2016) Transcriptomic changes drive physiological responses to progressive drought stress and rehydration in tomato. Front Plant Sci 7:371

Islam MS, Wang M-H (2009) Expression of dehydration responsive element-binding protein-3 (DREB3) under different abiotic stresses in tomato. BMB Rep 42(9):611–616

Janni M, Gullì M, Maestri E, Marmiroli M, Valliyodan B, Nguyen HT, Marmiroli N (2020) Molecular and genetic bases of heat stress responses in crop plants and breeding for increased resilience and productivity. J Exp Bot 71(13):3780–3802

Jenks MA, Wood AJ (2009) Genes for plant abiotic stress. Wiley

Ji T, Li S, Huang M, Di Q, Wang X, Wei M, Shi Q, Li Y, Gong B, Yang F (2017) Overexpression of cucumber phospholipase D alpha gene (CsPLDα) in tobacco enhanced salinity stress tolerance by regulating Na+–K+ balance and lipid peroxidation. Front Plant Sci 8:499

Jinek M, Chylinski K, Fonfara I, Hauer M, Doudna JA, Charpentier E (2012) A programmable dual-RNA-guided DNA endonuclease in adaptive bacterial immunity. Science (New York, NY) 337(6096):816–821

John R, Pandey R, Sopory S, Rajam M (2010) Engineering antioxidant enzymes for abiotic stress tolerance in plants. J Plant Biol 37:1–18

Joshi R, Wani SH, Singh B, Bohra A, Dar ZA, Lone AA, Pareek A, Singla-Pareek SL (2016) Transcription factors and plants response to drought stress: current understanding and future directions. Front Plant Sci 7:204078

Joshi S, Patil S, Shaikh A, Jamla M, Kumar V (2023) Modern omics toolbox for producing combined and multifactorial abiotic stress tolerant plants. Plant Stress 100301

Kahn TL, Fender SE, Bray EA, O'Connell MA (1993) Characterization of expression of drought- and abscisic acid-regulated tomato genes in the drought-resistant species Lycopersicon pennellii. Plant Physiol 103(2):597–605

Karkute SG, Ansari WA, Singh AK, Singh PM, Rai N, Bahadur A, Singh J (2021) Characterization of high-temperature stress-tolerant tomato (Solanum lycopersicum L.) genotypes by biochemical analysis and expression profiling of heat-responsive genes. 3 Biotech 11:1–10

Kazan K (2015) Diverse roles of jasmonates and ethylene in abiotic stress tolerance. Trends Plant Sci 20(4):219–229

Kelly G, Lugassi N, Belausov E, Wolf D, Khamaisi B, Brandsma D, Kottapalli J, Fidel L, Ben-Zvi B, Egbaria A (2017) The Solanum tuberosum KST1 partial promoter as a tool for guard cell expression in multiple plant species. J Exp Bot 68(11):2885–2897

Khan M, Ali S, Al Azzawi TNI, Saqib S, Ullah F, Ayaz A, Zaman W (2023) The key roles of ROS and RNS as a signaling molecule in plant–microbe interactions. Antioxidants 12(2):268

Kim JY, Kim S-K, Jung J, Jeong M-J, Ryu C-M (2018) Exploring the sound-modulated delay in tomato ripening through expression analysis of coding and non-coding RNAs. Ann Bot 122(7):1231–1244

Kissoudis C, Chowdhury R, van Heusden S, van de Wiel C, Finkers R, Visser RG, Bai Y, van der Linden G (2015) Combined biotic and abiotic stress resistance in tomato. Euphytica 202:317–332

Knapp S, Peralta IE (2016) The tomato (Solanum lycopersicum L., Solanaceae) and its botanical relatives. In: Causse M, Giovannoni J, Bouzayen M, Zouine M (eds) The tomato genome. Springer, Berlin/Heidelberg, pp 7–21

Kocsy G, Galiba G, Brunold C (2001) Role of glutathione in adaptation and signalling during chilling and cold acclimation in plants. Physiol Plant 113(2):158–164

Kopecká R, Kameniarová M, Černý M, Brzobohatý B, Novák J (2023) Abiotic stress in crop production. Int J Mol Sci 24(7):6603

Kotak S, Larkindale J, Lee U, von Koskull-Döring P, Vierling E, Scharf K-D (2007) Complexity of the heat stress response in plants. Curr Opin Plant Biol 10(3):310–316

Kramer MG, Redenbaugh K (1994) Commercialization of a tomato with an antisense polygalacturonase gene: the FLAVR SAVR™ tomato story. Euphytica 79(3):293–297

Krishna R, Ansari WA, Soumia P, Yadav A, Jaiswal DK, Kumar S, Singh AK, Singh M, Verma JP (2022) Biotechnological interventions in tomato (Solanum lycopersicum) for drought stress tolerance: achievements and future prospects. Biotech 11(4):48

Kushwaha BK, Singh S, Tripathi DK, Sharma S, Prasad SM, Chauhan DK, Kumar V, Singh VP (2019) New adventitious root formation and primary root biomass accumulation are regulated by nitric oxide and reactive oxygen species in rice seedlings under arsenate stress. J Hazard Mater 361:134–140

Lee S, Hirt H, Lee Y (2001) Phosphatidic acid activates a wound-activated MAPK in Glycine max. Plant J 26(5):479–486

Lee JH, Yoo SJ, Park SH, Hwang I, Lee JS, Ahn JH (2007) Role of SVP in the control of flowering time by ambient temperature in Arabidopsis. Genes Dev 21(4):397–402

Li J, Sima W, Ouyang B, Wang T, Ziaf K, Luo Z, Liu L, Li H, Chen M, Huang Y (2012) Tomato SlDREB gene restricts leaf expansion and internode elongation by downregulating key genes for gibberellin biosynthesis. J Exp Bot 63(18):6407–6420

Li D, Ma N-N, Wang J-R, Yang D-Y, Zhao S-J, Meng Q-W (2013) Overexpression of tomato enhancer of SOS3-1 (LeENH1) in tobacco enhanced salinity tolerance by excluding Na+ from the cytosol. Plant Physiol Biochem 70:150–158

Li R, Zhang L, Wang L, Chen L, Zhao R, Sheng J, Shen L (2018a) Reduction of tomato-plant chilling tolerance by CRISPR–Cas9-mediated SlCBF1 mutagenesis. J Agric Food Chem 66(34):9042–9051

Li Y, Wang H, Zhang Y, Martin C (2018b) Can the world's favorite fruit, tomato, provide an effective biosynthetic chassis for high-value metabolites? Plant Cell Rep 37:1443–1450

Lichtenthaler HK (1988) In vivo chlorophyll fluorescence as a tool for stress detection in plants. Applications of Chlorophyll Fluorescence in Photosynthesis Research, Stress Physiology, Hydrobiology and Remote Sensing: An introduction to the various fields of applications of the in vivo chlorophyll fluorescence also including the proceedings of the first International Chlorophyll Fluorescence Symposium held in the Physikzentrum, Bad Honnef, FRG, 6–8 June 1998, Springer

Liu JX, Howell SH (2016) Managing the protein folding demands in the endoplasmic reticulum of plants. New Phytol 211(2):418–428

Liu H, Zhou Y, Li H, Wang T, Zhang J, Ouyang B, Ye Z (2018a) Molecular and functional characterization of ShNAC1, an NAC transcription factor from Solanum habrochaites. Plant Sci 271:9–19

Liu M, Yu H, Zhao G, Huang Q, Lu Y, Ouyang B (2018b) Identification of drought-responsive microRNAs in tomato using high-throughput sequencing. Funct Integr Genomics 18:67–78

Liu D-D, Sun X-S, Liu L, Shi H-D, Chen S-Y, Zhao D-K (2019) Overexpression of the melatonin synthesis-related gene SlCOMT1 improves the resistance of tomato to salt stress. Molecules 24(8):1514

Liu X, Xie X, Zheng C, Wei L, Li X, Jin Y, Zhang G, Jiang C-J, Liang Z (2022) RNAi-mediated suppression of the abscisic acid catabolism gene OsABA8ox1 increases abscisic acid content and tolerance to saline–alkaline stress in rice (Oryza sativa L.). Crop J 10(2):354–367

Liu Y, Zhang L, Meng S, Zhang H, Wang S, Xu C, Liu Y, Xu T, He Y, Cui Y (2024) Galactinol regulates JA biosynthesis to enhance tomato cold tolerance. J Agric Food Chem 72:2547

Londo JP, Kovaleski AP, Lillis JA (2018) Divergence in the transcriptional landscape between low temperature and freeze shock in cultivated grapevine (Vitis vinifera). Hortic Res 5

Luo J, Butelli E, Hill L, Parr A, Niggeweg R, Bailey P, Weisshaar B, Martin C (2008) AtMYB12 regulates caffeoyl quinic acid and flavonol synthesis in tomato: expression in fruit results in very high levels of both types of polyphenol. Plant J 56(2):316–326

Lushchak VI (2014) Free radicals, reactive oxygen species, oxidative stress and its classification. Chem Biol Interact 224:164–175

Lv H, Wang X, Dong X, Gao M, Dong D, Li C, Jing S, Guo YD, Zhang N (2024) CRISPR/Cas9 edited *SlGT30* improved both drought resistance and fruit yield through endoreduplication. Plant Cell Environ

Maach M, Baghour M, Akodad M, Gálvez FJ, Sánchez ME, Aranda MN, Venema K, Rodríguez-Rosales MP (2020) Overexpression of LeNHX4 improved yield, fruit quality and salt tolerance in tomato plants (Solanum lycopersicum L.). Mol Biol Rep 47(6):4145–4153

Maach M, Rodríguez-Rosales MP, Venema K, Akodad M, Moumen A, Skalli A, Baghour M (2021) Improved yield, fruit quality, and salt resistance in tomato co-overexpressing LeNHX2 and SlSOS2 genes. Physiol Mol Biol Plants 27(4):703–712

Magnan F, Ranty B, Charpenteau M, Sotta B, Galaud JP, Aldon D (2008) Mutations in AtCML9, a calmodulin-like protein from Arabidopsis thaliana, alter plant responses to abiotic stress and abscisic acid. Plant J 56(4):575–589

Mahesh U, Mamidala P, Rapolu S, Aragao FJ, Souza M, Rao P, Kirti P, Nanna RS (2013) Constitutive overexpression of small HSP24. 4 gene in transgenic tomato conferring tolerance to high-temperature stress. Mol Breed 32:687–697

Martí R, Roselló S, Cebolla-Cornejo J (2016) Tomato as a source of carotenoids and polyphenols targeted to cancer prevention. Cancers 8(6):58

Martinez V, Nieves-Cordones M, Lopez-Delacalle M, Rodenas R, Mestre TC, Garcia-Sanchez F, Rubio F, Nortes PA, Mittler R, Rivero RM (2018) Tolerance to stress combination in tomato plants: new insights in the protective role of melatonin. Molecules 23(3):535

Martín-Pizarro C, Posé D (2018) Genome editing as a tool for fruit ripening manipulation. Front Plant Sci 9:414235

Meng X, Cai J, Deng L, Li G, Sun J, Han Y, Dong T, Liu Y, Xu T, Liu S (2020) SlSTE1 promotes abscisic acid-dependent salt stress-responsive pathways via improving ion homeostasis and reactive oxygen species scavenging in tomato. J Integr Plant Biol 62(12):1942–1966

Mittler R, Zandalinas SI, Fichman Y, Van Breusegem F (2022) Reactive oxygen species signalling in plant stress responses. Nat Rev Mol Cell Biol 23(10):663–679

Mujjassim NE, Mallik M, Rathod NKK, Nitesh S (2019) Cisgenesis and intragenesis a new tool for conventional plant breeding: a review. J Pharmacogn Phytochem 8(1):2485–2489

Munné-Bosch S, Villadangos S (2023) Cheap, cost-effective, and quick stress biomarkers for drought stress detection and monitoring in plants. Trends Plant Sci 28(5):527–536

Mushtaq N, Wang Y, Fan J, Li Y, Ding J (2022) Down-regulation of cytokinin receptor gene SlHK2 improves plant tolerance to drought, heat, and combined stresses in tomato. Plan Theory 11(2):154

Nath M, Bhatt D, Prasad R, Gill SS, Anjum NA, Tuteja N (2016) Reactive oxygen species generation-scavenging and signaling during plant-arbuscular mycorrhizal and Piriformospora indica interaction under stress condition. Front Plant Sci 7:219102

Nautiyal PC, Shono M, Egawa Y (2005) Enhanced thermotolerance of the vegetative part of MT-sHSP transgenic tomato line. Sci Hortic 105(3):393–409

Nawaz M, Wang Z (2020) Abscisic acid and glycine betaine mediated tolerance mechanisms under drought stress and recovery in Axonopus compressus: a new insight. Sci Rep 10(1):6942

Nehanjali Parmar NP, Singh K, Deepika Sharma DS, Lal Singh LS, Pankaj Kumar PK, Nanjundan J, Khan Y, Chauhan D, Thakur A (2017) Genetic engineering strategies for biotic and abiotic stress tolerance and quality enhancement in horticultural crops: a comprehensive review

Nie WF, Wang MM, Xia XJ, Zhou YH, Shi K, Chen Z, Yu JQ (2013) Silencing of tomato RBOH1 and MPK2 abolishes brassinosteroid-induced H2O2 generation and stress tolerance. Plant Cell Environ 36(4):789–803

Nievola CC, Carvalho CP, Carvalho V, Rodrigues E (2017) Rapid responses of plants to temperature changes. Temperature 4(4):371–405

Nivedithadevi D, Somasundaram R, Pannerselvam R (2012) Effect of abscisic acid, paclobutrazol and salicylic acid on the growth and pigment variation in Solanum trilobatum (l). Int J Drug Dev Res 4(3):236–246

Ohama N, Kusakabe K, Mizoi J, Zhao H, Kidokoro S, Koizumi S, Takahashi F, Ishida T, Yanagisawa S, Shinozaki K (2016) The transcriptional cascade in the heat stress response of Arabidopsis is strictly regulated at the level of transcription factor expression. Plant Cell 28(1):181–201

Ohama N, Sato H, Shinozaki K, Yamaguchi-Shinozaki K (2017) Transcriptional regulatory network of plant heat stress response. Trends Plant Sci 22(1):53–65

Orellana S, Yanez M, Espinoza A, Verdugo I, Gonzalez E, Ruiz-Lara S, Casaretto JA (2010) The transcription factor SlAREB1 confers drought, salt stress tolerance and regulates biotic and abiotic stress-related genes in tomato. Plant Cell Environ 33(12):2191–2208

Osakabe K, Osakabe Y, Toki S (2010) Site-directed mutagenesis in Arabidopsis using custom-designed zinc finger nucleases. Proc Natl Acad Sci USA 107(26):12034–12039

Osakabe Y, Yamaguchi-Shinozaki K, Shinozaki K, Tran L-SP (2013) Sensing the environment: key roles of membrane-localized kinases in plant perception and response to abiotic stress. J Exp Bot 64(2):445–458

Pailles Y, Awlia M, Julkowska M, Passone L, Zemmouri K, Negrão S, Schmöckel SM, Tester M (2020) Diverse traits contribute to salinity tolerance of wild tomato seedlings from the Galapagos Islands. Plant Physiol 182(1):534–546

Pan Y, Seymour GB, Lu C, Hu Z, Chen X, Chen G (2012) An ethylene response factor (ERF5) promoting adaptation to drought and salt tolerance in tomato. Plant Cell Rep 31:349–360

Pandey SK, Nookaraju A, Upadhyaya CP, Gururani MA, Venkatesh J, Kim DH, Park SW (2011) An update on biotechnological approaches for improving abiotic stress tolerance in tomato. Crop Sci 51(6):2303–2324

Pandey P, Ramegowda V, Senthil-Kumar M (2015) Shared and unique responses of plants to multiple individual stresses and stress combinations: physiological and molecular mechanisms. Front Plant Sci 6:153997

Pandey N, Iqbal Z, Pandey BK, Sawant SV (2017) Phytohormones and drought stress: plant responses to transcriptional regulation. Mech Plant Hormone Signal Under Stress 2:477–504

Panno S, Davino S, Caruso AG, Bertacca S, Crnogorac A, Mandić A, Noris E, Matić S (2021) A review of the most common and economically important diseases that undermine the cultivation of tomato crop in the mediterranean basin. Agronomy 11(11):2188

Perez Rojo F, Nyman RKM, Johnson AAT, Navarro MP, Ryan MH, Erskine W, Kaur P (2018) CRISPR-Cas systems: ushering in the new genome editing era. Bioengineered 9(1):214–221

Pessoa HP, Dariva FD, Copati MGF, de Paula RG, de Oliveira Dias F, Gomes CN (2023) Uncovering tomato candidate genes associated with drought tolerance using Solanum pennellii introgression lines. PLoS One 18(6):e0287178

Pua E-C, Davey MR (2007) Biotechnology in agriculture and forestry 59: transgenic crops IV. Springer

Qi S, Lin Q, Zhu H, Gao F, Zhang W, Hua X (2016) The RING finger E3 ligase SpRing is a positive regulator of salt stress signaling in salt-tolerant wild tomato species. Plant Cell Physiol 57(3):528–539

Qu A-L, Ding Y-F, Jiang Q, Zhu C (2013) Molecular mechanisms of the plant heat stress response. Biochem Biophys Res Commun 432(2):203–207

Quinet M, Angosto T, Yuste-Lisbona FJ, Blanchard-Gros R, Bigot S, Martinez J-P, Lutts S (2019) Tomato fruit development and metabolism. Front Plant Sci 10:475784

Raiola A, Rigano MM, Calafiore R, Frusciante L, Barone A (2014) Enhancing the health-promoting effects of tomato fruit for biofortified food. Mediat Inflamm. 2014

Raja MM, Vijayalakshmi G, Naik ML, Basha PO, Sergeant K, Hausman JF, Khan PSSV (2019) Pollen development and function under heat stress: from effects to responses. Acta Physiol Plant 41:1–20

Ranty B, Aldon D, Cotelle V, Galaud J-P, Thuleau P, Mazars C (2016) Calcium sensors as key hubs in plant responses to biotic and abiotic stresses. Front Plant Sci 7:186794

Rao KM, Raghavendra AS, Reddy KJ (2006) Physiology and molecular biology of stress tolerance in plants. Springer
Ratcliffe OJ, Riechmann JL (2002) Arabidopsis transcription factors and the regulation of flowering time: a genomic perspective. Curr Issues Mol Biol 4(3):77–91
Rehem BC, Bertolde FZ, de Almeida A-AF (2012) Regulation of gene expression in response to abiotic stress in plants. Cell Metabol Cell Homeo Stress Res:13–38
Reichel M, Li Y, Li J, Millar AA (2015) Inhibiting plant micro RNA activity: molecular SPONGE s, target MIMIC s and STTM s all display variable efficacies against target micro RNA s. Plant Biotechnol J 13(7):915–926
Romero-Aranda MR, González-Fernández P, Pérez-Tienda JR, López-Diaz MR, Espinosa J, Granum E, Traverso JÁ, Pineda B, Garcia-Sogo B, Moreno V (2020) Na+ transporter HKT1; 2 reduces flower Na+ content and considerably mitigates the decline in tomato fruit yields under saline conditions. Plant Physiol Biochem 154:341–352
Rommens CM, Haring MA, Swords K, Davies HV, Belknap WR (2007) The intragenic approach as a new extension to traditional plant breeding. Trends Plant Sci 12(9):397–403
Roşca M, Mihalache G, Stoleru V (2023) Tomato responses to salinity stress: from morphological traits to genetic changes. Front Plant Sci 14:1118383
Rothan C, Diouf I, Causse M (2019) Trait discovery and editing in tomato. Plant J 97(1):73–90
Ruggiero A, Landi S, Punzo P, Possenti M, Van Oosten MJ, Costa A, Morelli G, Maggio A, Grillo S, Batelli G (2019) Salinity and ABA seed responses in pepper: expression and interaction of ABA core signaling components. Front Plant Sci 10:434092
Sah SK, Reddy KR, Li J (2016) Abscisic acid and abiotic stress tolerance in crop plants. Front Plant Sci 7:190245
Saikia B, Debbarma J, Maharana J, Singha DL, Velmuruagan N, Dekaboruah H, Arunkumar KP, Chikkaputtaiah C (2020) SlHyPRP1 and DEA1, the multiple stress responsive eight-cysteine motif family genes of tomato (Solanum lycopersicum L.) are expressed tissue specifically, localize and interact at cytoplasm and plasma membrane in vivo. Physiol Mol Biol Plants 26(12):2553–2568
Saikia B, Debbarma RSJ, Maharana J, Sastry GN, Chikkaputtaiah C (2024) CRISPR/Cas9-based genome editing and functional analysis of SlHyPRP1 and SlDEA1 genes of Solanum lycopersicum L. in imparting genetic tolerance to multiple stress factors. Front Plant Sci 15:1304381–1304381
Scharf K-D, Höhfeld I, Nover L (1998) Heat stress response and heat stress transcription factors. J Biosci 23:313–329
Scharf K-D, Berberich T, Ebersberger I, Nover L (2012) The plant heat stress transcription factor (Hsf) family: structure, function and evolution. Biochimica et Biophysica Acta (BBA)-Gene Regul Mech 1819(2):104–119
Schouten HJ (2022) The origin of cisgenesis, and its evolving definition. Cisgenic crops: potential and prospects. Springer, pp 1–13
Shanmugavadivel P, Prakash C, Mithra SA (2019) Molecular approaches for dissecting and improving drought and heat tolerance in rice. In: Advances in rice research for abiotic stress tolerance. Elsevier, pp 839–867
Shi Y, Ding Y, Yang S (2018) Molecular regulation of CBF signaling in cold acclimation. Trends Plant Sci 23(7):623–637
Sikora P, Chawade A, Larsson M, Olsson J, Olsson O (2011) Mutagenesis as a tool in plant genetics, functional genomics, and breeding. Int J Plant Genomics 2011:314829–314829
Singh S, Tripathi DK, Singh S, Sharma S, Dubey NK, Chauhan DK, Vaculík M (2017a) Toxicity of aluminium on various levels of plant cells and organism: a review. Environ Exp Bot 137:177–193
Singh VP, Singh S, Tripathi DK, Prasad SM, Chauhan DK (2017b) Reactive oxygen species in plants: boon or bane-revisiting the role of ROS. Wiley
Singh H, Kumar P, Kumar A, Kyriacou MC, Colla G, Rouphael Y (2020) Grafting tomato as a tool to improve salt tolerance. Agronomy 10(2):263

Singh A, Banerjee A, Roychoudhury A (2022) Fluoride tolerance in rice is negatively regulated by the 'stress-phytohormone' abscisic acid (ABA), but promoted by ABA-antagonist growth regulators, melatonin, and gibberellic acid. Protoplasma 259(5):1331–1350

Slimestad R, Verheul M (2009) Review of flavonoids and other phenolics from fruits of different tomato (Lycopersicon esculentum Mill.) cultivars. J Sci Food Agric 89(8):1255–1270

Smith AF (2001) The tomato in America: early history, culture, and cookery. University of Illinois Press

Solanke AU, Sharma AK (2008) Signal transduction during cold stress in plants. Physiol Mol Biol Plants 14:69–79

Sun J, Qi L, Li Y, Chu J, Li C (2012) PIF4–mediated activation of YUCCA8 expression integrates temperature into the auxin pathway in regulating Arabidopsis hypocotyl growth. PLoS Genet 8(3):e1002594

Sun S, Wen D, Yang W, Meng Q, Shi Q, Gong B (2020) Overexpression of caffeic acid O-methyltransferase 1 (COMT1) increases melatonin level and salt stress tolerance in tomato plant. J Plant Growth Regul 39:1221–1235

Suresh BV, Roy R, Sahu K, Misra G, Chattopadhyay D (2014) Tomato genomic resources database: an integrated repository of useful tomato genomic information for basic and applied research. PLoS One 9(1):e86387

Suzuki N, Katano K (2018) Coordination between ROS regulatory systems and other pathways under heat stress and pathogen attack. Front Plant Sci 9:351917

Swetha B, Devi H, Kumar K (2024) Urban horticulture: a cutting-edge strategy and essential for the future. Int J Environ Clim Change 14(3):227–238

Thameur A, Ferchichi A, López-Carbonell M (2011) Quantification of free and conjugated abscisic acid in five genotypes of barley (Hordeum vulgare L.) under water stress conditions. S Afr J Bot 77(1):222–228

Thomson DW, Dinger ME (2016) Endogenous microRNA sponges: evidence and controversy. Nat Rev Genet 17(5):272–283

Tiwari P, Khare T, Shriram V, Bae H, Kumar V (2021) Plant synthetic biology for producing potent phyto-antimicrobials to combat antimicrobial resistance. Biotechnol Adv 48:107729

Tohge T, Fernie AR (2015) Metabolomics-inspired insight into developmental, environmental and genetic aspects of tomato fruit chemical composition and quality. Plant Cell Physiol 56(9):1681–1696

Tomato Genome Consortium (2012) The tomato genome sequence provides insights into fleshy fruit evolution. Nature 485(7400):635

Tran MT, Doan DTH, Kim J, Song YJ, Sung YW, Das S, Kim EJ, Son GH, Kim SH, Van Vu T (2021) CRISPR/Cas9-based precise excision of SlHyPRP1 domain (s) to obtain salt stress-tolerant tomato. Plant Cell Rep 40:999–1011

Tran MT, Son GH, Song YJ, Nguyen NT, Park S, Thach TV, Kim J, Sung YW, Das S, Pramanik D, Lee J, Son K-H, Kim SH, Vu TV, Kim J-Y (2023) CRISPR-Cas9-based precise engineering of SlHyPRP1 protein towards multi-stress tolerance in tomato. Front Plant Sci 14:1186932–1186932

Trebichalský P, Tóth T, Bajčan D, Harangozo L, Vollmannová A (2017) Hormonal changes in spring barley after triazine herbicide treatment and its mixtures of regulators of polyamine biosynthesis

Tripathi DK, Singh S, Singh S, Srivastava PK, Singh VP, Singh S, Prasad SM, Singh PK, Dubey NK, Pandey AC (2017) Nitric oxide alleviates silver nanoparticles (AgNps)-induced phytotoxicity in Pisum sativum seedlings. Plant Physiol Biochem 110:167–177

Turkan I (2017) Emerging roles for ROS and RNS–versatile molecules in plants, vol 68. Oxford University Press, pp 4413–4416

Tuteja N (2007) Abscisic acid and abiotic stress signaling. Plant Signal Behav 2(3):135–138

Van Vu T, Do VN (2017) Customization of artificial microRNA design. In: MicroRNA profiling: methods and protocols, pp 235–243

Van Zelm E, Zhang Y, Testerink C (2020) Salt tolerance mechanisms of plants. Annu Rev Plant Biol 71:403–433

Vishwakarma K, Upadhyay N, Kumar N, Yadav G, Singh J, Mishra RK, Kumar V, Verma R, Upadhyay R, Pandey M (2017) Abscisic acid signaling and abiotic stress tolerance in plants: a review on current knowledge and future prospects. Front Plant Sci 8:161

Viuda-Martos M, Sanchez-Zapata E, Sayas-Barberá E, Sendra E, Pérez-Álvarez J, Fernández-López J (2014) Tomato and tomato byproducts. Human health benefits of lycopene and its application to meat products: a review. Crit Rev Food Sci Nutr 54(8):1032–1049

Vogel E, Donat MG, Alexander LV, Meinshausen M, Ray DK, Karoly D, Meinshausen N, Frieler K (2019) The effects of climate extremes on global agricultural yields. Environ Res Lett 14(5):054010

von Born P, Bernardo-Faura M, Rubio-Somoza I (2018) An artificial miRNA system reveals that relative contribution of translational inhibition to miRNA-mediated regulation depends on environmental and developmental factors in Arabidopsis thaliana. PLoS One 13(2):e0192984–e0192984

Vurukonda SSKP, Vardharajula S, Shrivastava M, SkZ A (2016) Enhancement of drought stress tolerance in crops by plant growth promoting rhizobacteria. Microbiol Res 184:13–24

Waheed K, Nawaz H, Hanif MA, Rehman R (2020) Tomato. Medicinal plants of South Asia. Elsevier, pp 631–644

Wang Y, Wisniewski M, Meilan R, Cui M, Fuchigami L (2006) Transgenic tomato (Lycopersicon esculentum) overexpressing cAPX exhibits enhanced tolerance to UV-B and heat stress. J Appl Hortic 8:87–90

Wang J-Y, Lai L-D, Tong S-M, Li QL (2013) Constitutive and salt-inducible expression of SlBADH gene in transgenic tomato (Solanum lycopersicum L. cv. Micro-Tom) enhances salt tolerance. Biochem Biophys Res Commun 432(2):262–267

Wang L, Li Q-T, Lei Q, Feng C, Zheng X, Zhou F, Li L, Liu X, Wang Z, Kong J (2017) Ectopically expressing MdPIP1; 3, an aquaporin gene, increased fruit size and enhanced drought tolerance of transgenic tomatoes. BMC Plant Biol 17:1–13

Wang L, Ma R, Yin Y, Jiao Z (2018) Role of carbon ion beams irradiation in mitigating cold stress in Arabidopsis thaliana. Ecotoxicol Environ Saf 162:341–347

Wang X, Zhang H, Xie Q, Liu Y, Lv H, Bai R, Ma R, Li X, Zhang X, Guo Y-D (2020a) SlSNAT interacts with HSP40, a molecular chaperone, to regulate melatonin biosynthesis and promote thermotolerance in tomato. Plant Cell Physiol 61(5):909–921

Wang Z, Hong Y, Zhu G, Li Y, Niu Q, Yao J, Hua K, Bai J, Zhu Y, Shi H (2020b) Loss of salt tolerance during tomato domestication conferred by variation in a Na+/K+ transporter. EMBO J 39(10):e103256

Wang J, Chen C, Wu C, Meng Q, Zhuang K, Ma N (2023a) SlMYB41 positively regulates tomato thermotolerance by activating the expression of SlHSP90. 3. Plant Physiol Biochem 204:108106

Wang L, Wang L, Tan M, Wang L, Zhao W, You J, Wang L, Yan X, Wang W (2023b) The pattern of alternative splicing and DNA methylation alteration and their interaction in linseed (Linum usitatissimum L.) response to repeated drought stresses. Biol Res 56(1):12

Wang T, Ma X, Chen Y, Wang C, Xia Z, Liu Z, Gao L, Zhang W (2024) SlNAC3 suppresses cold tolerance in tomatoes by enhancing ethylene biosynthesis. Plant Cell Environ 47:3132

Wani SH, Sah SK, Hossain MA, Kumar V, Balachandran SM (2016) Transgenic approaches for abiotic stress tolerance in crop plants. In: Advances in plant breeding strategies: agronomic, abiotic and biotic stress traits, pp 345–396

Waszczak C, Carmody M, Kangasjärvi J (2018) Reactive oxygen species in plant signaling. Annu Rev Plant Biol 69:209–236

Watanabe CK, Hachiya T, Terashima I, Noguchi K (2008) The lack of alternative oxidase at low temperature leads to a disruption of the balance in carbon and nitrogen metabolism, and to an up-regulation of antioxidant defence systems in Arabidopsis thaliana leaves. Plant Cell Environ 31(8):1190–1202

Wei L, Feng L, Liu Y, Liao W (2022) Mitogen-activated protein kinase is involved in salt stress response in tomato (Solanum lycopersicum) seedlings. Int J Mol Sci 23(14):7645

Xin H, Li Q, Wu X, Yin B, Li J, Zhu J (2023) The Arabidopsis thaliana integrin-like gene AT14A improves drought tolerance in Solanum lycopersicum. J Plant Res 136(4):563–576
Xu T, Niu J, Jiang Z (2022) Sensing mechanisms: calcium signaling mediated abiotic stress in plants. Front Plant Sci 13:925863
Yadav S, Modi P, Dave A, Vijapura A, Patel D, Patel M (2020) Effect of abiotic stress on crops. Sustain Crop Prod 3(17):5–16
Yamada T, Spangenberg G, Yamaguchi-Shinozaki K, Shinozaki K (2009) DREB regulons in abiotic-stress-responsive gene expression in plants. Molecular breeding of forage and turf. Springer
Yan C, Xie D (2015) Jasmonate in plant defence: sentinel or double agent? Plant Biotechnol J 13(9):1233–1240
Yanagawa Y, Yoda H, Osaki K, Amano Y, Aono M, Seo S, Kuchitsu K, Mitsuhara I (2016) Mitogen-activated protein kinase 4-like carrying an MEY motif instead of a TXY motif is involved in ozone tolerance and regulation of stomatal closure in tobacco. J Exp Bot 67(11):3471–3479
Yang Y, Guo Y (2018) Unraveling salt stress signaling in plants. J Integr Plant Biol 60(9):796–804
Yang D-Y, Zhuang K-Y, Ma N-N (2023) Overexpression of SlGGP-LIKE gene enhanced the resistance of tomato to salt stress. Protoplasma 260(2):625–635
Yao P, Sun Z, Li C, Zhao X, Li M, Deng R, Huang Y, Zhao H, Chen H, Wu Q (2018) Overexpression of Fagopyrum tataricum FtbHLH2 enhances tolerance to cold stress in transgenic Arabidopsis. Plant Physiol Biochem 125:85–94
Yoshida T, Ohama N, Nakajima J, Kidokoro S, Mizoi J, Nakashima K, Maruyama K, Kim J-M, Seki M, Todaka D (2011) Arabidopsis HsfA1 transcription factors function as the main positive regulators in heat shock-responsive gene expression. Mol Gen Genomics 286:321–332
Yuan P, Wei Y, Gao X, Song C, Jiao J, Wang M, Zhang K, Song S, Zheng X, Bai T (2024) The RING-H2 type E3 ubiquitin ligase gene MdATL16 positively regulates salt tolerance in transgenic tomato and apple. Environ Exp Bot:105689
Zarka DG, Vogel JT, Cook D, Thomashow MF (2003) Cold induction of Arabidopsis CBF genes involves multiple ICE (inducer of CBF expression) promoter elements and a cold-regulatory circuit that is desensitized by low temperature. Plant Physiol 133(2):910–918
Zhang Z, Huang R (2010) Enhanced tolerance to freezing in tobacco and tomato overexpressing transcription factor TERF2/LeERF2 is modulated by ethylene biosynthesis. Plant Mol Biol 73:241–249
Zhang JZ, Creelman RA, Zhu J-K (2004) From laboratory to field. Using information from Arabidopsis to engineer salt, cold, and drought tolerance in crops. Plant Physiol 135(2):615–621
Zhang Y, Liu H, Li B, Zhang J-T, Li Y, Zhang H (2009) Generation of selectable marker-free transgenic tomato resistant to drought, cold and oxidative stress using the Cre/loxP DNA excision system. Transgenic Res 18:607–619
Zhang N, Zhang D, Chen SL, Gong B-Q, Guo Y, Xu L, Zhang X-N, Li J-F (2018) Engineering artificial microRNAs for multiplex gene silencing and simplified transgenic screen. Plant Physiol 178(3):989–1001
Zhang L, Guo X, Zhang Z, Wang A, Zhu J (2021) Cold-regulated gene LeCOR413PM2 confers cold stress tolerance in tomato plants. Gene 764:145097
Zhang H, Zhu J, Gong Z, Zhu J-K (2022a) Abiotic stress responses in plants. Nat Rev Genet 23(2):104–119
Zhang Z, Zhang Y, Yuan L, Zhou F, Gao Y, Kang Z, Li T, Hu X (2022b) Exogenous 5-aminolevulinic acid alleviates low-temperature injury by regulating glutathione metabolism and β-alanine metabolism in tomato seedling roots. Ecotoxicol Environ Saf 245:114112
Zhang Y, Peng Y, Liu J, Yan J, Zhu K, Sun X, Bu X, Wang X, Ahammed GJ, Liu Y (2023) Tetratricopeptide repeat protein SlREC2 positively regulates cold tolerance in tomato. Plant Physiol 192(1):648–665
Zhang Z, Yuan L, Dang J, Zhang Y, Wen Y, Du Y, Liang Y, Wang Y, Liu T, Li T (2024) 5-Aminolevulinic acid improves cold resistance through regulation of SlMYB4/SlMYB88-SlGSTU43 module to scavenge reactive oxygen species in tomato. Hortic Res 11(3):uhae026

Zhao C, Zhang H, Song C, Zhu J-K, Shabala S (2020) Mechanisms of plant responses and adaptation to soil salinity. The Innovation 1(1):1

Zhao T, Wu T, Pei T, Wang Z, Yang H, Jiang J, Zhang H, Chen X, Li J, Xu X (2021) Overexpression of SlGATA17 promotes drought tolerance in transgenic tomato plants by enhancing activation of the phenylpropanoid biosynthetic pathway. Front Plant Sci 12:634888

Zhen Y, Ungerer MC (2008) Relaxed selection on the CBF/DREB1 regulatory genes and reduced freezing tolerance in the southern range of Arabidopsis thaliana. Mol Biol Evol 25(12):2547–2555

Zhou J, Wang J, Li X, Xia X-J, Zhou Y-H, Shi K, Chen Z, Yu J-Q (2014) H2O2 mediates the crosstalk of brassinosteroid and abscisic acid in tomato responses to heat and oxidative stresses. J Exp Bot 65(15):4371–4383

Zhou R, Yu X, Zhao T, Ottosen C-O, Rosenqvist E, Wu Z (2019) Physiological analysis and transcriptome sequencing reveal the effects of combined cold and drought on tomato leaf. BMC Plant Biol 19(1):377–377

Zinn KE, Tunc-Ozdemir M, Harper JF (2010) Temperature stress and plant sexual reproduction: uncovering the weakest links. J Exp Bot 61(7):1959–1968

Climate-Proofing Tomatoes: Molecular Strategies for Adapting to Changing Climate

4

Zubaria Haakim, Yumna Ahmad, Javed Iqbal, Banzeer Ahsan Abbasi, Shumaila Ijaz, Shoaib Khan, Tariq Mahmood, and Sobia Kanwal

Abstract

Tomato (*Lycopersicon esculentum*), a critical member of the *Solanaceae* family, holds substantial nutritional and economic significance worldwide. This chapter delves into the global challenges posed by climate change on tomato cultivation, emphasizing its impact on yield, quality, and geographical distribution. Elevated temperatures, drought, salinity, flooding, and disease pressures significantly hinder tomato production, exacerbated by shifting climate patterns. The primary objective is to explore molecular strategies for enhancing tomato resilience to these stresses. Advanced biotechnological approaches, including genetic engineering, CRISPR/Cas9 technology, and marker-assisted breeding, are high-

Zubaria Haakim, Yumna Ahmad and Javed Iqbal contributed equally with all other contributors.

Z. Haakim · Y. Ahmad · T. Mahmood
Department of Plant Sciences, Faculty of Biological Sciences, Quaid-i-Azam University, Islamabad, Pakistan
e-mail: zubaria@bs.qau.edu.pk; tmahmood@qau.edu.pk

J. Iqbal (✉)
Department of Botany, Bacha Khan University, Charsadda, Khyber Pakhtunkhwa, Pakistan

B. A. Abbasi
Department of Botany, Rawalpindi Women University, Rawalpindi, Pakistan

S. Ijaz
Department of Burn and Plastic Surgery, Shenzhen Institute of Translational Medicine, The First Affiliated Hospital of Shenzhen University, School of Biomedical Engineering, Shenzhen University Medical School, Shenzhen University, Shenzhen, China

S. Khan · S. Kanwal
Department of Biology and Environmental Sciences, Allama Iqbal Open University, Islamabad, Pakistan
e-mail: sobia.kanwal@aiou.edu.pk

M. Waseem et al. (eds.), *Omics Approaches for Tomato Yield and Quality Trait Improvement*, https://doi.org/10.1007/978-981-96-3890-1_4

lighted for their potential to climate-proof tomatoes. By overexpressing heat shock proteins, manipulating abscisic acid (ABA) signaling pathways, and improving osmoprotectant biosynthesis, these strategies aim to bolster stress tolerance in tomatoes. The integration of multi-omics approaches and synthetic biology further enhances the understanding and development of stress-resilient tomato varieties. Addressing the socioeconomic and ethical considerations of GMOs is crucial for public acceptance and regulatory compliance. This comprehensive review underscores the importance of collaborative efforts in advancing sustainable tomato production, ensuring global food security amid the challenges of climate change.

Keywords

Molecular strategies · Biotechnological approaches · Genetic engineering · CRISPR/Cas9 technology · Marker-assisted breeding · Heat shock proteins · ABA signaling pathways · GMOs

4.1 Introduction

Tomato, scientifically known as *Lycopersicon esculentum*, is classified under the order *Solanales* and holds significant importance as a member of the *Solanaceae* family. This family encompasses approximately 1500–2000 species, establishing it as one of the most diverse families of flowering plants (Causse et al. 2020). The *Solanaceae* family includes key global food crops such as potatoes and tomatoes. Nightshades, eggplants, tamarillos, horse nettles, and a variety of attractive blooms and fruit plants are all members of this plant family, which is well-known for its nutritional benefits. The tomato, which has its roots in the South American Andes, was initially consumed as food in Mexico and later became popular globally following the Spanish conquest of the Americas (Spaldon et al. 2015). In terms of production and use, tomatoes rank second among *Solanaceae* crops, just after potatoes. Tropical and subtropical regions are ideal for tomato growth, with temperatures between 15 °C and 30 °C and moderate rainfall of around 1000 mm being the ideal conditions (Kissoudis et al. 2016).

The global tomato industry is a highly dynamic segment of horticulture, driven by extensive research into superior germplasm and innovative cultivation methods, such as protected environment production and processing. A significant issue in tomato farming is the overuse of agrochemicals, especially for controlling pests and diseases. Improper pesticide use has led to the accumulation of harmful residues, raising concerns about the safety of tomatoes amid increasing public awareness of health, environmental impact, and labor conditions (Majhi et al. 2022). To address these challenges, various strategies have been implemented to promote sustainable management of the tomato value chain, including the use of protected environments to ensure optimal growing conditions. Different countries have varying agronomic practices, processing methods, marketing strategies, and overall organization when it comes to the tomato industry. According to the FAO, tomato output worldwide has increased and is now at 123 million tons yearly with a production area of over 4.5 million hectares (Solankey et al. 2021). Leading tomato-producing nations include China, the European Union, the United States, and Turkey. However, production costs remain a major hurdle, particularly in developed countries. Developing nations encounter challenges like pests, diseases, inconsistent production systems, and chemical misuse. Currently, there is a global focus on reducing input costs for things like energy, water, pesticides, fertilizers, and other inputs in order to improve the value chain's sustainability. How safe and environmentally friendly production processes are will be crucial in deciding how important tomatoes will remain as a staple produce in the globe (La Pena and Hughes 2007).

Global crop losses primarily result from environmental stresses, which can lead to significant yield reductions of more than 50% in major crops. The production environment for tropical vegetables varies depending on seasonal and regional conditions. Climate change is predicted to worsen these pressures, as warmer temperatures, limited access to irrigation, more frequent flooding, and soil salinity all create obstacles to maintaining and boosting vegetable yields. Severe weather conditions can also harm soil quality and lead to erosion, requiring more fertilizer use or better nutrient management to support or even improve crop growth in the face of rising CO_2 levels in the atmosphere (Solankey et al. 2019).

Depending on their stage of growth, the intensity and duration of the environmental stress, and other factors, plants react to these factors in different ways. They use chemical and physical mechanisms to lessen the effects of stress, while interactions with the environment can make these reactions more complex. It is essential for the production of tropical vegetables to adapt to these pressures brought on by the climate. The majority of scientific studies being conducted so far have focused on tomatoes in order to better understand the effects of environmental challenges. This underscores the need for more studies to determine how other vegetable crops are impacted by the increasing abiotic stresses linked to climate change (Naik et al. 2017).

4.2 Objectives

- Emphasize the nutritional and economic significance of tomatoes as a global agricultural staple.
- Outline how climate change is posing unprecedented challenges to tomato cultivation, affecting yield, quality, and geographical distribution.
- Present the primary objective of exploring and elucidating molecular strategies to enhance the resilience of tomatoes to changing climatic conditions.
- Provide a comprehensive review that integrates current knowledge on the physiological impacts of climate stressors on tomatoes.
- Discuss advanced biotechnological approaches aimed at climate-proofing tomatoes, ensuring their sustainable production in the face of climate change.

4.3 Impact of Climate Change on Tomato Cultivation

4.3.1 High Temperatures

Temperature constraints significantly impact the range and yield of numerous crops. In tropical regions, elevated temperatures are prevalent throughout the growing season. Tomatoes grown in these regions are more susceptible to temperature-related stress when climatic patterns change. Temperature increases are predicted in tomato-growing regions based on the analysis of climate patterns, and over the next several decades, there will be an increase in the frequency of above-optimal temperature circumstances (Chapman et al. 2012).

Temperatures above 30 °C can negatively impact pollen viability, fruit set, and overall plant vigor. High temperatures can disrupt cellular processes such as photosynthesis and respiration, leading to reduced productivity. On the other hand, low temperatures have the potential to hinder the process of flowering and fruit ripening, making cultivation more challenging. Tomatoes' growth and reproduction can be significantly affected by temperature, either on its own or when combined with other environmental factors. High temperatures can disrupt crucial biochemical reactions needed for normal plant cell function, especially impacting photosynthesis. This can result in lower tomato yields, smaller fruits, and decreased fruit quality (Gonzalo et al. 2020).

Prior to flowering, temperature stress can lead to a number of developmental abnormalities, including abnormalities in the endothecium and epidermis, an inability to open the stomium, and insufficient pollen production. Bud drop, uneven flower development, limited pollen production, dehiscence, problems with viability, ovule abortion, decreased availability of carbohydrates, and other reproductive anomalies are signs of fruit set failure at high temperatures. Additionally, photosynthesis significantly decreases at temperatures above the optimal range, leading to a noticeable decline in potential productivity (Razzaq et al. 2021).

4.3.2 Drought

Unpredictable dry spells are the main issue affecting global food security and have historically caused significant famines. With the world's water resources being limited, the impact of consecutive droughts will only worsen as the population grows and competition for water resources increases. Climate change is anticipated to heighten vulnerability to water availability, with extreme water stress conditions expected to detrimentally impact crop yields, particularly for water-intensive crops like tomatoes (Gomès et al. 2021).

Significant water stress for crops might result from decreased irrigation water supply, increased evapotranspiration, and higher temperatures when combined with decreased precipitation. Tomatoes, renowned for their high water content exceeding 90%, depend heavily on water levels for both quantity and quality. During drought periods, tomato production is markedly affected. Drought stress results in soil solute

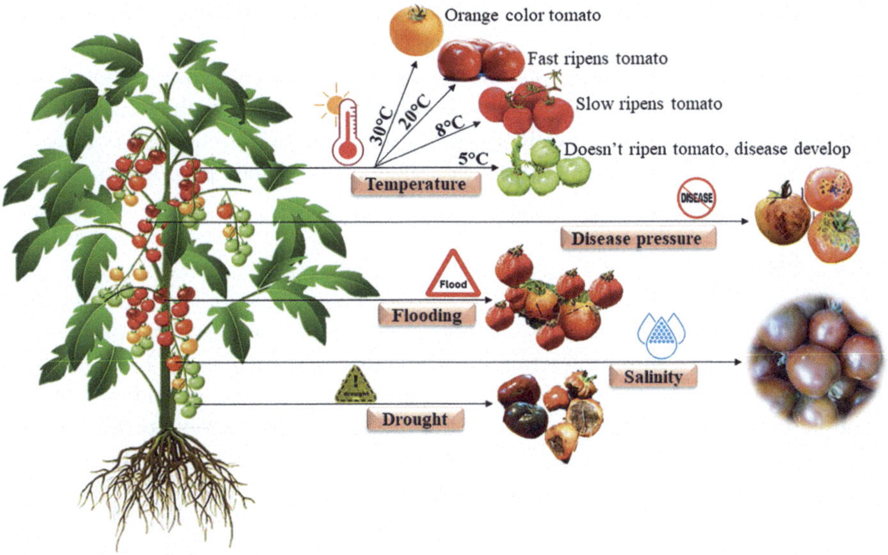

Fig. 4.1 Climate changing factors impacting tomato cultivation

accumulation, prompting water to exit plant cells through osmosis (Wani et al. 2022). This elevates solute concentration within plant cells, reduces water potential, and disrupts essential cellular processes such as photosynthesis. Variations exist in the timing, intensity, and length of dry periods that affect tomatoes (Fig. 4.1).

4.3.3 Salinity

Vegetable production is under serious threat due to rising soil salinity, especially in irrigated farmlands that contribute to 40% of the global food supply. High salt levels in the soil significantly reduce the yield of many crops, with tomatoes being especially sensitive at every stage of growth. According to the USDA, tomatoes are highly susceptible to the effects of saline soils. In hot and arid climates, the high rates of evapotranspiration cause significant water loss, leading to salt buildup around plant roots, which hinders their ability to absorb water.

Initially, salinity leads to water deficits by increasing the concentration of solutes in the soil, which disrupts the balance of potassium (K^+) and sodium (Na^+) ions, and raises the levels of harmful Na^+ and Cl^- ions for plants (Storey and Storey 2012). In tomatoes, the impact of salt stress is evident through symptoms such as loss of turgor pressure, stunted growth, wilting, leaf curling, abnormal leaf positioning (epinasty), leaf shedding, reduced photosynthesis, changes in respiration, cellular damage, tissue death, and, in severe cases, the death of the plant.

Salinity levels vary throughout the year, generally reaching their highest during dry seasons and declining during rainy periods due to the influx of freshwater.

Coastal areas are particularly vulnerable to saline-related disasters such as tsunamis, which can leave agricultural land barren (Raza et al. 2019). Even after seawater withdraws, the lingering contamination of groundwater and resulting osmotic stress can harm crops and degrade soil quality. In inland regions, the problem is worsened by the use of traditional water wells for irrigation, as salts from the bedrock elevate the salinity of well water, posing further challenges for tomato production.

4.3.4 Flooding

Vegetable production experiences challenges during the rainy season, particularly with excess moisture from heavy rainfall. Tomatoes, in particular, are highly vulnerable to flooding due to their limited genetic diversity in tolerating such conditions. Flooding reduces oxygen availability in the roots, affecting aerobic processes crucial for plant health (Abasi et al. 2023). When submerged, tomato plants internally release ethylene, a hormone that can damage them.

Low oxygen levels prompt roots to increase production of 1-aminocyclopropane-1-carboxylic acid (ACC), a precursor to ethylene. This often results in rapid leaf growth as a response to waterlogging, exacerbated by the buildup of ethylene. The impact of flooding is exacerbated in higher temperatures, often leading to rapid wilting and eventual death of tomato plants after a brief period of inundation during hot weather (Gao et al. 2023).

4.3.5 Disease Pressure

Disease pressure in tomato cultivation refers to the overall impact and prevalence of diseases affecting the crops, which can be influenced by a variety of factors. Tomatoes are susceptible to several types of diseases, including fungal, bacterial, and viral infections. Fungal diseases like late blight thrive in cool, wet conditions, causing dark lesions on leaves, stems, and fruits, while early blight and powdery mildew are more common in warm, humid, or dry climates.

Bacterial diseases such as bacterial wilt and bacterial spot lead to rapid wilting and water-soaked lesions, respectively, often flourishing in warm, humid environments. Viruses such as tomato mosaic virus (ToMV) and tomato yellow leaf curl virus (TYLCV) can cause mottling, leaf distortion, and stunted growth in tomatoes, with TYLCV being transmitted by whiteflies (Giovanna et al. 2022).

Environmental conditions play a significant role in disease pressure. Specific temperature ranges can favor different pathogens, and high humidity and frequent rainfall create ideal conditions for fungal and bacterial diseases. Overhead irrigation can spread waterborne pathogens, while poor soil health and inadequate drainage can promote soil-borne diseases. Breeding for disease resistance and planting a diverse range of varieties can mitigate disease risks. Pathogen dynamics, including mutation and evolution, can lead to resistance against control measures, complicating management efforts (Schouten et al. 2020).

4.4 Molecular Strategies for Climate-Proofing Tomatoes

As climate change continues to affect agriculture, it becomes crucial to develop tomato plants that can endure extreme weather conditions like high temperatures and droughts. Researchers are focusing on molecular techniques to create tomato varieties that are more resilient to these challenges. This section will explore the different molecular strategies being used to improve the climate resistance of tomatoes.

4.5 Genetic Engineering and CRISPR/Cas9 Technology

Genetic engineering, particularly with CRISPR/Cas9, allows precise modifications in tomatoes to enhance their resilience. By targeting specific genes, researchers aim to develop tomato varieties that are better suited to withstand extreme heat and other climate-related stresses (Table 1).

Table 1 CRISPR/Cas9 Driven Genetic Modifications in Tomato Exploring Target Genes Mutant Phenotypes and DNA Repair Mechanisms

Trait category	Target gene	Editing tool	Major mutant phenotype	Repair pathway	References
Growth habit	SIHPAT homolog	CRISPR/SpCas9	Multiple aspects of tomato reproductive development	cNHEJ	Brooks et al. (2014)
	SICLV1	CRISPR/SpCas9	Weak branching and fasciated flowers	cNHEJ	Xu et al. (2015)
	SILIN (long inflorescence)	CRISPR/SpCas9	Moderately branched inflorescences and increased flower production	cNHEJ	Roldan et al. (2017)
Fruit quality	SIRIN (ripening inhibitor)	CRISPR/SpCas9	Incomplete-ripening fruits, extended shelf life	cNHEJ	Ito et al. (2015)
	SIALC (alcobaca)	CRISPR/SpCas9	Long shelf life	cNHEJ	Yu et al. (2017)
	SITFM6 (tomato fruit malate on chromosome6)	CRISPR/SpCas9	Reduced fruit malate content	cNHEJ	Ye et al. (2017)
Biotic stress tolerance	SIDMR6-1 (downy mildew resistance 6–1)	CRISPR/SpCas9	Disease resistance against different pathogens, including *P. syringae* and *P. capsici*	cNHEJ	Paula de Toledo Thomazella et al. (2016)
	SIMLO1 (mildew locus O 1)	CRISPR/SpCas9	Powdery mildew disease resistance	cNHEJ	Nekrasov et al. (2017)
Abiotic stress tolerance	SIMAPK3 (mitogen-activated protein kinase 3)	CRISPR/SpCas9	Reduced drought tolerance	cNHEJ	Wang et al. (2017)
	SISOS1 (salt overly sensitive 1)	CRISPR/SpCas9	Increased salt sensitivity	cNHEJ	Wang et al. (2020)

4.6 Enhancing Heat Tolerance

4.6.1 Role of Heat Shock Proteins (HSPs)

Heat shock proteins (HSPs) are vital molecular chaperones that protect plants from heat stress by stabilizing proteins and cellular structures. These proteins are highly conserved and are classified based on their molecular weight, including HSP100, HSP90, HSP70, HSP60, and small HSPs (sHSPs) (Khan et al. 2021). When plants experience heat stress, HSPs are upregulated to help refold denatured proteins, prevent protein clumping, and maintain cellular balance (Reddy et al. 2016).

In tomatoes, the overexpression of HSPs, particularly HSP70, has been shown to significantly improve heat tolerance. For instance, transgenic tomatoes with elevated HSP70 levels exhibited increased thermotolerance, enhanced photosynthetic efficiency, and reduced cellular damage under heat stress conditions (Hoshikawa et al. 2021). These findings underscore the critical role of HSPs in the heat stress response and suggest that they could be promising targets for genetic improvements aimed at increasing heat tolerance in tomatoes.

4.7 CRISPR/Cas9-Mediated Gene Editing for Heat Tolerance

CRISPR/Cas9 technology offers a precise and efficient tool for genome editing, enabling targeted modifications in stress-responsive genes. This system uses a guide RNA (gRNA) to direct the Cas9 nuclease to specific DNA sequences, allowing for the targeted knockout or alteration of genes (Ahmad et al. 2022).

When it comes to enhancing heat tolerance, CRISPR/Cas9 can be applied to edit the promoter regions of heat-responsive genes, like those encoding heat shock proteins (HSPs), to boost their expression during heat stress (Ahmad et al. 2024). For instance, modifying the promoter of the HSP70 gene could increase its activity under high-temperature conditions, thereby improving the plant's ability to withstand heat (Divya et al. 2019). Furthermore, CRISPR/Cas9 can be used to knock out genes that act as negative regulators of heat stress responses, providing an additional way to enhance thermotolerance in tomato plants (Yu et al. 2019).

4.8 Improving Drought Resistance

4.8.1 ABA Signaling Pathways

Abscisic acid (ABA) is a crucial phytohormone that plays a central role in regulating how plants respond to drought stress. ABA signaling controls various physiological processes, including the closure of stomata, gene expression, and osmotic adjustment (Salvi et al. 2021). During drought conditions, ABA levels rise, initiating a series of signaling events that help the plant manage water scarcity (Mukherjee et al. 2023).

Key players in the ABA signaling pathway include ABA receptors (PYR/PYL/RCAR), protein phosphatases (PP2Cs), and SnRK2 kinases. When ABA binds to its receptors, it inhibits PP2Cs, which leads to the activation of SnRK2 kinases. These kinases then phosphorylate downstream targets, including transcription factors that regulate genes involved in stress responses (Saddhe et al. 2017).

Genetically manipulating components of the ABA signaling pathway can greatly enhance drought resistance in tomatoes. For example, overexpressing the transcription factor dehydration-responsive element-binding protein 1 (DREB1) has been shown to improve drought tolerance by activating ABA-responsive genes (Jiang et al. 2017). Additionally, using CRISPR/Cas9 to knock out negative regulators like PP2C phosphatases can increase ABA sensitivity, further boosting drought resistance (Xu et al. 2023).

4.9 Genetic Modifications for Enhanced Drought Response

Genetic engineering has been utilized to introduce drought-resistant traits into tomato plants by enhancing the expression of genes involved in osmoprotectant biosynthesis, such as proline and trehalose. Proline functions as an osmoprotectant by stabilizing proteins and membranes during osmotic stress, while trehalose helps protect against stress by preserving cellular structures and functions (Nawaz et al. 2022).

In transgenic tomatoes, overexpression of the pyrroline-5-carboxylate synthase (P5CS) gene, a crucial enzyme in proline biosynthesis, has led to improved drought tolerance. These tomatoes showed higher proline accumulation, better water retention, and increased resistance to drought (Liu et al. 2024). Similarly, overexpression of the trehalose-6-phosphate synthase-1 (TPS1) gene, which is involved in trehalose biosynthesis, has provided drought resistance by enhancing osmotic adjustment and reducing oxidative damage (Kosar et al. 2019).

4.10 Genomic Selection and Marker-Assisted Breeding

4.10.1 Identifying Climate-Resilient Traits

Genomic selection (GS) and marker-assisted breeding (MAB) represent potent methodologies for identifying and integrating climate-resilient traits into tomato breeding initiatives. GS employs genome-wide markers to forecast the performance of breeding lines, facilitating the selection of individuals possessing desired traits based on their genetic profiles (Sinha et al. 2023). This approach expedites the creation of stress-tolerant tomato varieties by amalgamating genomic insights with phenotypic data.

Quantitative trait loci (QTLs) mapping and genome-wide association studies (GWAS) have pinpointed several genetic loci linked to heat and drought tolerance in tomatoes (Albert et al. 2016). Additionally, GWAS have identified specific SNP markers associated with drought resistance, aiding in the selection of genotypes that are more resilient to drought conditions (Neelapu and Chaitanya 2024).

4.11 Marker-Assisted Breeding for Stress Tolerance

Marker-assisted breeding leverages molecular markers linked to desirable traits to facilitate the selection and incorporation of these traits into breeding lines (Salgotra and Stewart Jr 2020). By using markers associated with stress tolerance, breeders can efficiently develop tomato varieties that perform well under adverse climatic conditions.

Markers linked to QTLs for heat tolerance can be used to select breeding lines with improved fruit set and yield under high-temperature conditions (Golam et al. 2012). Similarly, markers associated with drought tolerance traits, such as root architecture and osmotic adjustment, can be employed to develop varieties with enhanced water-use efficiency and resilience to water deficit (Xiong et al. 2021).

4.12 Metabolic Engineering

4.12.1 Enhancing Osmoprotectants

Osmoprotectants like proline and glycine betaine are essential for preserving cellular integrity and function under osmotic stress. These compounds accumulate in response to stress, where they help protect cellular structures and stabilize proteins and membranes (Dikilitas et al. 2020).

Genetic engineering methods have been utilized to enhance the biosynthesis of osmoprotectants in tomatoes (Wai et al. 2020). Overexpression of P5CS, a key enzyme in proline synthesis, has been shown to enhance drought tolerance by increasing proline accumulation and improving water retention in plants. Similarly, genetic engineering aimed at boosting glycine betaine biosynthesis has proven effective in enhancing stress tolerance in diverse crops, including tomatoes (Krishna et al. 2019) (Fig. 4.2).

4.13 Genetic Engineering Approaches

Genetic engineering encompasses the incorporation of specific genes into the plant genome to bolster stress tolerance. Overexpressing genes responsible for osmoprotectant biosynthesis, such as those for proline and trehalose, have demonstrated enhanced drought tolerance in tomatoes. These genetic alterations augment the plant's capacity to uphold cellular homeostasis amid stressful conditions, thereby enhancing overall resilience to stress (Manna et al. 2021).

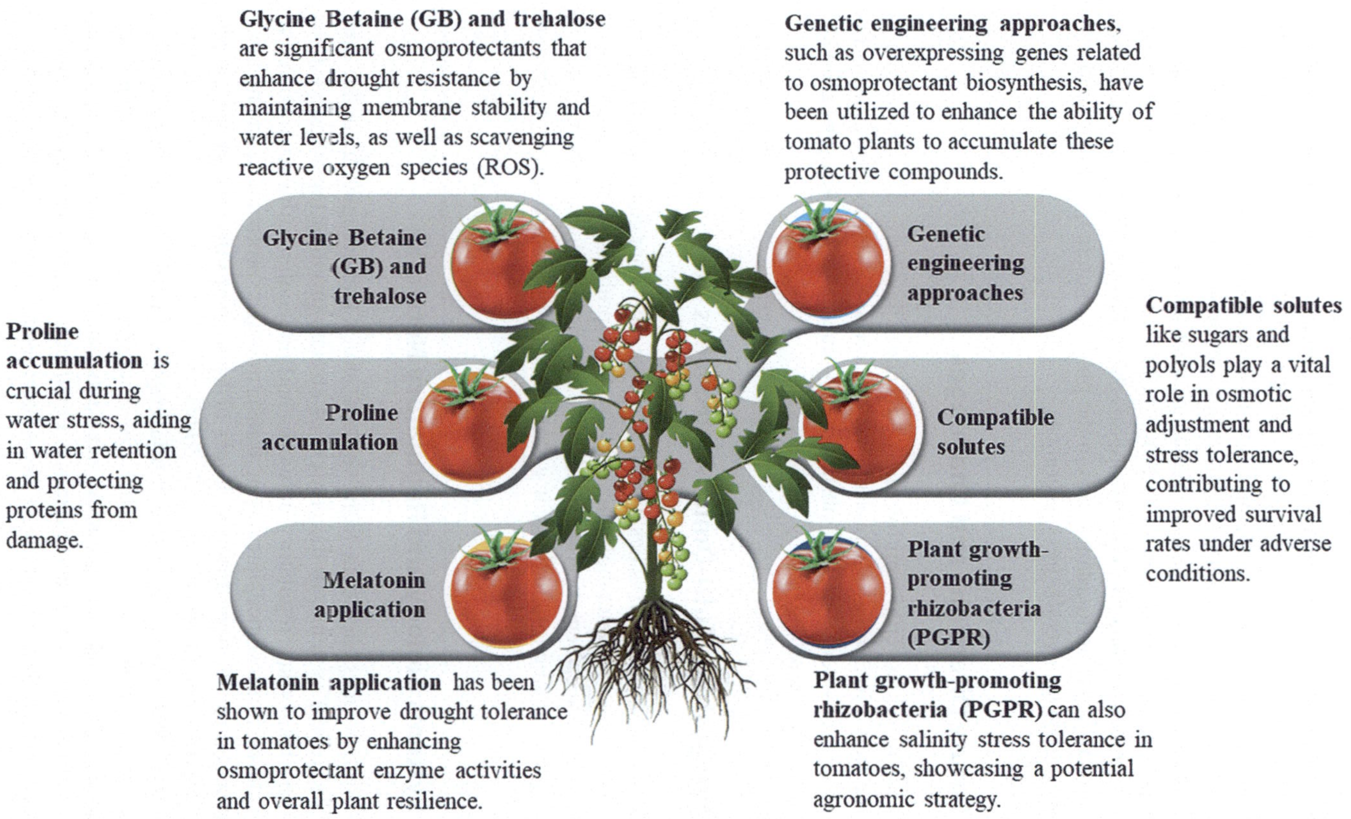

Fig. 4.2 Key points on enhancing osmoprotectants in tomatoes

4.14 Boosting Antioxidant Production

Antioxidants like ascorbate (vitamin C) and glutathione are vital for reducing oxidative damage caused by reactive oxygen species (ROS) during stress conditions (Gill and Tuteja 2010). Boosting the production of these antioxidants through metabolic engineering can significantly enhance stress resilience in tomatoes. For example, overexpressing genes involved in the ascorbate–glutathione cycle, such as ascorbate peroxidase (APX) and glutathione reductase (GR), has been shown to improve the plant's tolerance to various stresses (Krishna et al. 2021). These genetic modifications increase antioxidant levels in the plant, which helps reduce oxidative damage and enhances overall stress tolerance.

4.15 Future Directions and Challenges

As the field of tomato crop improvement advances, there are still challenges to address and new directions to explore. This section will outline the future paths researchers are considering and the obstacles they may face in developing climate-resilient tomato varieties.

4.16 Integrating Multi-Omics Approaches

Integrating multi-omics approaches, which combine data from genomics, transcriptomics, proteomics, and metabolomics, offers a comprehensive understanding of tomato biology. This integration helps identify key factors that can be targeted to enhance climate resilience, making it a crucial area for future research (Smyth et al. 2014).

4.17 Systems Biology and Regulatory Networks

Systems biology integrates data from various omics technologies to understand complex biological systems. In the context of climate-proofing tomatoes, systems biology can identify key regulatory networks and molecular pathways involved in stress responses. For instance, integrating transcriptomics, proteomics, and metabolomics data can reveal how different levels of biological information are coordinated under stress conditions (Raza et al. 2024).

Regulatory networks, which include transcription factors, signaling molecules, and their target genes, are essential for managing plant stress responses. Understanding and characterizing these networks can shed light on the molecular mechanisms that underpin stress tolerance. For instance, transcriptomic studies of tomatoes exposed to heat and drought stress have revealed differentially expressed genes involved in critical regulatory pathways, such as the ABA signaling pathway and heat shock proteins (HSPs). These insights help in unraveling how plants adapt to stressful conditions and can guide the development of more resilient crop varieties (Manna et al. 2021).

4.18 Omics Technologies in Stress Tolerance Research

Omics technologies, genomics, transcriptomics, proteomics, and metabolomics, provide a comprehensive approach to studying plant stress responses. Genomics offers the genetic blueprint of an organism, while transcriptomics reveals how gene expression changes in response to stress. Proteomics and metabolomics give insights into protein functions and metabolic shifts, respectively (Zhuang et al. 2014). For example, RNA-seq technology has been employed to investigate how tomatoes respond to drought stress, pinpointing key genes and pathways associated with drought tolerance (Bian et al. 2021). Additionally, proteomic studies have identified proteins linked to heat tolerance, such as heat shock proteins (HSPs) and antioxidant enzymes (Ding et al. 2020).

4.19 Enhancing Climate Resilience Through Synthetic Biology

4.19.1 Synthetic Gene Circuits and Regulatory Networks

Synthetic biology offers novel tools for engineering stress resilience in plants. By constructing synthetic gene circuits and regulatory networks, researchers can design plants with enhanced stress tolerance. Synthetic gene circuits are engineered pathways that can be programmed to respond to specific environmental cues, activating or repressing stress-responsive genes as needed (SinghMohan and BhallaPrem 2022). Synthetic promoters responsive to multiple stress signals can be designed to drive the expression of protective genes under adverse conditions. These synthetic constructs can provide a coordinated and robust defense mechanism against stress, enhancing the plant's resilience to climate change (Hasanuzzaman et al. 2020).

4.20 Potential Applications in Stress Tolerance

The application of synthetic biology in stress tolerance is vast. It includes the development of stress-responsive gene circuits, the engineering of biosynthetic pathways for osmoprotectants and antioxidants, and the creation of novel regulatory networks. These approaches can be used to enhance the production of stress-protective compounds, improve water-use efficiency, and increase overall stress resilience.

Synthetic biology can be used to engineer the biosynthesis of trehalose and proline, enhancing the plant's ability to cope with drought stress (Joshi et al. 2020). Additionally, synthetic regulatory networks can be designed to enhance the expression of antioxidant enzymes, reducing oxidative damage under heat stress (Raja et al. 2020).

4.21 Socioeconomic and Ethical Considerations

4.21.1 Public Perception and Acceptance of GMOs

Public perception and acceptance of genetically modified organisms (GMOs) are critical factors in the deployment of molecular strategies for climate-proofing tomatoes. Despite the potential benefits, there are concerns about the safety and environmental impact of GMOs. Addressing these concerns through transparent communication, rigorous safety assessments, and public engagement is essential for gaining public trust and acceptance (Cummings et al. 2024).

4.22 Regulatory Frameworks and Policies

The development and deployment of GMOs are governed by regulatory frameworks and policies designed to ensure their safety for human health and the environment. These regulations vary by country and can influence the adoption and commercialization of genetically engineered crops (Menz et al. 2020). Ensuring that regulatory frameworks are science-based, transparent, and adaptable to new technologies is crucial for the successful implementation of molecular strategies for climate-proofing tomatoes.

4.23 Conclusion

Adapting tomato cultivation to the challenges posed by climate change requires a multifaceted approach integrating genetic engineering, genomic selection, and metabolic engineering. Advances in molecular biology and biotechnology offer powerful tools to enhance stress tolerance and ensure sustainable tomato production. Strategies such as overexpressing heat shock proteins, manipulating ABA signaling pathways, and using CRISPR/Cas9 for targeted gene editing have shown promise in improving heat and drought tolerance. Continued research and development are essential for advancing our understanding of plant stress responses and developing effective climate-proofing strategies. Integrating multi-omics approaches, leveraging synthetic biology, and engaging in marker-assisted breeding can accelerate the development of stress-tolerant tomato varieties. Collaborative efforts among researchers, breeders, policymakers, and stakeholders are vital for addressing the challenges posed by climate change. Developing climate-resilient tomato varieties can significantly impact global food security. By ensuring stable and sustainable tomato production under changing climatic conditions, we can safeguard food supplies and support the livelihoods of millions of farmers worldwide. The molecular strategies discussed in this chapter offer promising avenues for enhancing the resilience of tomatoes and contributing to global efforts to mitigate the effects of climate change on agriculture.

References

Abasi F, Raja NI, Ehsan M, Ali H, Shahbaz M (2023) Adaptation strategies with respect to heat shock proteins and antioxidant potential; an era of food security and climate change. Int J Biol Macromol 128379

Ahmad Y, Haider S, Iqbal J, Abbasi BA, Yaseen T, Mahmood T (2022) The mechanisms of genome editing technologies in crop plants. In: Principles and practices of OMICS and genome editing for crop improvement. Springer, Cham, pp 295–313

Ahmad Y, Haakim Z, Iqbal J, Abbasi BA, Mahmood T, Kazi M (2024) Technological innovations for abiotic stress resistance in horticultural crops. In: OMICs-based techniques for global food security, pp 233–244

Albert E, Segura V, Gricourt J, Bonnefoi J, Derivot L, Causse M (2016) Association mapping reveals the genetic architecture of tomato response to water deficit: focus on major fruit quality traits. J Exp Bot 67(22):6413–6430

Bian Z, Wang Y, Zhang X, Grundy S, Hardy K, Yang Q, Lu C (2021) A transcriptome analysis revealing the new insight of green light on tomato plant growth and drought stress tolerance. Front Plant Sci 12:649283

Brooks C, Nekrasov V, Lippman ZB, Van Eck J (2014) Efficient gene editing in tomato in the first generation using the clustered regularly interspaced short palindromic repeats/CRISPR-associated9 system. Plant Physiol 166(3):1292–1297

Causse M, Zhao J, Diouf I, Wang J, Lefebvre V, Caromel B et al (2020) Genomic designing for climate-smart tomato. In: Genomic designing of climate-smart vegetable crops, pp 47–159

Chapman SC, Chakraborty S, Dreccer MF, Howden SM (2012) Plant adaptation to climate change – opportunities and priorities in breeding. Crop Pasture Sci 63(3):251–268

Cummings C, Selfa T, Lindberg S, Bain C (2024) Identifying public trust building priorities of gene editing in agriculture and food. Agric Hum Values 41(1):47–60

de Toledo Thomazella DP, Brail Q, Dahlbeck D, Staskawicz BJ (2016) CRISPR-Cas9 mediated mutagenesis of a DMR6 ortholog in tomato confers broad-spectrum disease resistance. BioRxiv 064824. https://doi.org/10.1101/064824

Dikilitas M, Simsek E, Roychoudhury A (2020) Role of proline and glycine betaine in overcoming abiotic stresses. In: Protective chemical agents in the amelioration of plant abiotic stress: biochemical and molecular perspectives, pp 1–23

Ding H, Mo S, Qian Y, Yuan G, Wu X, Ge C (2020) Integrated proteome and transcriptome analyses revealed key factors involved in tomato (Solanum lycopersicum) under high temperature stress. Food Energy Secur 9(4):e239

Divya K, Bhatnagar-Mathur P, Sharma KK, Reddy PS (2019) Heat shock proteins (Hsps) mediated signalling pathways during abiotic stress conditions. In: Plant signaling molecules. Woodhead Publishing, pp 499–516

Gao L, Kantar MB, Moxley D, Ortiz-Barrientos D, Rieseberg LH (2023) Crop adaptation to climate change: an evolutionary perspective, vol 16. Mol Plant, p 1518

Gill SS, Tuteja N (2010) Reactive oxygen species and antioxidant machinery in abiotic stress tolerance in crop plants. Plant Physiol Biochem 48(12):909–930

Giovanna F, Floriana A, Simona B, Irene B, Monica C, Chiara DA et al (2022) The smart-breed project: innovative molecular technologies for the adaptation of horticultural species to climate change through precision breeding. Poster Communications Siga

Golam F, Prodhan ZH, Nezhadahmadi A, Rahman M (2012) Heat tolerance in tomato. Life Sci J 9(4):1936–1950

Gomès É, Maillot P, Duchêne É (2021) Molecular tools for adapting viticulture to climate change. Front Plant Sci 12:633846

Gonzalo MJ, Li YC, Chen KY, Gil D, Montoro T, Nájera I et al (2020) Genetic control of reproductive traits in tomatoes under high temperature. Front Plant Sci 11:326

Hasanuzzaman M, Bhuyan MB, Zulfiqar F, Raza A, Mohsin SM, Mahmud JA et al (2020) Reactive oxygen species and antioxidant defense in plants under abiotic stress: revisiting the crucial role of a universal defense regulator. Antioxidants 9(8):681

Hoshikawa K, Pham D, Ezura H, Schafleitner R, Nakashima K (2021) Genetic and molecular mechanisms conferring heat stress tolerance in tomato plants. Front Plant Sci 12:786688

Ito Y, Nishizawa-Yokoi A, Endo M, Mikami M, Toki S (2015) CRISPR/Cas9-mediated mutagenesis of the RIN locus that regulates tomato fruit ripening. Biochem Biophys Res Commun 467(1):76–82

Jiang L, Wang Y, Zhang S, He R, Li W, Han J, Cheng X (2017) Tomato SlDREB1 gene conferred the transcriptional activation of drought-induced gene and an enhanced tolerance of the transgenic Arabidopsis to drought stress. Plant Growth Regul 81:131–145

Joshi R, Sahoo KK, Singh AK, Anwar K, Pundir P, Gautam RK et al (2020) Enhancing trehalose biosynthesis improves yield potential in marker-free transgenic rice under drought, saline, and sodic conditions. J Exp Bot 71(2):653–668

Khan S, Jabeen R, Deeba F, Waheed U, Khanum P, Iqbal N (2021) Heat shock proteins: classification, functions and expressions in plants during environmental stresses. J Bioresour Manag 8(2):9

Kissoudis C, Van De Wiel C, Visser RG, Van Der Linden G (2016) Future-proof crops: challenges and strategies for climate resilience improvement. Curr Opin Plant Biol 30:47–56

Kosar F, Akram NA, Sadiq M, Al-Qurainy F, Ashraf M (2019) Trehalose: a key organic osmolyte effectively involved in plant abiotic stress tolerance. J Plant Growth Regul 38:606–618

Krishna R, Karkute SG, Ansari WA, Jaiswal DK, Verma JP, Singh M (2019) Transgenic tomatoes for abiotic stress tolerance: status and way ahead. 3 Biotech 9:1–14

Krishna R, Ansari WA, Jaiswal DK, Singh AK, Prasad R, Verma JP, Singh M (2021) Overexpression of AtDREB1 and BcZAT12 genes confers drought tolerance by reducing oxidative stress in double transgenic tomato (Solanum lycopersicum L.). Plant Cell Rep 40(11):2173–2190

La Pena RD, Hughes J (2007) Improving vegetable productivity in a variable and changing climate

Liu W, Wei JW, Shan Q, Liu M, Xu J, Gong B (2024) Genetic engineering of drought-and salt-tolerant tomato via Δ1-pyrroline-5-carboxylate reductase S-nitrosylation. Plant Physiol 195(2):1038–1052

Majhi PK, Raza B, Behera PP, Singh SK, Shiv A, Mogali SC et al (2022) Future-proofing plants against climate change: a path to ensure sustainable food systems. In: Biodiversity, functional ecosystems and sustainable food production. Springer, Cham, pp 73–116

Manna M, Thakur T, Chirom O, Mandlik R, Deshmukh R, Salvi P (2021) Transcription factors as key molecular target to strengthen the drought stress tolerance in plants. Physiol Plant 172(2):847–868

Menz J, Modrzejewski D, Hartung F, Wilhelm R, Sprink T (2020) Genome edited crops touch the market: a view on the global development and regulatory environment. Front Plant Sci 11:586027

Mukherjee A, Dwivedi S, Bhagavatula L, Datta S (2023) Integration of light and ABA signaling pathways to combat drought stress in plants. Plant Cell Rep 42(5):829–841

Naik PS, Singh M, Ranjan JK (2017) Impact of climate change on vegetable production and adaptation measures. In: Abiotic stress management for resilient agriculture, pp 413–428

Nawaz M, Hassan MU, Chattha MU, Mahmood A, Shah AN, Hashem M et al (2022) Trehalose: a promising osmo-protectant against salinity stress – physiological and molecular mechanisms and future prospective. Mol Biol Rep 49(12):11255–11271

Neelapu NRR, Chaitanya KV (2024) Bioinformatics tools for assessing drought stress tolerance in crops. Bioinformatics for Plant Research and Crop Breeding

Nekrasov V, Wang C, Win J, Lanz C, Weigel D, Kamoun S (2017) Rapid generation of a transgene-free powdery mildew resistant tomato by genome deletion. Sci Rep 7(1):482

Raja V, Qadir SU, Alyemeni MN, Ahmad P (2020) Impact of drought and heat stress individually and in combination on physio-biochemical parameters, antioxidant responses, and gene expression in Solanum lycopersicum. 3 Biotech 10:1–18

Raza A, Razzaq A, Mehmood SS, Zou X, Zhang X, Lv Y, Xu J (2019) Impact of climate change on crops adaptation and strategies to tackle its outcome: a review. Plan Theory 8(2):34

Raza A, Salehi H, Bashir S, Tabassum J, Jamla M, Charagh S et al (2024) Transcriptomics, proteomics, and metabolomics interventions prompt crop improvement against metal (loid) toxicity. Plant Cell Rep 43(3):80

Razzaq A, Wani SH, Saleem F, Yu M, Zhou M, Shabala S (2021) Rewilding crops for climate resilience: economic analysis and de novo domestication strategies. J Exp Bot 72(18):6123–6139

Reddy PS, Chakradhar T, Reddy RA, Nitnavare RB, Mahanty S, Reddy MK (2016) Role of heat shock proteins in improving heat stress tolerance in crop plants. In: Heat shock proteins and plants, pp 283–307

Roldan MVG, Périlleux C, Morin H, Huerga-Fernandez S, Latrasse D, Benhamed M, Bendahmane A (2017) Natural and induced loss of function mutations in SlMBP21 MADS-box gene led to jointless-2 phenotype in tomato. Sci Rep 7(1):4402

Saddhe AA, Kundan K, Padmanabh D (2017) Mechanism of ABA signaling in response to abiotic stress in plants. Mech Plant Hormone Signal Under Stress 1:173–195

Salgotra RK, Stewart CN Jr (2020) Functional markers for precision plant breeding. Int J Mol Sci 21(13):4792

Salvi P, Manna M, Kaur H, Thakur T, Gandass N, Bhatt D, Muthamilarasan M (2021) Phytohormone signaling and crosstalk in regulating drought stress response in plants. Plant Cell Rep 40:1305–1329

Schouten HJ, Tikunov Y, Visser RGF (2020) Tomato diversified by modern plant breeding. Prophyta Annu 2020:14–19

Singh MB, Bhalla PL (2022) Biological parts for engineering abiotic stress tolerance in plants. BioDesign Res

Sinha D, Maurya AK, Abdi G, Majeed M, Agarwal R, Mukherjee R et al (2023) Integrated genomic selection for accelerating breeding programs of climate-smart cereals. Genes 14(7):1484

Solankey SS, Prakash S, Pawan S, Randhir K (2019) Vulnerability of vegetable crops to climate change. In: National Seminar on "recent advances in agriculture for sustainable rural development (RAASRD-2019)". VKSCoA, Dumraon (BAU, Sabour), Bihar, pp 64–70

Solankey SS, Kumari M, Akhtar S, Singh HK, Ray PK (2021) Challenges and opportunities in vegetable production in changing climate: mitigation and adaptation strategies. Adv Res Vegetable Prod Under Chang Clim 1:13–59

Spaldon S, Samnotra RK, Chopra S (2015) Climate resilient technologies to meet the challenges in vegetable production. Int Res Curr Acad Rev 3(2):28–47

Storey KB, Storey JM (2012) Strategies of molecular adaptation to climate change: the challenges for amphibians and reptiles. In: Temperature adaptation in a changing climate: nature at risk. CABI, Wallingford, pp 98–119

Wai AH, Naing AH, Lee DJ, Kim CK, Chung MY (2020) Molecular genetic approaches for enhancing stress tolerance and fruit quality of tomato. Plant Biotechnol Rep 14:515–537

Wang L, Chen L, Li R, Zhao R, Yang M, Sheng J, Shen L (2017) Reduced drought tolerance by CRISPR/Cas9-mediated SlMAPK3 mutagenesis in tomato plants. J Agric Food Chem 65(39):8674–8682

Wang Z, Hong Y, Zhu G, Li Y, Niu Q, Yao J et al (2020) Loss of salt tolerance during tomato domestication conferred by variation in a Na+/K+ transporter. EMBO J 39(10):e103256

Wani AK, Akhtar N, Sher F, Navarrete AA, Américo-Pinheiro JHP (2022) Microbial adaptation to different environmental conditions: molecular perspective of evolved genetic and cellular systems. Arch Microbiol 204(2):144

Xiong R, Liu S, Considine MJ, Siddique KH, Lam HM, Chen Y (2021) Root system architecture, physiological and transcriptional traits of soybean (Glycine max L.) in response to water deficit: a review. Physiol Plant 172(2):405–418

Xu C, Liberatore KL, MacAlister CA, Huang Z, Chu YH, Jiang K et al (2015) A cascade of arabinosyltransferases controls shoot meristem size in tomato. Nat Genet 47(7):784–792

Xu L, Gao Q, Feng J, Xu Y, Jiang J, Deng L et al (2023) Physiological and phosphoproteomic analyses revealed that the NtPOD63 L knockout mutant enhances drought tolerance in tobacco. Ind Crop Prod 193:116218

Ye J, Wang X, Hu T, Zhang F, Wang B, Li C et al (2017) An InDel in the promoter of Al-ACTIVATED MALATE TRANSPORTER9 selected during tomato domestication determines fruit malate contents and aluminum tolerance. Plant Cell 29(9):2249–2268

Yu QH, Wang B, Li N, Tang Y, Yang S, Yang T et al (2017) CRISPR/Cas9-induced targeted mutagenesis and gene replacement to generate long-shelf life tomato lines. Sci Rep 7(1):11874

Yu W, Wang L, Zhao R, Sheng J, Zhang S, Li R, Shen L (2019) Knockout of SlMAPK3 enhances tolerance to heat stress involving ROS homeostasis in tomato plants. BMC Plant Biol 19:1–13

Zhuang J, Zhang J, Hou XL, Wang F, Xiong AS (2014) Transcriptomic, proteomic, metabolomic and functional genomic approaches for the study of abiotic stress in vegetable crops. Crit Rev Plant Sci 33(2–3):225–237

Nutrigenomics in Tomato: Enhancing Health Benefits Through Molecular Breeding

5

Nida Firdous, Umar Farooq, Aliza Batool, Fareena Jamil, and Areeba Aslam

Abstract

Nutrigenomics is the branch of nutrition that uses molecular methods to identify, access, and interpret the range of responses that individuals or groups may experience while following a certain diet. Since humans learned long ago that a person's diet and environment may both impact their health, they have used food and plants as medicine. It seeks to elucidate the ways in which the components of a particular meal (bioactive substance) may affect gene expression, either upregulating or downregulating its potential. It is viable to increase tomatoes' biological make up of important nutrients and bioactive substances including lycopene, vitamins, and antioxidants by using molecular breeding techniques. These improvements not only boost tomatoes' nutritional content but also assist in reducing the risk of chronic illnesses like cancer and heart disease. This chapter attempts to give a thorough review of the most recent developments in tomato molecular breeding, emphasizing the ways in which these methods might be applied to produce more wholesome, nutrient-dense food crops. Genetic advancements impact on public health, taking into account the possible advantages and moral dilemmas associated with genetically engineering food plants.

Keywords

Bioactive · Boost · Ethical · Nutritious · Potential · Population · Risk · Vitamins

N. Firdous (✉) · U. Farooq · A. Batool · F. Jamil · A. Aslam
Faculty of Food and Home Sciences, Muhammad Nawaz Shareef University of Agriculture, Multan, Pakistan
e-mail: nida.firdous@mnsuam.edu.pk

M. Waseem et al. (eds.), *Omics Approaches for Tomato Yield and Quality Trait Improvement*, https://doi.org/10.1007/978-981-96-3890-1_5

5.1 Introduction

It has long been known that each person responds differently to the meals they eat. Food-related phenotypic variation can range from something as delicate as being sensitive to resentment, as demonstrated by reactions to substances like phenylthiocarbamide, to something as obvious as obesity, as demonstrated by variations in calorie consumption . "Nutrigenomics" or "nutritional genomics" refers to the scientific study of how genes interact with diet or consumption to alter phenotypes (Davis and Milner 2004; Trujillo et al. 2006). The field of nutrigenomics is beginning to bring the genetic pathways and related molecular targets that explain food components' capacity to elicit a physiologically appropriate response to more clarity. The hereditary components that impact a person's eating habits are starting to be identified by researchers. Many new analytical methods, such as genome-wide linkage scans and association studies like candidate gene markers and those examining eating behavior, are starting to emerge, despite the fact that this field of study is still quite young (Rankinen and Bouchard 2006).

What particular foods tell your genes is revealed by the emerging field of nutrigenomics. Your diet has a direct impact on the genetic signals that reach your body. Your body uses these communications to tell which molecules to burn or retain as calories. In turn, these messages regulate every molecule that makes up your metabolism. You may significantly alter the way that food interacts with your body, lose weight, and improve your health if you can understand the language of your DNA and take control of the signals and directives they provide to your body and metabolism (Mark 2006). The study of gene products, their physiological roles, and their interactions is known as nutrigenomics. It highlights how nutrients affect the genome, proteome, and metabolome and describes how nutrient regimes and particular nutrients relate to human health.

There is an immense need to Investigates the relationship between the things we eat and our DNA (Fenech et al. 2011; Kaput 2005). It specifically investigates the effects of various meals on individuals with various genetic composition. This technology seeks to pair individuals with the most appropriate foods. For designing functionsthat would enhance health as claimed individual's needs, and to 1 day apply individualized nutrition guidance, this growing subject brings together scientists, legislators, and health experts. Our eating habits, our parents' eating habits, and the way in which they affect the development of our cells and biological systems all merge. Moreover, it is similar to how our genetic composition determines the varied pathways that our food will take in regulating our anatomical growth (Pavlidis et al. 2015; Uthpala et al. 2020). All things considered, a nutrigenomic approach offers a snapshot illustrating genes that are either activated or inactive (the genetic potential) at any given time, an understanding of how gene/protein networks may work together to produce the observed response, and a technique to ascertain how nutrients affect gene/protein expression (Ramachandran 2006; Kandaswamy 2011; Pathak et al. 2000).

Following the human genome project (HGP), efforts and conversation focused on the "post-genomic era," a new biological era in which gains in "omics" science research were made possible by the development of bioinformatics. These fields of study employ biotechnology to separate and analyze a larger variety of biomolecules from the same family, including metabolites, proteins, RNA, and DNA. As a result, other omics sciences including proteomics, transcriptomics, and metabolomics emerged as groundbreaking instruments after genomics (Dalmiel et al. 2013; Koutsos et al. 2009). As a result, the original concept of nutrigenomics mainly covered research on how nutrients and bioactive foods affect an individual's gene expression. These days, this concept has broadened, and studies on nutritional components that function to safeguard the genome are now included in the field of nutrigenomics. Consequently, the goal of the new research is to comprehend how nutrition affects the transcriptome, proteome, metabolome, and genome (Ronteltap et al. 2009; Cozzolino 2012).

Nutrigenomics can produce a variety of results in a single experiment; thus, creative thinking is required in this field. Furthermore, nutrigenomics can glean biologically significant information from the gathered data (Cozzolino and Cominetti 2013). The ultimate response can only be obtained via a number of studies or polls conducted across various teams and groups. As a result, there is an increasing collaboration in the fields of nutrition, biology, medicine, genomics, and bioinformation among nations, teams, and research organizations (Ronteltap et al. 2009; Costa and Rosa 2011; Dauncey 2012).

5.2 Importance of Tomatoes in Diet

Tomatoes (*Lycopersicon esculentum*) are major agricultural products that are consumed by humans and are valued as commodities globally. One of the most adaptable and extensively utilized food plants, it can be eaten raw or as a component of other foods and goods. As a significant source of carotenoids, especially lycopene, ascorbic acid, and phenolic compounds, tomato-based products are abundant in antioxidant chemicals (Kaur and Kapoor 2002; George et al. 2004; Sahlin et al. 2004). One of the most important foods for humans is tomatoes, which are abundant in bioactive substances as well (Erge and Karadeniz 2011). Research has indicated a connection between high tomato consumption and a lower risk of some cancers (Chan et al. 2005; Rao and Ali 2007).

5.3 Health Benefits of Tomatoes

Residents in the Mediterranean region are at a lower risk of developing an assortment of serious chronic illnesses, such as coronary heart disease and various cancers (breast, colon, and prostate). A traditional diet rich in fruits, vegetables, and cooked tomatoes, in particular, together with copious amounts of olive oil seems to be associated with this lower risk. Public health advice to eat more veggies, particularly cooked tomatoes with olive oil, has been promoted by these findings (Weisburger 2002). Regular consumption of tomatoes and products made from them has been linked to a lower risk of cardiovascular diseases (CVDs) and a variety of malignancies (Borguini and Ferraz 2009).

A prospective study found that eating tomatoes for 2–3 days (as opposed to zero days) was linked to a considerable drop in mortality (48%) and a lower chance of dying from diarrhea. The study examined the association between 28, 753 children between the ages of six and 60 months and tomato consumption in terms of illness and death. Tomato consumption was also strongly and inversely correlated with respiratory illnesses (Fawzi et al. 2000).

Though it can be found in a variety of fruits and vegetables, tomatoes and processed tomato products are the main sources of lycopene in the diet of North Americans. Lycopene is a carotenoid that lacks provitamin-A activity. Eating more tomatoes and tomato-based foods can lower your risk of developing chronic illnesses like cancer and heart disease, according to several recent studies. Additionally, there is an inverse relationship between serum and tissue lycopene levels and the risk of chronic illnesses. The beneficial effects of lycopene are mainly attributed to its antioxidant properties, but mounting evidence indicates that additional mechanisms, including metabolic pathways of hormone ,immune system regulation and intercellular gap junction communication, may also be at play (Rao and Agarwal 2000).

The additional carotenoids found in tomatoes are vital parts of the human body's defense system; they include phytofluene, beta-carotene, neurosporene, lutein, and zeaxanthin. Furthermore, lutein and zeaxanthin are believed to be the primary guardians against cataracts and age-related macular degeneration (Mares et al. 2001). One of lycopene's most important modes of action is undoubtedly its strong antioxidant activity. It can lessen mutagenesis and trap singlet oxygen. Without causing toxicity or cell death, it can, at physiological concentrations, suppress the growth of human cancer cells by blocking growth factor receptor signaling and the progression of the cell cycle, particularly in prostate cancer cells (Heber 2002). A vital component of all living things is oxygen. Reactive oxygen species (ROSs) are produced as a necessary byproduct of the body's aerobic metabolism. ROSs have the ability to alter gene expression (Lee et al. 2004). L-ascorbic acid and phenolic chemicals are also found in tomatoes as natural antioxidants. Free radical chain reactions' electron morbidity is suppressed by phenolic antioxidants, which also function as electron or hydrogen donors. Ascorbic acid is well known for its ability to scavenge oxygen (Blokhina et al. 2003), break DNA strands and modify bases, oxidize lipids and proteins, and more in cells. Unprecedented strides have been achieved in the past few years in identifying and comprehending the part ROSs play in numerous illnesses. Arteriosclerosis, cancer, rheumatoid arthritis, and cardiovascular disorders are a few of these (Melo and Guerra 2002). It is believed that these tomato phytochemicals lower the chances of human illnesses like prostate cancer and cardiovascular diseases (CVDs). It seems that making the right dietary decisions can both prevent and delay the start of prostate cancer (Wilkinson and Chodak 2003; Cohen 2002).

5.4 Bioactive Compounds in Tomatoes

5.4.1 Phenolic Compounds

Tomato fruits include a wide variety of flavonoids and phenolic substances that have been investigated and identified. With a concentration of 309.7 mg per 100 g dry weight, naringenin chalcone was found to be the most prevalent phenolic component in tomatoes. Subsequently, the level of 3-caffeoylquinic acid and quercetin-3-rutinoside in tomatoes were found to be the second and third highest, at 71.7 mg and 60 mg per 100 g dry weight, respectively (Choi et al. 2014). Rutin is the most prevalent flavonoid found, with naringenin (0.65–1.19 mg/100 g fresh weight), quercetin (0.048–0.141 mg/100 g fresh weight), and myricetin (0.017–0.286 mg/100 g fresh weight) following in order of abundance. Furthermore, at 0.75–1.38 mg of phenolic acid per 100 g of fresh weight, chlorogenic acid was the most prevalent phenolic acid constituent in tomatoes (Kaur et al. 2013).

5.4.2 Carotenoids

Tomato fruits contain a variety of carotenoids. For example, up to 88% of carotenoids can be found in tomato peels and other by-products (Perveen et al. 2015; Viuda-Martos et al. 2014). The tissues that carry out the photosynthetic activities of immature green tomatoes are the primary source of beta-carotene, which is found in fewer than 1% of tomatoes. (Rizk et al. 2014).

5.4.3 Lycopene

The most abundant group of carotenoid pigments in tomatoes, lycopene (alpha-carotene), is a vivid red pigment that signifies fruit maturity (Daood et al. 2014; Ilahy et al. 2016). Many factors, both hereditary and environmental, can have a significant influence on tomatoes' lycopene levels (Brandt et al. 2006). From a genetic perspective, fruit color accurately indicates the amount of lycopene present; orange cultivars exhibit lower lycopene content than most red cultivars, while red cultivars contain more lycopene content than yellow cultivars (Coyago-Cruz et al. 2019). Some black tomato varieties, such as the Black Krim tomatoes, however, contain more lycopene than red varieties (Choi and Lee 2015). Factors affecting the Lycopene contents in tomatoes are illustrated in (Fig. 5.1).

5.4.4 Beta-Carotene

Carotenoids are found in tomatoes and are regarded as the second most prevalent carotenoid. Their product, beta-carotene, has a content level of 0.23–2.83 mg per 100 g of fresh weight, which is significantly lower than lycopene. From a structural

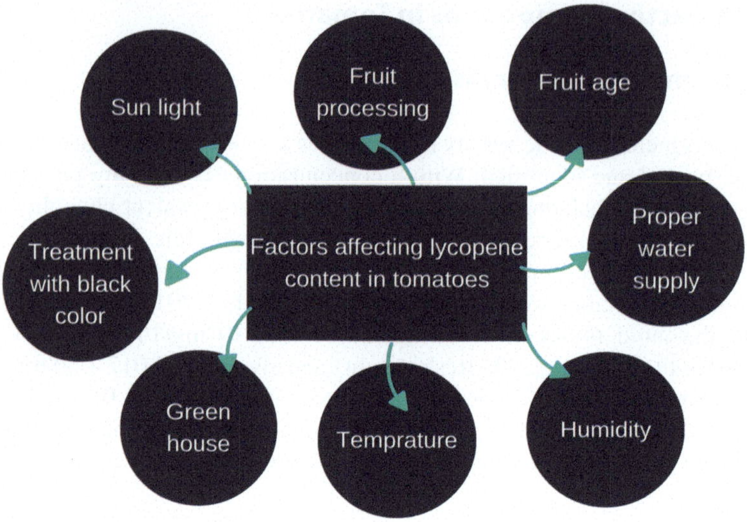

Fig. 5.1 Factors affecting the lycopene content in tomatoes

standpoint, beta-carotene exhibits provitamin-A activity due to its two retinyl groups and is in charge of fruit pigmentation, which ranges from yellow to orange (Baranska 2006).

5.4.5 Vitamins and Minerals

Tomatoes are good sources of micronutrients and also contain minerals like K, Na, Mg, and S, as well as vitamins like A, B, and E. Potassium is the most prevalent mineral in fresh tomatoes, with 237 mg per unit. Phosphorus is next with 24 mg; magnesium is at 11 mg; and sodium is at 5 mg. Moreover, the tomato-processing industry's wastes are a great source of nutrients in addition to a variety of phytochemicals (Zhivkova 2020). It was found that With a concentration level of up to 2238 mg/unit, the potassium element was the most concentrated in wasted tomatoes. S, Mg, and Na were next with 135, 134, and 132 mg/kg, respectivelyillustrated in the research carried out by (Asquer et al. 2013). Potassium constituted the greatest mineral level (9.97 mg/kg) in tomatoes that were discarded. Tomatoes are one of the primary food sources of vitamin C, among other nutrients. The properties of vitamin C include being soluble in water, being able to be stored in the body, and having the ability to perform a variety of biological tasks (Pehlivan 2017). This vitamin concentration in fresh tomatoes ranges from 8 to 16.3 mg per 100 g of fresh weight. However, a number of variables, including fruit growth, genotype, variety, and environmental condition, play a major role (Marti et al. 2018). Vitamin C is one of the most potent antioxidants and has a wide range of health benefits. When it comes to vitamin E, it is a kind of fat-soluble substance made up of several chemical structures, such as four tocopherols (alpha, beta, gamma, and delta) and four tocotrienols (alpha, beta, gamma, and delta). Since alpha-tocopherol is the only form of vitamin E that is preserved in human plasma, it is defined as the standard recommended dietary amount of vitamin E (Traber et al. 2006).

5.4.6 Proteins and Amino Acids

Tomato seeds have a practical purpose and may be recycled, making them essential components of fresh tomatoes. In addition to having a greater concentration of bioactive chemicals, tomato seeds were also found to contain a higher amount of crude protein—roughly 24.5%—with the highest concentration of aspartic and glutamic acids (Persia et al. 2003). The seeds of tomato fruits contain the majority of the fruit's protein. Studies conducted recently suggest that the protein content of tomato seeds may be approximately twice as high as other cereals including wheat (Sogi et al. 2005). It has been claimed that the specific protein content of tomatoes and tomato-based meals can exceed 21.9% on a dry weight basis.

5.4.7 Carbohydrates

Soluble sugars like glucose and fructose and non-soluble dietary fibers make up the two main components of tomatoes' carbohydrate content. Higher level of sugar content in tomatoes is responsible for Better sweetness and texture of the finished products that give the sector a more successful market b. Still, given that sugar metabolism appears to be regulated by multiple quantitative characteristics, tomatoes and their associated products tend to have relatively low carbohydrate contents that are nevertheless effective in terms of flavor and texture. Dietary fiber has been demonstrated to have several advantages for the digestive system. It is made up of a variety of materials, including cellulose, hemicellulose, lignin, pectic, and gums.

5.4.8 Fatty Acids

Tomato peels and seeds are the main sources of fat. Fresh tomatoes have a total fat content of about 0.2 g/kg, but the tomato pomace—which contains 12–18% of the fat from the tomato seeds—is high in fat (King and Zeidler 2004). According to reports, unsaturated fatty acids make up 77.04% of the fatty acids in tomato wastes, while saturated fatty acids make up 22.72%. Furthermore, linoleic acid accounted for up to 51.91% of the total fat content in tomatoes, with oleic and palmitic acids following at 18.5% and 16.2%, respectively (Nour et al. 2018). It is important to note that tomato wastes had a dangerous ratio of 12.56:1, which is quite mild when compared to other criteria, for n-6:n-3 polyunsaturated fatty acids (Rossell 2001).

5.4.9 Glycoalkaloids

The glycoalkaloid tomatine, which was initially identified, is actually a combination of two glycoalkaloids: dehydrotomatine and R-tomatine. Every portion of the tomato plant has both these chemicals (Friedman 2002) (Fig. 5.2).

5.5 Health Benefits of Tomatoes

The body's aerobic metabolism will inevitably produce reactive oxygen species (ROSs), which include both free radicals and non-radicals. ROSs have the ability to alter gene expression, break DNA strands and modify bases, and oxidize lipids and proteins in cells (Lee et al. 2004). Superoxide dismutase, glutathione peroxidase, and catalase are just a few of the enzymatic and nonenzymatic processes that aerobic organisms must develop in order to neutralize those ROSs (Davey et al. 2000). The term "antioxidants" often refers to a substrate's lowest level of capacity to impede the body's oxidation response. Both synthetic and natural ones are divided into two types (Borguini and Ferraz 2009).

5.5.1 Antioxidants

Antioxidants have a crucial role in keeping free radicals from donating electrons, which, in turn, prevents all diseases. Frequent consumption of tomatoes supplies naturally occurring antioxidants necessary for oxidation inhibition and disease prevention. Reactive oxygen species (ROSs) cause oxidative damage to cells and develop conditions that are directly related to oxidative stress, which include neoplasia, atherosclerosis, and neurological diseases (Shahidi and Ambigaipalan 2015). Tomatoes are a great source of bioactive substances including flavonoids, carotenoids, and polyphenols. They are linked to inhibitions of the oxidizing chain reaction's initiation and propagation phases, as well as the ability to combine with metal ions and inactivate the Fenton reaction (Faraga et al. 2010). Furthermore, a high quantity of vitamins like Vitamin D and citric acid in tomatoes are found to lessen the incidence of various cancers and cardiovascular disorders (Sahlin et al. 2004).

5.5.2 Cardioprotective Actions

The interest in tomato diet and phenolic compounds has increased since dietary phenolics have demonstrated multiple health benefits, such as lowering the risk of cardiovascular diseases, Alzheimer's disease, and some types of cancer (Chaudhary et al. 2018).

In addition, it has been demonstrated that tomato wastes generated during processing contain bioactive phenolic extracts with antiproliferative activity, which further affects the cell lines HeLa (cervix epithelioid carcinoma), MCF7 (breast

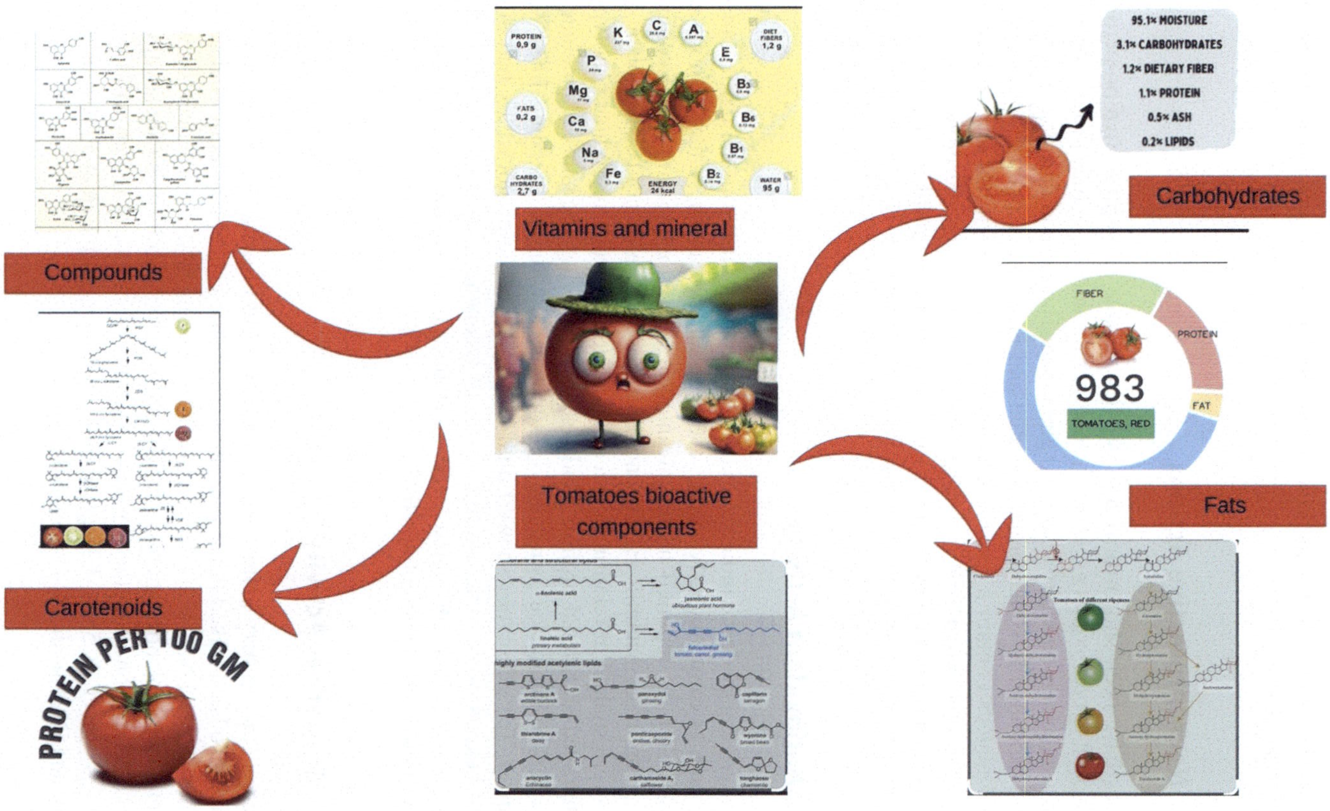

Fig. 5.2 Bioactive components present in tomatoes. These components provide us with health benefits

adenocarcinoma), and MRC-5 (fetal lungs) (Gorecka et al. 2020). The antioxidant benefits of vitamin C, which is also a source of bioactivity, are often achieved by electron donation, which shields proteins and lipid membranes from oxidative damage (Pehlivan 2017).

5.5.3 Neuroprotective Actions

Tomatoes have been shown in several studies to have neuroprotective properties. The leaves of *Lycopersicon esculentum* contain substances that may provide protection against the neurotoxicity caused by glutamate. The method depends on controlling the mitochondria's membrane potential and nicotinic receptors (Taveira et al. 2014). Tomatoes contain a substance called lycopene, which has several positive health impacts particularly on neurodegenerative illnesses. Furthermore, it acts as a safeguard against harmful materials drifting through the atmosphere. Lycopene has been shown to significantly lessen neurotoxicity caused by toxins (Paul et al. 2020).

5.5.4 Antioxidant Activity

Epidemiological evidence indicates that the presence of numerous antioxidant molecules, such as carotenoids, lycopene, ascorbic acid, vitamin E, and flavonoids, is responsible for the reported health advantages. Many tomato nutritional components—vitamin C, vitamin E, flavonoids, lycopene, carotenoids, and phenolic acids—were studied in relation to breeding initiatives meant to produce genotypes with improved nutritional profiles. Under strict monitoring, the antioxidant content of six open-pollinated cultivars and 12 sophisticated tomato breeding lines was ascertained. Ten genotypes showed high levels of total carotenoids; six had high levels of beta-carotene; nine had high levels of lycopene; 15 had high levels of flavonoids; and two had substantial levels of vitamin E among the 18 genotypes that were examined (Frusciante et al. 2007).

5.5.5 Anticancer Activity

Lycopene is the primary antioxidant in tomatoes, but oligosaccharides from tomatoes have also been linked to the prevention of gastric cancer (Aydemir et al. 2013; Bhuvaneswari and Nagini 2005). Tomato-derived nanoparticles have been shown to have anticancer potential against cervical and colorectal cell lines (Shejawal et al. 2021). Consumption of dietary tomatoes significantly reduced the growth factor-1 that causes cancer (IC50 of 3.4 μg/mL and significant inhibition at 30 μg/mL dose) (Kapoor and Dharmesh 2017) (Fig. 5.3).

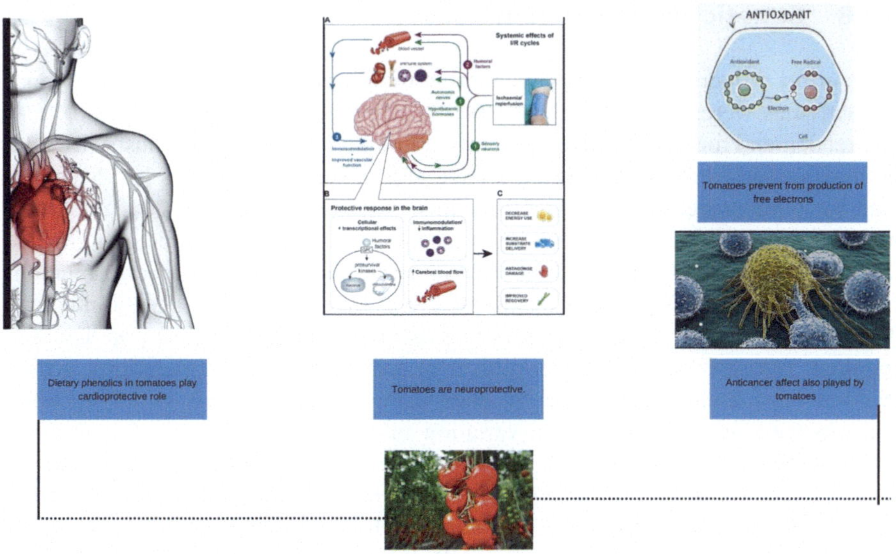

Fig. 5.3 Health benefits of tomatoes

5.5.6 Genetic Engineering

The most emphasis has been paid to the genetic engineering (GE) of higher plants since many of these crops are staple foods that are used to provide vital nutrients like carotenoids. Furthermore, compared to microbial production, they are inexpensive renewable resources that are environmentally favorable from an economic standpoint (Paine et al. 2005).

5.5.7 Current and Prospective Methods for Increasing Tomato Fruit's Carotenoid Content

It is evident that ripe tomato fruits are significant dietary sources of health-promoting carotenoids. Now consumer is more aware about nutritional aspects of food commodities and is demanding nutritionally enriched fresh produces. . The creation of fruits and vegetables with enhanced consumer health quality attributes has attracted significant industry interest in the quest for more nutrient-dense food products. Elite tomato varieties now process numerous significant features that were added by conventional plant breeding. In recent times, there has been significant progress in utilizing tomato introgression (IL) and recombinant inbred line (RIL) populations to introduce traits into peak varieties through marker-assisted selection (MAS) (Fraser et al. 2009).

5.5.8 Genetic Markers in Tomatoes

In an effort to increase tomatoes' general horticultural qualities, systematic breeding activities were initiated in the 1930s. Breeding aims grew more specialized as market demand for more particular features required by the tomato industry for both processing and fresh markets grew. By the 1950s, superior varieties had been created for both processing and fresh markets (Foolad and Panthee 2012). Genetic markers need to fulfill a new condition in order to be valuable in breedingfor the trait of interest to exist. The second step is to ascertain the genetic linkage relationship between relevant genes or quantitative trait loci (QTLs) and genetic markers. Third, for the linked genetic markers to exist, polymorphism in the population is required. Because marker polymorphism in tomato breeding populations is rare, this has been a significant concern when using molecular markers for breeding because there is so little genetic variation in the tomato cultigen (Stevens and Robbins 2007; Commemte et al. 2004; Arens et al. 2010). Owing to the minimal genetic diversity within the tomato cultigen, tomato variety creation and marker discoveries are making use of novel marker technologies that can identify minute genetic variations. For many reasons, SNPs have emerged as the preferred marker in tomato genetics and breeding programs, for instance (Yang et al. 2004; Labate and Baldo 2005).

5.5.9 Use of Marker-Assisted Selection in Tomato Breeding

Molecular markers were first proposed as an indirect selection criterion for breeding purposes in the tomato. The real application of MAS in tomato breeding dates back to 30 years, when the acid phosphatase (Aps-11 locus) isozyme marker was used as an indirect selection criterion for nematode resistance in tomato breeding. Paradoxically, tomato breeding efforts, both public and private, continue to employ this isozyme marker to select for resistance to nematodes. However, more recently, MAS has become a standard procedure in many tomato breeding programs, especially in the corporate sector, for a number of reasons, including the following three. This is due to the introduction of new molecular markers and maps in tomatoes (Foolad and Panthee 2012). First, by using a panel of molecular markers to screen seed batches, MAS is frequently used to evaluate the quality of hybrids produced overseas. The tools applied for this aim are very diverse; SNPs and PCR-based markers are frequently used, and in certain situations, even well-known isozyme markers are enlisted. Furthermore, MAS is an efficient method for rapid screening of germplasm for disease resistance or fruit quality, provided trustworthy markers are strongly connected to resistance genes or specific fruit quality loci. Breeding populations are frequently "indexed" by applying a panel of linked markers to individual selections, pools of seed, or tissues from early-generation populations.

Telling the breeder which disease resistances or fruit quality traits are fixed or segregated in a particular population helps with breeding efforts. To authenticate the linkage—or lack thereof—between markers and the trait(s) of interest, as well as to confirm the results of MAS, organism screening may frequently still be necessary. Third, MAS is used for marker-assisted backcrossing (MAB) following the discovery of trustworthy associations between markers and basic qualities of interest. These characteristics can include but are not limited to resistance to disease, fruit color, and carotenoid content (like lycopene and beta-carotene) (Brouwer and St. Clair 2004; Collard et al. 2005).

5.5.10 CRISPR/Cas9 and Genome Editing

Following the initial discoveries of CRISPR/Cas9-mediated gene editing, a swift series of publications showcased the successful editing of numerous organisms, including plants. It may be said that genes of interest caused a revolution in the study of gene function in all living things (Doudna and Charpentier 2014). Given all the resources at hand, genomic sequences for over 500 genotypes, wild species, transcriptome data, and mutant populations and a transformation methodology applied to numerous genotypes by agrobacterium, the tomato provided an excellent test model to illustrate the viability in a significant food crop (Sato and Tabata 2016; Van Eck et al. 2018). CRISPR/Cas9 has been used in numerous studies, ranging from proof-of-concept demonstrations to gene editing operations to analyze fruit biology features, change plant architecture, and explore metabolic engineering techniques. It was shown to be an effective instrument for furthering research on gene function and tomato trait modification (Ron et al. 2014; Van 2017). Apart from the discovery of transgenic lines harboring the targeted insertion/deletion mutations (indels), it was demonstrated that indels are passed down to the next generations. Furthermore, segregation revealed the existence of lines that did not have T-DNA but still carried the indels (Brooks et al. 2014; Pan et al. 2016). Beyond investigations into gene function, the capacity to precisely target a gene to modify its expression is valuable. In other cases, such as those where strong genetic ties could make trait change impossible, gene editing will be crucial in enabling breeding efforts to enhance programs, as is the case with significant food crops like tomatoes (Belhaj et al. 2013; Nekrasov et al. 2017; Brooks et al. 2014; Reem and Van 2019).

5.5.11 Enhancing Health Benefits Through Nutrigenomics

Numerous nutrients and bioactive phytochemicals have been demonstrated in basic research to function as signaling molecules. They bind to cellular sensors like transcription factors or directly to promoters that affect gene and protein expression and, in turn, metabolite production. For instance, eating a diet high in carbohydrates causes the glycolytic and lipogenesis enzymes to be upregulated. But it suppresses the expression of the enzymes involved in gluconeogenesis. Furthermore, the utilization of the genomic approach has promoted the clarification of individual genetic variations that indicate varying responses to diet and increased susceptibility to nutrition-related illnesses. People with phenylketonuria, for example, are affected by a mutation in the phenylalanine hydroxylase coding gene, so they may avoid phenylalanine containing diets(Blau et al. 2010). Furthermore, it has been demonstrated that genetic variants or polymorphisms in the genes encoding proteins or enzymes (5,10- methylenetetrahydrofolatereductase (MTHFR)) have an impact on the catalytic activity (Reddy et al. 2018), which, in turn, affects the requirements for nutrients and metabolism of each individual. While nutrigenetics examines the impact of genetic variations, such as SNPs, mutations, copy-number variation, and epigenetic modifications, on nutrition biology, nutrigenomics focuses on dietary or nutrient-induced changes in the transcriptome, proteome, and metabolome (Muller and Kersten 2003) (Table 5.1).

5.6 Today's Condition of Tomato Breeding

5.6.1 Standard Methods

The Andes area of Central America is said to be the birthplace of the domestication of cultivated tomatoes (*Solanum lycopersicum var. lycopersicum*). *S. pimpinellifolium* and *S. lycopersicum var. cerasiforme*, which are regarded as the direct progenitors of the cultivated tomato, were the two intermediary phases in this process. A key evolutionary force during domestication was the selection for larger fruit size (Lin et al. 2014). Additional features that set domesticated tomatoes apart from their wild counterparts are development factors and different fruit attributes (Bai and Lindhout 2007). It is possible that cultivated tomatoes originated in Mexico before making its way to Europe and subsequently other regions of the world (Jenkins 1948). However, genetic drift greatly decreased the genetic diversity of domesticated tomatoes when tomato varieties were selected to adapt to certain geographic locations (Bai and Lindhout 2007; Lin et al. 2014).

According to Bai and Lindhout (2007), conventional tomato breeding techniques include pedigree, hybrid, and backcross breeding, each of which focuses on a distinct feature to satisfy a range of consumer and market demands. Individual farmers or institutional breeders may have distinct objectives for their breeding operations. Cross-hybridization between inbred tomato populations is an easy and rapid way to produce genetic variants because of the decreased genetic diversity among these populations caused by the long period of selective domestication (Ranc et al. 2008). This makes it possible to choose new types with advantageous characteristics. The pedigree approach helps select hybrids with novel qualities obtained from the gene pools of the parental populations by meticulously recording the performance of many generations of hybrids from genetically diverse parents. Breeders utilize the backcross breeding strategy to transfer novel alleles from wild relatives into cultivated lines for qualities like resistance to biotic or abiotic stress, which may not be present in the cultivated gene pool.

In order to reestablish the genetic background, this technique entails repeatedly backcrossing the children with the parental cultivated lines. These conventional breeding techniques need thorough monitoring and selection over several

Table 5.1 Important components of plants that have epigenetic changes

Sr. No.	Fundamental plants	Bioactive constituent	Epigenetic roles	References
1.	Apples	Phloretin	HDAC restriction, reexpression of TSGs, and demethylation	Orlikova et al. (2012)
2.	Broccoli	Isothiocyanates	DNA methylation, p21 gene activation, chromatin remodeling, and gene reactivation (cyclin D2)	Ho et al. (2009)
3.	Cashew nuts	Anacardic acid	HATi and DNA repair	Balasubramanyam et al. (2003)
4.	Fenugreek	Rhaponticin	DNA repair and gene reactivation	Wani and Kumar (2016)
5.	Citrus	Hesperidin	DNMT inhibitor	Fang et al. (2007)
6.	Cinnamon	Coumaric acid	DNMT inhibitor increases p53 acetylation	Olaharski et al. (2005)
7.	Coffee	Caffeic acid	DNMT inhibitor	Lee and Zhu (2006)
8.	Garlic	Allyl mercaptan	Histone acetylation, HDAC inhibitor	Nian et al. (2008)
9.	Grapes	Resveratrol	A DNMT inhibitor causes gene reactivation and DNA demethylation	Howitz et al. (2003)
10.	Soyabean	Genistein	DNMT inhibitor raises acetylation at TSG and reduces HDAC activity	Bosviel et al. (2012)
11.	Tea	Epigallocatechin gallate (EGCG)	DNMT blocker	Fang et al. (2003)
12.	Tomatoes	Lycopene	Causes the reactivation of genes (GSTP1, RARbeta, and HIN-1) and DNA demethylation	King-Batoon et al. (2008)
13.	Red pepper	Capsaicin	DNMT blocker	Srinivasan (2013)
14.	Turmeric	Curcumin	DNMT blocker	Shu et al. (2011)
15.	Ginger	Gingerol	Decreases the atherogenic changes of LDL and platelet aggregation	Tchombe et al. (2012)
16.	Mango, blackberry	Gallic acid	HATi	Choi et al. (2009b)
17.	Frequently found in fruits and vegetables	Quercetin	DNMTi stimulates gene expression, demethylation, and the p16 promoter	Tan et al. (2009)
18.	Red onion with red cabbage	Cyaniding	DNMTi	Paluszczak et al. (2010)
19.	Walnuts	Myricetin	DNMTi	Paluszczak et al. (2010)

generations, which is time-consuming and labor-intensive. But because of the use of molecular markers and marker-assisted selection (MAS) in the genomics age, these techniques are much more effective (Collard and Mackill 2008). PCR or sequencing is employed to monitor the existence of QTLs in hybrid progeny by identifying DNA markers that are closely associated with the relevant QTLs. The effectiveness of traditional plant breeding has greatly increased as a result of developments in genome sequencing technologies.

5.6.2 Novel Strategies in Plant Breeding

The term "new plant breeding techniques" (NPBTs) refers to the most recent molecular methods utilized in plant breeding, especially when it comes to genome editing and genomics, such as CRISPR/Cas nucleases. NPBTs concentrate on precisely manipulating plant genomes. Molecular-level plant breeding is now much more accurate and efficient because of the introduction and broad use of CRISPR/Cas systems (Chen et al. 2019). From single base adjustments using base editors and prime editors to bigger alterations spanning several kilobases using techniques like homology-directed knock-in (HKI), the CRISPR/Cas complexes provide a variety of approaches for precise gene sequence modification (Van Vu et al. 2019).

5.6.3 CRISPR/Cas-Based Editing of Genomes

According to Barrangou et al. (2007), prokaryotes use CRISPR/Cas-based DNA interference as a defensive mechanism against viral phages. Usually, a complex formed by a single CRISPR guide RNA (sgRNA) is what activates a single-unit Cas nuclease such as SpCas9. Next, this Cas–sgRNA complex "scans" for a target double-stranded DNA with a corresponding protospacer sequence. It binds and cleaves both DNA strands when it recognizes a protospacer adjacent motif (PAM), where N might be A, T, G, or C, next to the target (Jinek et al. 2012). Because CRISPR/Cas nucleases may cause double-strand breaks (DSBs), they are frequently utilized for precise genomic alterations (Belhaj et al. 2013; Li et al. 2013; Nekrasov et al. 2013; Shan et al. 2013). To preserve genomic integrity, these DSBs trigger the cell's repair systems, which is essential because even one unrepaired DSB has the potential to be fatal.

Nonhomologous end joining (NHEJ) and the cell cycle–dependent homologous recombination (HR) pathways are the main mechanisms for double-strand break (DSB) repair (Puchta 2005; Lieber 2010; Chapman et al. 2012). DNA ligase IV is essential for NHEJ's direct ligation of the two damaged ends, which fixes double-strand breaks. Nevertheless, NHEJ can occasionally be prone to mistakes, which can result in minor alterations like insertions or deletions. The HR pathway replaces the sequences close to the breaks with homologous sequences from the donor DNA in plant somatic cells in order to repair double-strand breaks (DSBs). Sub-pathways of HR include synthesis-dependent strand annealing (SDSA) and single-strand annealing (SSA) (Puchta 2005; Van Vu et al. 2019).

New developments have brought about a variety of CRISPR/Cas systems that have the ability to modify RNA (Cox et al. 2017) or cause particular base changes by means of deaminases, resulting in transversions (Kurt et al. 2020; Zhao et al. 2020) or single base transitions (Komor et al. 2016; Nishida et al. 2016; Gaudelli et al. 2017). The prime editor is another cutting-edge method of genome editing that avoids causing double-strand breaks (DSBs). It works by using reverse transcriptase to transcribe genetic material from an extended sgRNA. By attaching to the sequence upstream of the nicked site on the nontarget strand, this priming extended gRNA (pegRNA) directs the Cas9-fused RT. Reverse transcription is then started from the 3′-OH of the nicked end using pegRNA as a template (Anzalone et al. 2019). Although the primary editor system has been effectively used in crops like wheat and rice (Lin et al. 2020), its present performance in plants indicates that there is potential for additional development and study in this field.

By 2050, current food production technologies are predicted to be insufficient to feed the world's fast-expanding population (Ray et al. 2013). Strategic breeding initiatives have been started to create sustainable agricultural techniques in response to these catastrophic forecasts, with a primary focus on producing optimal or "super" crops (Zsogon et al. 2017). The issues of feeding the world's population by 2050 may be resolved by the engineering of novel crops through re-domestication or de novo domestication of wild relatives or semidomesticated plants (Fernie and Yan 2019; Hickey et al. 2019). Plant breeding can benefit greatly from the use of CRISPR/Cas technology, especially in terms of expediting the introduction of new alleles into superior cultivars and streamlining the de novo domestication of wild plants.

The accuracy and potency of CRISPR/Cas-based technologies are critical to the success of these breeding projects. There have been reported cases of de novo domestication of orphan crops and wild tomatoes, indicating a radical change in tomato breeding (Lemmon et al. 2018; Li et al. 2018; Zsogon et al. 2018). Certain critical features, such as stress tolerance and fruit quality, have been lowered by traditional domestication and breeding techniques (Liu et al. 2020; Wang et al. 2020). De novo domestication, which involves carefully introducing certain features from domesticated cultivars into wild or semi-cultivated plants, has the potential to produce novel crops and provide other means of supplying food. This procedure is made possible by the CRISPR/Cas technology, which allows for the modification of almost any genomic location, including those found in semi-cultivated and wild relatives of plants. Thousands of semi-cultivated plant species are considered orphan crops and are used locally for food, even though just 15 plant species are now exploited for large-scale food production (Lemmon et al. 2018).

For the purpose of de novo domestication of elite features in tomatoes, *S. pimpinellifolium*, a wild cousin, has been utilized as a parent plant (Zsogon et al. 2018). Growth habit (SELF PRUNING (SP) and SP5G), fruit setting (OV, CLV3, FW2.2, MULT, and WUS), and fruit quality (CycB and SlGGP1) were among the features that were being targeted for de novo domestication. By carefully inserting knock-out mutations into the chosen genes within the *S. pimpinellifolium* genome, these features were obtained. Lemmon and colleagues employed an alternative de novo domestication technique in which they inserted selected mutations (orthologs of domesticated tomato alleles) into the genome of the orphan crop *Physalis pruinosa*, also known as the sweet ground cherry, which is native to Central and South America. The resultant plants have traits resembling those of their tomato counterparts, including determinate growth, a rise in fruit quantity, and a larger number of locules (Lemmon et al. 2018). By working together, these investigations have paved the way for the quick de novo domestication and re-domestication of plants, with the goal of breeding adaptable tomatoes that can adjust to changing environmental conditions without sacrificing or improving fruit quality or yield.

5.7 Challenges and Considerations

5.7.1 Genetic Complexity and Variation

The intricate genetic architecture of features linked to health benefits is one of the main obstacles in the field of tomato nutrigenomics. It can be challenging to identify certain genetic targets for improvement since traits like lycopene content, flavonoid levels, and antioxidant capacity are frequently regulated by several genes. Furthermore, breeding efforts are made more difficult by the limited genetic variety among cultivated tomato varieties, which forces researchers to look for wild species that possess desired features. The well-optimized genetic background of the current cultivars may be disturbed, though, if these features are introduced into cultivated types (Vu et al. 2020).

5.7.2 Precision in Genetic Modifications

Precision in the genetic alteration of tomatoes has been transformed by molecular breeding techniques like CRISPR/Cas9. However, there are obstacles involved in using these technologies. Off-target effects, or unintentional genetic modifications, can lead to undesirable traits. In addition, there is ongoing controversy around the acceptability and control of genetically modified organisms (GMOs), especially in areas with strict regulatory policies. The timely distribution of improved tomato varieties is significantly hampered by the unclear regulations around genome-edited crops (Chaudhary et al. 2019).

5.7.3 Environmental and Stress Factors

When breeding tomatoes to increase their nutritional value, environmental and stress-related aspects must also be taken into account. Abiotic (drought and temperature extremes) and biotic (diseases and pests) factors that affect tomatoes can affect the expression of genes crucial to nutrition (Chaudhary et al. 2019). It is difficult to develop cultivars that retain their nutritional advantages in these circumstances; it calls for a careful balancing act between stress resistance and nutrient increase.

5.7.4 Consumer Acceptance and Marketability

Customer acceptability is a factor in the nutrigenomics of tomatoes as well. The marketability of these goods may be impacted by public distrust regarding genetically modified foods, even if molecular breeding can increase the health advantages of tomatoes. For nutrigenomically modified tomatoes to succeed in the market, breeding practices must be transparent, and consumers must be informed of the advantages of these alterations (Ge and Zhang 2023).

5.7.5 Ethical and Legal Considerations

An important factor is also the ethical issues surrounding the alteration of food crops for improved health benefits. Concerns about food sovereignty and access are brought up by the possibility of generating genetically modified crops that are patent-protected, especially in underdeveloped nations (Ge and Zhang 2023). Laws must strike a balance between public access to nutritionally improved crops and innovation.

5.8 Upcoming Perceptions about Tomato Breeding

The main obstacle for hybrid breeding programs that utilize marker-assisted selection (MAS) in the future to battle tomato viruses is finding novel antiviral genes that can target newly developing strains of viruses. Based on the arms race pathogen–host model of coevolution, we propose a more comprehensive integration of the ecological, biogeographical, and genetic disciplines of both tomatoes and tomato viruses. This strategy will help pinpoint resistance sources and comprehend how locally adaptable tomatoes are. Understanding the heritability of antiviral properties will also need the use of monogenetic techniques in gene regulatory networks (Guevara-Escudero et al. 2021).

Breeding tactics should leverage genome-wide association studies (GWAS), genomic prediction (GP), and quantitative trait loci (QTLs) mapping to identify putative novel resistance genes. New virus resistance genes have already been successfully identified using these techniques (Qi et al. 2022). To address the problem of building broad-spectrum resistance against many viruses, researchers advise utilizing the gene pyramiding strategy through MAS to quickly integrate several resistance genes into a single plant. The main obstacles to RNAi-based transgenic breeding for virus-resistant tomatoes are the strict laws regulating genetically modified (GM) crops and public opposition. Subsequent investigations have to concentrate on developing transgenic technology in methods that tackle social issues (Sáez et al. 2020).

Although CRISPR/Cas-based genome editing has a lot of potential for creating tomato varieties resistant to viruses, there are a few issues that need to be resolved before this technique can be effectively applied to antiviral breeding. These difficulties include the creation of multi-target gene editing systems, DNA-free gene editing methods, increased gene editing efficiency, and decreased off-target effects (Lv et al. 2022).

5.9 Conclusion

Cultivated tomatoes have undergone selective breeding to maintain traits that are attractive for processing and eating. However, the hereditary variety of the contemporary, extensively grown tomato varieties remain significantly condensed as a result of domestication and subsequent breeding. Particularly in light of regional and global climate change, tomato producers must balance maintaining production with diversifying their product line to include tomatoes with more enticing insignia, tastes, or nutritional and fitness advantages. Historically, mixture passage with a contributor source—typically a desolate relative—has been necessary to introduce a useful gene into a cultivated variety. The creation of a hybrid genome containing the desired allele may be hampered by the overview of undesirable inherited material or relation drag from the giver maternal due to this process. Most undesirable features may be eliminated by repeatedly backcrossing to the elite maternal stroke and selecting the suitable allele in each group though linkage drag complicates this procedure. It frequently takes years using conventional breeding to produce a new tomato variability that is acceptable for farming, even using marker-assisted selection (MAS).

CRISPR/Cas technology has been used recently in tomato breeding to speed up "de novo domestication." By introducing elite qualities that have arisen via domestication, this method selectively introduces new traits into wild cousins such as *S. pimpinellifolium*. The resultant plants preserve important natural qualities like stress tolerance, especially those impacted by several genes or alleles, while also displaying enhanced traits like bigger fruits, quicker development, and greater quality and production. The creation of improved tomato cultivars resistant to climate change might be greatly accelerated by using this technique. Another method is toward de novo control tomato-like foundling harvests using CRISPR/Cas-based introgression of orthologous genes or alleles from farmed tomatoes, which will increase the crops' growth and yield.

It would be stimulating to relate this method toward a range of latent tomato-like orphan crops in order to increase the harvest of native gardeners. Compared to random mutagenesis techniques employing radiation or chemicals, targeted mutagenesis based on CRISPR/Cas9 is significantly more accurate. On the other hand, complex alleles requiring precise base/sequence alterations encode many important tomato traits. The availability of more precise editing methods like BE, PE, and HKI has increased with the advent of CRISPR/Cas apps. These tools are constantly being improved to become more precise and effective as tomato breeding picks up speed.

References

Anzalone AV, Randolph PB, Davis JR, Sousa AA, Koblan LW, Levy JM et al (2019) Search-and-replace genome editing without double-strand breaks or donor DNA. Nature 576:149–157. https://doi.org/10.1038/s41586-019-1711-4

Bai Y, Lindhout P (2007) Domestication and breeding of tomatoes: what have we gained and what can we gain in the future? Ann Bot 100:1085–1094. https://doi.org/10.1093/aob/mcm150

Balasubramanyam K, Swaminathan V, Ranganathan A, Kundu TK (2003) Small molecule modulators of histone acetyltransferase p300. J Biol Chem 278(21):19134–19140. https://doi.org/10.1074/jbc.M301580200

Barrangou R, Fremaux C, Deveau H, Richards M, Boyaval P, Moineau S et al (2007) CRISPR provides acquired resistance against viruses in prokaryotes. Science 315:1709–1712. https://doi.org/10.1126/science.1138140

Belhaj K, Chaparro-Garcia A, Kamoun S, Nekrasov V (2013) Plant genome editing made easy: targeted mutagenesis in model and crop plants using the CRISPR/Cas system. Plant Methods 9:39. https://doi.org/10.1186/1746-4811-9-39

Bosviel R, Dumollard E, Déchelotte P, Bignon YJ, Bernard-Gallon D (2012) Can soy phytoestrogens decrease DNA-methylation in BRCA1 and BRCA2 oncosuppressorgenes in breast cancer? Omics 16(5):235–244. https://doi.org/10.1089/omi.2011.0105

Chapman JR, Taylor MR, Boulton SJ (2012) Playing the end game: DNA double-strand break repair pathway choice. Mol Cell 47:497–510. https://doi.org/10.1016/j.molcel.2012.07.029

Chaudhary J, Alisha A, Bhatt V, Chandanshive S, Kumar N, Mir Z et al (2019) Mutation breeding in tomato: advances, applicability and challenges. Plants 8(5):128

Chen K, Wang Y, Zhang R, Zhang H, Gao C (2019) CRISPR/Cas genome editing and precision plant breeding in agriculture. Annu Rev Plant Biol 70:667–697. https://doi.org/10.1146/annurev-arplant-050718-100049

Collard BC, Mackill DJ (2008) Marker-assisted selection: an approach for precision plant breeding in the twenty-first century. Philos Trans R Soc Lond B Biol Sci 363:557–572. https://doi.org/10.1098/rstb.2007.2170

Cox DBT, Gootenberg JS, Abudayyeh OO, Franklin B, Kellner MJ, Joung J et al (2017) RNA editing with CRISPR-Cas13. Science 358:1019–1027. https://doi.org/10.1126/science.aaq0180

Fang MZ, Wang Y, Ai N, Hou Z, Sun Y, Lu H et al (2003) Tea polyphenol (−)-epigallocatechin-3-gallate inhibits DNA methyltransferase and reactivates methylation-silenced genes in cancer cell lines. Cancer Res 63(22):7563–7570

Fang M, Chen D, Yang CS (2007) Dietary polyphenols may affect DNA methylation. J Nutr 137(1 Suppl):223S–228S. https://doi.org/10.1093/jn/137.1.223S

Fernie AR, Yan J (2019) De novo domestication: an alternative route toward new crops for the future. Mol Plant 12:615–631. https://doi.org/10.1016/j.molp.2019.03.016

Ge P, Zhang Y (2023) Genes, genetics and breeding of tomato. Horticulturae 9(11):1208

Guevara-Escudero M, Osorio AN, Cortés AJ (2021) Integrative pre-breeding for biotic resistance in forest trees. Plants 10:2022

Hickey LT, Robinson H, Jackson SA, Leal-Bertioli SCM, Tester M (2019) Breeding crops to feed 10 billion. Nat Biotechnol 37:744–754. https://doi.org/10.1038/s41587-019-0152-9

Ho E, Clarke JD, Dashwood RH (2009) Dietary sulforaphane, a histone deacetylase inhibitor for cancer prevention. J Nutr 139(12):2393–2396. https://doi.org/10.3945/jn.109.113332

Howitz KT, Bitterman KJ, Cohen HY, Lamming DW, Lavu S, Wood JG et al (2003) Small molecule activators of sirtuins extend *Saccharomyces cerevisiae* lifespan. Nature 425(6954):191–196. https://doi.org/10.1038/nature01960

Jenkins JA (1948) The origin of the cultivated tomato. Econ Bot 2:379–392. https://doi.org/10.1007/BF02859492

Jinek M, Chylinski K, Fonfara I, Hauer M, Doudna JA, Charpentier E (2012) A programmable dual-RNA-guided DNA endonuclease in adaptive bacterial immunity. Science 337:816–821. https://doi.org/10.1126/science.1225829

King-Batoon A, Leszczynska JM, Klein CB (2008) Modulation of gene methylation by genistein or lycopene in breast cancer cells. Environ Mol Mutagen 49(1):36–45. https://doi.org/10.1002/em.20363

Komor AC, Kim YB, Packer MS, Zuris JA, Liu DR (2016) Programmable editing of a target base in genomic DNA without doublestranded DNA cleavage. Nature 533:420–424. https://doi.org/10.1038/nature17946

Kurt IC, Zhou R, Iyer S, Garcia SP, Miller BR, Langner LM et al (2020) CRISPR C-to-G base editors for inducing targeted DNA transversions in human cells. Nat Biotechnol 39:41. https://doi.org/10.1038/s41587-020-0609-x

Lee WJ, Zhu BT (2006) Inhibition of DNA methylation by caffeic acid and chlorogenic acid, two common catechol-containing coffee polyphenols. Carcinogenesis 27:269–277

Lemmon ZH, Reem NT, Dalrymple J, Soyk S, Swartwood KE, Rodriguez-Leal D et al (2018) Rapid improvement of domestication traits in an orphan crop by genome editing. Nat Plants 4:766–770. https://doi.org/10.1038/s41477-018-0259-x

Lieber MR (2010) The mechanism of double-strand DNA break repair by the nonhomologous DNA end-joining pathway. Annu Rev Biochem 79:181–211. https://doi.org/10.1146/annurev.biochem.052308.093131

Lin T, Zhu G, Zhang J, Xu X, Yu Q, Zheng Z et al (2014) Genomic analyses provide insights into the history of tomato breeding. Nat Genet 46:1220–1226. https://doi.org/10.1038/ng.3117

Lv J, Deng M, Jiang S, Zhu H, Li Z, Wang Z, Li J, Yang Z, Yue Y, Xu J et al (2022) Mapping and functional characterization of the tomato spotted wilt virus resistance gene SlCHS3 in Solanum lycopersicum. Mol Breed 42:55

Nekrasov V, Staskawicz B, Weigel D, Jones JD, Kamoun S (2013) Targeted mutagenesis in the model plant Nicotiana benthamiana using Cas9 RNA-guided endonuclease. Nat Biotechnol 31:691–693. https://doi.org/10.1038/nbt.2655

Nian H, Delage B, Pinto JT, Dashwood RH (2008) Allyl mercaptan, a garlic-derived organosulfur compound, inhibits histone deacetylase and enhances Sp3 binding on the P21WAF1 promoter. Carcinogenesis 29(9):1816–1824. https://doi.org/10.1093/carcin/bgn165

Olaharski AJ, Rine J, Marshall BL et al (2005) The flavouring agent dihydrocoumarin reverses epigenetic silencing and inhibits sirtuin deacetylases. PLoS Genet 1(6):e77

Orlikova B, Schnekenburger M, Zloh M, Golais F, Diederich M, Tasdemir D (2012) Natural chalcones as dual inhibitors of HDACs and NF-kappaB. Oncol Rep 28(3):797–805. https://doi.org/10.3892/or.2012.1870

Paluszczak J, Krajka-Kuzniak V, Baer-Dubowska W (2010) The effect of dietary polyphenols on the epigenetic regulation of gene expression in MCF7 breast cancer cells. Toxicol Lett 192(2):119–125. https://doi.org/10.1016/j.toxlet.2009.10.010

Puchta H (2005) The repair of double-strand breaks in plants: mechanisms and consequences for genome evolution. J Exp Bot 56:1–14. https://doi.org/10.1093/jxb/eri025

Qi S, Shen Y, Wang X, Zhang S, Li Y, Islam MM, Wang J, Zhao P, Zhan X, Zhang F et al (2022) A new NLR gene for resistance to Tomato spotted wilt virus in tomato (Solanum lycopersicum). Theor Appl Genet 135:1493–1509

Ranc N, Munos S, Santoni S, Causse M (2008) A clarified position for Solanum lycopersicum var. cerasiforme in the evolutionary history of tomatoes (solanaceae). BMC Plant Biol 8:130. https://doi.org/10.1186/1471-2229-8-130

Ray DK, Mueller ND, West PC, Foley JA (2013) Yield trends are insufficient to double global crop production by 2050. PLoS One 8:e66428. https://doi.org/10.1371/journal.pone.0066428

Sáez C, Martínez C, Montero-Pau J, Esteras C, Sifres A, Blanca J, Ferriol M, López C, Picó B (2020) A major QTL located in chromosome 8 of Cucurbita moschata is responsible for resistance to tomato leaf curl New Delhi virus. Front Plant Sci 11:207

Shan Q, Wang Y, Li J, Zhang Y, Chen K, Liang Z et al (2013) Targeted genome modification of crop plants using a CRISPR-Cas system. Nat Biotechnol 31:686–688. https://doi.org/10.1038/nbt.2650

Shu L, Khor TO, Lee JH, Boyanapalli SS, Huang Y, Wu TY et al (2011) Epigenetic CpG demethylation of the promoter and reactivation of the expression of Neurog1 by curcumin in prostate LNCaP cells. AAPS J 13(4):606–614. https://doi.org/10.1208/s12248-011-9300-y

Srinivasan K (2013) Dietary spices as beneficial modulators of lipid profile in condition of metabolic disorders and disease. Food Funct 4(4):503–521. https://doi.org/10.1039/c2fo30249g

Tan S, Wang C, Lu C, Zhao B, Cui Y, Shi X et al (2009) Quercetin is able to demethylate the p16INK4a gene promoter. Chemotherapy 55(1):6–10. https://doi.org/10.1159/000166383

Tchombe NL, Louajri A, Benajiba MN (2012) Therapeutical effects of ginger. ISESCO J Sci Technol 8:64–69

Van Vu T, Sung YW, Kim J, Doan DTH, Tran MT, Kim JY (2019) Challenges and perspectives in homology-directed gene targeting in monocot plants. Rice 12:95. https://doi.org/10.1186/s12284-019-0355-1

Vu TV, Das S, Tran MT, Hong JC, Kim JY (2020) Precision genome engineering for the breeding of tomatoes: recent progress and future perspectives. Front Genome Ed 2:612137

Zhao D, Li J, Li S, Xin X, Hu M, Price MA et al (2020) Glycosylase base editors enable C-to-A and C-to-G base changes. Nat Biotechnol 39:35. https://doi.org/10.1038/s41587-020-0592-2

Zsogon A, Cermak T, Voytas D, Peres LE (2017) Genome editing as a tool to achieve the crop ideotype and de novo domestication of wild relatives: case study in tomato. Plant Sci 256:120–130. https://doi.org/10.1016/j.plantsci.2016.12.012

Unraveling the Transcriptome: Transcriptomics for Understanding Tomato Development and Traits

6

Muhammad Saeed, Faiz Ahmad Joyia, Ghulam Mustafa, Aneela Shadab, Ehtisham Hussain, Muhammad Asif Raheem, and Aqsa Parvaiz

Abstract

Transcriptomics has emerged as a powerful tool for dissecting the molecular mechanisms underlying tomato development and trait formation. In this chapter, we delve into the application of transcriptomics to unravel the complexities of tomato biology. We explore how gene expression profiling has provided insights into the regulatory networks orchestrating key developmental processes, such as fruit ripening, flower development, and response to environmental stimuli. By integrating high-throughput sequencing technologies with bioinformatic tools, we highlight the identification of candidate genes and pathways governing important agronomic traits in tomatoes, including fruit quality, disease resistance, and stress tolerance. The transcriptome analysis not only aids in understanding the genetic basis of these traits but also offers potential targets for crop improvement through breeding or genetic engineering. Furthermore, we discuss the challenges and future directions of transcriptomics in tomato research, emphasizing the need for comprehensive datasets and advanced analytical methods to fully exploit the potential of transcriptome data. Ultimately, this chapter underscores the pivotal role of transcriptomics in advancing our knowledge of tomato biology and enhancing the sustainability and productivity of tomato agriculture.

M. Saeed · F. A. Joyia · G. Mustafa · E. Hussain · M. A. Raheem
Centre of Agricultural Biochemistry and Biotechnology (CABB), University of Agriculture, Faisalabad, Pakistan

A. Shadab · A. Parvaiz (✉)
Department of Biochemistry and Biotechnology, The Women University Multan, Multan, Pakistan
e-mail: aqsa.6497@wum.edu.pk

M. Waseem et al. (eds.), *Omics Approaches for Tomato Yield and Quality Trait Improvement*, https://doi.org/10.1007/978-981-96-3890-1_6

Keywords

Transcriptome · Tomato · Signal transduction · Abiotic stress · CRISPR

6.1 Introduction

The production of tomatoes is made more difficult by the fact that they are grown in a variety of climate zones, despite the fact that tomatoes are crops that thrive in temperate regions. Changing climatic conditions and abiotic stressors, such as drought, salt, and heat, can have a significant impact on the productivity and yield of crops. Abiotic stresses grow more complicated when they are subjected to field settings, which often include the coexistence of many stressors (Barone et al. 2009). For this reason, the creation of sustainable, high-yielding cultivars that have better tolerance to a variety of abiotic stressors is a necessity for producing enough food to satisfy the need for food on a worldwide scale. Despite the fact that several attempts have been made to address specific stress features under controlled settings, this strategy is not always viable. This is due to the fact that the plant response is different in the field, where various variables and stressors are applied concurrently. Conventional breeding has resulted in considerable advancements for a wide range of characteristics throughout the course of the past 10 years. These characteristics include resistance to biotic and abiotic stress, yield components, and quality-related characteristics. Nevertheless, traditional types are vulnerable to a number of pressures that may be found in a variety of settings. Therefore, taking into consideration the genetic complexity and the interactions between the environment, the implementation of methodologies that are more extensive and interdisciplinary gives a better strategy to increase the stress tolerance of current crops (Hajheidari et al. 2007).

6.2 Overview of Transcriptomics and Its Significance in Plant Research

All living creatures are dependent on the availability of water, and it is essential in many regions of the earth. Water is of crucial significance to all living beings. Taking into account the anticipated increase in the global population as well as the rising need for food, the situation is becoming direr. In light of this, it is essential for any industrial system, including agriculture, to make efficient use of this resource. A significant goal of breeding programs all over the globe is to enhance the performance of plants when they are subjected to drought stress. In order to accomplish this objective, it is not a straightforward operation since conventional plant breeding requires a large number of selection cycles that include a variety of experiments conducted in a variety of locales and over a period of many years (Wang et al. 2004). In addition, a summary of transgeny and patents pertaining to drought-resistant plants is included below. In addition, this study offers breeders a foundation around which they may build their breeding programs as well as conduct more research and make decisions (Mochida and Shinozaki 2011).

6.3 Importance of Tomatoes in Agriculture and as a Model System for Study

In comparison to its ancestors, which were found in the wild and were not edible, the tomato that is produced today and is eaten by a large number of people has accomplished a great deal. The differences in the availability of desired foods, such as tomatoes, for human consumption may be attributed to variations in breeding tactics. On the other hand, the shelf life of tomatoes, like that of other horticultural products, is rather short. This short shelf life leads to losses that may reach approximately 50% of the produce (Ranjan et al. 2012). This is especially true in underdeveloped nations as opposed to countries that have more modern technology and stronger infrastructure. Food security worries are quite serious, particularly when one considers the fact that the population boom that is expected to occur by the year 2050 would need an increase in food production as well as the development of food that is more nutrient-dense. This is something that applies to the tomato crop just as much as it does to the other crops. The consumer of today is more knowledgeable and is seeking for foods that are high in nutrients in order to live a long and healthy life. Up until quite recently, there was not much effort made to manufacture food, including fruits and vegetables, that was nutritionally enriched. On top of that, harsh conditions contribute to plant stress, which, in turn, has an effect on the output and nutritional quality of products (Roychowdhury et al. 2023).

Major tomato-cultivating nations that are responsible for 61% of the world's tomato resources are China (33%), India (11%), Turkey (7%), the United States (6%), and Egypt (4%). Iran, Italy, Spain, Mexico, Brazil, Nigeria, and Russia are some of the other nations that produce a significant amount of goods. Although the total tomato output in the globe reached 182.3 million tonnes from 4.8 million hectares in 2017, the worldwide export of fresh and prepared tomatoes earned profits of $12.8 billion in the United States in 2016. There were three nations that were the leading exporters of tomatoes: Mexico, the Netherlands, and Spain. Their respective export shares were 22, 13, and 12% (Fig. 6.1).

The United States, Morocco, Turkey, Jordan, India, France, and Belgium are among the other main nations that export fresh tomatoes. Not only is the United States placed among the top nations that import fresh tomatoes, but it is also ranked among the top countries that export tomatoes, making it the tenth-most important country in this category. According to the trade data for 2016, Norway, Finland, Switzerland, Ireland, Sweden, Germany, Denmark, Austria, and Canada were among the top ten nations that imported fresh tomatoes. The price of imported fresh tomatoes ranged from $2475 to $1611 per tonne for these countries. Because of this, the tomato is a crop that is commercially significant all over the globe. As a result of its demand, there is a need for the development of novel cultivars that are high-yielding and have improved nutritional quality (Mousavi and Hotta 2005).

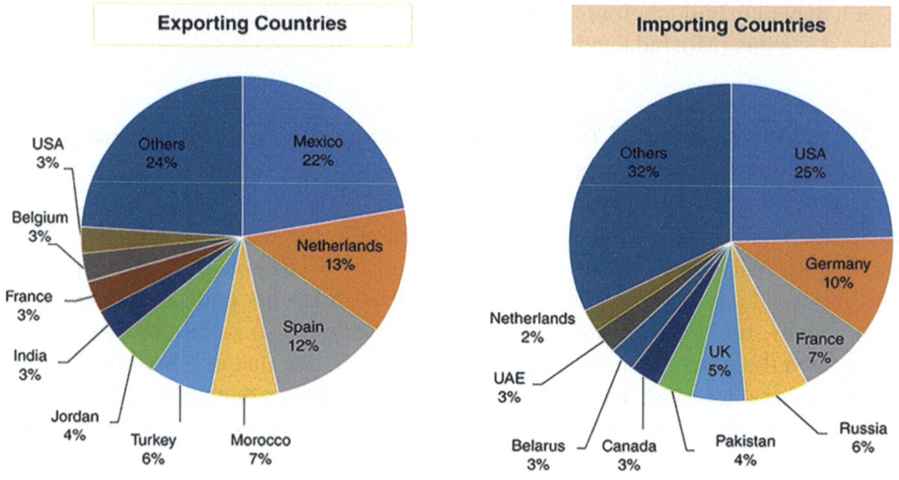

Fig. 6.1 Export rate of tomatoes

6.4 Objectives of Applying Transcriptomics in Tomato Research

With the help of transcriptomics, researchers in the tomato industry want to get a thorough understanding of the molecular mechanisms that underlie important developmental processes and features, which will ultimately lead to improvements in agricultural techniques (Anwar et al. 2019). In addition, this research makes a contribution to the study of gene function, the field of evolutionary biology, and the development of significant public databases and resources (Rai et al. 2021). After everything is said and done, the goals of using transcriptomics in tomato research focus on expanding basic understanding and pushing practical improvements in order to increase the sustainability and productivity of tomato agriculture (Møller et al. 2007).

6.5 Transcriptomics Techniques

This is an essential component of crop management and pest control. Higher-resolution insights into the ecology of the rhizosphere may be obtained by the use of other omics methods, such as metagenomics and metatranscriptomics. It is possible that they will bring to light important elements that influence the biodiversity or activities that occur below the earth, hence influencing the expression of stressful pathways. A relatively new area of application, the application of microRNAs and other short RNAs, has immense promise for the selective activation of defense pathways. This field of application is quite new. Nevertheless, there are constraints that need to be addressed, including those pertaining to the stability of the RNA molecules and of their efficient transport (Ambrosino et al. 2020).

6.6 High-Throughput Sequencing Technologies

Between 2011 and 2020, 45 new tomato virus species were identified. The bulk of these findings occurred during the past few years in nations located in the Neotropics and Palearctic. Among them, 14 were found by high-throughput sequencing (HTS). In 2020 alone, three investigations utilizing HTS—Ciuffo. In particular, HTS was used to uncover 7 out of 13 RNA viruses, whereas a non-HTS technique was used to discover 24 out of 32 DNA viruses, including species of satellite viruses. In general, methods that do not rely on HTS are still preferred for finding viruses in certain plant species, but HTS has evolved into a tool for finding viruses in a variety of species, even when no previous information on such viruses is known. This section provides a comprehensive background on the impact of High-Throughput Sequencing (HTS) as a technique in tomato viromics. It begins with the identification of viruses in individual plants, extends to the post-discovery characterization of novel viruses, and culminates in the holistic study of plant viromes within agroecosystems (Ahammed et al., 2013). Additionally, it highlights the role of HTS in tomato virus discovery and generic detection (El-Esawi et al., 2020)...

In the past several years, HTS has identified 14 novel tomato viruses or viral satellites. Sequences submitted in GenBank (accession number MK517476) demonstrated that tomatoes cultivated in Italy tested positive for Tomato matilda virus (TMaV) (*Iflaviridae*), a family of viruses first identified in Australia using high-throughput sequencing (HTS). To this day, we still do not know how ToFBV and TMaV affect tomato output, how infectious they are, or any other relevant biological factors (Møller et al. 2007).

It was also possible to employ HTS to identify previously identified viruses in tomatoes, in addition to discovering novel viruses. The first reports of five known viruses in tomatoes came from China: potato virus A, *Betaflexiviridae*, turnip yellows, potato virus S, and tobacco vein banding mosaic (*Potyviridae*), all of which are found in potatoes (Miller et al. 2010). Tomatoes from Slovenia were the first to be found using HTS; this particular virus, henbane mosaic virus (HMV), was initially identified in henbane in 1932 in England and was thereafter seldom seen. Since only partial genomic sequences of the virus were previously accessible, HTS was also used to reconstruct whole genomic sequences of the tomato isolate and historical isolates from viral collections in the same research (Mubarok et al. 2023).

Research in the fields of plant sciences, biomedical sciences, veterinary medicine, and pharmacology may all benefit greatly from the use of DNA microarrays, which are very effective techniques. The detection of microorganisms, clinical diagnosis, gene expression analysis, host–pathogen interactions, food safety testing, and environmental monitoring are all possible with this system. It is also capable of handling enormous volumes of samples. Due to the practical problems involved, microbial ecology presents a challenge when it comes to detecting and estimating the populations of microorganisms. When it comes to identifying bacteria and yeasts at the species level, as well as recognizing many species at the same time, conventional approaches such as dilution plating and biochemical tests are inadequate. Not only are these methods inaccurate and time-consuming, but they are also incapable of identifying species that cannot be cultivated.

DNA microarrays have been employed since 1995 in several scientific areas. These strategies are very effective and beneficial for conducting high-throughput experiments in parallel within the field of molecular biological research (Rinaldi et al. 2007). One of the approaches that may be used is the CustomArray®, which was created by CombiMatrix. This particular method has been shown to be very suited for the simultaneous and multiple detection of ten important tomato viruses. This demonstrates that the method is suitable for use in routine diagnostic examinations.

6.7 Transcriptomics and Stress Tolerance in Tomatoes

The introduction of more sophisticated technology has led to the development of a number of multidimensional omics disciplines. These disciplines include genomics, transcriptomics, proteomics, and metabolomics, among others. The breakthroughs that have been made in genomic research have been significant in the identification of several gene families and pathways that influence tolerance to abiotic stresses and, as a result, increase yield (Wang et al. 2022).

6.8 Major Abiotic Stress in Tomatoes

Variations in temperature, osmotic pressure, drought, salt, waterlogging, bruising, ozone, poisonous chemicals, excessive light, and ultraviolet radiation are all examples of abiotic stress, which is a wide word that embraces a range of circumstances. As a consequence of these major concerns, a chain reaction of morphological, physiological, molecular, and biochemical alterations is triggered, which ultimately leads to the destruction of cells and the suppression of glucose metabolism in general. A variety of environmental stressors, which function as ecological barriers and shake cellular integrity, are included in this category of environmental stressors. Their presence has a detrimental effect on the growth of plants and the production of agricultural goods. Because of the complexity of stress and the genesis of stress, it is uncommon to detect specific abiotic stress; thus, in response to a wide variety of stressors, a large number of stress-responsive components and pathways are widespread (Wang et al. 2022).

6.9 Physiological Transformation Due to Abiotic Stress in Tomatoes

Plants go through several stages of stress, including alarm, acclimatization, resistance, weariness, and recovery. These stages are not mutually exclusive. A synopsis of the many proteome responses that are seen in each phase is presented below. In spite of this, it is possible to see other physiological, cytosolic, and molecular reactions, which will be discussed in further detail in the following sections.

6.10 Genome Sequencing of Tomatoes

Tomatoes grown in open fields are susceptible to a variety of abiotic stressors, which reduce crop output and quality, which, in turn, leads to reduced prices and an inability to meet market demand. Farmers often grow tomato plants in greenhouses to avoid the problems caused by abiotic stress, such as heat and flooding, in open fields. In addition to increasing production costs, growing crops in greenhouses has other negative effects, such as the quick accumulation of salt, NO_3, and PO_4 in the soil, which leads to soil deterioration and surface and groundwater pollution. Hence, it is more sustainable and economically beneficial to increase tomato cultivars' tolerance to abiotic stress (Kido et al. 2016). The primary factors that hinder tomato growth include abiotic stress conditions caused by extremes of water and temperature, an uneven distribution of nutrients in the soil, elemental toxicity, and excessive salt. Abiotic pressures become more complicated in agricultural settings due to the coexistence of several stressors (Aslam et al. 2022).

To address both biological and environmental stresses, this investigation examines the roles of SDG33 and SDG34, two histone H3 lysine methyltransferases, found in tomatoes (Suzuki et al. 2012). The SDG33 and SDG34 mutants changed the H3K36 and H3K4 methylations, the gene expression involved in many processes, and responses to both biotic and abiotic stimuli. While single mutants retained drought resistance, the double mutant showed improved tolerance, in line with distinct and supplementary roles. After the drought ended, mutants were better able to heal and survive because they kept their water status high. Particularly, mutants are better able to endure stress because stressed plants produce negative regulators and reduce trimethylation in H3K4 and H3K36. A first-of-its-kind genome-wide analysis identified 19 PDI genes in tomatoes. These genes were found to be unevenly distributed across eight of the 12 chromosomes in tomatoes, and three pairs of paralogous genes had segmental duplications. Expression profiling studies found that most PDI genes exhibited tissue-specific and development-specific expression patterns. Furthermore, treatments with high temperatures, salt, and abscisic acid robustly activated most PDI genes, while freezing and stressors, such as nutrition and water deficiencies, stimulated only a limited subset of these genes. The PDI gene family, which includes SlPDI1-1, SlPDI1-3, SlPDI1-4, SlPDI2-1, SlPDI4-1, and SlPDI5-1, seems to have a role in the management of abiotic challenges in tomatoes, as shown by their dominant expression in response to ABA treatment and abiotic stresses (Gerszberg and Hnatuszko-Konka 2017).

6.11 Genomic Approaches to Combat Abiotic Stress

Plants adapt to high temperatures and sustain life through several processes, such as transcription; translation; control of calcium, phytohormone, and sugar; and lipid signaling, as well as primary and secondary metabolism. Various nonliving factors, such as exposure to high temperatures during the growth of buds, may affect the positioning and development of the male (anther cone) and female (style) reproductive organs in tomato flowers. This phenomenon, known as stigma exertion, has been studied. Heat stress resistance is influenced by genetic diversity. The researchers showed that *SlGRAS10*, a stress-tolerant transcription factor, plays a key role in a specific kind of abiotic stress tolerance by enhancing the tomato plant's osmotic potential, flavonoid biosynthesis, and ROS scavenging mechanism.

6.12 The Fruit Proteome of Tomatoes

Much of what we know about the anatomy and biology of *Solanaceae* species comes from proteomics research. Most research on the *Solanaceae* family has focused on three members—tomato, potato, and tobacco—because of their economic importance.

A study of the cherry tomato pericarp proteome found an upregulation of proteins involved in the synthesis of proteins or the degradation of amino acids as the cells divide. Fruit development was associated with an increase in proteins involved in glucose metabolism and oxidative processes, in contrast to the transient expression of proteins involved in photosynthesis and cytoskeleton creation during the cell growth stage (García-Sánchez et al. 2021).

6.13 Metabolomic Applications in Plant Stress Responses

Primary metabolites play a crucial role in the growth and development of plants and are highly consistent in their molecular structures and amounts throughout the plant kingdom (Li et al. 2019).

6.14 Metabolomics of Plant–Microbe Interactions

This process also activates pathways connected to defense mechanisms and increases the production of different phenylpropanoid precursors (Caicedo and Peralta 2013).

6.15 Conclusions and Future Recommendations

The cultivation of tomatoes is widespread around the globe, and it holds the third position in terms of global production, behind potatoes and sweet potatoes. Additionally, tomatoes top the list of canned vegetables. The plant is developed specifically for its edible fruits, which may be consumed either in their raw state or after undergoing various processing methods. It provides a rich supply of vitamins A, B, and C, along with minerals and the antioxidant lycopene. The worldwide demand for tomatoes is strong, but the poor production is mostly caused by restrictions such as biotic and abiotic factors acting alone or in combination. The most effective solutions to overcoming these restrictions are the approaches linked with omics. Various abiotic stressors regularly impact the development and production of crops. Various omics-based methodologies, including transcriptomics, proteomics, metabolomics, and others, have been effectively used either alone or in conjunction to comprehend how plants respond to stress. When stress circumstances begin, plants adapt by altering their metabolic pathways and genetic controls, resulting in the activation of new genes. Therefore, it is crucial to clarify the gene and its function in order to uncover the plant's reaction to stress. This achievement is solely attributable to the advancements in biotechnology and the use of omics tools and methodologies. The tomato is a very significant agricultural crop on a worldwide scale. This review presents a variety of omics research and integrated management techniques, which provide insights into the current state and prospective future approaches for effectively managing abiotic stress in tomatoes. By using the aforementioned technologies in tomato cultivation, we anticipate that agriculture will undergo a second Green Revolution. This revolution has the potential to satisfy the increasing food requirements of growing populations and provide nutritional security. Furthermore, it can significantly enhance crop output to fulfill the demand for food.

References

Ahammed GJ, Zhou YH, Xia XJ, Mao WH, Shi K, Yu JQ (2013) Brassinosteroid regulates secondary metabolism in tomato towards enhanced tolerance to phenanthrene. Biol Plant 57:154 158

Ambrosino L, Colantuono C, Diretto G, Fiore A, Chiusano ML (2020) Bioinformatics resources for plant abiotic stress responses: state of the art and opportunities in the fast evolving-omics era. Plan Theory 9(5):591

Anwar R, Fatima T, Mattoo AK (2019) Tomatoes: a model crop of solanaceous plants. In: Oxford research encyclopedia of environmental science. Oxford University Press, pp 1–50

Aslam HMU, Riaz H, Killiny N, Zhou XG, Thomashow LS, Peters NT, Chanda AK (2022) Microarray technology for detection of plant diseases. In: Trends in plant disease assessment, pp 203–223

Barone A, Matteo AD, Carputo D, Frusciante L (2009) High-throughput genomics enhances tomato breeding efficiency. Curr Genomics 10(1):1–9

Caicedo A, Peralta I (2013) Basic information about tomatoes and the tomato group. In: Liedl BE, Labate JA, Stommel JR, Slade A, Kole C (eds) Genetics, genomics and breeding of tomato, vol 40. CRC Press, Boca Raton, pp 1–36

El-Esawi MA, Elkelish A, Soliman M, Elansary HO, Zaid A, Wani SH (2020) Serratia marcescens BM1 enhances cadmium stress tolerance and phytoremediation potential of soybean through modulation of Osmolytes, leaf gas exchange, antioxidant machinery, and stress-responsive genes expression. Antioxidants 9:43

García-Sánchez S, Gala M, Žoldák G (2021) Nanoimpact in plants: lessons from the transcriptome. Plan Theory 10(4):751

Gerszberg A, Hnatuszko-Konka K (2017) Tomato tolerance to abiotic stress: a review of most often engineered target sequences. Plant Growth Regul 83:175–198

Hajheidari M, Eivazi A, Buchanan BB, Wong JH, Majidi I, Salekdeh GH (2007) Proteomics uncovers a role for redox in drought tolerance in wheat. J Proteome Res 6:1451–1460

Kido ÉA, Ferreira-Neto JRC, Pandolfi V, de Melo Souza AC, Benko-Iseppon AM (2016) Drought stress tolerance in plants: insights from transcriptomic studies. In: Hossain M, Wani S, Bhattacharjee S, Burritt D, Tran LS (eds) Drought stress tolerance in plants, vol 2: molecular and genetic perspectives. Springer, Cham, pp 153–185

Li R, Liu C, Zhao R, Wang L, Chen L, Yu W, Zhang S, Sheng J, Shen L (2019) CRISPR/Cas9-Mediated SlNPR1 mutagenesis reduces tomato plant drought tolerance. BMC Plant Biol 19:38

Miller G, Suzuki N, Ciftci-Yilmaz S, Mittler R (2010) Reactive oxygen species homeostasis and signalling during drought and salinity stresses. Plant Cell Environ 33:453–467

Mochida K, Shinozaki K (2011) Advances in omics and bioinformatics tools for systems analyses of plant functions. Plant Cell Physiol 52(12):2017–2038

Møller IM, Jensen PE, Hansson A (2007) Oxidative modifications to cellular components in plants. Annu Rev Plant Biol 58:459–481

Mousavi A, Hotta Y (2005) Glycine-Rich proteins: a class of novel proteins. Appl Biochem Biotechnol 120:169–174

Mubarok S, Qonit MAH, Rahmat BPN, Budiarto R, Suminar E, Nuraini A (2023) An overview of ethylene insensitive tomato mutants: advantages and disadvantages for postharvest fruit shelf-life and future perspective. Front Plant Sci 14:1079052

Rai G, Parveen A, Jamwal G, Basu U, Kumar R, Rai P, Sharma J, Alalawy A, Al-Duais M, Hossain M et al (2021) Leaf proteome response to drought stress and antioxidant potential in tomato (Solanum lycopersicum L.). Atmos 12:1021

Ranjan A, Ichihashi Y, Sinha NR (2012) The tomato genome: implications for plant breeding, genomics and evolution. Genome Biol 13:167

Rinaldi M, Ventrella D, Gagliano C (2007) Comparison of nitrogen and irrigation strategies in tomato using CROPGRO model. A case study from Southern Italy. Agric Water Manag 87(1):91–105

Roychowdhury R, Das SP, Gupta A, Parihar P, Chandrasekhar K, Sarker U et al (2023) Multi-omics pipeline and omics-integration approach to decipher plant's abiotic stress tolerance responses. Genes 14(6):1281

Suzuki N, Koussevitzky S, Mittler R, Miller G (2012) ROS and redox signalling in the response of plants to abiotic stress. Plant Cell Environ 35:259–270

Wang W, Vinocur B, Shoseyov O, Altman A (2004) Role of plant heat-shock proteins and molecular chaperones in the abiotic stress response. Trends Plant Sci 9:244–252

Wang Z-Y, Zhao S, Liu J-F, Zhao H-Y, Sun X-Y, Wu T-R, Pei T, Wang Y, Liu Q-F, Yang H-H et al (2022) Genome-wide identification of Tomato Golden 2-Like transcription factors and abiotic stress related members screening. BMC Plant Biol 22:82

Postharvest Metabolomics: Prolonging Shelf Life and Quality of Tomatoes

7

Nida Firdous, Umar Farooq, Aliza Batool, Areeba Aslam, Aliza Erum, Shabbir Ahmad, Muhammad Sibt-e-Abbas, Muhammad Usman, and Salman Khalil

Abstract

Postharvest metabolomics is a complex process and rapidly growing field in which metabolomic techniques are used to identify and understand the biochemical changes that occurred during the postharvest period of crops. Understanding postharvest metabolomics helps enhance the shelf life, develop targeted strategies for enhancing quality, decrease food waste, and enhance the sustainability of crops. This chapter illustrates postharvest metabolomic techniques for prolonging the shelf life and quality of tomatoes. Tomatoes are economically and nutritionally vital crops, which are widely used in many forms worldwide. Due to the perishable nature of tomatoes, their shelf life is low, and this results in postharvest losses. Tomatoes' shelf life is limited due to the metabolomic changes that occurred after harvesting. This chapter also focuses on postharvest treatments related to postharvest metabolomics for maintaining the shelf life and quality of tomatoes such as modified atmospheric packaging, edible coatings, ethylene management, and disease and pest control. These techniques help in enhancing the sensory attributes and shelf life of tomatoes. This chapter also focuses on utilizing techniques like mass spectrometry (MS) and nuclear magnetic resonance (NMR) spectroscopy to profile and quantify metabolites to use them as a potent tool for extending the shelf life and quality of tomatoes.

N. Firdous (✉) · U. Farooq · A. Batool · A. Aslam · A. Erum · S. Ahmad · M. Sibt-e-Abbas · M. Usman · S. Khalil
Faculty of Food and Home Sciences, Muhammad Nawaz Shareef University of Agriculture, Multan, Pakistan
e-mail: nida.firdous@mnsuam.eud.pk

M. Waseem et al. (eds.), *Omics Approaches for Tomato Yield and Quality Trait Improvement*, https://doi.org/10.1007/978-981-96-3890-1_7

Keywords

Metabolomics · Postharvest changes · Food waste · Tomato quality · Mass spectrometry · Nuclear magnetic resonance

7.1 Introduction

Tomatoes (*Lycopersicon esculentum*) are valuable products of agriculture that are consumed all over the world. Tomatoes are versatile foods that may be eaten raw or cooked and still provide significant nutritional benefits to a diet. Tomatoes farmed commercially account for more than 80% of the tomatoes used in processed forms including ketchup, soup, and juice. Many of the health advantages of eating a diet high in tomatoes and tomato-based products are known to be related to their high antioxidant content (Collins et al. 2022). Tomatoes (*Solanum lycopersicum*) are an excellent source of minerals, carotenoids, and phytochemicals, including lycopene, potassium, iron, folate, vitamin C, phenolic components, flavonoids, hydroxycinnamic acid, chlorogenic acid, homovanillic acid, and ferulic acid (Alexander and Grierson 2002). The fruit's genetic makeup, level of ripeness, and growing environment all affect its chemical makeup. Frequent use of tomato products and varieties is associated with a decreased risk of cardiovascular diseases and a variety of malignancies. Antioxidants, namely carotenoids (lycopene and beta-carotenoids) and phenolic substances, are responsible for this beneficial effect. Tomatoes change dramatically in terms of color, texture, flavor, and scent as they ripen. Along with the synthesis and accumulation of several components, including carotenoids, the fruit's color changes due to the breakdown of chlorophyll, which reveals underlying pigments (Baldwin et al. 2011; Zhang et al. 2016). Postharvest losses are especially common with tomatoes, as they are with other juicy fruits. Fruits' degree of softening, shriveling, and rotting determines the shelf life of food, which is defined as the amount of time that a product that has been stored stays fit for consumption. One of the most crucial characteristics of commercially grown tomatoes is their postharvest shelf life, which can be reduced by an abundance of factors, including temperature fluctuations and pathogen exposure, causing rapid over-ripening (Thole et al. 2020).

As demonstrated by the idea of systems biology, there is currently a growing interest in thoroughly comprehending not only the biological mechanisms at the molecular level (biological components) but also the impacts of a continuing biological process in the organism as a whole (biological functioning). In this regard, metabolomics, which is regarded as a phenotyping tool, is among the most potent bioanalytical techniques that enable the acquisition of an image of an organism's metabolite during a biological process. In a nutshell, the metabolomic method is recognizing and figuring out which set of metabolites or particular metabolites are present (Klassen et al. 2017).

7.2 Nutritional Elements in Tomatoes

Fruits and vegetables provide a variety of health-promoting phytochemicals, including fiber, natural antioxidants, and other bioactive compounds. There is a strong body of evidence from epidemiological studies suggesting that diets high in these foods are linked to a decreased risk of several diseases. Nevertheless, the kind of consumption, raw or processed, that determines whether fruit and vegetables have the potential to improve health is the only factor that matters. Processing of vegetables is thought to impact their bioactive components, contents, activity, and bioavailability (Borguini and Torres 2009). Phenolic substances, which are phytochemicals with potential biological action, are abundant in fruits and vegetables and their by-products. There are two main categories of phenolics found in both commercial goods and raw tomatoes (Perveen et al. 2015). These categories include flavonoids, which include kaempferol, quercetin, rutin, and naringenin, and phenolic acids. Tomato fruits contain an abundance of phenolic chemicals and flavonoids that have been investigated and identified. As can be observed, the concentration of naringenin chalcone, which was thought to be the most prevalent phenolic component in tomatoes, was 309.7 mg per 100 grams of dry weight. Subsequently, tomatoes exhibited the second- and third-highest concentrations of 3-caffeoylquinic acid and quercetin-3 rutinoside, with 71.1 mg and 60 mg of each per 100 g dry weight, respectively. Rutin is the most prevalent flavonoid found, with naringenin (0.65–1.19 mg per 100 g fresh weight), quercetin (0.048–0.141 mg per 100 g fresh weight), and myricetin (0.017–0.286 mg per 100 fresh weight) following in the order of abundance. Furthermore, at 0.75–1.38 mg of phenolic acid per 100 g of fresh weight, chlorogenic acid was the most prevalent phenolic acid constituent in tomatoes (Wang et al. 2023).

7.3 Therapeutic Effects of Bioactive Compounds Present in Tomatoes

Tomatoes and tomato-based products are thought to be a significant source of carotenoids and are high in antioxidant chemicals, particularly lycopene, ascorbic acid, and phenolic compounds. Regular tomato and tomato product consumption has been linked to a decreased risk of cardiovascular diseases and a variety of malignancies (Tan et al. 2010). Antioxidants, especially phenolic chemicals and carotenoids (lycopene and beta-carotene), are responsible for this beneficial effect. The hues of red, orange, and yellow found in the leaves, fruits, and flowers are attributed to carotenoids. Pigments called carotenoids are primarily responsible for shielding plants from the photooxidative process. Carotenoids work in harmony with other antioxidants to strengthen the human body's defense mechanisms against free radical damage (Mein et al. 2008). The primary antioxidant found in both raw and processed tomatoes is lycopene. The chemical lycopene, which is mostly present in tomato skins, can be found in tomatoes up to 5 mg per 100 g. Lycopene's antioxidant qualities, which shield lipoproteins from oxidation, contribute to its potential function in the prevention of cardiovascular illnesses. Flavonoids and phenolic compounds are a significant class of nonessential dietary chemicals that have been linked to the prevention of cancer and atherosclerosis (Agarwal and Rao 2000). A major factor in ascorbic acid's high effectiveness in biological systems as an antioxidant is how easily its electrons can be lost. Numerous reactive substances are reduced by the acid by giving an electron (Borguini and Torres 2009). Lycopene in tomatoes increases fertility by increasing sperm quality and speed while lowering the quantity of aberrant sperm in men. Tomato consumption helps fend off age-related illnesses such as Alzheimer's, Parkinson's, osteoporosis, and dementia. Vitamins C and A are abundant in tomatoes and play a crucial role in preventing muscle degeneration and enhancing vision. It is also thought to be an effective blood purifier and is helpful in treating urinary tract infections. Because they include a lot of fiber, tomatoes can help with weight loss and simple digestion.

7.4 Physiological Changes in Tomatoes

The postharvest quality of fresh tomatoes is significantly influenced by physiological factors, including respiration, ethylene production, and transpiration. Just like any other living thing, tomato cells display a variety of biological activities even after harvest. These processes carry on from maturity to ripening, senescence, and death, which is the final stage. The reason for quality degradation is that when tomato cells are separated from their father plant, they are unable to recover stored elements and energy that had been lost to biological processes. The normal physiological processes that cause fruits and vegetables to deteriorate are transpiration, respiration, and the production of C_2H_2 gas between 20–25 °C. Tomatoes ripen and degrade quickly as a result. Knowing this, tomatoes are kept in cold storage (below 12 °C) to slow down the physiological processes that lead to faster deterioration (Nunes et al. 2009). One of the primary issues with tomatoes losing their freshness is chilling damage. These physiological flaws reduce longevity, which restricts the amount of cold temperature storage that can be done. After harvesting, the quality of fresh tomatoes is greatly influenced by physiological factors such as respiration, ethylene generation, and water loss (Uchenna and Abiodun 2024; Vijay et al. 2011).

Tomatoes are highly perishable; there are a number of issues with storage, marketing, and transportation. The fruits suffer a significant postharvest loss in addition to losing quality due to a lack of knowledge regarding proper postharvest handling, packaging, and temperature (Mrema and Rolle 2002). Studies have contributed to the rise in growth in tomato production, but the goal of making the most money will only be achieved if this growth is complemented by comparable measures to reduce postharvest losses and prolong shelf life (Idah et al. 2007). Fruits and vegetables are cleaned with potassium permanganate or chlorine before being packaged in numerous countries throughout the world. Fresh produce is kept in what is known as "modified atmosphere packaging (MAP)." MAP involves wrapping them in plastic sheets that limit the flow of gases that cause respiratory problems. This causes carbon dioxide to build up and oxygen to be depleted surrounding the crop, potentially extending its shelf life. Packaging broccoli in perforated polyethylene bags with 1% holes extended its shelf life.

A fruit's flavor, texture, and scent are all part of its complex organoleptic qualities. The mix of sugars, organic acids, volatile chemicals, and free amino acids determines the flavor, which is experienced via a combination of odor, taste, and mouthfeel. Changes in the overall nutritional quality can also result from changes in any one of these substances (Babalola et al. 2010; Akter and Khan 2012). The process of fruit ripening initiates pathways that affect the amounts of volatile fragrance compounds, sugars, acids, and pigments (carotenoids and flavonoids) that give the fruit its color and softness (Sablani et al. 2006). During climacteric ripening, ethylene is produced autocatalytically and controls many of these physiological processes. In the context of commerce, early harvesting, managing the postharvest storage environment, and genetic selection for slow- or late-ripening cultivars are frequently used to regulate fruit ripening. Nonetheless, a better knowledge of the metabolic alterations that underpin fruit development and ripening should aid in the creation of fresh tactics to extend fruits' shelf life and enhance their quality after harvest (Lee and Kader 2000).

7.5 Basics of Metabolomics

The advent of improved technology has led to the emergence of multiple multidimensional omics disciplines that focus on the molecular constituents of living organisms, such as transcriptomics, proteomics, metabolomics, and genomics. The progress made in genomic investigation has contributed to the identification of several gene families and processes that influence yield-boosting tolerance to abiotic stresses (Khoso et al. 2022). The application of proteomic and metabolomic technologies has grown in significance as a means of holistically investigating the mechanisms behind plant growth and development as well as the responses to biotic and abiotic challenges. The phenylpropanoid pathway was crucial in fruit ripening, particularly in skin tissue, and protein expression altered differently in tomato fruit flesh and skin tissues during fruit development. In all three tomato fruit–shape typologies, there were notable differences in the amounts of amino acids, glycoalkaloid chemicals, and total phenolics. Variations in amino acid levels during tomato fruit ripening were observed and it was discovered that aspartate content grew visibly and gamma-aminobutyric acid levels gradually reduced, whereas organic acid did not exhibit any discernible changes under salt stress. Because of this, omics technology is a very useful tool for comprehensively elucidating the metabolic pathways and levels of protein expression in various developmental stages or environments (Tang et al. 2020). Factors affecting food quality and role of metabolomics are illustrated in (Fig. 7.1).

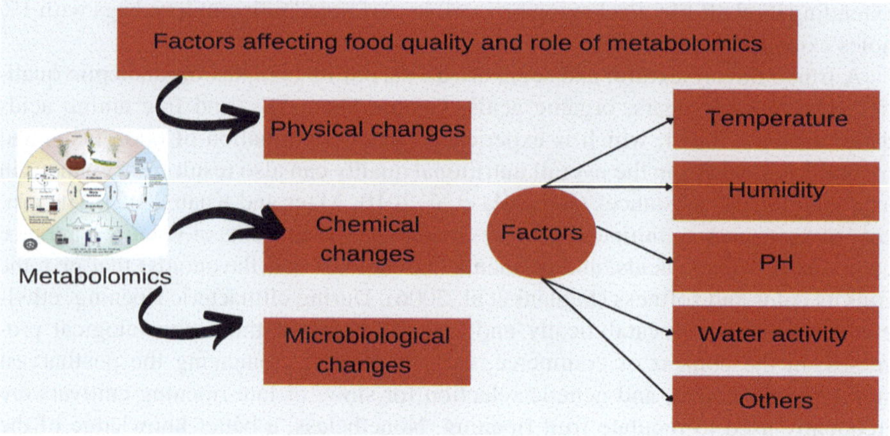

Fig. 7.1 Factors affecting the quality of tomatoes

7.6 Role of Metabolites

Metabolites play a crucial role in plant metabolism because of their impact on biomass and architecture. With the advent of metabolomics, precise metabolite profiling in microorganisms, plants, and mammals has been possible. Metabolomics has recently become recognized as a significant advance (Wishart et al. 2007). Metabolism is a technique that enables the rapid and accurate study of metabolites by detecting a wide variety of compounds from a single extract. Put differently, metabolomics provides a comprehensive understanding of cellular metabolites, or tiny organic substances, which take part in many cellular processes and so reflect a cell's true physiological state. Given the speed at which metabolomics is developing, metabolite analysis of transgenic lines and mutants offers promise for deciphering metabolomics networks and identifying the underlying candidate gene or genes (Hanhineva and Aharoni 2009).

Identification and measurement of plant metabolomics seek to comprehend the connection between pathogenic, genetic, and/or environmental factors in terms of how the metabolism expresses itself differently. Plant diagnostics has made extensive use of the development of proteomics, metabolomics, and genomic techniques in recent years. Due to its beneficial composition, which includes volatiles, primary and secondary metabolites, and other nutraceutical substances, fruit consumption is crucial for human health and nutrition (Raiola et al. 2014). Fruit farmers, consumers, and producers all benefit from their valuable agronomic features and metabolic content. Fruits are necessary component of our diet, and breeding efforts in terms of metabolite content need to be thoroughly analyzed for nutrient-rich cultivars. To date, the fruit has yielded hundreds of compounds, including primary and secondary metabolites. Within certain plant cells, multiple processes related to anabolism, catabolism, and amphibolism take place. Secondary metabolites are produced from primary metabolites and are needed for fruit set and fertilization, which are the hallmarks of living cells. Primary metabolites are produced first (Tang et al. 2020).

7.6.1 Metabolite Profiling Applied to Tomatoes to Uncover Valuable Features

For plant phenotyping and diagnostic analysis, metabolite profiling is a quickly developing method that attempts to measure the whole metabolome, or metabolic complement, of biological materials. Recently this approach, which has enormous potential to improve the compositional quality of crops, has been utilized to evaluate the natural variables in metabolite content by using these tools (Osorio et al. 2009).

7.7 Techniques and Technologies in Metabolomics

Introduced by Jeremy Nicholson in 1999, the metabolomic analysis first came into being. As many metabolites as feasible in a biological system could be identified and quantified using this approach, which was first presented as a complete analysis (Fiehn et al. 2000; Zhang et al. 2020). Studying molecular weight compounds (usually less than 1500 Da) in organisms, cells, or tissues at a given time and in a given environment is a quantitative, qualitative, and dynamic process (Feng et al. 2020). New developments in technology have made it possible to quantify more metabolites at once, and the outcomes of metabolic profiling are not influenced by the method of data collection. The lack of a distinct approach is one of the primary issues with metabolomics (León et al. 2018).

Targeted and untargeted techniques are the two primary categories of metabolomic approaches. The targeted strategy is hypothesis-driven, concentrating on a select group of chemicals and typically permitting measurement. Although this approach is more constrained and only reaches a limited number of metabolites, it enables the correlation of material changes with well-established, well-characterized analytes (Begou et al. 2017; León et al. 2018; Pezzatti et al. 2020). The hypothesis-driven focused strategy concentrates on a group of molecules and typically permits the quantification of those compounds. Since it covers a larger range of compounds and can be helpful in identifying other metabolites, the untargeted technique is more popular. It is exploratory and generates a hypothesis (Liu and Locasale 2017; Pezzatti et al. 2020).

7.7.1 Untargeted Analysis

The untargeted analysis can be divided into two types such as profiling and fingerprinting. Without requiring identification or quantification, fingerprinting entails analyzing as many substances as feasible in a matrix, including their detection and statistical processing of the findings. In order to produce a typical sample fingerprinting that is quicker and more comfortable, it allows the identification of sample components in different locations based on metabolite patterns. However, profiling concentrates on examining a particular class of metabolites, which are often detected and measured using databases or software. Effective statistical procedures, such as multivariate analysis, are required to reduce the data complexity as a result of enormous datasets produced by untargeted metabolomics. Additionally, even with offering a thorough examination of metabolite in the sample, it can still be challenging to validate and combine the data, necessitating frequently confirmatory, ideally quantitative targeted analysis (Castro-Puyana et al. 2017; Ibáñez et al. 2013; López-Ruiz et al. 2019).

Fingerprinting and profiling are the two categories under which untargeted analysis falls. In fingerprinting, the greatest number of substances in a matrix are analyzed; their detection is made; and the results are statistically processed without requiring identification or quantification. It is a quicker and more practical way to identify sample components according to distinct regions based on metabolite patterns in order to achieve a typical sample fingerprinting. However, profiling is centered on analyzing a particular class of metabolites, which are often detected and measured using software tools or databases (Lacalle-Bergeron et al. 2021).

Multivariate analysis is one of the most effective statistical approaches that is required to reduce the complexity of the data, as untargeted metabolomics generate massive datasets. However, even with offering a thorough examination of metabolites in the sample, it can still be challenging to compile and validate the results, necessitating frequently confirmatory, ideally quantitative targeted analysis (Begou et al. 2017)..

7.7.2 Tomato Postharvest Quality Change Revealed by Measuring the Metabolites of Organic Acids Using HPLC Mass Spectrometry

Organic acid separation, detection, and quantification have been achieved through the rapid development of a variety of analytical techniques and instruments, such as gas chromatography coupled with mass spectrometry (GC-MS), capillary electrophoresis (CE), high-performance liquid chromatography (HPLC), enzymatic biosensors, and infrared (IR) spectroscopy. In recent years, the detection of organic acids has also made use of liquid chromatography combined with tandem mass spectrometry (MS), or LC-MS/MS (Feizi et al. 2021; Fraga-Corral et al. 2022; Mielko et al. 2021). However, the majority of studies only pay attention to the organic acids such as fumaric, citric, pyruvic, malic, and succinic that are a part of the tricarboxylic acid (TCA) cycle. Ascorbic, citric, fumaric, malic, pyruvic, and oxalic acids were the only main acids that were quantified in the majority of earlier studies on the organic. Nonetheless, because of their strong polarity and comparable structures, different organic acids are challenging to separate using typical columns and mobile phases. Even using LC-MS/MS, co-eluting organic acids would interfere with the measurement of individual acids. In the meantime, the target compounds' ionization efficiency was highly impacted by the composition of the mobile phase, and appropriate additions might significantly improve the MS signal (Meng et al. 2023).

7.8 Analytical Techniques

At the molecular level, analytical techniques like NMR and MS reveal details on the chemical makeup of biological materials. These two methods are often applied independently in plant metabolomics to profile plant systems. In terms of sensitivity and detection, LC-MS and NMR are two different analytical methods. LC-MS is a quick and accurate method, on the one hand. However, molecular elucidation has certain inherent limits, such as the resolution of isomers, and the separation of metabolites depends on the chromatographic column used. The detection is also determined by the analytes' ability to ionize. On the other hand, because the analytes are soluble, NMR is indiscriminate with regard to matrix properties. When it comes to identifying molecular structures, NMR is a very selective method but not as sensitive as MS (Moco et al. 2008).

7.9 NMR/HRMS Spectroscopy

Natural product structure determination in metabolomics research is made possible by the special, potent, and trustworthy capabilities of NMR spectroscopy. Its intrinsic quantitative character, simple sample preparation, and high-throughput analysis of many metabolites are combined with its strong structural elucidation capabilities (Pontes et al. 2017; Lee et al. 2009). Furthermore, NMR spectroscopy is noninvasive, produces extremely repeatable results, and does not need the isolation of substances from biological mixtures. The sensitivity of NMR may have limited its application in the past, especially when compared to mass spectrometry (MS), another popular technique. However, NMR has advanced with recent technological advancements, such as upgrades in hardware (e.g., high-resolution NMR and the use of cryoprobes, microprobes, and in-flow probes), pulse sequence, and spectral acquisition and post-processing protocols. Consequently, it is not unexpected that NMR has emerged as the standard untargeted metabolomic profiling technique in a number of plant science fields, such as pathology, pharmacology, chemotaxonomy, plant functional genomics, plant stress physiology, and green biotechnology (Giraudeau 2020). The application of NMR metabolomics has yielded extensive and intricate datasets by profiling the effects of time, stress, nutritional status, and environmental perturbations on complex combinations of plant metabolites (Dixon et al. 2006; Consonni and Cagliani 2019).

High-resolution mass spectrometry (HRMS) and nuclear magnetic resonance (NMR) are the two primary analytical platforms utilized for the metabolomic study. The NMR technology is more useful for targeted approaches because it gives quantitative data and uses an intrinsic magnetic property of atomic nuclei to detect molecular properties. The approach offers the benefits of being noninvasive, very repeatable, and of excellent quality. Furthermore, it is thought to be an environmentally beneficial analytical technology because it permits the use of green solvents, streamlined extraction techniques, and process automation. It also detects all organic chemical classes and offers accurate compound identification (Utpott et al. 2022).

7.10 Improvement of Phenolic Component Extraction from Tomatoes with Microwave Assistance

7.10.1 FTIR and HPLC Characterization and Comparison with Traditional Solvent Extraction

Fruit and vegetable bioactive components are typically extracted, especially by conventional solvent extraction (CSE). Nevertheless, CSE pollutes the environment because it takes a long time to process and needs a lot of solvent. Thus, as more efficient, cutting-edge, and quick alternatives to conventional procedures, innovative techniques that lower the quality of organic solvent and boost extraction efficiency have been created recently (Baltacoğlu et al. 2021).

7.10.2 Microwave-Assisted Extraction

Microwave-assisted extraction (MAE) is one of the most sophisticated techniques for removing bioactive substances from fruits and vegetables. Foods can be heated both indoors and outdoors without creating a temperature differential during MAE. There is minimal phenolic component extraction from tomatoes using MAE. Antioxidant activities can be monitored to identify the ideal extraction conditions, which they then to employed to analyze the phenolic content. HPLC and Fourier transform infrared (FTIR) spectroscopy are mostly used for phenolic components' identification (Fig. 7.2).

7.11 Metabolic Alterations During Tomato Fruit Ripening

One of the most crucial characteristics of fruit quality is pigmentation, which changes significantly due to increased carotenoid production and chlorophyll breakdown. This gives the fruit its distinctive color (Klee and Giovannoni 2011).

1. Cell Wall

Rearranging the connections between the cellulose and hemicelluloses, solubilizing and depolymerizing the pectin polysaccharides, and hydrolyzing the cellulose and hemicelluloses are all intricate processes that occur during the ripening stage of

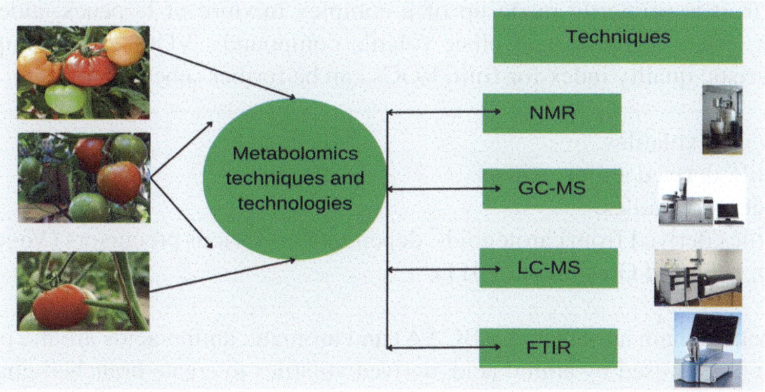

Fig. 7.2 Different metabolomic techniques used in postharvest tomatoes. Among these techniques are *NMR* nuclear magnetic resonance, *GC-MS* gas chromatography/mass spectrometry, *FTIR* Fourier transform infrared spectroscopy, and *LC-MS* liquid chromatography/mass spectrometry

fruit cell wall remodeling, which is one of the main characteristics of fruit texture and a major factor in determining fruit shelf life and commercial value (Goulao and Oliveira 2008). Fruit texture is primarily determined by cell wall remodeling, a complex process that involves the hydrolysis of cellulose and hemicellulose, the solubilization and depolymerization of pectin polysaccharide, and the rearrangements of their connections. During the ripening stage of fruit development cell wall remodeling occurs (Payasi et al. 2009).

2. Central Carbon Metabolites

Central carbon metabolites are the main factors that determine the flavor and quality of fruit. They also serve as crucial carbon skeletons for other metabolites in addition to having a direct impact on the sour-sweet taste (Malundo et al. 1995). Fruit ripening involves closely regulated levels of sugar and organic acid content to balance importation from the source organ and utilization through glycolysis and the tricarboxylic acid (TCA) cycle (Carrari and Fernie 2006).

3. Secondary Metabolism

Alkaloids, volatile organic compounds (VOCs), and polyphenols, which function as bioactive substances against cancer, heart disease, and inflammation, make up tomato fruit's secondary metabolism (Andersen and Markham 2005). Flavonoid metabolites, the most significant component class of polyphenols are produced from phenylalanine through phenylpropanoid and polyketide pathways (Perez de Souza et al. 2020).

The fruit is primarily made up of a complex mixture of terpenes, aldehydes, alcohols, esters, ketones, and other volatile compounds. VOCs are an important characteristic quality index for fruit. VOCs can be further subclassified into:

- Fatty acid volatiles.
- Volatile-derived amino acids.
- Terpenoid volatiles.
- Volatiles derived from carotenoids, depending on various precursors (Vogel et al. 2010; Klee and Giovannoni 2011).

Branched-chain amino acids (BCAA) and aromatic amino acids are the primary building blocks used by amino acid–derived volatiles to create branched-chain and phenylpropanoid volatiles when the tomato fruit ripens; the release of phenylpropanoid volatile is mostly dependent on the glycoconjugation reaction, while the alpha-keto acid intermediate in the catabolism of BCCAs serves as a direct precursor for branched-chain flavor volatiles (Zhu et al. 2022).

4. Colors

Pigmentation is the most crucial characteristic of fruit appearance quality, which changes significantly as a result of increased carotenoid biosynthesis and chlorophyll breakdown during the ripening process, giving the fruit its distinctive color.

7.12 Postharvest Variables

The following postharvest variable might impact the quality of harvested produce: temperature is the most crucial element to preserve the quality of harvested produce and extend its shelf life. Intervals between harvesting and cooling or processing can cause indirect losses like off-flavor and degradation in nutritional content, as well as direct losses like water loss and decomposition.

7.12.1 Ethylene Management

Ethylene (C_2H_4), the natural plant hormone, has two carbons and has no smell or color. It is released when climacteric fruits reach maturity. It is also referred to as the natural aging, death, or ripening hormone. It is present in trace amounts, and when it builds up, it can cause fruit deterioration and waste while the harvested product is in the postharvest stage. Numerous physiological processes related to plant growth, germination, development, ripening, maturity, and senescence are governed by it. It contributes to the abscission of plant organs as well. Gas regulation, physical handling, chemical treatments, radiations, respiration control, and relative humidity (RH) management are a few techniques that can be applied to minimize postharvest changes. Fruits' physiological and biochemical characteristics are influenced by a variety of pre- and postharvest elements including agricultural practices, ripening stages at harvest, and cultivation. Tomatoes' physiological and biochemical characteristics are largely impacted throughout the maturity and ripening stages (Trong et al. 2019).

7.12.2 Intervention and Strategies for the Prolonged Shelf Life of Tomatoes

- Preharvest treatments (e.g., irrigation system deficit).
- Postharvest treatments (e.g., temperature, relative humidity, and pH).
- Optimum storage conditions.
- Ethylene management.

Temperature, light, water, weather, and soil nutrients affect the metabolite composition of tomatoes. High temperatures can increase the level of amino acids in tomatoes. Genetic makeup can determine the metabolite composition of tomatoes (Klee and Giovannoni 2011). Different varieties can have distinct metabolite profiles and breeding programs, which target the specific metabolites to enhance the quality of tomatoes. Postharvest handling such as storage and handling practices can impact the metabolite composition of tomatoes with changes in enzyme activity and metabolite degradation (Zhang et al. 2018). Sucrose, glucose, and fructose can influence the biochemical composition of tomatoes. Organic acids such as citric acid, tartaric acid, and malic acid can influence the metabolite composition of tomatoes. Amino acids such as glutamic acid, aspartic acid, and alanine can influence the metabolite composition of tomatoes. For proper storage of tomatoes, it is necessary to maintain temperature, relative humidity, light, and atmospheric conditions because these affect the metabolite composition of tomatoes (Carrari and Fernie 2006).

7.13 Strategies for Prolonging the Shelf Life of Tomatoes

- Metabolic engineering.
- Postharvest treatments.
- Genetic breeding for enhanced metabolite profile.

7.13.1 Metabolic Engineering

Metabolic engineering is a powerful approach for enhancing the shelf life of tomatoes. By manipulating the metabolic pathways involved in ripening and senescence, scientists can develop tomatoes with improved shelf life. (Klee and Giovannoni 2011). Ethylene is a hormone that promotes ripening and senescence. By reducing ethylene production, tomatoes can be slowed down in their ripening process.

Antioxidants help protect tomatoes from oxidative stress, which can contribute to spoilage. Enhancing antioxidant activity can help extend shelf life (Waseem et al. 2019). Cell wall modifications can improve texture and firmness, reducing the likelihood of spoilage. Sugars play a crucial role in tomato ripening. Manipulating sugar metabolism can help regulate ripening and extend shelf life. Tomatoes that are more stress-tolerant are less likely to be spoiled quickly. Metabolic engineering can be used to enhance stress tolerance (Gapper and Giovannoni 2013).

7.13.2 Postharvest Treatments

Maintenance of optimum temperature plays an important role in postharvest treatments (Fig. 7.3).

Another physiological condition that arises from exposing fruits to low temperatures for extended periods of time is chilling injury (CI). While ripe tomatoes should normally be stored between 7 and 10°C, green, mature tomatoes should be stored at around 13° (Suslow and Cantwell 2009; Biswas et al. 2016). Green, mature tomatoes are more susceptible to CI than ripened tomatoes because they take longer to ripen and produce greater CI alterations when stored at low temperatures. The primary signs of CI in tomatoes include uneven or incomplete ripening, browning, surface pitting, areas that are saturated with water, mealy texture, and heightened susceptibility to microbial deterioration. The rate at which ions leak out of tomatoes' tissue and the permeability of the cell membrane are associated with the severity of CI (Aghdam et al. 2014; Biswas et al. 2016).

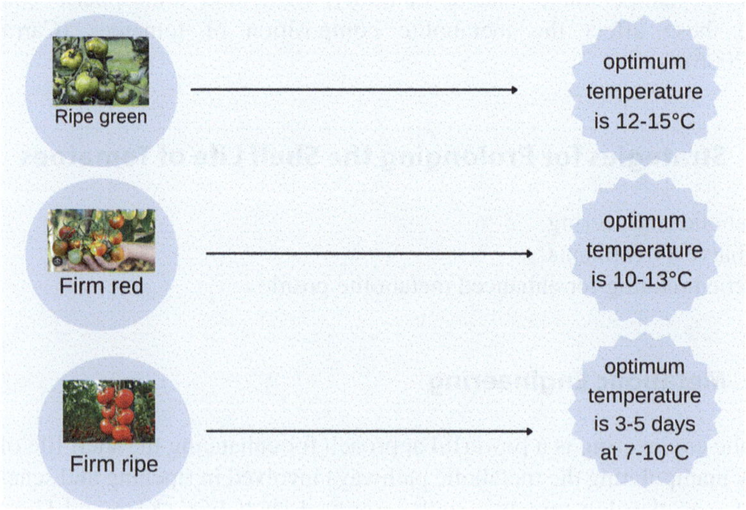

Fig. 7.3 This figure explains the optimum temperature requirements for tomatoes. Stages of tomatoes with temperature ranges

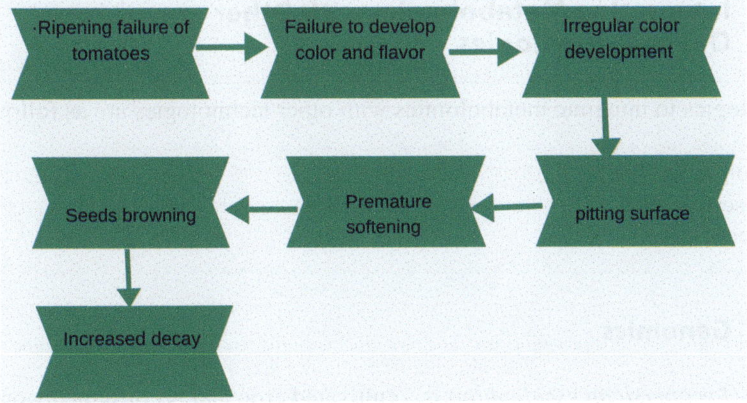

Fig. 7.4 Changes that occur if the specific required conditions are not provided to fruits

- For standard ripening of tomatoes, the temperature is about 18–21 °C, and RH is 90–95%.
- For slow ripening of tomatoes, the temperature is about 14–16 °C.

Tomatoes are sensitive to chilling temperatures. If kept for more than 2 weeks at 10 °C and more than 6–8 days at 5 °C, chilling injury problems may develop. Chilling injuries may cause all changes that are illustrated below: (Fig. 7.4).

7.13.3 Genetic Breeding for Enhanced Metabolite Profile

Genome editing technology refers to the process of making precise changes in the DNA, and it is a rapid breeding method that disrupts the genes in the carotenoid biosynthesis pathway to develop yellow and orange tomatoes.

Genome editing technology can be used in various processes such as:

- Crop improvement in agriculture
- Gene therapy
- Basic scientific research
- Cancer improvement

The benefits of genome editing are as follows:

- Allowing for precise changes in the genome
- More efficient results than traditional breeding technologies
- More flexible process used in plants, animals, and microbes

Multiplex editing technology is used in carotenoid metabolism to enhance lycopene content and nutritional content in tomatoes.

7.14 Integrating Metabolomics with Other Omics Technologies

Technologies to integrate metabolomics with other technologies are as follows:

- Genomics
- Transcriptomics
- Proteomics

7.14.1 Genomics

Tomato (*Lycopersicon esculentum*) is a cultivated crop and is consumed worldwide and considered as the second-most consumable vegetable. It is the most studied crop in genomics and genomes. It is the first crop in which genetic linkage mapping was constructed.

Genomic technology is used in:

- Genome sequencing
- Gene expression
- Genomic variations

7.14.2 Transcriptomics

Measurement of all mRNA transcripts in a single cell or a population of cells is possible through a technique called transcriptomics, which is the study of gene expression or genome-wide expression profiling (Wang et al. 2009). Understanding the functional pieces of the genome requires transcriptome analysis. Transcriptome sequencing technology offers a quick and affordable method for generating a wide variety of expressed sequence tags (ESTs), which may aid in the computational identification of molecular markers in vegetable crops like SNPs and SSRs (Barbazuk et al. 2007).

Transcriptomic technology is used in:

- Gene expression profiling
- Regulatory network analysis
- Transcriptome assembly
- Transcriptome quantification
- Gene regulation

7.14.3 Proteomics

Along with transcriptomics and genomics, proteomics is the large-scale examination of the profiles and activities of proteins. It is regarded as one of the scientific research tools of the future (Liu and Chen 2012).

Proteomics has many applications in tomato research such as:

- Improvement of shelf life and quality of tomatoes
- Production of disease-free tomato varieties
- Help in the ripening and development of tomatoes

Genomic, transcriptomic, and proteomic technologies can be used to improve resistance to biotic and abiotic stress in tomatoes to prolong their shelf life..

7.14.4 Future Directions for Enhancing the Shelf Life of Tomatoes

Future directions for enhancing the shelf life of tomatoes are as follows (Baptista et al. 2020) :

- Edible coating of tomatoes
- Optimum storage conditions
- Optimum temperature
- Greenhouse tomatoes
- Controlled atmosphere
- Proper packaging

Tomatoes are perishable; they have a short shelf life and are prone to significant changes in physiological quality when stored. Tomatoes' qualitative characteristics can be effectively preserved by edible coatings. Different techniques are used to generate composite coatings, and nearly all of them have drawbacks. Choosing a coating process presents another challenge for the composite edible coating (Ali et al. 2010). Generally speaking, the dipping method is used to coat tomatoes, although it does not guarantee a consistent thickness of coating material throughout the surface of the tomato. The most difficult challenge of coating tomatoes for eating is choosing the right coating. Quality loss will continue to be a significant problem for tomato-growers and handlers until both aspects are appropriately managed. It is believed that the quality and storage life of tomatoes after harvest depend not just on the postharvest factors alone but also on several preharvest factors during production (Shehata et al. 2021).

7.15 Factors Influencing Tomato Postharvest Quality

The fruit keeps its life after harvesting and continues to operate as a living tissue in all respects. Fruit senescence and subsequent ripening are initiated by the climacteric burst of ethylene that renders the fruit appetizing too. In order to ensure that the fruit reaches consumers at its best eating quality, the aim of any postharvest management procedure or treatment is to control the amount and timing of ethylene production. Inadequate management of the following variables can have an impact on the quality of tomatoes after harvest (Beckles 2012).

1. Temperature

The best technique to preserve quality has been discovered to be the proper regulation of the temperature between the time of harvesting and consumption. Harvested fruits can slow down many of the metabolic processes that contribute to ripening if they are kept cool at low temperatures of around 20 °C. This will give you more time to handle the produce after harvest. The shelf life of a crop will typically decrease by 1 day for every hour that passes between harvesting and cooling. The rate of respiration, or the creation of CO_2, in harvested or stored fruit products can be accelerated by high temperatures. Oxygen or CO_2 levels, exposure duration, and ripening stage are some of the additional variables that might cause ethylene generation in stored climacteric items like tomatoes (De Wild et al. 2003). Texture, nutrition, flavor, aroma, and nutrition can all be preserved with low-temperature preservation. Tomatoes, being tropical fruits, are also negatively impacted by exposure to very low temperatures. Tomato fruits that are kept below 10 °C may suffer chilling damage.

2. Relative Humidity

Relative humidity, a measure of how much moisture is in the surrounding air, is what mostly causes water loss from harvested fruit products. Harvested fruits preserve their nutritional value, weight flavor, and appearance at extremely high relative humidity levels while slowing down the rate of withering, softening, and juiciness. Tomato fruits can shrink after being harvested and have a very high water content. Any little amount of moisture loss might cause the fruit to shrink. Relative humidity levels between 85 and 95% (v/v) are ideal for mature, green tomatoes but between 90 and 95% (v/v) for firmer, ripe fruits. When evapotranspiration rises above the ideal range, fruits become shriveled. Tomato fruits may shrink if they are stored at lower relative humidity (Suslow and Cantwell 2009).

3. Combination Gases

The way that various gases mix in a storage environment affects how long tomato fruits may be stored. The ideal atmosphere required to prevent senescence in all kinds of tomatoes mature, green and ripe fruit is 3–5% (v/v) of oxygen, but only

1–3% (v/v) of carbon dioxide, respectively. Whereas 94–66% (v/v) of nitrogen gas is necessary (Sandhya. 2010). While enhancing several tomato quality attributes, carbon monoxide reduces postharvest pathogenic infections (Beckles 2012).

4. Application of Calcium Chloride Postharvest

Calcium-based fertilizers can help minimize plant calcium shortage. Tomato plants that get insufficient calcium during fertilization may be more vulnerable to several calcium-related diseases, such as blossom end rot (Prudent et al. 2009; Saure, 2001). When applied as a fertilizer, calcium has been shown to boost tomato yields. It has also been demonstrated that adding calcium after harvest improves a number of fruit and vegetable storage characteristics. When fresh food, like tomatoes, is treated with calcium chloride after harvest, it slows down respiration, produces less ethylene, and delays senescence (Hussain et al. 2012).

7.16 Physical Management

Harvested fruits' postharvest quality or life can be significantly impacted by physical treatment. Inadequate handling, both during and after harvesting, might cause mechanical damage that degrades the quality (Miller 2002). Common industrial tomato production techniques include mechanical harvesting, long-distance shipping, crate packaging, sorting, grading, and washing. The fruits run a significant risk of suffering mechanical damage during any of these stages, which could cause them to be cut, bruised, scarred, scuffed, or punctured. The use of improper harvesting containers and packaging materials can lead to mechanical injury in small-scale tomato cultivation (Arah et al. 2015a).

7.17 Effects of Postharvest Storage on the Principal Metabolic Pathways: Texture and Taste of Fruits

Different species and storage circumstances have a major impact on the behavior of sugar content, which is generally measured by the soluble solid content (SSC), during the postharvest period. The major sugar profiles found in mature fruits—glucose, fructose, and sucrose—usually alter in tandem with the SSC trend.

7.17.1 Sugar and Sugar Derivatives

A crucial physiological process that results in mature fruit firmness during postharvest storage is softening; however, firmness loss due to over-ripening can cause physical damage and microbial attack, which, in turn, results in a significant reduction in fruit quality. According to GC-MS and NMR research conducted on tomatoes stored at room temperature, some species have a tendency for their major carbs to decrease. However, sugar intake during fruit ripening may impact postharvest water loss by disrupting the growth of cuticles. As a matter of fact, transgenic tomato fruits stored at ambient temperatures for a period of 12 days showed less water loss and wrinkling following stable silencing of the cell wall invertase LIN5, a crucial mediator of SSC content (Vallarino et al. 2017).

7.17.2 Organic Acids

Fruit postharvest metabolism is associated with a number of organic acids. Unexpectedly, the most prevalent organic acid has an effect on fruits' shelf life, especially in tomatoes. GC-MS measurements show that the malate content of a number of tomato genotypes, such as non-ripening and ripening inhibitor, delayed fruit deterioration and in genotypes used in commercial applications due to their delayed maturation and senescence, malate content decreased during ripening and storing food after harvest at room temperature. Remarkably, it was also demonstrated that ripe tomatoes that developed on the vine had lower malate levels than fruits that matured off the vine (Osorio et al. 2020).

A distinct postharvest behavior was seen at room temperature in tomato fruits with altered malate concentrations than in fruits of the wild type, which were caused by downregulating the expression of malate dehydrogenase (MDH) and fumarase, two TCA cycle enzymes. Fascinatingly, fruits with the MDH-deficient genotype contained more malate and behaved worse after harvest compared to non-transformed fruits, losing more water and being more vulnerable to *Botrytis cinerea* spores and opportunistic fungal infections. However, the genotype lacking in fumarase showed less water loss and a lower vulnerability to opportunistic fungus while having a relatively low malate level (Centeno et al. 2011).

7.17.3 Amino Acids

Due to the fact that these chemicals are implicated in several pathways produced during fruit ripening, postharvest storage also has a significant impact on antioxidant concentration (Uarrota et al. 2019). For instance, amino acid catabolism can offset the decrease in TCA cycle–derived electron supply during senescence as seen by NMR-based metabolite profiling of bananas kept at ambient temperatures (Zhang et al. 2011). The abrupt emergence of salsolinol, which has been shown to be derived from both dopamine and acetaldehyde, the latter of which is created from ethanol, coincided with the elimination of dopamine (Yuan et al. 2017).

7.18 Repercussions on Secondary Metabolites After Harvest

Fruits contain two major groups of secondary metabolites: terpenoid and polyphenol chemicals, which are significant for their nutritional value and organoleptic properties, as well as for giving fruits their attractive color. Tomato lines rich in anthocyanins are able to store their fruits for extended periods of time without losing fruit quality, mostly because they are less vulnerable to *Botrytis cinerea*.

7.18.1 Carotenoids

Several fruits and vegetables have a visually appealing color due to a class of terpenoids called carotenoids. Temperature seems to affect postharvest accumulation in tomatoes and citrus, despite the fact that many staple crops have been shown to have little stability during postharvest storage, mostly because of beta-carotene's fast turnover (Carmona et al. 2012; Javanmardi and Kubota 2006).

7.18.2 Variable Profiles in the Postharvest Period and Their Effects on Fruit Aroma

Three main kinds of metabolites, sugars, acids, and volatiles, are in charge of giving fruits their flavor. Fruit flavor is mostly determined by the ratio of sugar to acid, although volatiles also contribute to the distinctive flavor of individual fruits. Mature fruits include mostly volatile compounds that are either derivatives of fatty and amino acids or come from terpenoid and phenylpropanoid pathways. The usual method for achieving volatile profiling is to extract them from the headspace (HS), or the airspace surrounding the fruit, and then use GC-MS to detect them. Solid-phase microextraction (SPME), the process of sampling from headspace, is most frequently carried out by the volatiles' coating on fused silica fiber for adsorption on a stationary phase (Vallarino et al. 2018). Since not all volatiles affect fruit scent, GC-olfactometry, a supplementary method, can be used to identify molecules that are odor-active. Significant changes in fruit volatile profiles have been shown to occur during postharvest, and these changes are frequently the cause of the reduced sensory acceptability following extended storage. For example, general tendencies identified by GC-olfactometry indicate a decrease in "green" or "fresh" notes and a corresponding rise in musty, overripe, and fruity scents (Pott et al. 2020).

7.19 Impact of Edible Coatings on Tomatoes' Quality Attributes and Physiological Reactions

7.19.1 Content of Moisture

An essential factor in evaluating tomato quality is its moisture level, which also affects other factors including firmness, freshness, and weight loss throughout storage. Following harvest, tomatoes typically have a moisture level of 90–95% (Correia et al. 2015; Rajkumar et al. 2007). As they are stored, however, the natural process of evaporation causes the moisture content to gradually decrease. Tomatoes that transpire from their scarred stems lose moisture to the surrounding air, which can cause physiological problems like wilting, shrinking, and loss of turgidity (Kabir et al. 2020).

7.19.2 Loss of Weight and Firmness

Tomatoes' textural qualities are significantly influenced by firmness and weight loss. Because tomato enzymes that break down cell walls, such as polygalacturonase and pectin methylesterase, are more active when tomatoes are stored, their firmness decreases, and their cell wall strength decreases as a result of decreased cell-to-cell adhesion (Dijik et al. 2006; Kabir et al. 2020). Ethanol content and respiration rate are the two main factors that affect an enzyme's activity (Sinha et al. 2019). As a result of transpiration from the stem scars, the tomato loses water through a decrease in turgor pressure (Fich et al. 2020).

7.19.3 Rate of Respiration

Due to their climacteric nature, tomatoes continue to breathe even after harvest. In tomato plants, the rate of respiration rises to a designated climacteric peak, which signifies the start of senescence (Li et al. 2017). By altering the gaseous control surrounding the whole fresh tomato, edible coatings seek to lower the respiration rate and thus postpone or prevent the climacteric peak from occurring (Zhang et al. 2019; Fitch-Verges et al. 2019). When compared to uncoated samples kept at 20 °C and 85% relative humidity for 14 days, a 1.5% chitosan coating effectively decreased the respiration rate of tomatoes by 21%. The extra surface coating extended the tomato's shelf life by 3 to 4 days by suppressing the rate of respiration (Pagno et al. 2018).

7.19.4 Ethylene Production

A significant factor in quickening the ripening process is ethylene, sometimes referred to as the "ripening hormone," which increases the activity of ripening enzymes such as ACC synthase and oxidase (Alexander and Grierson 2002; Paul and Pandey 2014; Waseem et al. 2019). One important substrate for ripening enzymes' activity is thought to be oxygen (Cruz et al. 2018). Tomatoes coated with edible coatings exhibit an oxygen barrier that inhibits the action of enzymes and postpones the ripening-related changes in the fruit. Edible coatings have been shown in numerous studies to be successful in lowering tomato ethylene production (Safari et al. 2021).

7.19.5 Total Soluble Solids

When a tomato ripens, total soluble solids (TSS), a measure of its sweetness, progressively rises and then peaks before trending downward (Nath et al. 2012; Tigist et al. 2013). The breakdown of complex carbohydrates into simple sugars and the drying out of tomatoes during storage are the two main causes of the rise in TSS (Mahfoudhi and Hamdi 2015; Ruelas-Chacon et al. 2017; Mohammed et al. 2021).

7.19.6 Titratable Acidity

Another crucial factor in fruits and vegetables is their titratable acidity (TA), which affects how well-liked the product is by consumers (Zhang et al. 2019). The fruit's total acidity is expressed as an approximate number that typically drops with storage. Citric acid, which serves as a respiration substrate, that is the organic acid found in tomatoes. Tomato storage causes a drop in TA because respiration that uses the organic acid persists during that time (Azmai et al. 2018; Ruiz-Martínez et al. 2020; Safari et al. 2021).

7.19.7 Lycopene

The carotenoid ingredient called lycopene, which gives tomatoes their red color, is also thought to have certain health advantages (Story et al. 2010). The variety, harvesting time, and maturation stage of fresh tomatoes all affect their lycopene levels (Górecka et al. 2020). Due to a reduction in the rate of metabolic respiration that abruptly decrease when senescence occurs. Due to a decrease in the rate of metabolic respiration that slows down the ripening-related metabolic alterations, edible coatings have shown that lycopene can be better retained or sustained in tomatoes during the storage period (Dávila-Aviña et al. 2014).

7.19.8 Ascorbic Acid

Chrysargyris et al. (2016) reported that when tomatoes were stored for 14 days, the ascorbic acid level rose in the 10% aloe vera–gel edible coating; however, there was no discernible change in the ascorbic acid content between the control samples and coatings made with 5 and 15% aloe vera gel (Yadav et al. 2022; Zhu et al. 2022).

7.20 Conclusion

Tomatoes' quality control after harvest begins from the field and continues till the end user. The preharvest activities undertaken during production partially determine the postharvest quality status of the fruits. No postharvest treatment or handling methods can improve any fruit's quality after harvest but can simply maintain it. For postharvest quality management, it is necessary to manage and understand the preharvest factors such as pruning of tomatoes, maturity stage of tomatoes, and the selection of fertilizers, cultivars, and irrigation methods for tomatoes. These preharvest factors play an important role in producing good-quality tomatoes. Due to the perishable nature and short shelf life of tomatoes, more quality losses occur at the time of harvesting. For good-quality tomatoes, it is mandatory to use good postharvest practices or harvesting factors such as optimum relative humidity, optimum temperature conditions, proper storage conditions, and use of proper physical handling practices (Yadav et al. 2022).

When tomatoes are coated with edible natural coatings, the average weight loss is reduced, the shelf life is extended, and the sensory attributes of tomatoes are somewhat improved. Therefore, it can be said that applying a natural coating made of aloe vera has a lot of potential to lessen the physiological changes brought on by ripening and increase the shelf life of tomatoes and other fruits and vegetables. In order to protect the country and climate, natural coatings could replace dangerous chemicals (Arefin et al. 2020).

All things considered, postharvest metabolomics has been shown to be a revolutionary method for increasing tomato quality and shelf life. Key biomarkers and metabolic pathways involved in ripening, senescence, and spoiling can be found through metabolic profiling. With this information, focused mitigation techniques can be developed, such as modifying metabolic pathways, sorting materials optimally, or using metabolic inhibitors. Technological developments such as high-throughput metabolomic platforms and bioinformatic tools have made a substantial contribution to our comprehension of tomato postharvest physiology. These realizations make it easier to create targeted treatments that preserve the fruit's nutritional content, firmness, and flavor throughout distribution and storage.

References

Agarwal S, Rao AV (2000) Tomato lycopene and its role in human health and chronic diseases. CMAJ 163(6):739–744

Aghdam MS, Asghari M, Khorsandi O, Mohayeji M (2014) Alleviation of postharvest chilling injury of tomato fruit by salicylic acid treatment. J Food Sci Technol 51(10):2815–2820

Akter H, Khan SA (2012) Effect of gamma irradiation on the quality (colour, firmness and total soluble solid) of tomato (Lycopersicon esculentum Mill.) stored at different temperature. Asian J Agric Res 6(1):12–20

Alexander L, Grierson D (2002) Ethylene biosynthesis and action in tomato: a model for climacteric fruit ripening. J Exp Bot 53(377):2039–2055

Ali A, Maqbool M, Ramachandran S, Alderson PG (2010) Gum Arabic as a novel edible coating for enhancing shelf life and improving postharvest quality of tomato (Solanum lycopersicum L.) fruit. Poatharvest Biol Technol 58(1):42–47

Andersen OM, Markham KR (2005) Flavonoids: chemistry, biochemistry and applications. CRC press

Arah IK, Amaglo H, Kumah EK, Ofori H (2015a) Preharvest and postharvest factors affecting the quality and shelf life of harvested tomatoes: a mini review. Int J Agron 2015(1):478041

Arah IK, Kumah EK, Anku EK, Amaglo H (2015b) An overview of post-harvest losses in tomato production in Africa: causes and possible prevention strategies. J Biol Agric Healthcare 5(16):78–88

Arefin P, Amin R, Sadiq MZA, Dey SS, Boby F, Parvaz MS (2020) A study of effectiveness of natural coatings on the shelf-life extension of tomatoes by the observation of TomloxC gene expression. Biosci Biotechnol Res Asia 17(3):459–465

Azmai MW, Latif AN, Zain MN (2018) The efficacy of combined coatings of chitosan and cinnamic acid on tomatoes. Int Food Res J 25:185–194

Babalola DA, Makinde YO, Omonona BT, Oyekanmi MO (2010) Determinants of post-harvest losses in tomato production: a case study of Imeko-Afon local government area of Ogun state. Acta Satech 3(2):14–18

Baldwin E, Plotto A, Narciso J, Bai J (2011) Effect of 1-methylcyclopropene on tomato flavour components, shelf life and decay as influenced by harvest maturity and storage temperature. J Sci Food Agric 91(6):969–980

Baltacoğlu H, Baltacıoğlu C, Okur I, Tanrıvermiş A, Yalıç M (2021) Optimization of microwave-assisted extraction of phenolic compounds from tomato: Characterization by FTIR and HPLC and comparison with conventional solvent extraction. Vib Spectrosc 113:103204

Baptista RC, Horita CN, Sant'Ana, A.S. (2020) Edible coatings: Challenges and opportunities. Food Res Int 137:109612

Barbazuk WB, Emrich SJ, Chen HD, Li L, Schnable PS (2007) SNP discovery via 454 transcriptome sequencing. Plant J 51(5):910–918

Beckles DM (2012) Factors affecting the postharvest soluble solids and sugar content of tomato (Solanum lycopersicum L.) fruit. Biol Technol 63(1):129–140

Begou O, Gika HG, Wilson ID, Theodoridis G (2017) Hyphenated MS-based targeted approaches in metabolomics. Analyst 142(17):3079–3100

Biswas P, East AR, Hewett EW, Heyes JA (2016) Chilling injury in tomato fruit. Hortic Rev 44:229–278

Borguini RG, Torres EAFDS (2009) Tomatoes and tomato products as dietary sources of antioxidants. Food Rev Int 25(4):313–325

Carmona L, Zacarías L, Rodrigo MJ (2012) Stimulation of coloration and carotenoid biosynthesis during postharvest storage of 'Navelina'orange fruit at 12 C. Postharvest Biol Technol 74:108–117

Carrari F, Fernie AR (2006) Metabolic regulation underlying tomato fruit development. J Exp Bot 57(9):1883–1897

Castro-Puyana M, Pérez-Míguez R, Montero L, Herrero M (2017) Reprint of: Application of mass spectrometry-based metabolomics approaches for food safety, quality and traceability. TrAC Trends Anal Chem 96:62–78

Centeno DC, Osorio S, Nunes-Nesi A, Bertolo AL, Carneiro RT, Araujo WL, Fernie AR (2011) Malate plays a crucial role in starch metabolism, ripening, and soluble solid content of tomato fruit and affects postharvest softening. Plant Cell 23(1):162–184

Chrysargyris A, Nikou A, Tzortzakis N (2016) Effectiveness of Aloe vera gel coating for maintaining tomato fruit quality. N Z J Crop Hortic Sci 44(3):203–217

Collins EJ, Bowyer C, Tsouza A, Chopra M (2022) Tomatoes: An extensive review of the associated health impacts of tomatoes and factors that can affect their cultivation. Biol 11(2):239
Consonni R, Cagliani LR (2019) The potentiality of NMR-based metabolomics in food science and food authentication assessment. Magn Reson Chem 57(9):558–578
Correia AFK, Loro AC, Zanatta S, Spoto MHF, Vieira TMFS (2015) Effect of temperature, time, and material thick-ness on the dehydration process of tomato. Int J Food Sci 2015:970724
Cruz AB, Bianchetti RE, Alves FRR, Purgatto E, Peres LEP, Rossi M, Freschi L (2018) Light, ethylene and auxin signaling interaction regulates carotenoid biosynthesis during tomato fruit ripening. Front Plant Sci 9:1370
Dávila-Aviña JE, Villa-Rodríguez JA, Villegas-Ochoa MA, Tortoledo-Ortiz O, Olivas GI, Ayala-Zavala JF, González-Aguilar GA (2014) Effect of edible coatings on bioactive compounds and antioxidant capacity of tomatoes at different maturity stages. J Food Sci Technol 51(10):2706–2712
De Wild HP, Otma EC, Peppelenbos HW (2003) Carbon dioxide action on ethylene biosynthesis of preclimacteric and climacteric pear fruit. J Exp Bot 54(387):1537–1544
Dixon RA, Gang DR, Charlton AJ, Fiehn O, Kuiper HA, Reynolds TL, Seiber JN (2006) Applications of metabolomics in agriculture. Agric Food Chem 54(24):8984–8994
Feizi N, Hashemi-Nasab FS, Golpelichi F, Saburouh N, Parastar H (2021) Recent trends in application of chemometric methods for GC-MS and GCŨGC-MS-based metabolomic studies. TrAC Trends Anal Chem 138:116239
Feng Z, Ding C, Li W, Wang D, Cui D (2020) Applications of metabolomics in the research of soybean plant under abiotic stress. Food Chem 310:125914
Fich EA, Fisher J, Zamir D, Rose JK (2020) Transpirationfrom tomato fruit occurs primarily via trichome-associated transcuticular polar pores. Plant Physiol 184(4):1840–1852
Fiehn O, Kopka J, Dörmann P, Altmann T, Trethewey RN, Willmitzer L (2000) Metabolite profiling for plant functional genomics. Nat Biotechnol 18(11):1157–1161
Fraga-Corral M, Carpena M, Garcia-Oliveira P, Pereira AG, Prieto MA, Simal-Gandara J (2022) Analytical metabolomics and applications in health, environmental and food science. Crit Rev Anal Chem 52(4):712–734
Gapper NE, Giovannoni JJ (2013) Molecular and genetic regulation of fruit ripening. Plant Mol Biol 82(6):575–591
Giraudeau P (2020) NMR-based metabolomics and fluxomics: developments and future prospects. Analyst 145(7):2457–2472
Górecka D, Wawrzyniak A, Jędrusek-Golińska A, Dziedzic K, Hamułka J, Kowalczewski PŁ, Walkowiak J (2020) Lycopene in tomatoes and tomato products. Open Chem 18(1):752–775
Goulao LF, Oliveira CM (2008) Cell wall modifications during fruit ripening: when a fruit is not the fruit. Trends Food Sci Technol 19(1):4–25
Hanhineva K, Aharoni A (2009) Metabolomics in fruit development. In: Molecular techniques in crop improvement, 2nd edn, pp 675–693
Hussain PR, Meena RS, Dar MA, Wani AM (2012) Effect of post-harvest calcium chloride dip treatment and gamma irradiation on storage quality and shelf-life extension of Red delicious apple. J Food Sci Technol 49:415–426
Ibáñez C, García-Cañas V, Valdés A, Simó C (2013) Novel MS-based approaches and applications in food metabolomics. TrAC Trends Anal Chem 52:100–111
Idah, P. A., E. S. Ajisegiri, and M. G. Yisa, 2007. An assessment of impact damage to fresh tomato fruits
Javanmardi J, Kubota C (2006) Variation of lycopene, antioxidant activity, total soluble solids and weight loss of tomato during postharvest storage. Postharvest Biol Technol 41(2):151–155
Kabir M, Nur S, Ali M, Lee WH, Cho SI, Chung SO (2020) Physicochemical quality changes in tomatoes during delayed cooling and storage in a controlled chamber. Agriculture 10(6):196
Khoso MA, Hussain A, Ritonga FN, Ali Q, Channa MM, Alshegaihi RM, Manghwar H (2022) WRKY transcription factors (TFs): molecular switches to regulate drought, temperature, and salinity stresses in plants. Front Plant Sci 13:1039329

Klassen A, Faccio AT, Canuto GAB, da Cruz PLR, Ribeiro HC, Tavares MFM, Sussulini A (2017) Metabolomics: definitions and significance in systems biology. Adv Exp Med Biol 965:3–17
Klee HJ, Giovannoni JJ (2011) Genetics and control of tomato fruit ripening and quality attributes. Annu Rev Genet 45(1):41–59
Lacalle-Bergeron L, Izquierdo-Sandoval D, Sancho JV, López FJ, Hernández F, Portolés T (2021) Chromatography hyphenated to high resolution mass spectrometry in untargeted metabolomics for investigation of food (bio) markers. TrAC Trends Anal Chem 135:116161
Lee SK, Kader AA (2000) Preharvest and postharvest factors influencing vitamin C content of horticultural crops. Postharvest Biol Technol 20(3):207–220
Lee EJ, Shaykhutdinov R, Weljie AM, Vogel HJ, Facchini PJ, Park SU, Yang TJ (2009) Quality assessment of ginseng by 1H NMR metabolite fingerprinting and profiling analysis. J Agric Food Chem 57(16):7513–7522
León C, Cifuentes A, Valdés A (2018) Foodomics applications. In: Comprehensive analytical chemistry, vol 82. Elsevier, pp 643–685
Li C, Tao J, Zhang H (2017) Peach gum polysaccharides-based edible coatings extend shelf life of cherry tomatoes. 3Biotech 7(3):1–5
Liu ZP, Chen L (2012) Proteome-wide prediction of protein-protein interactions from high-throughput data. Protein Cell 3(7):508–520
Liu X, Locasale JW (2017) Metabolomics: a primer. Trends Biochem Sci 42(4):274–284
López-Ruiz R, Romero-González R, Frenich AG (2019) Metabolomics approaches for the determination of multiple contaminants in food. Curr Opin Food Sci 28:49–57
Mahfoudhi N, Hamdi S (2015) Use of almond gum and gumarabic as novel edible coating to delay postharvest ripening and to maintain sweet cherry (Prunus avium) quality during storage. J Food Process Preserv 39(6):1499–1508
Malundo TMM, Shewfelt RL, Scott JW (1995) Flavor quality of fresh tomato (Lycopersicon esculentum Mill.) as affected by sugar and acid levels. Postharvest Biol Technol 6(1–2):103–110
Mein JR, Lian F, Wang XD (2008) Biological activity of lycopene metabolites: implications for cancer prevention. Nutr Rev 66(12):667–683
Meng, F., Liu, H., Hu, S., Jia, C., Zhang, M., Li, S., ... & Wang, Q. (2023). The brassinosteroid signaling component SlBZR1 promotes tomato fruit ripening and carotenoid accumulation. Journal of Integrative Plant Biology, 65(7), 1794-1813.
Mielko KA, Pudełko-Malik N, Tarczewska A, Młynarz P (2021) NMR spectroscopy as a "green analytical method" in metabolomics and proteomics studies. Sustain Chem Pharm 22:100474
Miller AR (2002) Harvest and handling injury: physiology, biochemistry, and detection. In: Postharvest physiology and pathology of vegetables. CRC Press, pp 215–249
Moco S, Forshed J, De Vos RC, Bino RJ, Vervoort J (2008) Intra-and inter-metabolite correlation spectroscopy of tomato metabolomics data obtained by liquid chromatography-mass spectrometry and nuclear magnetic resonance. Metabolomics 4:202–215
Mohammed, I. A., & Al Dulaimi, M. A. K. (2021, November). An Economic analysis of the costs of producing tomato under greenhouse in Anbar Governorate for the agricultural season 2019-2020. In IOP Conference Series: Earth and Environmental Science (Vol. 904, No. 1, p. 012061). IOP Publishing
Mrema GC, Rolle RS (2002) Status of the postharvest sector and its contribution to agricultural development and economic growth. 9th JIRCAS International Symposium 2002 – 'Value-addition to agricultural products', 13–20
Nath A, Deka BC, Singh A, Patel RK, Paul D, Misra LK, Ojha H (2012) Extension of shelf life of pear fruits using different packaging materials. J Food Sci Technol 49(5):556–563
Nunes MCN, Emond JP, Rauth M, Dea S, Chau KV (2009) Environmental conditions encountered during typical consumer retail display affect fruit and vegetable quality and waste. Postharvest Biol Technol 51(2):232–241
Osorio S, Tohge T, Fernie AR (2009) Application of metabolomic profiling for identifying valuable traits in tomato. CABI Rev 2009:1–9

Osorio S, Carneiro RT, Lytovchenko A, McQuinn R, Sørensen I, Vallarino JG, Rose JK (2020) Genetic and metabolic effects of ripening mutations and vine detachment on tomato fruit quality. Plant Biotechnol J 18(1):106–118
Pagno CH, Castagna A, Trivellini A, Sodi A, Ranieri A, Ferreira EA, Flôres SH (2018) The nutraceutical quality of tomato fruit during domestic storage is affected by chitosancoating. J Food Process Preserv 42(1):13326
Paul V, Pandey R (2014) Role of internal atmosphere on fruitripening and storability—a review. J Food Sci Technol 51(7):1223–1250
Payasi A, Mishra NN, Chaves ALS, Singh R (2009) Biochemistry of fruit softening: an overview. Physiol Mol Biol Plants 15(2):103–113
Perez de Souza L, Garbowicz K, Brotman Y, Tohge T, Fernie AR (2020) The acetate pathway supports flavonoid and lipid biosynthesis in Arabidopsis. Plant Physiol 182(2):857–869
Perveen R, Suleria HAR, Anjum FM, Butt MS, Pasha I, Ahmad S (2015) Tomato (Solanum lycopersicum) carotenoids and lycopenes chemistry; metabolism, absorption, nutrition, and allied health claims—A comprehensive review. Crit Rev Food Sci Nutr 55(7):919–929
Pezzatti J, Boccard J, Codesido S, Gagnebin Y, Joshi A, Picard D, Rudaz S (2020) Implementation of liquid chromatography–high resolution mass spectrometry methods for untargeted metabolomic analyses of biological samples: a tutorial. Anal Chim Acta 1105:28–44
Pontes JGM, Brasil AJM, Cruz GC, de Souza RN, Tasic L (2017) NMR-based metabolomics strategies: plants, animals and humans. Anal Methods 9(7):1078–1096
Pott DM, Vallarino JG, Osorio S (2020) Metabolite changes during postharvest storage: Effects on fruit quality traits. Meta 10(5):187
Prudent M, Causse M, Génard M, Tripodi P, Grandillo S, Bertin N (2009) Genetic and physiological analysis of tomato fruit weight and composition: influence of carbon availability on QTL detection. J Exp Bot 60(3):923–937
Raiola A, Rigano MM, Calafiore R, Frusciante L, Barone A (2014) Enhancing the health-promoting effects of tomato fruit for biofortified food. Mediat Inflamm 2014(1):139873
Rajkumar P, Kulanthaisami S, Raghavan GSV, Gariépy Y, Orsat V (2007) Drying kinetics of tomato slices in vacuum assisted solar and open sun drying methods. Dry Technol 25(7–8):1349–1135
Ruelas-Chacon X, Contreras-Esquivel JC, Montañez J, Aguilera-Carbo AF, Reyes-Vega ML, Peralta-Rodríguez RD, Sanchéz-Brambila G (2017) Guar gum as an edible coating forenhancing shelf-life and improving postharvest quality of romatomato (Solanum lycopersicum L.). J Food Qual 2017:1–9
Ruiz-Martínez J, Aguirre-Joya JA, Rojas R, Vicente A, Aguilar-González MA, Rodríguez-Herrera R, Aguilar CN (2020) Candelilla wax edible coating with Flourensia cernua bioactives toprolong the quality of tomato fruits. Food Secur 9(9):130
Sablani SS, Opara LU, Al-Balushi K (2006) Influence of bruising and storage temperature on vitamin C content of tomato fruit. J Food Agric Environ 4(1):54
Safari ZS, Ding P, Nakasha JJ, Yusoff SF (2021) Control-ling Fusarium oxysporum tomato fruit rot under tropical condition using both chitosan and vanillin. Coatings 11(3):367
Sandhya. (2010) Modified atmosphere packaging of fresh produce: Current status and future needs. LWT-Food Sci Technol 43(3):381–392
Saure, M. C. (2001). Blossom-end rot of tomato (Lycopersicon esculentum Mill.)—a calcium-or a stress-related disorder?. Scientia horticulturae, 90(3-4), 193–208
Shehata SA, Abdelrahman SZ, Megahed MM, Abdeldaym EA, El-Mogy MM, Abdelgawad KF (2021) Extending shelf life and maintaining quality of tomato fruit by calcium chloride, hydrogen peroxide, chitosan, and ozonated water. Horticulturae 7(9):309
Sinha, S. R., Singha, A., Faruquee, M., Jiku, M. A. S., Rahaman, M. A., Alam, M. A., & Kader, M. A. (2019). Post-harvest assessment of fruit quality and shelf life of two elite tomato varieties cultivated in Bangladesh. Bulletin of the National Research Centre, 43, 1–12
Story EN, Kopec RE, Schwartz SJ, Harris GK (2010) Anupdate on the health effects of tomato lycopene. Annu Rev Food Sci Technol 1:189–210

Suslow TV, Cantwell M (2009) Tomato recommendations for maintaining postharvest quality. Produce facts. Davis: Postharvest Technology Research & Information Center. Available at: http://postharvest.ucdavis.edu/files/259455.pdf. Accessed on 18 Aug 2021

Tan HL, Thomas-Ahner JM, Grainger EM, Wan L, Francis DM, Schwartz SJ, Clinton SK (2010) Tomato-based food products for prostate cancer prevention: what have we learned? Cancer Metastasis Rev 29:553–568

Tang H, Zhang X, Gong B, Yan Y, Shi Q (2020) Proteomics and metabolomics analysis of tomato fruit at different maturity stages and under salt treatment. Food Chem 311:126009

Thole V, Vain P, Yang RY, da Silva JAB, Enfissi EM, Nogueira M, Martin C (2020) Analysis of tomato post-harvest properties: fruit color, shelf life, and fungal susceptibility. Curr Protoc Plant Biol 5(2):e20108

Tigist M, Workneh TS, Woldetsadik K (2013) Effects of varietyon the quality of tomato stored under ambient conditions. J Food Sci Technol 50(3):477–448

Trong LV, Tuong LQ, Thinh BB, Khoi NT, Trong VT (2019) Physiological and biochemical changes in tomato fruit (Solanum lycopersicum L.) during growth and ripening cultivated in Vietnam. Biosci Res 16(2):1736–1744

Uarrota VG, Fuentealba C, Hernández I, Defilippi-Bruzzone B, Meneses C, Campos-Vargas R, Pedreschi R (2019) Integration of proteomics and metabolomics data of early and middle season Hass avocados under heat treatment. Food Chem 289:512–521

Utpott M, Rodrigues E, de Oliveira Rios A, Mercali GD, Flôres SH (2022) Metabolomics: An analytical technique for food processing evaluation. Food Chem 366:130685

Vallarino JG, Yeats TH, Maximova E, Rose JK, Fernie AR, Osorio S (2017) Postharvest changes in LIN5-down-regulated plants suggest a role for sugar deficiency in cuticle metabolism during ripening. Phytochemistry 142:11–20

Vallarino JG, Erban A, Fehrle I, Fernie AR, Kopka J, Osorio S (2018) Acquisition of volatile compounds by gas chromatography–mass spectrometry (GC-MS). Methods Mol Biol 1778:225–239

Vogel JT, Tieman DM, Sims CA, Odabasi AZ, Clark DG, Klee HJ (2010) Carotenoid content impacts flavor acceptability in tomato (Solanum lycopersicum). J Sci Food Agric 90(13):2233–2240

Wang C, Li M, Duan X, Abu-Izneid T, Rauf Z, Khan A, Suleria HA (2023) Phytochemical and nutritional profiling of tomatoes; impact of processing on bioavailability-a comprehensive review. Food Rev Int 39(8):5986–6010

Wang, H., Schauer, N., Usadel, B., Frasse, P., Zouine, M., Hernould, M., ... & Bouzayen, M. (2009). Regulatory features underlying pollination-dependent and-independent tomato fruit set revealed by transcript and primary metabolite profiling. The Plant Cell, 21(5), 1428–1452

Waseem M, Li N, Su D, Chen J, Li Z (2019) Overexpression of a basic helix–loop–helix transcription factor gene, SlbHLH22, promotes early flowering and accelerates fruit ripening in tomato (Solanum lycopersicum L.). Planta 250:173–185

Wishart DS, Tzur D, Knox C, Eisner R, Guo AC, Young N, Querengesser L (2007) HMDB: the human metabolome database. Nucleic Acids Res 35(suppl_1):D521–D526

Yadav A, Kumar N, Upadhyay A, Sethi S, Singh A (2022) Edible coating as postharvest management strategy for shelf-life extension of fresh tomato (Solanum lycopersicum L.): an overview. J Food Sci 87(6):2256–2290

Yuan Y, Zhao Y, Yang J, Jiang Y, Lu F, Jia Y, Yang B (2017) Metabolomic analyses of banana during postharvest senescence by 1H-high resolution-NMR. Food Chem 218:406–412

Zhang L, Yu Z, Jiang L, Jiang J, Luo H, Fu L (2011) Effect of post-harvest heat treatment on proteome change of peach fruit during ripening. J Proteome 74(7):1135–1149

Zhang B, Tieman DM, Jiao C, Xu Y, Chen K, Fei Z, Klee HJ (2016) Chilling-induced tomato flavor loss is associated with altered volatile synthesis and transient changes in DNA methylation. Proc Natl Acad Sci 113(44):12580–12585

Zhang J, Ren Y, Ma Q (2018) Effect of temperature and light on metabolic composition on tomato fruit. Food Chem 245:652–659

Zhang X, Zhang X, Liu X, Du M, Tian Y (2019) Effect of polysaccharide derived from Osmunda japonica Thunb-incorporated carboxymethyl cellulose coatings on preservation of tomatoes. J Food Process Preserv 43(12):14239

Zhu F, Wen W, Cheng Y, Fernie AR (2022) The metabolic changes that effect fruit quality during tomato fruit ripening. Mol Hortic 2(1):2

Zhuang J, Zhang J, Hou XL, Wang F, Xiong AS (2014a) Transcriptomic, proteomic, metabolomic and functional genomic approaches for the study of abiotic stress in vegetable crops. Crit Rev Plant Sci 33(2–3):225–237

Insights from Genome Editing into Tomato Germplasm: Current Status and Future Perspective

8

Shumaila Ijaz, Javed Iqbal, Banzeer Ahsan Abbasi, Zakir Ullah, Rooma Waqar, Muhammad Usman, Tabassum Yaseen, Shoaib Khan, Rashid Iqbal, Ghulam Murtaza, Sadaf Arbab, Farishta Zarshan, Madiha Imtiaz, and Tariq Mahmood

Abstract

Genome editing (GE) has become a potent method for precisely enhancing tomato features, allowing genetic factors to be modified with specificity to increase nutritional quality, disease resistance, and yield. Recent research has revealed the effective use of clustered regularly interspaced short palindromic repeat (CRISPR/Cas9) technology to modify certain genes linked to fruit quality and stress responses, resulting in improved fruit qualities including larger fruit and better flavor profiles. Furthermore, studies show that genome editing can

Authors Shumaila Ijaz and Javed Iqbal have been equally contributed to this chapter.

S. Ijaz
Department of Burn and Plastic Surgery, Shenzhen Institute of Translational Medicine, The First Affiliated Hospital of Shenzhen University, School of Biomedical Engineering, Shenzhen University Medical School, Shenzhen University, Shenzhen, China

J. Iqbal (✉) · T. Yaseen · S. Arbab · F. Zarshan · M. Imtiaz
Department of Botany, Bacha Khan University, Charsadda, Khyber Pakhtunkhwa, Pakistan
e-mail: tabassumyaseen@bkuc.edu.pk

B. A. Abbasi
Department of Botany, Rawalpindi Women University, Rawalpindi, Pakistan

Z. Ullah · R. Waqar · T. Mahmood
Department of Plant Sciences, Faculty of Biological Sciences, Quaid-i-Azam University, Islamabad, Pakistan
e-mail: roomawaqar@bs.qau.edu.pk; tmahmood@qau.edu.pk

M. Usman
Department of Plant Pathology, University of Agriculture, Faisalabad, Pakistan

M. Waseem et al. (eds.), *Omics Approaches for Tomato Yield and Quality Trait Improvement*, https://doi.org/10.1007/978-981-96-3890-1_8

help produce tomato types that are more resistant to infections, which will lessen the need for chemical pesticides. But there are still issues that might prevent these technologies from being widely used, such as public acceptance of genetically modified organisms (GMOs) and regulatory barriers. Even though genome editing has a lot of potential to improve tomato farming, more study and discussion are needed to resolve these issues and realize the full range of advantages. With the ability to make precise genetic changes that improve desired features, CRISPR technology is essential to the modification of tomato properties. Studies show that CRISPR may be used to enhance tomato characteristics including disease resistance, fruit quality, and stress tolerance. Promising outcomes have been observed in the use of genetic modification in tomato breeding. For instance, gamma-aminobutyric acid (GABA) concentration, domestication qualities, and jointless-2 phenotype have all been effectively edited by researchers, leading to tomatoes with higher nutritional value, improved domestication traits, and changed fruit characteristics. Moreover, quantitative trait variation in tomatoes has been engineered through the use of genome editing tools. Researchers have created a variety of cis-regulatory alleles using promoter editing, which offers useful quantitative diversity for breeding. This technique enables the rapid evaluation of a large number of promoter variations for genes controlling crucial tomato production factors such as fruit size, inflorescence branching, and plant architecture. The chapter describes how to change certain genes in the tomato genome using precise techniques such as the CRISPR/Cas9 system and its derivatives. With the use of these instruments, researchers can improve the effectiveness, accuracy, and success of breeding programs, which have the potential to completely transform crop breeding.

Keywords

Genome editing · Genetic modification · CRISPR/Cas9 technology · Chemical pesticides · Genetically modified organisms (GMOs) · Regulatory barriers · Precise genetic changes · Cis-regulatory alleles

---

S. Khan
Department of Chemistry, Abbottabad University of Science and Technology (AUST), Abbottabad, Pakistan

R. Iqbal
Department of Agronomy, Faculty of Agriculture and Environment, The Islamia University of Bahawalpur, Bahawalpur, Pakistan

Department of Life Sciences, Western Caspian University, Baku, Azerbaijan
e-mail: rashid.iqbal@iub.edu.pk

G. Murtaza
School of Agriculture, Yunnan University, Kunming, Yunnan, China

8.1 Introduction

The development of clustered regularly interspaced short palindromic repeat (CRISPR/Cas9) technology has transformed genetic engineering by providing previously unheard-of levels of genome editing (GE) efficiency and precision. With the use of this innovative instrument, which comes from a bacterial defensive mechanism, scientists may precisely alter the DNA of many creatures, including plants. The use of CRISPR/Cas9 is particularly important in the field of agriculture because it provides new opportunities to improve resilience, enhance agricultural features, and address issues related to food security and climate change. A crop that will greatly benefit from this technique is the tomato, which is a mainstay in diets all over the world and an essential component of many culinary traditions (Singh et al. 2018; Macarrón Palacios et al. 2024). In addition to their nutritional worth, tomatoes are prized for their economic significance in the world market. But conventional breeding techniques sometimes fail to produce the quick improvements required to create new cultivars resistant to pests, illnesses, and environmental challenges. Researchers can now precisely target genes linked to desirable qualities like disease resistance, fruit quality, and yield potential in tomato breeding programs, thanks to the incorporation of CRISPR/Cas9. In order to produce tomato varieties that can flourish in shifting climates and satisfy customers' increasing needs, precise trait improvement is essential (Bergougnoux 2014; Ahmad et al. 2022).

It is evident as we investigate the possibilities of CRISPR/Cas9 and its uses in tomato genome editing that this technology may promote sustainable farming methods in addition to increasing agricultural output. Through the enhancement of crop resilience and decreased dependence on chemical inputs, CRISPR/Cas9 opens the door to a more sustainable agricultural future. Furthermore, the ethical questions raised by genome editing are becoming more pressing, sparking conversations about how these technologies may affect environmental health, biodiversity, and food security. This introduction lays the groundwork for a more in-depth look at how CRISPR/Cas9 will influence tomato farming going forward and what more the field of genome editing has to offer (Liu et al. 2021, 2022).

In contemporary agriculture, genome editing has been shown to be a revolutionary method, especially for improving tomato precision traits. Researchers may create tomato cultivars with enhanced resistance, increased nutritional content, and superior overall quality by making targeted alterations to the tomato genome using technologies like CRISPR/Cas9 (Nagamine and Ezura 2022; Haider et al. 2021a). Conventional breeding techniques, which frequently depend on the sluggish process of cross-breeding and selection, can be unreliable and ineffective. Genome editing, on the other hand, enables researchers to precisely modify the genes that confer desired features, greatly speeding up the breeding process and increasing the tomato crop's potential for innovation. For example, by focusing on genes linked to disease resistance, such as the MLO gene, scientists have created tomato cultivars that are resistant to diseases like powdery mildew, which eliminates the need for chemical fungicides. Using fewer chemicals in agriculture not only increases tomato output and quality but also supports ecologically friendly practices (Ahmar et al. 2020; Ahmad 2023).

Furthermore, genome editing is used for purposes other than resistance to illnesses. Heat and drought resistance are essential characteristics as climate change continues to impact agricultural output worldwide. Through the use of CRISPR/Cas9, scientists have been able to change genes that control stress responses, such as SlDREB2 and SlHSP. As a result, tomato varieties that can thrive under harsh environmental circumstances have been created. These improvements are especially important for areas where there is a shortage of water since they allow growers to grow tomatoes with less irrigation, which conserves water. Additionally, by raising the quantities of vitamins, antioxidants, and other healthy substances, genome editing might enhance the nutritional profile of tomatoes. For example, certain changes to the Sly-miR156 gene can result in higher levels of carotenoid content, which is vital for human health. These developments support the battle against hunger in a variety of demographics in addition to satisfying consumer demand for better food alternatives (Ahmad et al. 2022; Ndudzo et al. 2024).

Utilizing genome editing tools like CRISPR/Cas9 is not without its difficulties and moral implications. Concerns about biodiversity, unexpected repercussions, and the effects of releasing genetically modified organisms (GMOs) into the environment must all be taken into consideration as technology advances. Regulations must change to keep up with these advancements, guaranteeing thorough and transparent safety evaluations and promoting an atmosphere that encourages research and development (R&D). Furthermore, important to the acceptability of genome-edited crops is public perception. Campaigns to raise consumer knowledge of the advantages and security of these technologies are essential for closing the acceptability gap between scientific innovation and the general population. To sum up, genome editing is a revolutionary step forward in the search for precisely engineered tomato traits. Through the utilization of CRISPR/Cas9 and other genome editing technologies, scientists may produce tomato varieties that possess enhanced nutritional value and increased resistance to environmental stressors and illnesses. The advantages might go beyond specific crops to include more general agricultural sustainability and food security. In order to ensure that these potent tools are used responsibly and effectively for the benefit of agriculture and society at large, it is crucial that stakeholders from scientists to legislators and consumers engage in candid conversations about the ethical implications and regulatory considerations of genome editing as the field continues to develop (Ahmad et al. 2024a; Ezin and Symonds 2023).

The potential for genome editing in tomatoes is enormous because of ongoing technological developments and a growing comprehension of plant biology. Improving precision editing tools is one of the most important possible development topics. Even though CRISPR/Cas9 has already completely changed the sector, research is still being done to improve these instruments' specificity and efficiency. The advent of novel Cas proteins, including Cas12 and Cas13, presents the possibility of more adaptable editing features like the capacity to target several genes at once or carry out precise alterations without creating double-strand breaks in the

DNA. By carefully modifying many genes involved in these qualities, tomato cultivars with complex attributes, such as enhanced taste, texture, and nutritional value, may be developed. Combining genome editing with cutting-edge breeding methods is another exciting area. The introduction of desired features from wild tomato cousins into cultivated varieties may occur more quickly if CRISPR technology is used with conventional breeding techniques like marker-assisted selection. This strategy not only increases genetic variety but also makes it possible to quickly create new cultivars that are more suited to shifting environmental circumstances. To increase the genetic pool accessible for breeding programs, scientists are investigating the possibility of de novo domestication, in which wild animals are altered to express characteristics that make them suited for agriculture (Ahmad et al. 2024b; Nagamine et al. 2023; Haider et al. 2021a).

Furthermore, major breakthroughs are anticipated when genome editing and synthetic biology are combined. Through the design and construction of novel biological components, tools, and systems, scientists may engineer tomatoes with specific features not present in the wild. This may involve creating tomatoes that can naturally resist pests or perhaps altering metabolic pathways to increase the production of healthy substances like vitamins and antioxidants. These developments might enhance tomatoes' nutritional value while lowering the need for chemical pesticides and encouraging more environmentally friendly farming methods. Enhancement of stress tolerance is an additional crucial topic of attention. It will be crucial to produce tomato varieties that can resist abiotic conditions like heat, salt, and drought as climate change continues to present difficulties for agriculture. In the future, genome editing may be used to precisely control the expression of genes linked to stress reactions, allowing tomatoes to flourish in less-than-ideal environments. This might make tomato crops far more resilient, assuring food security in areas that are becoming more susceptible to climatic fluctuation (Ali et al. 2022; Dayma et al. 2023).

Last, how genome editing is regulated will have a significant impact on how it is used in the future. Clear and encouraging regulatory frameworks that promote the safe and ethical use of genome editing technologies are becoming more and more necessary as public opinion about genetically modified organisms (GMOs) changes. Establishing criteria that distinguish conventional genetically modified organisms (GMOs) from genome-edited crops, which frequently lack foreign DNA and could be more palatable to consumers, is part of this. Regulatory agencies can assist in realizing the full promise of genome editing in tomato breeding and other fields by creating an atmosphere that promotes creativity while guaranteeing safety. There will undoubtedly be many developments in the field of genome editing for the purpose of precisely enhancing tomato traits in the future. The potential to completely transform tomato farming is enormous, ranging from higher stress tolerance and supporting regulatory frameworks to better editing tools and the use of synthetic biology. These developments promise to enhance the resilience and quality of tomatoes while also advancing more general objectives of food security and sustainability in a world that is changing constantly (Ijaz et al. 2023, 2024; Haider et al. 2021b, 2022; Ullah et al. 2023a).

In conclusion, CRISPR/Cas9's development and use in genome editing to improve certain traits in tomatoes represent an important turning point in agricultural biotechnology. The potential to create tomato varieties with enhanced nutritional content and quality, as well as increased resistance to diseases and environmental challenges, is presented by the capacity to precisely change the tomato genome. The creation of tomatoes that can withstand the difficulties of a changing environment and the increasing needs of customers will surely be aided by the combination of cutting-edge editing technologies, synthetic biology, and creative breeding tactics in the future. Furthermore, the broad adoption and application of these technologies will depend on how well the regulatory environment and public perception around genome editing are handled. By embracing these developments, we can ensure that agriculture remains stable and sustainable in the future and that tomatoes continue to be an essential part of the world's food chain. This is an exciting time for scientists, farmers, and consumers alike since CRISPR/Cas9 has the ability to improve food security and encourage environmental stewardship in addition to particular crops.

8.2 Tools for Editing Genomes

Variations in the genetic pool provide the foundation for biological diversification. In plant breeding, genetic variety is essential for producing new cultivars and serves as the foundation for enhancing an organism's features. Mutations are spontaneous genetic differences that happen throughout evolution as a result of damage to DNA or mistakes made during the replication process called mutagenesis. It is not feasible to depend solely on natural variants for crop breeding since natural mutations are sporadic, random, and spontaneous (Manova and Gruszka 2015). Therefore, increasing genetic variation requires intentional mutagenesis, which is a necessary stage in the breeding procedure.

Methods for genome editing (GE) have revolutionized biology by enabling accurate, effective, and focused changes to the genomes of a wide range of species, including humans, animals, plants, and microbes. The basic mechanism of GE is DNA double-strand breaks (DSBs), which are caused by nucleases that are carefully engineered. After that, the double-stranded DNA (dsDNA) break is repaired via either homology-dependent recombination (HDR) or nonhomologous end joining (NHEJ). With very low editing frequencies, HDR makes it possible to create exact point mutations, deletions, or beneficial gene knock-ins for crop breeding. In contrast, NHEJ is prone to errors, allowing for inadvertent small insertions or deletions alongside substitutions, which ultimately may lead to gene knockout (Veillet et al. 2019). A variety of genome editing techniques, including induced mutagenesis, transposable elements, zinc finger nucleases (ZFNs), transcriptional activator-like effector nucleases (TALENs), oligonucleotide-directed mutagenesis (ODM), epigenome editing, and more recently, clustered regularly interspaced short palindromic repeat (CRISPR/Cas9) systems, can be employed to elucidate the functions

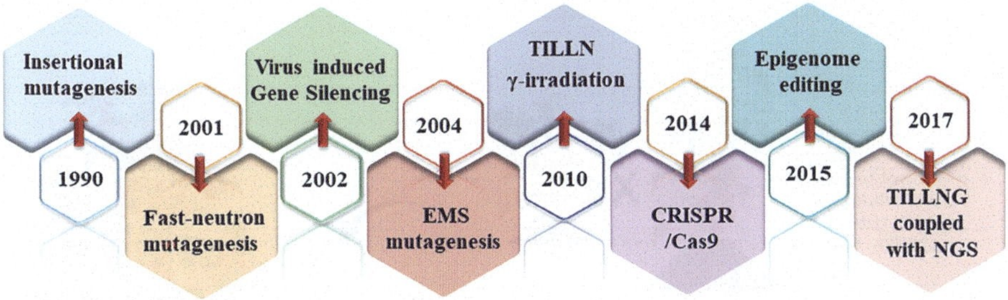

Fig. 8.1 The timeline of the breakthrough in tomato genome editing. Insertional mutagenesis, γ-irradiation, CRISPR/Cas9, fast-neutron mutagenesis, targeted induced local lesions in genomes (TILLING), TALENs, virus-induced gene silencing (VIGS), TILLING in conjunction with next-generation sequencing (NGS), ethyl methanesulfonate (EMS) mutagenesis, and other mutagenesis techniques have all been outlined

of unknown and uncharacterized genes (Vats et al. 2019). The historical progression of genome editing technology is illustrated in Fig. 8.1.

8.2.1 Induced Mutagenesis

The process of induced mutagenesis makes it possible to introduce new genetic alleles and produce novel genetic properties for the purpose of improving crops and understanding gene functions. Genetic breeding techniques have advanced remarkably since Stadler first reported on mutant breeding in 1928 (Stadler 1928). It has been extensively employed in crop plants that are resistant to conventional breeding methods and have limited genetic variability (Fig. 8.2a). Once created, the mutagenized population is a perpetual resource. Since mutagenic plants have a long history of safety and do not contain any foreign DNA, they are exempt from the EU GMO restrictions (Fig. 8.2b). On the other hand, creating a mutant population requires a lot of work and time. Physical mutagenesis and chemical mutagenesis are further classifications of mutagenesis based on the inciting agent (Fig. 8.2c).

8.2.2 Chemical Mutagenesis

The chemicals employed to induce plant mutagenesis consist of epoxides, sulfonic esters, oxidizing agents, purine analogs, and alkylating agents (Leitao 2012). Chemical mutagenesis offers the benefit of not requiring expensive equipment, as opposed to physical mutagenesis. This mutagenesis results in point mutations or single base alterations rather than large deletions or chromosomal rearrangements, which typically result in the acquisition or loss of functionality and give birth to novel allelic mutants. Many alkylating compounds are used for mutagenesis, including azides, ethyleneimine (EI), ethyl nitro urethane (ENU), diethyl sulfate (DES), and 1-methyl-1-nitrosourea (MNU) (Saito et al. 2011). It mostly results in random

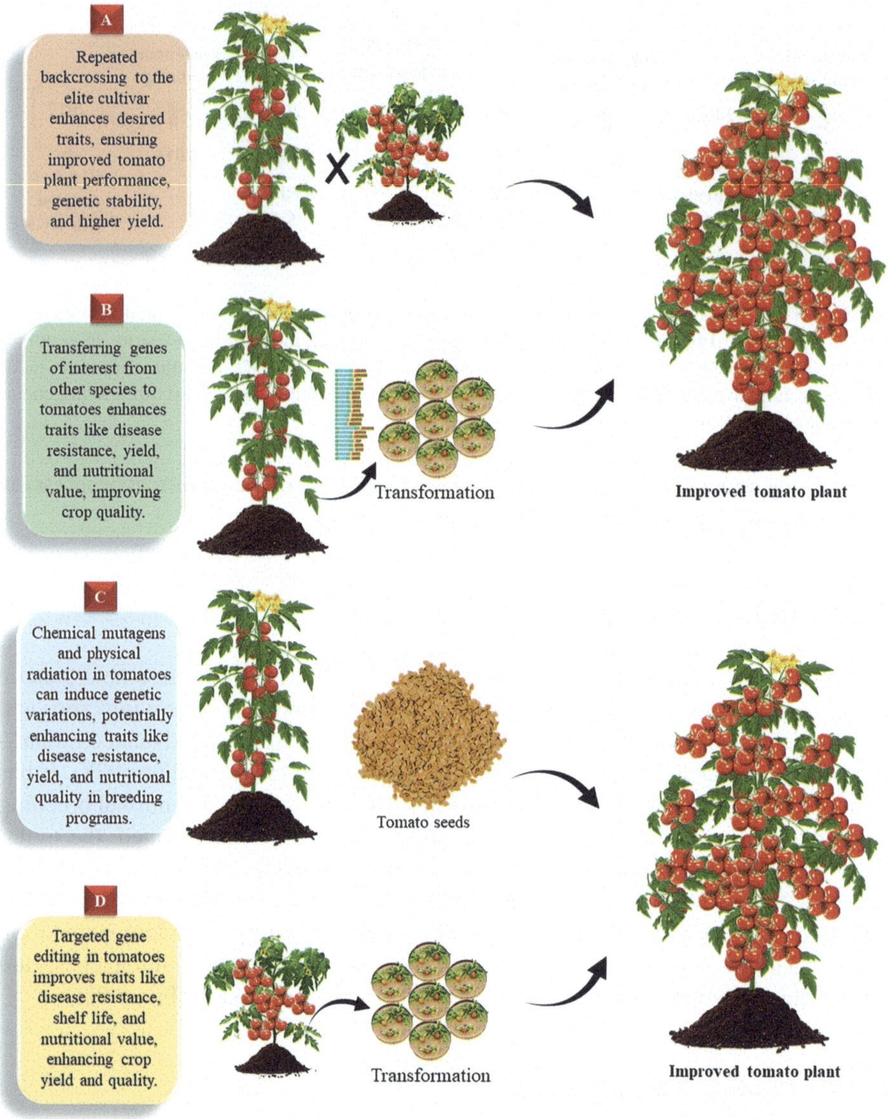

Fig. 8.2 The breeding methods used to improve crops. (**a**) In conventional breeding, two tomato species with desired features are selectively bred. For instance, a tall plant that is tolerant of stress but produces little is crossed with a little plant that produces more fruits. To produce plants with desired qualities, backcrossing to the elite cultivar is done repeatedly. The selection process in conventional breeding often results in a loss of genetic variety. (**b**) Transgenic breeding entails using the transformation process to transfer a desired gene into the selected plant from another species. (**c**) To cause mutations, physical radiation or a chemical mutagen is used in mutation breeding. In order to lessen chimerism, M2 or higher populations are created after the mutant populations (M1) are established. (**d**) Targeted genome editing provides a schematic of the process used to produce a wide range of improved plant features. Following the creation of modified plants, phenotypic and genotypic screening is done to identify the plants that have the desired features

point mutations, the majority of which are base pair transitions from G/C to A/T in species genome (Till et al. 2007). Like EMS, MNU mostly results in G/C to A/T transitions, along with lower frequencies of inversions and translocations (Maluszynski et al. 2003).

ENU strongly generates A/T to G/C transitions as well as transversions, in contrast to MNU. Furthermore, nitrogenous base analogs including 2-aminopurine, 5-bromouracil, and maleic hydrazide have mutagenic properties. Plants, however, hardly ever employ these bases (Maluszynski et al. 2003). Tomato cultivars including Moneymaker, M82, *Lycopersicon esculentum* Mill. (Kostov et al. 2007), Red Setter (Mazzucato et al. 2015), and *Arka Vikas* have all benefited from the effective use of EMS (Kostov et al. 2007). In 2007, the research group led by Kostov established an ethyl methanesulfonate (EMS) population at a concentration of 1.5% in *Lycopersicon esculentum* Mill. to investigate the resistance of tomato plants to broomrape in breeding programs. Their research identified 16 plants demonstrating resistance to *Orobanche ramosa* (Kostov et al. 2007). In a separate investigation, a total of 5508 lines of the Red Setter cultivar were subjected to treatments with EMS concentrations of 0.7 and 1%, analyzing seven distinct fruit quality traits to pinpoint 66 point mutations. This mutant population was cultivated for the purposes of both forward and reverse genetic screening, as outlined by Minoia et al. (2010). To facilitate functional genomic research within tomatoes, two EMS doses of 1% and 0.5% were utilized to generate a mutant population from the diminutive dwarf tomato cultivar Micro-Tom, with the 1% concentration yielding the most favorable results (Just et al. 2013).

A mutant population can be subjected to screening for mutations in specific genes of interest via forward genetics, which correlates phenotypic traits to specific genes, or reverse genetics, which establishes a relationship between genes and phenotypes. A prominent method employed to investigate gene functionality and locate point mutations within a mutagenized population is known as targeted induced local lesions in genomes (TILLING). This technique serves as a high-throughput reverse genetic approach (Minoia et al. 2010). There are three essential criteria that must be fulfilled for the TILLING methodology to yield optimal outcomes: first, the implementation of effective pooling strategies; second, a comprehensive gene model and protein conservation framework that aids in pinpointing gene regions with the highest likelihood of variation; and third, the identification of appropriate PCR primer pairs designed to amplify products in the range of approximately 1000–2000 base pairs. Ultimately, sequencing efforts will be employed to confirm and assess the mutations that have been detected (Maghuly and Laimer 2017). This process ultimately provides an allelic sequence comprising splice site, non-sense, mis-sense, and silent mutations to investigate the implications of diverse gene alterations. TILLING has been applied to a number of plant species, including *Arabidopsis*, maize, barley, wheat, tomato, and lotus, among others (Perry et al. 2003).

Due to its capacity to facilitate the rapid and reliable screening of a substantial number of amplicons within a condensed time frame through the sequencing of smaller amplicons (approximately 300 bp), next-generation sequencing (NGS)–based mutation detection is progressively supplanting traditional TILLING methodologies (Zhu et al. 2012, 2018). In the initial application of NGS to tomato research, Rigola et al. identified two novel alleles of SlelF4E utilizing a three-dimensional pooling approach, referred to as "keypoint" technology. When TILLING and NGS technologies were used in 2017 to screen a population of 2300 tomato lines for EMS, 64 mutations were found at a frequency of one mutation every 367 kilobases (Gupta et al. 2017).

8.2.3 Physical Mutagenesis

It has been determined that ionizing radiations are the physical mutagenic agents most efficient in inducing mutations (Smith 1958). Although physical mutagenesis has demonstrated a high degree of success in causing significant DNA fragment deletions, its main uses in mutagenesis induction are for gene rearrangement and the production of knockout mutants (Naito et al. 2005). γ- and X-rays are the most often utilized kind. When plants are exposed to γ-radiation, they can experience damage such as double-strand breaks and other DNA damage because free radicals are produced. According to Shikazono et al. (2005), γ-ray irradiation can result in significant deletions and chromosomal rearrangements. Fast-neutron radiation, on the other hand, mostly causes deletion mutations that affect a few base pairs to a few kilobases.

Furthermore, it is well known that ion beams made up of helium, protons, and heavy charged particles are extremely mutagenic. Larger DNA lesions are caused by their increased acceleration and high linear energy transfer (Abe et al. 2012). According to Tanaka et al. (2010), these ion beam radiations cause single- and double-stranded breaks that result in point mutations, translocations, deletions, and inversions. Using high-speed carbon ions, several new mutations in *Arabidopsis* were found (Shikazono et al. 2005). Although physical mutagenesis can arise from several sources, most mutations caused by physical mutagens are caused by γ-irradiation.

More than 3200 distinct kinds of induced mutant lines are available at the Joint FAO/IAEA, a joint venture between the United Nations Food and Agriculture Organization and the International Atomic Energy Agency, located in Vienna (Mba 2013). Using physical mutagenesis, several mutants are produced in a variety of plant species, including rice (Sato et al. 2006), tomatoes (Matsukura et al. 2007), and *Arabidopsis* (Ijaz et al. 2024).

8.3 Importance of Precision Trait Enhancement in Tomatoes

8.3.1 Cas-Based CRISPR/Targeted Mutagenesis

Strong precision plant breeding is now feasible with the use of molecular tools, thanks to the CRISPR/Cas revolution. Beginning in 2013, early studies on CRISPR/Cas-based targeted mutagenesis in tomatoes were published, with a focus on how

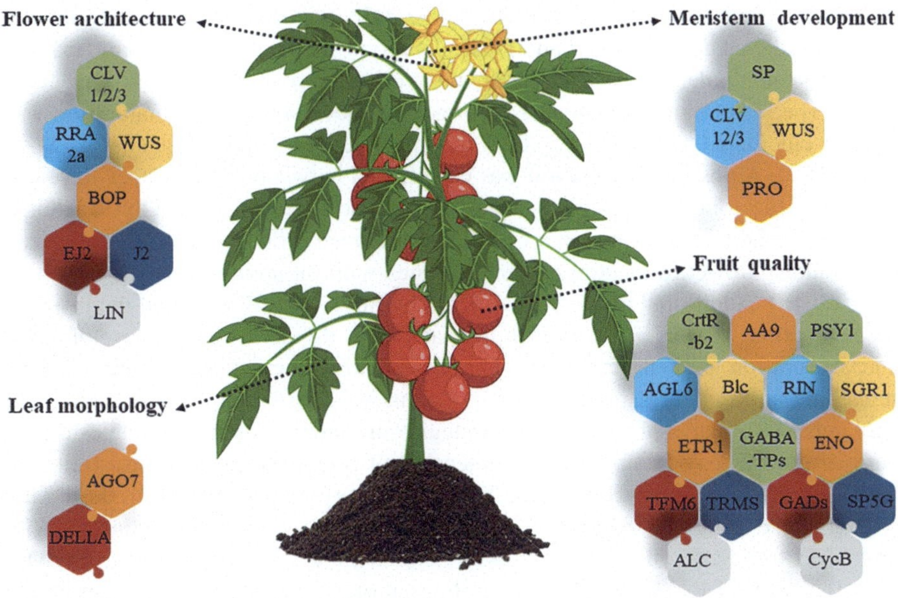

Fig. 8.3 Tomato primary genes targeted using CRISPR/Cas-based targeted mutagenesis. A normal tomato plant in its vegetative development stage was sketched. The genes altered by CRISPR/Cas are shown in relation to the tissues they impact. The gene set associated with abiotic stress is displayed around the plant, signifying their worldwide effects on it

practical and helpful the method was for analyzing gene function using knockout methods (Fig. 8.3).

8.3.2 Engineering to Foster Habits of Growth

In 2014, Brooks and associates published the first CRISPR/Cas9-based tomato genome editing data, showcasing incredibly successful targeted CRISPR/Cas9 changes of four different loci (Brooks et al. 2014). It was also demonstrated that the genotype and phenotype were passed down to the next generations, underscoring the enormous promise of CRISPR/Cas technology in tomato breeding and genetic engineering. Controlling plant growth and inflorescence development has been a top priority throughout tomato domestication and breeding history in an effort to boost fruit output and yield. Using CRISPR/Cas9, a functional investigation of novel genes implicated in the regulation of shoot meristem development in the Clavata (CLV)–Wushel (WUS) circuit was carried out in 2015. After the purposeful mutation of tomato CLV homologs, an arabinosyltransferase generated mutant plants with traits like those of naturally occurring mutants. The study advanced our knowledge of the regulation of the CLV pathway, which may be important for customizing mutant alleles for tomato breeding (Xu et al. 2015).

The targeted knockout of SlPDS successfully yielded strains with biallelic knockout alleles that exhibited an albino phenotype as an observable marker; however, this manipulation also hindered the proliferation of the strains (Pan et al. 2016). The genetic investigation of inflorescence development and maturation was further elucidated through CRISPR/Cas9-mediated targeted mutagenesis (Soyk et al. 2017). A positive correlation was established between inflorescence development and the tomato transcriptional cofactors known as BLADE-ON-PETIOLE (SlBOPs). The knockout mutants of SlBOP produced inflorescences characterized by a solitary flower, indicating potential issues with blooming, as reported by Xu et al. (2016). Additional research conducted by Soyk et al. illuminated the involvement of other MADS-box transcription factors in the architecture, development, and maturation of inflorescences. The quantitative regulation of inflorescence branching and the corresponding flower counts was attributed to the J2, EJ2, and LIN genes (Roldan et al. 2017). The strategic combination of mutant alleles across various groups may facilitate the engineering of preferred inflorescence traits for agricultural applications (Soyk et al. 2017). The morphology of the fruit during the setting phase is influenced by the activities of several proteins, including OVATE and SUPPRESSOR OF OVATE1 (SOV1), both of which belong to the OVATE FAMILY PROTEIN (OFP) class. The elimination of TRM5, a component of the TONNEAU1 Recruiting Motif (TRM) family that operates downstream and in close proximity to OVATE and SOV1, could preserve the elongated fruit shapes (oval or pear-shaped) produced by the mutant alleles of OVATE and SOV1. The knockout mutant of EXCESSIVE NUMBER OF FLORAL ORGANS (SlENO) exhibited an increase in both floral organs and locules. The application of CRISPR/Cas9-mediated targeted mutagenesis at the SlENO locus proved to be exceptionally effective. Gibberellins (GAs) engage with DELLA proteins, leading to their degradation, which can have detrimental effects on plant growth. Consequently, GAs serve as regulators of vegetative plant development. A mutation affecting the interaction domain of the DELLA protein PROCERA induced tomato dwarfism by obstructing GA binding and consequently retarding plant growth (Tomlinson et al. 2019; Yuste-Lisbona et al. 2020).

8.4 Enhancement of Fruit Quality

In contemporary markets, there is a growing interest in tomato fruit quality, particularly for fresh applications (Fig. 8.4). However, due to the industrialization and domestication of tomato production focused on yield, many qualitative traits of the fruit, including flavors, have been lost in farmed tomatoes (Tieman et al. 2017). In order to customize tomato fruit using CRISPR/Cas9 technology for greater quality, it is vital to study the activities of the genes and their mutations during fruit establishment and maturity (Fig. 8.2). Fruit shelf life is a critical determinant of fruit quality and preservation potential. An essential MADS-box transcription factor gene was eliminated by CRISPR/Cas9, resulting in an incomplete ripening stage and extended shelf life in fruit (Ito et al. 2015). Nonetheless, the rin mutants may

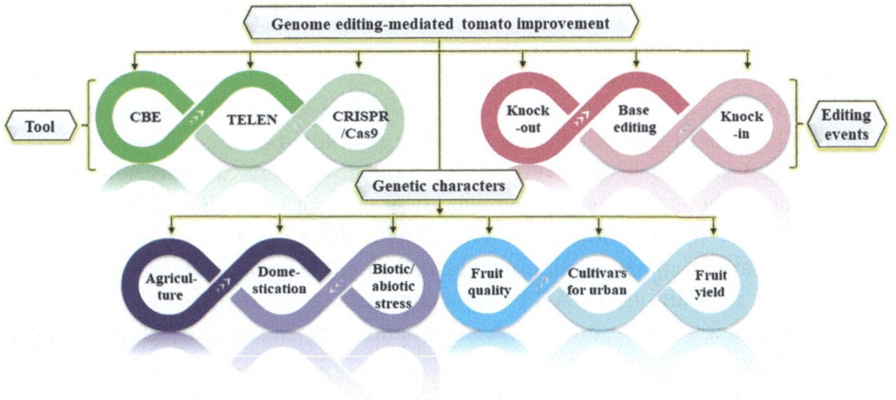

Fig. 8.4 Genome editing applications for tomato trait enhancement. In tomatoes, genome editing improves characteristics including production, shelf life, disease resistance, and nutritional value. Precision changes are made possible by methods like CRISPR, which speed up agricultural innovation and breeding

cause tomato fruit to contain less lycopene, even in the heterozygous form (Herner and Sink 1973). Yu et al. (2017) state that alcobaca (alc) is another well-characterized mutant allele in tomato breeding for extended shelf life and, maybe, fewer adverse effects. The seedless tomato, which may be used for processing as well as fresh eating, is one interesting example. The tomato AGAMOUS-LIKE 6 (SlAGL6) gene was mutated by EMS to produce parthenocarpic tomato lines. This method encouraged fruit growth without fertilization, particularly in situations of high heat. CRISPR/Cas9 complexes may potentially be used to precisely produce the SLAGL6 allele. It is interesting to note that seeds continue to develop normally if fertilization is successful under typical circumstances, indicating that the features may have applications in practical breeding (Klap et al. 2017).

A different method of parthenocarpic tomato breeding involved specifically targeting the AUXIN-INDUCED 9 (SlIAA9) mutation, which is a repressor of unfertilized fruit growth. A SlIAA9 knockout mutant produced parthenocarpic fruits, but it also resulted in an aberrant leaf shape, which may have an impact on yield (Ueta et al. 2017). An intermediate metabolite of C4 plants, malic acid (also known as malate salt), is involved in root Al detoxification, fruit quality, and plant development. It has been demonstrated that tomato fruit malate content is related to tomato fruit malate on chromosome 6 (TFM6). Fruit malate content decreased after the TFM6 gene was deleted using CRISPR/Cas9. However, a 3-bp deletion in the WRKY transcription repressor SlWRKY42's binding site in the TFM6 promoter enhanced the quantity of malate that accumulated in the fruit (Ye et al. 2017). Gamma-aminobutyric acid (GABA) is a neurotransmitter that reduces hypotension and suppresses behavior, both of which are good for human health (Li et al. 2018a, b) (Nonaka et al. 2017; Lee et al. 2018). Li and associates employed an approach, but they concentrated on five additional genes that are involved in the GABA shunt's

GABA conversions. Red fruits exhibited a 3.5-fold higher GABA accumulation than WT fruits.

Fruits (11–12 folds, transgenic plants over-expressing C-terminally shortened SlGAD3 in fruits) and leaves (~6–20 folds) exhibit higher GABA buildup (Li et al. 2018a) on the other hand, resulted in altered fruit contents, longer blooming times, and decreased growth, respectively. As of right now, there are two nonexclusive reasons that might apply. First, glutamate, the precursor of GABA, is deficient as a result of GABA overaccumulation. The second theory is that excessive GABA buildup is harmful to plant development because it plays a role in the plant signaling system. Thus, in addition to GABA overaccumulation, future approaches such as it could be required to sequester GABA into vacuoles or enhance the glutamate synthesis pathway.

A significant amount of the colors and volatiles found in tomato fruits as by-products of their metabolism (metabolites) are derived from carotenoids, which are formed throughout the fruit's growth and maturation. Tomato fruit contains the pigment lycopene, which gives the fruit its color and may provide some health benefits (Story et al. 2010). Therefore, it is intriguing to breed tomatoes to produce fruit with higher lycopene content. Efforts to boost lycopene production were directed at five genes (LCY-B1, LCY-E, Blc, SlSGR1, and LCY-B2) involved in lycopene metabolism in the early processes and lycopene cyclization stages. The single mutation of SlSGR1 resulted in a 5.1-fold increase in lycopene content; nevertheless, the quantity of lycopene decreased relative to the SlSGR1 mutant, even if the level was still significantly greater than that of non-edited fruits. Another method for modifying fruit color is the control of carotenoid accumulation in tomato fruit. The main strategy for using CRISPR/Cas9 complexes is to target the genes that encode the enzymes that play an important role in the carotenoid pathway. Tomato fruits with yellow flesh were produced when phytoene synthase 1 (SlPSY1) was specifically knocked out, which eliminated the synthesis of lycopene (D'Ambrosio et al. 2018).

8.4.1 Tomato Genome Editing Techniques

Most genome editing methodologies, specifically TALEN, ZFN, and particularly CRISPR/Cas systems, alongside the cytidine base editor (CBE)—a variant of DNA base editing—have been employed in tomatoes for fundamental research and precision breeding (Shimatani et al. 2017). However, ZFN and TALEN are infrequently applied to tomatoes in contrast to the widely adopted CRISPR/Cas systems. Both technologies are constructed by merging the DNA cleavage domain of the endonuclease FokI with zinc finger proteins or transcriptional activator-like effectors to achieve site-specific cleavage, relying on protein–DNA interactions to ensure sequence specificity (Xu et al. 2019). In comparison to ZFN and TALEN, the third-generation genome editing technologies, namely CRISPR/Cas systems, exhibit remarkable simplicity and user-friendliness. They were first publicized in 2012 and rapidly gained prominence in the realm of genome editing instruments. Initially, CRISPR/Cas was demonstrated to serve as a potent acquired immune system in bacteria, providing defense against viral infections, as reported by Zhang et al. (2018). The specificity of genome editing is achieved through RNA–DNA interactions. The CRISPR/Cas systems can be categorized into five principal groups based on various RNA and Cas protein combinations; type II encompasses the most extensively researched CRISPR/Cas9 system (Makarova et al. 2015). When contrasted with ZFN and TALEN, the third-generation genome editing technologies, or CRISPR/Cas systems, are exceptionally straightforward and accessible. They were first introduced to the scientific community in 2012 and swiftly became recognized in the domain of genome editing techniques. The initial findings indicated that CRISPR/Cas functions as an efficient acquired immunity mechanism in bacteria, safeguarding against viral attacks (Vu et al. 2020; Haider et al. 2022), indicating promising future applications for CRISPR/Cpf1 in tomato genome editing. Although CRISPR/Cas9 has been extensively employed across various species, numerous technical advancements have been implemented to enhance the efficacy of CRISPR/Cas9-mediated genome editing in plants, particularly in tomatoes. These innovations can facilitate mutations in designated cell types, tissues, or organs; augment editing efficiency; diminish the probability of off-target effects; and expedite the generation of genome-edited and transgene-free plants. In tomatoes, the efficacy of homology-directed repair mediated genome editing was nearly tripled (Vu et al. 2020) when CRISPR/Cpf1 was integrated with Gemini viral multi-replicons in contrast to a Cas9-based single-replicon system. This indicates that CRISPR/Cpf1 holds promising potential for future applications in tomato genome editing. Despite the widespread utilization of CRISPR/Cas9 across a diverse array of species, the efficacy of CRISPR/Cas9-mediated genome editing in plants, particularly in tomatoes, has been augmented by a multitude of technical advancements (Doyle et al. 2019; Maher et al. 2020).

8.5 CRISPR/Cas-Based Accurate DNA Modifications and Replacements

8.5.1 Changes to the Base in Tomatoes

In tomato genome editing, the use of CRISPR/Cas complexes has progressed beyond targeted mutagenesis to encompass precise base-by-base alterations of extended DNA sequences. The Liu group at Harvard University has worked extensively on the direct progression of nucleotide deaminases for base editing (BE) by fitting them with CRISPR/Cas9 guides. For a detailed examination (Rees and Liu 2018). To address BE in tomatoes, *Petromyzon marinus* cytidine deaminase (PmCDA1) was linked to CRISPR/Cas9. Effective base substitutions led to a loss-of-function mutation in the DELLA gene that produced a procera phenotype with fewer serrated leaflets. Shimatani et al. (2017) reported that comparable instruments were employed to generate ethylene-insensitive strains by nucleotide changes in the ethylene receptor SlETR1. Another study produced a tomato that was resistant to herbicides using BE-based GE. When proline-197 in *Arabidopsis* acetolactate synthase (ALS) was substituted for another a.a., it was demonstrated to be the critical for imparting resistance to chlorsulfuron. Consequently, nCas9-PmCDA1 altered the equivalent proline-186 in tomato ALS1 to different types of amino acids. Chlorsulfuron treatment caused high resistance in strains having substituted bases (C to G or C to T) (Veillet et al. 2019). The all-in-one precision technique known as homologous recombination (HR)–based knock-in (HKI) is used to change long DNA sequences or single nucleotides (SNPs) in plants (Merker et al. 2020; Danilo et al. 2019).

8.5.2 Accelerated Allele Introgression and De Novo Domestication

It is predicted that food production using current technologies will not be able to meet the needs of a rapidly growing population by 2050 (Ray et al. 2013). To avert these dire predictions, the field has launched strategic breeding projects, and integrating sustainable agricultural solutions will be essential to removing obstacles to food production. One of the main objectives should be to generate perfect or super crops (Zsogon et al. 2017, 2018). If newly domesticated plants or their wild relatives were used for re-domestication or de novo domestication to develop new crops, more options would be available to meet the issue of feeding humanity by 2050 (Hickey et al. 2019). The most effective use of CRISPR/Cas technology in plant breeding would be to expedite the de novo domestication of wild plants for culture and the introgression of novel genes into elite cultivars (Fig. 8.5). For these treatments to be successful, the CRISPR/Cas-based technologies must be precise and efficient.

Some significant quality and stress-tolerance features have decreased as a result of the drawn-out domestication and traditional breeding procedures (Wang et al. 2017, 2020). Therefore, by carefully introducing certain features from their domesticated cultivars, it may be possible to de novo domesticate wild cousins or semi-cultivated plants, potentially producing novel crops as other possibilities for food supply. CRISPR/Cas technology, which makes it easier to modify practically any genomic area in a genome of interest, even those of wild relatives and semi-cultivated plants, strongly supports the concepts of re-domestication and de novo domestication. Just 15 plant species are employed to create food; nevertheless, hundreds more semi-cultivated species are used to provide food locally and are regarded as orphan crops (Fernie and Yan 2019). For the purpose of de novo domestication of elite features in tomatoes, the wild-related *S. pimpinellifolium* was employed as a parental plant. The most significant characteristics chosen for de novo domestication in the research were fruit quality (CycB and SlGGP1), fruit setting, and growth habit. When precise knockout mutations were incorporated in the chosen genes within the *S. pimpinellifolium* genome, the most desired features were acquired (Zsogon et al. 2018).

Physalis pruinosa, sweet ground cherry of the *Solanaceae* family, which is a native of Central and South America, is an orphan crop. Lemmon and colleagues

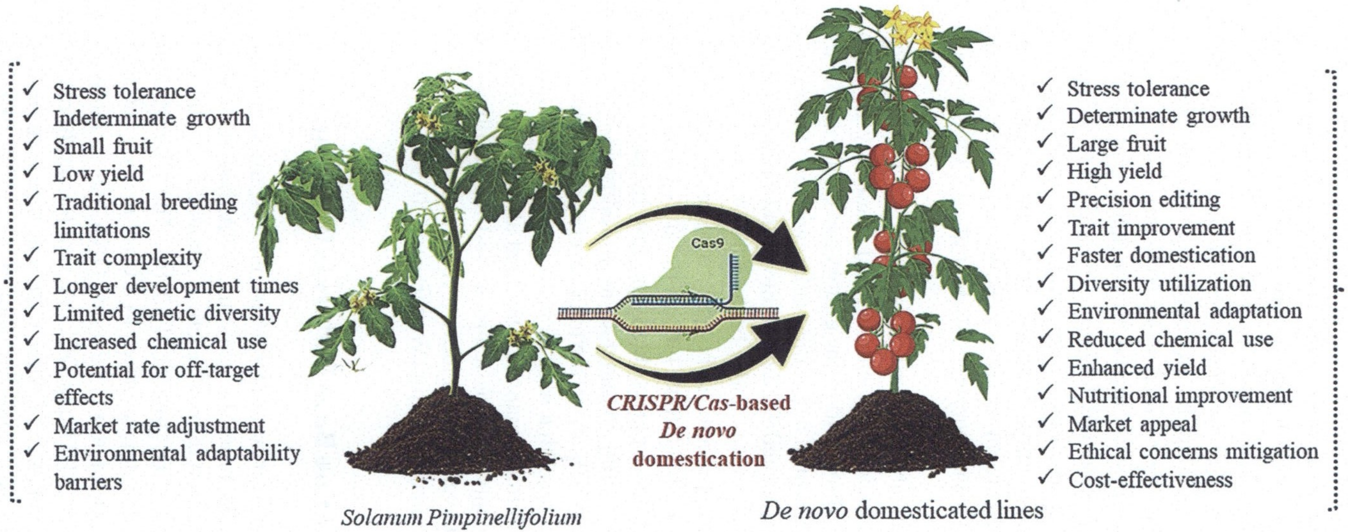

Fig. 8.5 Using CRISPR/Cas technology, *S. pimpinellifolium* is domesticated from the ground up. A typical plant named *S. pimpinellifolium* and its de novo domesticated version are shown, and the CRISPR/Cas-mediated domesticated traits are indicated. The illustrations of the plants lack size

suggested an alternative method of de novo domestication by introducing targeted mutations into its genome. In order to achieve determinate growth, more fruit, and more locules, the plant underwent rigorous breeding, producing characteristics that were similar to those of their tomato orthologs. When considered collectively, the aforementioned research has opened up a new avenue for the quickest possible de novo domestication or re-domestication of plants, enabling the development of robust tomatoes that can adapt to environmental changes due to their wild genetic heritage while maintaining or increasing fruit quality and yield (Lemmon et al. 2018). Targeted mutagenesis is incredibly adaptable and effective using CRISPR/Cas, allowing for the potential targeting of any desired location. Its accuracy, though, is controlled at the locus/gene level rather than at each and every base. Consequently, the methodology is easily adaptable to targeted coding sequence knockouts or haphazard alterations of noncoding sequences, including cis-elements, for haphazard promoter engineering (Ullah et al. 2023b; Mishra et al. 2020; Tran et al. 2020).

Prime editor's applicability in dicots is still restricted and requires refinement, despite its apparent efficiency in animals and suitability for monocot plants (Lu et al. 2020a, b). Base editors and prime editors can create an allele by substituting a small number of SNPs within a specific editing window; however, it becomes more challenging and sophisticated for base editors and prime editors to edit properly when more base modifications and a broader DNA window are required (Mishra et al. 2020). The HKI approach can be used to correctly edit most, if not all, of the necessary base/DNA modifications; however, it may be the final option for reliable gene editing of agricultural plants due to its low frequency and complex design. HKI-mediated precise editing in tomatoes may modify a single base up to hundreds of base pairs (Vu et al. 2020). From the trace level seen in nature, HKI frequency has steadily increased to a level that may be utilized for agricultural precision breeding in a practical and economical manner (Danilo et al. 2019). By fine-tuning CRISPR/Cas complex components, such as temperature-tolerant LbCas12a, HKI might be further enhanced and eventually achieve a real "speed-of-light" precision genome editing method for tomato breeding (Wolter et al. 2019).

8.6 Use of Cytidine Base Editors Is Growing

Since CRISPR/Cas9 gave rise to DNA base editors, these tools have the ability to precisely modify single bases in genomic DNA without causing double-strand breaks. These editors come in two varieties: adenine base editors (ABEs) and cytidine base editors (CBEs) (Komor et al. 2016). The CBE genome editing system consists of a cytidine deaminase linked to a Cas9 nickase (nCas9) that facilitates a specific C-to-T base conversion in genomic DNA and an inhibitor of uracil glycosylase (Eid et al. 2018). ETR1 and DELLA, two tomato hormone signaling genes, were effectively changed in 2017 when CBE was first used (Shimatani et al. 2017). The basic edition's efficacy increased from 26.2 to 53.8%. Branching-chain amino acid production is facilitated by the acetolactate synthase (ALS) gene, and resistance to chlorsulfuron can result from mutation of the tomato ALS1 proline-186 residue. Tomato plants resistant to chlorsulfuron were effectively developed using the CBE-mediated mutation of the ALS1 gene. Notably, accurate base edition efficiency reached 71.4% at the tomato ALS1 Pro186 codon in 12.9% of these transgene-free, chlorsulfuron-resistant tomato plants (Veillet et al. 2019). These findings suggest that CBE may be used to enhance tomato genetics.

8.7 Genome Editing Applications for Precision Trait Enhancement in Tomatoes

Genome editing has made a significant contribution to tomato genetic improvement (Kwon et al. 2020). Genome editing has primarily been applied to tomato breeding using the CRISPR/Cas9 technique to produce precise knockout mutations that have improved fruit output and quality, boosted stress tolerance, accelerated the domestication of wild tomatoes, and produced tomato cultivars specifically suited for urban agriculture. Tomato yield is one of the most crucial agronomic traits in crop breeding; it is primarily determined by fruit setting rate, flowering speed, fruit size, and quantity at maturity, and specific genes associated with tomato production have been found via genetic study (Zsogon et al. 2017). According to Rodriguez-Leal et al. (2017), CRISPR/Cas9 caused changes in the promoters of three tomato genes: *COMPOUND INFLORESCENCE* (S), *SELF PRUNING* (SP), and CLV3, which code for the tomato signaling peptide. A few of these CRISPR/Cas9 cis-regulatory modifications that enhanced the size or number of floral organs resulted in a higher yield of tomato fruit. CRISPR/Cas9-driven promoter mutagenesis will be employed as a revolutionary plant precision breeding technique to boost agricultural production, particularly for generating quantitative trait variation. The demand from consumers for higher-quality fruit has increased along with living standards; tomato fruit quality has been genetically enhanced by genome editing. Numerous elements, such as color, size, shape, nutrition, sweetness, acidity, scent, and shelf life, affect fruit quality. While red tomatoes are the most widely available, other varieties are as vital to satisfy a range of consumer preferences. Zhu et al. (2018) report that pink tomatoes are more popular than red tomatoes in Asia. Yang et al. (2019) suggest that pink tomatoes might be produced by purposefully using CRISPR/Cas9 to interfere with tomato MYB12, which is a master regulator of tomato flavonoid production.

Orange and yellow tomatoes were created via CRISPR/Cas9-mediated deletion of the tomato's phytoene synthase 1 (PSY1) and carotenoid isomerase (CRTISO) genes, which are crucial to the carotenoid biosynthesis pathway (Dahan-Meir et al. 2018). But while cultivating these pink, orange, and yellow tomatoes, some minerals, such as flavonoids and carotenoids, were lost. Because purple tomatoes have a high anthocyanin content, consumers adore them, especially those in the recently launched 'Sun Black' tomato brand. Purple tomatoes were produced by TALEN-mediated or CRISPR/Cas9-targeted insertion of a strong promoter upstream of the endogenous anthocyanin biosynthesis gene SlANT1 in red-fruited tomatoes, which is another superb example of genome editing–mediated targeted gene insertion in plant breeding (Cermak et al. 2015). Genome editing has been utilized to enhance tomato fruit's fundamental qualities in addition to altering the fruit's apparent color. Rich in nutrients, tomato fruit provides a significant amount of lycopene, vitamins, and minerals to our daily diet (Zhu et al. 2018; Bachtiar et al. 2015; Nonaka et al. 2017). These results suggest that by regulating the metabolic and synthesis pathways of plants, nucleotide substitutions or genome editing–mediated knockouts can be effectively employed to increase plant nutrients. A vital feature of horticultural crops is parthenocarpy, as the seedless fruit has a more flavorful pulp and may be processed into many other products. The main phytohormone that controls parthenocarpy is auxin (Hu et al. 2018; Klap et al. 2017).

SlAGL6 is a new and helpful gene for the production of seedless fruit; nevertheless, more research is needed to identify the precise regulatory function of SlAGL6 in parthenocarpy. Another crucial factor in fruit quality is shelf life, particularly when considering the perspective of the consumer. Tomato fruit had a long shelf life thanks to CRISPR/Cas9-induced mutagenesis and replacement of tomato ALC, which was found in the tomato spontaneous mutant alcobaca (alc) and proved to be an allele of the NOR gene (Bota et al. 2014). The fruit's shelf life was significantly increased by the CRISPR/Cas9-induced mutation of the tomato cell wall–modifying enzyme-encoding gene PL, which increased fruit firmness (Yang et al. 2017).

Selective breeding and tomato domestication reduced the genetic diversity of tomatoes grown today (Lin et al. 2014). The cultivated tomato has a compact design and a high yield, but it is more susceptible to environmental assaults from both bacterial and physical agents. Growing tomatoes has gotten increasingly challenging, particularly in light of the changing global environment. Thus, new methods and trait margins need to be carefully examined in the new tomato breeding situation. Compared to their wild counterparts, contemporary tomatoes have lost many of the genes that govern their biotic and abiotic stress-tolerant characteristics or fruit flavor (Bai et al. 2018). Nevertheless, conventional breeding methods are arduous and time-consuming, particularly when it comes to pyramiding polygenic characteristics or several desirable monogenic features (Fernie and Yan 2019). Furthermore, in a timely breeding effort, it may be possible to generate numerous monogenic phenotypes or polygenic features based on CRISPR/Cas. Random mutagenesis is another widely used method in tomato breeding. It involves breaking apart the tomato genome using both chemical and physical agents and, then, looking for novel alleles and viable mutant lines for subsequent breeding. Because the induced mutations are random and broad, the mutant lines show extremely defective traits that render them unfit for agricultural production. The allele may have undergone significant genetic alteration from its maternal ancestry, even in the unlikely event that we are able to isolate it. As a result, the process of choosing useful alleles from a large number of mutations is time-consuming and difficult. Once more, it seems that CRISPR/Cas-based technologies have revolutionized methods for obtaining valuable alleles for tomato breeding.

Based on the widely available omics datasets, the gathered data demonstrated the great potential and effective advancements of multiplexed editing that may be employed for the discovery of new alleles (Pramanik et al. 2020). High locus-specific accuracy may also be achieved in the creation of an entire metabolic pathway by multiplexed editing. Additionally, it is fantastic that haploid inducer–based genome editing can produce homozygous altered alleles in the first generation of genome-edited events (Kelliher et al. 2019). Selection and regeneration of altered events still face challenges, particularly in the situation of low-efficiency HKI where allele-associated marker-free circumstances are present (Maher et al. 2020) or the use of nanoparticles for the genome editing process might be implemented. These methods might considerably cut down on the amount of time and labor required for tomato breeding, which a small business could afford to contribute to the field.

8.8 Perspectives on Genome Editing–Mediated Precision Trait Enhancement in Tomatoes

Precision plant breeding is greatly accelerated by the novel genome editing technique, whereas conventional breeding takes a long time and typically results in a loss of genetic variety and fitness (Chen et al. 2019). It is interesting to note that the majority of these genome editing applications employed CRISPR/Cas9 technologies to create precise knockout mutations in tomatoes. Prime editing, a breakthrough in precision genome editing that may induce targeted substitutions, insertions, and deletions, has also been proven in recent studies on tomatoes (Lu et al. 2020a, b; Gonatopoulos-Pournatzis et al. 2020). Though it has been widely employed in tomato breeding to boost many agronomic characteristics including yield, nutritional quality, and stress responses, genome editing–mediated precision breeding still has a long way to go. Consumer complaints suggest that there is a significant issue with contemporary commercial tomato cultivars' declining flavor quality when compared to heritage tomato types. Previous domestication and breeding techniques of tomatoes that overemphasized output and outward appearance can be partly blamed for this (Xu et al. 2019). The tomato pan-genome, which was created by sequencing the genomes of 725 phylogenetically and geographically representative accessions, and the identification of the panSV (structural variant) genome for 100 different tomato lines have all made it possible to identify a large number of genes that are likely regulating various aspects of fruit quality, including flavor (Alonge et al. 2020). Precision breeding in tomatoes based on genome editing requires these rich and priceless gene resources.

8.9 Conclusion

In conclusion, genome editing technologies such as the CRISPR/Cas9 system hold considerable potential for precisely modifying tomato traits. By altering certain genes in the tomato genome, scientists can improve traits related to nutritional value, tomato yield, and tolerance to biotic and abiotic stresses. In tomato breeding, the use of genome editing methods has already shown encouraging results.

Researchers have effectively improved tomato domestication qualities, changed the fruit's properties, and increased nutritional value by focusing on and changing particular genes. Additionally, desirable qualities including fruit size, inflorescence branching, and plant architecture may be quickly evaluated and selected, thanks to the generation of quantitative trait variation by genome editing. In order to solve the issues associated with global food security, it is imperative that tomato varieties that are more durable, nutritious, and high-yielding be developed. A potent method for quickening the breeding process and introducing specific genetic alterations that improve desirable features is genome editing. To guarantee its appropriate and long-term application, it is crucial to take into account the safety, legal, and ethical issues surrounding genome editing in agriculture. In the future, trait improvement in tomatoes will be further advanced by ongoing research and development in genome editing techniques. The production of enhanced tomato varieties that can satisfy the changing needs of farmers, consumers, and the environment is very promising when cutting-edge technologies are combined with conventional breeding approaches. Researchers may add advantageous genetic changes that enhance characteristics including nutritional content, inflorescence branching, fruit size, plant architecture, and resistance to biotic and abiotic challenges by using genome editing methods. All things considered, genome editing has enormous potential for enhancing precise tomato traits. Through the use of CRISPR/Cas9 and other genome editing instruments, scientists may implement specific genetic alterations to enhance diverse characteristics in tomatoes, culminating in the creation of tomato cultivars that are more robust, nourishing, and prolific.

References

Abe T, Ryuto H, Fukunishi N (2012) Ion beam radiation mutagenesis. In: Plant mutation breeding and biotechnology, pp 99–106

Ahmad M (2023) Plant breeding advancements with "CRISPR-Cas" genome editing technologies will assist future food security. Front Plant Sci 14:1133036

Ahmad Y, Haider S, Iqbal J, Abbasi BA, Yaseen T, Mahmood T (2022) The mechanisms of genome editing technologies in crop plants. In: Principles and practices of OMICS and genome editing for crop improvement. Springer, Cham, pp 295–313

Ahmad Y, Haakim Z, Iqbal J, Abbasi BA, Mahmood T, Kazi M (2024a) Technological innovations for abiotic stress resistance in horticultural crops. In: OMICs-based techniques for global food security, pp 233–244

Ahmad Y, Haider S, Iqbal J, Naseer S, Attia KA, Mohammed AA, Mahmood T (2024b) In-silico analysis and transformation of OsMYB48 transcription factor driven by CaMV35S promoter in model plant–Nicotiana tabacum L. conferring abiotic stress tolerance. GM Crop Food 15(1):130–149

Ahmar S, Gill RA, Jung KH, Faheem A, Qasim MU, Mubeen M, Zhou W (2020) Conventional and molecular techniques from simple breeding to speed breeding in crop plants: recent advances and future outlook. Int J Mol Sci 21(7):2590

Ali I, Salah KBH, Sher H, Ali H, Ullah Z, Ali A, Mahmood T (2022) Drought stress enhances the efficiency of floral dip method of agrobacterium-mediated transformation in Arabidopsis thaliana. Braz J Biol 84:e259326

Alonge M, Wang XG, Benoit M, Soyk S, Pereira L, Zhang L, Suresh H, Ramakrishnan S et al (2020) Major impacts of widespread structural variation on gene expression and crop improvement in tomato. Cell 182:145–161
Bachtiar V, Near J, Johansen-Berg H, Stagg CJ (2015) Modulation of GABA and resting state functional connectivity by transcranial direct current stimulation. elife 4:e08789
Bai Y, Kissoudis C, Yan Z, Visser RGF, van der Linden G (2018) Plant behaviour under combined stress: tomato responses to combined salinity and pathogen stress. Plant J 93:781–793. https://doi.org/10.1111/tpj.13800
Bergougnoux V (2014) The history of tomato: from domestication to biopharming. Biotechnol Adv 32(1):170–189
Bota J, Conesa MA, Ochogavia JM, Medrano H, Francis DM, Cifre J (2014) Characterization of a landrace collection for Tomatiga de Ramellet (*Solanum lycopersicum* L.) from the Balearic Islands. Genet Resour Crop Ev 61:1131–1146
Brooks C, Nekrasov V, Lippman ZB, Van Eck J (2014) Efficient gene editing in tomato in the first generation using the clustered regularly interspaced short palindromic repeats/CRISPR-associated9 system. Plant Physiol 166:1292–1297
Cermak T, Baltes NJ, Cegan R, Zhang Y, Voytas DF (2015) High-frequency, precise modification of the tomato genome. Genome Biol 16:232
Chen K, Wang Y, Zhang R, Zhang H, Gao C (2019) CRISPR/Cas genome editing and precision plant breeding in agriculture. Annu Rev Plant Biol 70:667–697
Dahan-Meir T, Filler-Hayut S, Melamed-Bessudo C, Bocobza S, Czosnek H, Aharoni A, Levy AA (2018) Efficient in planta gene targeting in tomato using geminiviral replicons and the CRISPR/Cas9 system. Plant J 95:5–16
D'Ambrosio C, Stigliani AL, Giorio G (2018) CRISPR/Cas9 editing of carotenoid genes in tomato. Transgenic Res 27:367–378. https://doi.org/10.1007/s11248-018-0079-9
Danilo B, Perrot L, Mara K, Botton E, Nogue F, Mazier M (2019) Efficient and transgene-free gene targeting using agrobacterium-mediated delivery of the CRISPR/Cas9 system in tomato. Plant Cell Rep 38:459–462. https://doi.org/10.1007/s00299-019-02373-6
Dayma J, Koushik P, Dayma A (2023) Nano-boosted tomatoes: unravelling the potential of nanotechnology in hydroponics. Int J Nanobiotechnol 9(2):1–9p
Doyle C, Higginbottom K, Swift TA, Winfield M, Bellas C, Benito-Alifonso D, Fletcher T, Galan CM, Edwards K, Whitney HM (2019) A simple method for spray-on gene editing in planta. BioRxiv 805036
Eid A, Alshareef S, Mahfouz MM (2018) CRISPR base editors: genome editing without double-stranded breaks. Biochem J 475:1955–1964
Ezin V, Symonds RC (2023) Genome editing of tomato crop with CRISPR/Cas systems: past, present, and future. In: CRISPR/Cas-mediated genome editing in plants. Apple Academic Press, pp 191–238
Fernie AR, Yan J (2019) *De novo* domestication: an alternative route toward new crops for the future. Mol Plant 12:615–631. https://doi.org/10.1016/j.molp.2019.03.016
Gonatopoulos-Pournatzis T, Aregger M, Brown KR, Farhangmehr S, Braunschweig U, Ward HN, Ha KCH, Weiss A, Billmann M, Durbic T, Myers CL, Blencowe BJ, Moffat J (2020) Genetic interaction mapping and exon-resolution functional genomics with a hybrid Cas9-Cas12a platform. Nat Biotechnol 38:638–648
Gupta P, Reddaiah B, Salava H, Upadhyaya P, Tyagi K, Sarma S, Datta S, Malhotra B, Thomas S, Sunkum A (2017) Next-generation sequencing (NGS)-based identification of induced mutations in a doubly mutagenized tomato (Solanum lycopersicum) population. Plant J 92:495–508
Haider S, Iqbal J, Naseer S, Yaseen T, Shaukat M, Bibi H, Mahmood T (2021a) Molecular mechanisms of plant tolerance to heat stress: current landscape and future perspectives. Plant Cell Rep 40(12):2247–2271
Haider S, Iqbal J, Shaukat M, Naseer S, Mahmood T (2021b) The epigenetic chromatin-based regulation of somatic heat stress memory in plants. Plant Gene 27:100318

Haider S, Iqbal J, Naseer S, Shaukat M, Abbasi BA, Yaseen T, Mahmood T (2022) Unfolding molecular switches in plant heat stress resistance: a comprehensive review. Plant Cell Rep 41(3):775–798

Herner RC, Sink KC (1973) Ethylene production and respiratory behavior of the rin tomato mutant. Plant Physiol 52:38–42. https://doi.org/10.1104/pp.52.1.38

Hickey LT, Robinson H, Jackson SA, Leal-Bertioli SCM, Tester M (2019) Breeding crops to feed 10 billion. Nat Biotechnol 37:744–754. https://doi.org/10.1038/s41587-019-0152-9

Hu JH, Israeli A, Ori N, Sun TP (2018) The interaction between DELLA and ARF/IAA mediates crosstalk between gibberellin and auxin signaling to control fruit initiation in tomato. Plant Cell 30:1710–1728

Ijaz S, Iqbal J, Abbasi BA, Tufail A, Ullah Z, Sharifi-Rad J, Uddin S (2023) Biofortification: lessons from the past and strategies for future food security. In: Legumes biofortification. Springer, Cham, pp 521–545

Ijaz S, Iqbal J, Abbasi BA, Yaseen T, Rehman S, Kazi M, Mahmood T (2024) Role of OMICS-based technologies in plant sciences. In: OMICs-based techniques for global food security, pp 45–66

Ito Y, Nishizawa-Yokoi A, Endo M, Mikami M, Toki S (2015) CRISPR/Cas9-mediated mutagenesis of the RIN locus that regulates tomato fruit ripening. Biochem Biophys Res Commun 467:76–82. https://doi.org/10.1016/j.bbrc.2015.09.117

Just D, Garcia V, Fernandez L, Bres C, Mauxion J-P, Petit J, Jorly J, Assali J, Bournonville C, Ferrand C (2013) Micro-tom mutants for functional analysis of target genes and discovery of new alleles in tomato. Plant Biotechnol. 30:225

Kelliher T, Starr D, Su X, Tang G, Chen Z, Carter J et al (2019) One-step genome editing of elite crop germplasm during haploid induction. Nat Biotechnol 37:287–292. https://doi.org/10.1038/s41587-019-0038-x

Klap C, Yeshayahou E, Bolger AM, Arazi T, Gupta SK, Shabtai S et al (2017) Tomato facultative parthenocarpy results from SlAGAMOUS-LIKE 6 loss of function. Plant Biotechnol J 15:634–647. https://doi.org/10.1111/pbi.12662

Komor AC, Kim YB, Packer MS, Zuris JA, Liu DR (2016) Programmable editing of a target base in genomic DNA without double-stranded DNA cleavage. Nature 533:420–424

Kostov K, Batchvarova R, Slavov S (2007) Application of chemical mutagenesis to increase the resistance of tomato to Orobanche ramosa L. Bulg J Agric Sci 13:505–513

Kwon CT, Heo J, Lemmon ZH, Capua Y, Hutton SF, Van Eck J, Park SJ, Lippman ZB (2020) Rapid customization of Solanaceae fruit crops for urban agriculture. Nat Biotechnol 38:182–188

Lee J, Nonaka S, Takayama M, Ezura H (2018) Utilization of a genome-edited tomato (*Solanum lycopersicum*) with high gamma aminobutyric acid content in hybrid breeding. J Agric Food Chem 66:963–971. https://doi.org/10.1021/acs.jafc.7b05171

Leitao J (2012) Chemical mutagenesis. In: Shu QY, Forster BP, Nakagawa H (eds) Plant mutation breeding and biotechnology. CABI Publishing, Wallingford, pp 135–158

Lemmon ZH, Reem NT, Dalrymple J, Soyk S, Swartwood KE, Rodriguez-Leal D et al (2018) Rapid improvement of domestication traits in an orphan crop by genome editing. Nat Plants 4:766–770. https://doi.org/10.1038/s41477-018-0259-x

Li XD, Wang YN, Chen S, Tian HQ, Fu DQ, Zhu BZ, Luo YB, Zhu HL (2018a) Lycopene is enriched in tomato fruit by CRISPR/Cas9-mediated multiplex genome editing. Front Plant Sci 9:559

Li R, Li R, Li X, Fu D, Zhu B, Tian H et al (2018b) Multiplexed CRISPR/Cas9-mediated metabolic engineering of gamma-aminobutyric acid levels in Solanum lycopersicum. Plant Biotechnol J 16:415–427. https://doi.org/10.1111/pbi.12781

Lin T, Zhu G, Zhang J, Xu X, Yu Q, Zheng Z et al (2014) Genomic analyses provide insights into the history of tomato breeding. Nat Genet 46:1220–1226. https://doi.org/10.1038/ng.3117

Liu Q, Yang F, Zhang J, Liu H, Rahman S, Islam S, She M (2021) Application of CRISPR/Cas9 in crop quality improvement. Int J Mol Sci 22(8):4206

Liu H, Chen W, Li Y, Sun L, Chai Y, Chen H et al (2022) CRISPR/Cas9 technology and its utility for crop improvement. Int J Mol Sci 23(18):10442

Lu Y, Tian Y, Shen R, Yao Q, Wang M, Chen M, Dong J, Zhang T, Li F, Lei M, Zhu JK (2020a) Targeted, efficient sequence insertion and replacement in rice. Nat Biotechnol 38:1402–1407

Lu Y, Tian Y, Shen R, Yao Q, Zhong D, Zhang X et al (2020b) Precise genome modification in tomato using an improved prime editing system. Plant Biotechnol J 19:415. https://doi.org/10.1111/pbi.13497. [Epub ahead of print]

Macarrón Palacios A, Korus P, Wilkens BG, Heshmatpour N, Patnaik SR (2024) Revolutionizing in vivo therapy with CRISPR/Cas genome editing: breakthroughs, opportunities and challenges. Front Genome Ed 6:1342193

Maghuly F, Laimer M (2017) Forward and reverse genetics for the improvement of Jatropha. In: The Jatropha genome. Springer, Cham, pp 131–148

Maher MF, Nasti RA, Vollbrecht M, Starker CG, Clark MD, Voytas DF (2020) Plant gene editing through de novo induction of meristems. Nat Biotechnol 38:84–89

Makarova KS, Wolf YI, Alkhnbashi OS, Costa F, Shah SA, Saunders SJ, Barrangou R, Brouns SJJ, Charpentier E, Haft DH, Horvath P, Moineau S, Mojica FJM, Terns RM, Terns MP, White MF, Yakunin AF, Garrett RA, van der Oost J, Backofen R, Koonin EV (2015) An updated evolutionary classification of CRISPR-Cas systems. Nat Rev Microbiol 13:722–736

Maluszynski M, Szarejko I, Maluszynska J (2003) Encyclopedia of applied plant sciences. Academic, New York, NY, pp 186–201

Manova V, Gruszka D (2015) DNA damage and repair in plants—from models to crops. Front Plant Sci 6:885

Matsukura C, Yamaguchi I, Inamura M, Ban Y, Kobayashi Y, Yin Y-G, Saito T, Kuwata C, Imanishi S, Nishimura S (2007) Generation of gamma irradiation-induced mutant lines of the miniature tomato (*Solanum lycopersicum* L.) cultivar 'micro-tom'. Plant Biotechnol 24:39–44

Mazzucato A, Cellini F, Bouzayen M, Zouine M, Mila I, Minoia S, Petrozza A, Picarella ME, Ruiu F, Carriero F (2015) A TILLING allele of the tomato Aux/IAA9 gene offers new insights into fruit set mechanisms and perspectives for breeding seedless tomatoes. Mol Breed 35:22

Mba C (2013) Induced mutations unleash the potentials of plant genetic resources for food and agriculture. Agronomy 3:200–231

Merker L, Schindele P, Huang TK, Wolter F, Puchta H (2020) Enhancing in planta gene targeting efficiencies in Arabidopsis using temperature-tolerant CRISPR/LbCas12a. Plant Biotechnol J 18:2382–2384. https://doi.org/10.1111/pbi.13426

Minoia S, Petrozza A, D'Onofrio O, Piron F, Mosca G, Sozio G, Cellini F, Bendahmane A, Carriero F (2010) A new mutant genetic resource for tomato crop improvement by TILLING technology. BMC Res Notes 3:69

Mishra R, Joshi RK, Zhao K (2020) Base editing in crops: current advances, limitations and future implications. Biotechnol J 18:20–31. https://doi.org/10.1111/pbi.13225

Nagamine A, Ezura H (2022) Genome editing for improving crop nutrition. Front Genome Ed 4:850104

Nagamine A, Takayama M, Ezura H (2023) Genetic improvement of tomato using gene editing technologies. J Hortic Sci Biotechnol 98(1):1–9

Naito K, Kusaba M, Shikazono N, Takano T, Tanaka A, Tanisaka T, Nishimura M (2005) Transmissible and nontransmissible mutations induced by irradiating *Arabidopsis thaliana* pollen with γ-rays and carbon ions. Genetics 169:881–889

Ndudzo A, Makuvise AS, Moyo S, Bobo ED (2024) CRISPR-Cas9 genome editing in crop breeding for climate change resilience: implications for smallholder farmers in Africa. J Agric Food Res 16:101132

Nonaka S, Arai C, Takayama M, Matsukura C, Ezura H (2017) Efficient increase of gamma-aminobutyric acid (GABA) content in tomato fruits by targeted mutagenesis. Sci Rep 7:7057. https://doi.org/10.1038/s41598-017-06400-y

Pan C, Ye L, Qin L, Liu X, He Y, Wang J et al (2016) CRISPR/Cas9-mediated efficient and heritable targeted mutagenesis in tomato plants in the first and later generations. Sci Rep 6:24765. https://doi.org/10.1038/srep24765

Perry JA, Wang TL, Welham TJ, Gardner S, Pike JM, Yoshida S, Parniske M (2003) A TILLING reverse genetics tool and a web-accessible collection of mutants of the legume Lotus japonicus. Plant Physiol 131:866–871

Pramanik D, Shelake RM, Kim MJ, Kim JY (2020) CRISPR-mediated engineering across the central dogma in plant biology for basic research and crop improvement. Mol Plant 14:127. https://doi.org/10.1016/j.molp.2020.11.002

Ray DK, Mueller ND, West PC, Foley JA (2013) Yield trends are insufficient to double global crop production by 2050. PLoS One 8:e66428. https://doi.org/10.1371/journal.pone.0066428

Rees HA, Liu DR (2018) Base editing: precision chemistry on the genome and transcriptome of living cells. Nat Rev Genet 19:770–788. https://doi.org/10.1038/s41576-018-0059-1

Rodriguez-Leal D, Lemmon ZH, Man J, Bartlett ME, Lippman ZB (2017) Engineering quantitative trait variation for crop improvement by genome editing. Cell 171:470–480

Roldan MVG, Perilleux C, Morin H, Huerga-Fernandez S, Latrasse D, Benhamed M et al (2017) Natural and induced loss of function mutations in SlMBP21 MADS-box gene led to jointless-2 phenotype in tomato. Sci Rep 7:4402. https://doi.org/10.1038/s41598-017-04556-1

Saito T, Ariizumi T, Okabe Y, Asamizu E, Hiwasa-Tanase K, Fukuda N, Mizoguchi T, Yamazaki Y, Aoki K, Ezura H (2011) TOMATOMA: a novel tomato mutant database distributing micro-tom mutant collections. Plant Cell Physiol 52:283–296

Sato Y, Shirasawa K, Takahashi Y, Nishimura M, Nishio T (2006) Mutant selection from progeny of gamma-ray-irradiated rice by DNA heteroduplex cleavage using brassica petiole extract. Breed Sci 56:179–183

Shikazono N, Suzuki C, Kitamura S, Watanabe H, Tano S, Tanaka A (2005) Analysis of mutations induced by carbon ions in *Arabidopsis thaliana*. J Exp Bot 56:587–596

Shimatani Z, Kashojiya S, Takayama M, Terada R, Arazoe T, Ishii H et al (2017) Targeted base editing in rice and tomato using a CRISPR-Cas9 cytidine deaminase fusion. Nat Biotechnol 35:441–443. https://doi.org/10.1038/nbt.3833

Singh V, Gohil N, Ramirez Garcia R, Braddick D, Fofié CK (2018) Recent advances in CRISPR-Cas9 genome editing technology for biological and biomedical investigations. J Cell Biochem 119(1):81–94

Smith HH (1958) Radiation in the production of useful mutations. Bot Rev 24:1–24

Soyk S, Lemmon ZH, Oved M, Fisher J, Liberatore KL, Park SJ et al (2017) Bypassing negative epistasis on yield in tomato imposed by a domestication gene. Cell 169:e1112. https://doi.org/10.1016/j.cell.2017.04.032

Stadler LJ (1928) Genetic effects of X-rays in maize. Proc Natl Acad Sci USA 14:69

Story EN, Kopec RE, Schwartz SJ, Harris GK (2010) An update on the health effects of tomato lycopene. Annu Rev Food Sci Technol 1:189–210. https://doi.org/10.1146/annurev.food.102308.124120

Tanaka A, Shikazono N, Hase Y (2010) Studies on biological effects of ion beams on lethality, molecular nature of mutation, mutation rate, and spectrum of mutation phenotype for mutation breeding in higher plants. J Radiat Res 51:223–233

Tieman D, Zhu G, Resende MF Jr, Lin T, Nguyen C, Bies D et al (2017) A chemical genetic roadmap to improved tomato flavor. Science 355:391–394. https://doi.org/10.1126/science.aal1556

Till BJ, Cooper J, Tai TH, Colowit P, Greene EA, Henikoff S, Comai L (2007) Discovery of chemically induced mutations in rice by TILLING. BMC Plant Biol 7:19

Tomlinson L, Yang Y, Emenecker R, Smoker M, Taylor J, Perkins S et al (2019) Using CRISPR/Cas9 genome editing in tomato to create a gibberellin-responsive dominant dwarf DELLA allele. Plant Biotechnol J 17:132–140. https://doi.org/10.1111/pbi.12952

Tran MT, Doan DTH, Kim J, Song YJ, Sung YW, Das S et al (2020) CRISPR/Cas9-based precise excision of SlHyPRP1 domain(s) to obtain salt stress-tolerant tomato. Plant Cell Rep 40:999. https://doi.org/10.1007/s00299-020-02622-z. [Epub ahead of print]

Ueta R, Abe C, Watanabe T, Sugano SS, Ishihara R, Ezura H et al (2017) Rapid breeding of parthenocarpic tomato plants using CRISPR/Cas9. Sci Rep 7:507. https://doi.org/10.1038/s41598-017-00501-4

Ullah Z, Iqbal J, Abbasi BA, Akhtar W, Kanwal S, Ali I, Mahmood T (2023a) Assessment of Gus expression induced by anti-sense Os PPO gene promoter and antioxidant enzymatic assays in response to drought and heavy metal stress in transgenic Arabidopsis thaliana. Sustain For 15(17):12783

Ullah Z, Iqbal J, Abbasi BA, Ijaz S, Sharifi-Rad J, Yaseen T, Mahmood T (2023b) How genome editing can be helpful in the biofortification of legumes. In: Legumes biofortification. Springer, Cham, pp 207–232

Vats S, Kumawat S, Kumar V, Patil GB, Joshi T, Sonah H, Sharma TR, Deshmukh R (2019) Genome editing in plants: exploration of technological advancements and challenges. Cells 8:1386

Veillet F, Perrot L, Chauvin L, Kermarrec MP, Guyon-Debast A, Chauvin JE et al (2019) Transgene-free genome editing in tomato and potato plants using agrobacterium-mediated delivery of a CRISPR/Cas9 cytidine base editor. Int J Mol Sci 20:402. https://doi.org/10.3390/ijms20020402

Vu TV, Sivankalyani V, Kim EJ, Doan DTH, Tran MT, Kim J, Sung YW, Park M, Kang YJ, Kim JY (2020) Highly efficient homology-directed repair using CRISPR/Cpf1-geminiviral replicon in tomato. Plant Biotechnol J 18:2133–2143

Wang L, Chen L, Li R, Zhao R, Yang M, Sheng J et al (2017) Reduced drought tolerance by CRISPR/Cas9-mediated SlMAPK3 mutagenesis in tomato plants. J Agric Food Chem 65:8674–8682. https://doi.org/10.1021/acs.jafc.7b02745

Wang WQ, Allan AC, Yin XR (2020) Small RNAs with a big impact on horticultural traits. Crit Rev Plant Sci 39:30–43

Wolter F, Schindele P, Puchta H (2019) Plant breeding at the speed of light: the power of CRISPR/Cas to generate directed genetic diversity at multiple sites. BMC Plant Biol 19:176. https://doi.org/10.1186/s12870-019-1775-1

Xu C, Liberatore KL, MacAlister CA, Huang Z, Chu YH, Jiang K et al (2015) A cascade of arabinosyltransferases controls shoot meristem size in tomato. Nat Genet 47:784–792. https://doi.org/10.1038/ng.3309

Xu C, Park SJ, Van Eck J, Lippman ZB (2016) Control of inflorescence architecture in tomato by BTB/POZ transcriptional regulators. Genes Dev 30:2048–2061. https://doi.org/10.1101/gad.288415.116

Xu JM, Hua K, Lang ZB (2019) Genome editing for horticultural crop improvement. Hortic Res 6:113

Yang L, Huang W, Xiong FJ, Xian ZQ, Su DD, Ren MZ, Li ZG (2017) Silencing of SlPL, which encodes a pectate lyase in tomato, confers enhanced fruit firmness, prolonged shelf-life and reduced susceptibility to grey mould. Plant Biotechnol J 15:1544–1555

Yang TX, Deng L, Zhao W, Zhang RX, Jiang HL, Ye ZB, Li CB, Li CY (2019) Rapid breeding of pink-fruited tomato hybrids using the CRISPR/Cas9 system. J Genet Genomics 46:505–508

Ye J, Wang X, Hu T, Zhang F, Wang B, Li C et al (2017) An InDel in the promoter of Al-ACTIVATED MALATE TRANSPORTER9 selected during tomato domestication determines fruit malate contents and aluminum tolerance. Plant Cell 29:2249–2268. https://doi.org/10.1105/tpc.17.00211

Yu QH, Wang B, Li N, Tang Y, Yang S, Yang T et al (2017) CRISPR/Cas9-induced targeted mutagenesis and gene replacement to generate long-shelf life tomato lines. Sci Rep 7:11874. https://doi.org/10.1038/s41598-017-12262-1

Yuste-Lisbona FJ, Fernández-Lozano A, Pineda B, Bretones S, Ortíz-Atienza A, García-Sogo B et al (2020) ENO regulates tomato fruit size through the floral meristem development network. Proc Natl Acad Sci USA 117:8187–8195. https://doi.org/10.1073/pnas.1913688117

Zhang Y, Massel K, Godwin ID, Gao CX (2018) Applications and potential of genome editing in crop improvement. Genome Biol 19:210

Zhu Q, Smith SM, Ayele M, Yang L, Jogi A, Chaluvadi SR, Bennetzen JL (2012) High-throughput discovery of mutations in tef semi-dwarfing genes by next-generation sequencing analysis. Genetics 192:819–829

Zhu GT, Wang SC, Huang ZJ, Zhang SB, Liao QG, Zhang CZ, Lin T, Qin M, Peng M, Yang CK, Cao X, Han X, Wang XX, van der Knaap E, Zhang ZH, Cui X, Klee H, Fernie AR, Luo J, Huang SW (2018) Rewiring of the fruit metabolome in tomato breeding. Cell 172:249–261

Zsogon A, Cermak T, Voytas D, Peres LEP (2017) Genome editing as a tool to achieve the crop ideotype and de novo domestication of wild relatives: case study in tomato. Plant Sci 256:120–130

Zsogon A, Cermak T, Naves ER, Notini MM, Edel KH, Weinl S et al (2018) *De novo* domestication of wild tomato using genome editing. Nat Biotechnol 36:1211–1216. https://doi.org/10.1038/nbt.4272

CRISPR/Cas9 and Beyond: Genome Editing for Precision Trait Enhancement in Tomatoes

9

Samra Irum, Nazia Rehman, Safeena Inam,
Muhammad Zain Farid Khan, Muhammad Uzair,
Amna Muhammad, and Muhammad Ramzan Khan

Abstract

Genome editing, particularly through clustered regularly interspaced short palindromic repeats (CRISPR)–based technology, has revolutionized tomato biology and breeding, offering precise and efficient means for trait enhancement. This technology addresses various challenges in tomato cultivation, from improving quality, yield, and fruit ripening to enhancing abiotic and biotic stress resistance. By targeting specific genes associated with desirable traits, CRISPR/Cas9 enables rapid and precise modifications, overcoming the narrow genetic base of tomatoes. CRISPR/Cas9 technology's advantages include ease of experimental design, minimal off-targets mutations, and cost-effectiveness, making it ideal for enhancing productivity and yields in tomatoes. Traits such as plant architecture, flower characteristics, fruit composition, and stress resilience have been successfully edited using CRISPR/Cas9, showcasing its versatility and efficacy in tomato breeding. Innovations in CRISPR/Cas9 technology, such as the availability of plant-specific Cas variants, online design tools, and efficient transformation protocols, further enhance its utility in tomato research and breeding. Comparative analysis has highlighted the CRISPR/Cas9 superiority over other genome editing platforms, emphasizing its potential to revolutionize agricultural practices. CRISPR genome editing technology holds immense potential for accelerating applied research and precision breeding in tomatoes, contributing to global food security. It represents a transformative tool for precision trait enhancement in tomatoes, offering a pathway to address environmental challenges and meet the demands of a growing population. Further advancements are

S. Irum · N. Rehman (✉) · S. Inam · M. Z. F. Khan · M. Uzair · A. Muhammad · M. R. Khan
Functional Genomics and Bioinformatics Labs, National Institute for Genomics and Advanced Biotechnology (NIGAB), National Agricultural Research Centre, Islamabad, Pakistan

M. Waseem et al. (eds.), *Omics Approaches for Tomato Yield and Quality Trait Improvement*, https://doi.org/10.1007/978-981-96-3890-1_9

essential to fully realize its potential in designing elite tomato varieties to combat the current and future challenges.

Keywords

CRISPR · Tomatoes · Genome editing · Abiotic stress · Growth

9.1 Introduction

Tomato (*Solanum lycopersicum* L.) belongs to the *Solanaceae* family and is grown globally due to its nutritional and commercial values. In 2020, the world's tomato production was 41.52 million tons, and it has been assumed that in 2026, the production will increase to 51.93 million tons. Tomato is consumed both as fresh salad and processed food due to its nutritional importance for health, having vitamins A and C, minerals, and antioxidant compounds like beta-carotene (β-carotene) and lycopene (Cambiaso et al. 2019). Tomato breeding to develop improved high-yielding and high-quality varieties to meet the public demands is very essential. In this scenario, exploring the natural wild resources, which are rich in genetic diversity for desirable traits is a good and valuable source for developing elite tomato varieties. Conventional breeding and chemically and physically induced mutagenesis for breeding purposes are time-consuming and laborious. The availability of tomato genome sequencing (Tomato Genome Consortium 2012) and advancements in next-generation sequencing have resolved these issues. Currently, advanced breeding and multi-omics methods have accelerated the breeding process by unraveling the mystery of complex traits. Recently, tomato pan-genome studies identified new genes involved in fruit flavor (Gao et al. 2019) and tomato super-pan-genome (Li et al. 2023a), unraveling the whole picture of structural variations in both wild and cultivated tomato species. The effective utilization of this genetic variation is beneficial for the improvement of tomato cultivars. The advancements in genomic-assisted breeding have been reported by Tiwari et al. (2023) for speedy tomato improvement. In continuation of innovation, the conventional methods of breeding and mutagenesis methods are being rapidly substituted by genetic engineering technologies to overcome the associated issues. Currently, genomes of diverse plant species are frequently being modified by creating targeted insertion/deletions (Indels) and replacements through site direct mutagenesis (SDN) including meganucleases, transcription activator-like effector nucleases (TALENs), zinc finger nucleases (ZFNs), and the more recently emerging method, clustered regularly interspaced short palindromic repeats (CRISPR) associated with its Cas protein (Voytas and Gao 2014). CRISPR technology has revolutionized genome editing technology and enabled scientists to generate precise and trait-specific targeted mutations and functional studies of genes for desired traits (Larriba et al. 2024).

In tomatoes, till now CRISPR/Cas9 and other Cas protein variants have been utilized to edit numerous traits (Vu et al. 2020a; Chandrasekaran et al. 2021; Xia et al. 2021; Salava et al. 2021; Chaudhuri et al. 2022). CRISPR-based editing is greatly fruitful in tomatoes as several genes in tomatoes have been edited and functionally characterized and have contributed to the tomato crop improvement not only as a model crop species but also as a model for providing insight into plant biology. Notably, Japan is the first country to introduce CRISPR-edited GABA-rich tomatoes, the first vegetable crop that arrived in the food market (Waltz 2022).

This book chapter outlines the structure and mechanism of the CRISPR system as the current background of gene editing and its applications in tomato crop research intended to enhance resistance to environmental stresses, improve fruit quality (aroma, flavor, and nutritional enrichment), and modify plant architecture and breeding objectives in tomato cultivation. Additionally, it investigates the encountered future challenges and offers possible ways to establish the CRISPR technology as a speedy precision breeding technique for tomatoes.

9.2 CRISPR/Cas Mechanism of Action in Prokaryote

CRISPR/Cas-based system mediates the adaptive immune system in prokaryotes against the phages and plasmids. CRISPR system is a versatile, simple, precise, and cost-effective technology that has revolutionized molecular biology and biotechnology. The scientists have adopted this genome editing DNA targeting platform, which is guided by designed guide RNA (gRNA) to modify the genome of various species (Jinek et al. 2012).

The prokaryotic adaptive immunity systems have CRISPR loci that encode guide RNAs, and the loci are linked with CRISPR-associated (Cas) genes for immunization against bacteriophage (Fig. 9.1) (Heler et al. 2014). The whole process of attaining immunity is based on the steps of integration of phage DNA (as new spacers) into bacterial CRISPR repeat-spacer array, thus providing a memory record of former infections that allows the host to prevent upcoming infections of the same phage (Barrangou et al. 2007). The next step is CRISPR RNA (crRNA) biogenesis by transcription of the CRISPR array and tracrRNA, which leads to the processing of precursor-crRNA through enzymatic cleavage into mature CRISPR RNAs (crRNAs). In the third step of DNA interference through the combined activity of crRNAs, tracrRNA, and Cas9 protein, a complementary foreign target sequence (protospacer) triggers the destruction of entering nucleic acids upon a second-time infection (Garneau et al. 2010).

9.3 CRISPR/Cas System in Tomato Breeding

CRISPR technology is derived from the adaptive immunity system of the bacteria *Streptococcus pyogenes* and is extensively being utilized as an editing tool for the genome of target organisms. In literature the classification of Cas proteins categorized with Roman numbers as Class I (Type I, III, IV), Class II (Type II, V, VI). Cas proteins belonging to class I are more complex and multi-subunit complexes, whereas, the Cas proteins of class II systems are single and comparatively simple.

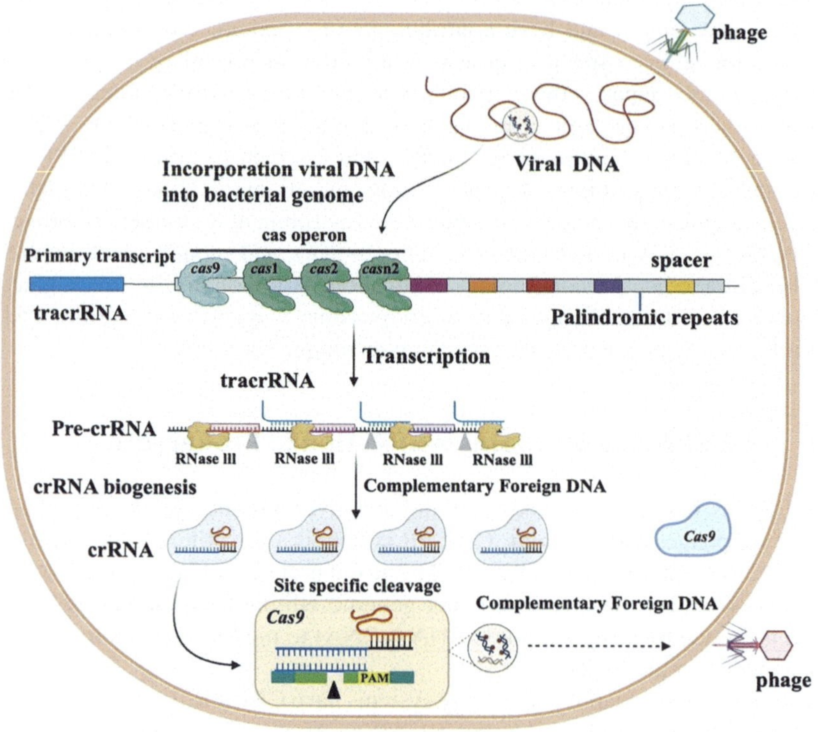

Fig. 9.1 CRISPR/Cas9-mediated mechanism of adaptive immunity in prokaryote. A generalized type II CRISPR locus in the CRISPR/Cas system includes repetitive sequences (repeats, light gray) interspaced by spacer (protospacer) sequences (spacers, colored rectangular boxes), linked with CRISPR-associated (Cas) genes (Cas operon) and is preceded by tracrRNA having homology with the repeats. Invading phage incorporates a spacer into the CRISPR array through associated Cas proteins such as Cas1, Casn2, and Cas2. Following the addition of a spacer element, the entire CRISPR region is transcribed to produce precursor CRISPR RNA. tracrRNA is transcribed separately and then hybridized to the pre-crRNA repeats to mature the crRNA by RNase III cleavage. During DNA interference, when the same phage infects a second time, the mature crRNA–tracrRNA makes a complex with Cas9 endonuclease and further leads it to the cleavage of foreign DNA of the same phage having 20-nt sequences homologous with crRNA sequence preceding the protospacer adjacent motif (PAM) sequence

So proteins of class II have been well-studied and extensively used in the genome editing of organisms (Liu et al. 2020a). The two key parts: single guide RNA (sgRNA) and CRISPR-associated (Cas9) proteins are essential for the system's functionality. Cas9 is a type II protein having two lobes called the recognition lobe (REC) and the nuclease lobe (NUC). The domains REC1 and REC2 of REC are authorized for binding gRNA, whereas the NUC lobe comprises a protospacer adjacent motif (PAM), HNH, and RuvC interacting domain (PI-Domain) (Nishimasu et al. 2014). Generally, CRISPR/Cas-mediated editing is a three-step process including recognition, cleavage, and repair (Shao et al. 2016). This is attained by designing an sgRNA that is highly specific to the target sequence; an RNA–Cas9 ribonucleoprotein complex is formed, which gets activated on the target site, causing double-stranded breaks (DBSs) in the three base pairs (bps) upstream to the PAM (Ceasar et al. 2016). PAM is a 2–5-bp short sequence located downstream to the site of cleavage and variable in size due to variants of Cas protein (Fig. 9.2). The most frequently used Cas-9 protein identifies the 5′-NGG-3′ PAM sequence (N is any nucleotide base). On recognition of a suitable target and PAM sites, Cas9 starts denaturation of target DNA and forms the RNA–DNA hybrid, leading to the generation of a double-strand break (DSB) by the NHN and RuvC domains of the nuclease (NUC) lobe. Last, the DSB is repaired by the host cellular apparatus either by non-homologous end joining (NHEJ) or homology-directed repair (HDR) pathways (Eid et al. 2018). The basic structural knowledge about the CRISPR complex is vital for further genome editing implementation. It allows for a comprehensive insight into the vital components dealing with the whole mechanisms, including target recognition, DNA–gRNA hybrid formation, DNA co-coupling and DNA break. This

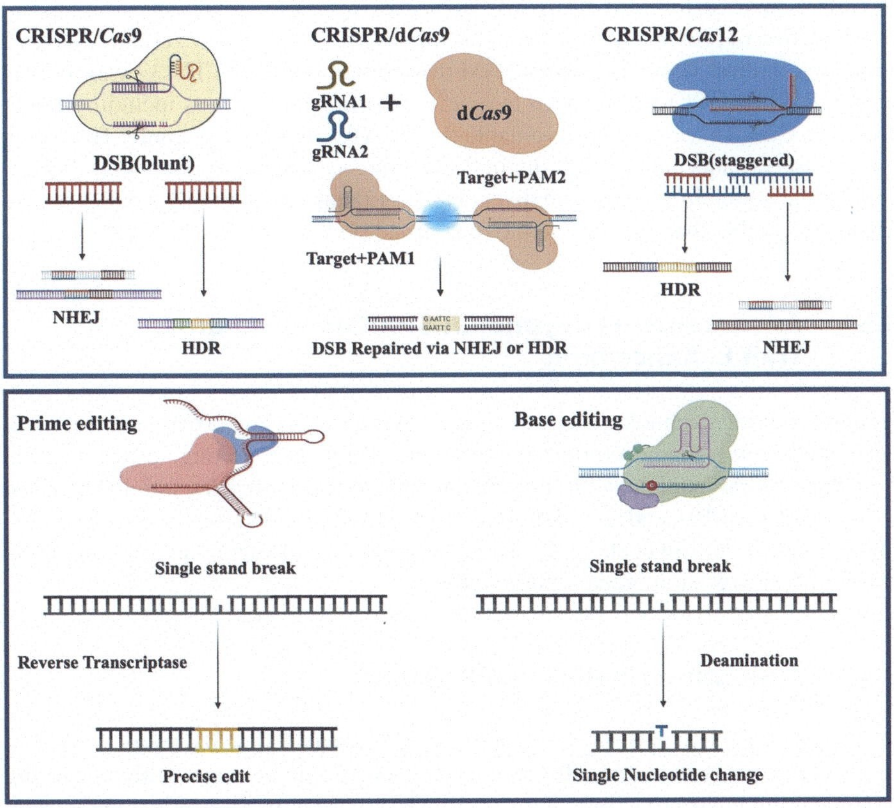

Fig. 9.2 Working models of CRISPR/Cas variants: CRISPR/Cas9, CRISPR/dCas9, CRISPR/Cas12, base editing, and prime editing

complete analysis can aid in enhancing target efficiencies by identifying and optimizing the key components for improved performance (Nishida et al. 2016).

9.4 CRISPR-Based Cas Variants

CRISPR/Cas editing platforms emerged as the third-generation gene editing technology in 2012 and rapidly developed as a megastar in genome editing systems due to its simplicity, efficiency, and usability in comparison with other genome editors (Zhang et al. 2018). In 2014, the CRISPR/Cas9 tool was used for the first time in tomatoes for targeting the tomato homolog of *Arabidopsis ARGONAUTE7* (*SlAGO7*), and it has now become the significant genome editing tool for the functional characterization of genes and precision plant breeding (Brooks et al. 2014; Xu et al. 2019). The availability of an "NGG" PAM sequence specific to CRISPR editing is lacking in the target gene sequence of an organism; in this case, the CRISPR system is not an appropriate choice for editing. Therefore, diversity in PAM sequence specificities is an important necessity of CRISPR/Cas9. Initially, the scientists have customized PAM-free nucleases by orthologous data mining and manipulation of protein. To overcome this limitation, another Cas9 variant (Cas12a), a member of class II, type V, with a PAM recognition site (5′-TTTN-3′) was discovered. CRISPR/Cpf1 tool has been regularly used in various crops including tomatoes. The Cas12 variant is quite valuable for the AT-rich tomato genome. Therefore, novel Cas9 variants having diversity in PAM sequence specificities would be beneficial for the restoration of genetic diversity. Cas12a protein is highly smaller in size and thermostable than Cas9 protein (Vu et al. 2021).

9.5 Advancements Beyond CRISPR/Cas in Tomato Trait Enhancement

Genome editing technology is advancing (Anzalone et al. 2020) with recently developed new genome editing systems: including base editing, prime editing, mitochondria-targeted transcription activator-like effector nucleases (mitoTALENs) (Kazama et al. 2019), and APOBEC–Cas9 fusion-induced deletion systems (AFIDs) (Wang et al. 2020; Lin et al. 2020). These systems have the potential to significantly revolutionize precision breeding in tomatoes.

9.5.1 Base Editors in the CRISPR System

DNA editors, such as cytosine base editors (CBEs) and adenine base editors (ABEs) can generate programmable substitutions (C:T or A:G) in genomes without causing double-strand breaks (Nishida et al. 2016). ABEs change bases A:T into G:C, while CBEs switch G:C into A:T (Wada et al. 2020). The CBE genome editing specifically promoted C-to-T base conversions in genomic DNA (Fig. 9.2). It involves combining a cytidine deaminase with a Cas nickase (referred to as nCas) and a uracil glycosylase inhibitor (Eid et al. 2018). CBE was effectively applied to tomatoes and modified genes *ETR1* and *DELLA* genes responsible for hormone signaling with base editing efficiencies ranging from 26.2% to 53.8% (Shimatani et al. 2017). Among them, the acetolactate synthase (*ALS*) gene, which has a role in the biosynthesis of branched amino acid chains (McCourt and Duggleby 2006), and proline-186 residue mutation can confer chlorsulfuron resistance (Yu et al. 2010). Overexpression of the *ALS1* gene targeted by CBEs in tomatoes conferred resistance to chlorsulfuron. In the *ALS1* Pro186 codon, the base editing efficiency reached 71.4% in tomato plants, and up to above one-tenth of the detected chlorsulfuron-resistant tomatoes regenerated were harboring no transgene copies (Veillet et al. 2019). These findings demonstrate that CBEs could be used for genetic improvement in tomatoes.

9.5.2 Prime Editing

Prime editing (PE) is a new DSB-free genome editing strategy that is capable of supporting base conversions, insertions, and deletions at designated locations. It is also known as a search and replace tool that offers user-specified sequence modification to other desired locations without creating DSBs. Moreover, there is no need to deliver a DNA repair template (Anzalone et al. 2019). PE comprises the Cas9 nickase protein (H840A) joined with reverse transcriptase. A special gRNA is developed and designated as prime editing guide RNA, known as pegRNA. The pegRNA consists of a 3′ primer binding site (PBS) and a spacer sequence at the t 5′ end that identify the target location, which has a 5′-spacer sequence that specifies the target site and a 3′-extension of primer binding site (PBS) sequence and a template for reverse transcriptase (RT) for incorporating the desired edits in the target genome (Fig. 9.2). This PE tool activity is based on the fusion protein (Cas9 nickase, H840A) guided by pegRNA to the rds target location for the induction of nick in the opposite nontarget strand (Tingting et al. 2023). Prime editing consists of two elements: a modified Cas9 nickase (nCas9), with an H840A alteration, and a reverse transcriptase, in addition to a customized gRNA (pegRNA) (Anzalone et al. 2019). The pegRNA includes guide RNA for targeting genetic locations, a reverse transcription template for introducing predetermined modifications, and a PBS that enables the nicked DNA strands 3′ end to bind with the pegRNA (Huang and Puchta 2021).

In prime editing, pegRNA locates the site of the target and employs the transcriptase along with PBS and RT templates for making precise edits on the nicked DNA strand (Zhan et al. 2021). Prime editing enables a range of 12 base substitutions and facilitates base substitutions, insertions, and deletions, providing more flexibility compared to conventional DNA base editors (Lin et al. 2020). Prime editing systems have been developed for crop improvement by codon optimization, promoter enhancement, and addition of localization signals, enabling efficient and precise genome editing (Molla et al. 2021). Notably, prime editing has not shown off-target effects in plants (Li et al. 2020) even though the editing efficiency may vary. Factors such as the length of the PBS or RT template in pegRNA and the positioning of the nicking sgRNA play roles in influencing editing efficiency in crops (Lin et al. 2020). Recent computational models, like PE_type, PE_position, and DeepPE, are utilized to evaluate PE efficiency (Kim et al. 2021), showcasing advancements. Identifying functionally important gene sites and comprehending the effects of single nucleotide polymorphisms in crops are greatly facilitated by site-directed mutations (Schilbert et al. 2020). When compared to transgenic approaches, prime editing emerges as an effective method for generating targeted mutations in tomatoes. This notable advantage positions prime editing as a tool for progressing plant research and precise breeding, hastening the integration of alleles into commercially grown varieties and enhancing crop enhancement efforts.

9.5.3 mitoTALENs-Based Gene Editing

In addition to nucleus, organelles such as mitochondria and chloroplasts also carrying their own DNA which harbor numerous essential genes involved in regulation of various cellular processes within these organelles (Bock and Knoop 2012). While the stable transformation of mitochondrial genomes remains a challenge, recent progress has successfully introduced heritable and stable modifications of mitochondria to genes in rice, rapeseed, and *Arabidopsis* by combining nuclear transformation with mitoTALENs. These mitoTALENs are made up of a TALEN nuclease connected to a mitochondrial localization signal for targeting the mitochondria (Kazama et al. 2019), for instance, making edits to the gene *ATP6-1* encoding an ATP synthase subunit that also has a counterpart in the nucleus, resulting in mutations exclusively within the mitochondrial gene version demonstrating the precision of mitoTALEN-based editing of mitochondrial genomes (Arimura et al. 2020). To optimize the effectiveness of mitoTALEN-mediated editing, different promoters (CaMV35S, Ubiquitin1, and RPS5A) and two types of TALEN (compact) were compared, with the RPS5A promoter and conventional TALEN demonstrating high efficiency (Kazama et al. 2019). Plant mitochondrial genomes contain around 60 genes and significant noncoding DNA content (Gualberto and Newton 2017), presenting mitoTALEN as a valuable tool for investigating the roles of mitochondria in plant cellular processes. Moreover, the manipulation of plastid genomes has proven effective in improving economic and agronomic characteristics in plants (Zhang et al. 2020a). This suggests that utilizing mitoTALEN for editing plant genomes could offer benefits for enhancing crop improvements.

9.5.4 APOBEC–Cas9 Fusion-Induced Deletion System (AFIDs) Editing

The CRISPR/Cas9 gene editing often leads to alterations to the DNA sequence like insertions and deletions. These modifications do not interfere with the regulatory elements and regions within the genome (Wang et al. 2020). An innovative method named the APOBEC-Cas9 fusion-induced deletion systems (AFIDs) comprising AFID1, AFID2, and AFID has been developed to effectively generate deletions in the genomes of rice and wheat (Wang et al. 2020). Among these systems, AFID3 is noteworthy for its capacity to produce deletions efficiently. This system incorporates elements such as DNA glucosidase from *Escherichia coli* and apurinic or apyrimidinic site lyase (AP lyase) from *E. coli* that collectively account for approximately one-third of its predictable deletions (Li et al. 2023b). These AFIDs facilitate nucleotide deletions in plant genomes, which provide a faster method to explore small functional regulatory elements and domains within genomic DNA. Furthermore, there is increasing evidence suggesting that functional regulatory components and domains within DNA like open reading frames (ORFs) and other cis-elements hold potential as targets for enhancing crops (Zhang et al. 2020a; Zhou et al. 2017). Additionally, certain transcription factors play varying roles by attaching to distinct cis-elements (Xiao et al. 2013). Therefore, AFIDs will expedite plant breeding through the adjustment of the target expression of genes primarily at the transcriptional and translational levels.

9.5.5 Target Activation–Induced Cytidine Deaminase (Target-AID) Editing

Target-AID editing technology combines the enzyme *Petromyzon marinus* cytidine deaminase with nuclease-deactivated Cas9 (dCas9), which can cleave stranded DNA molecules. Using dCas9 for target-AID technology produces effective and precise replacement of C with T; whereas nCas9 results in both base substitution and insertion–deletion at specified sites with enough efficiency to generate point mutations (either C to T or A to G), without causing any off-target effects elsewhere (Kawaguchi et al. 2021).

9.6 Application of CRISPR/Cas System in Tomato Gene Editing

Tomatoes are grown on a large scale around the globe because of their health benefits and economic value. The CRISPR/Cas technology has revolutionized the way tomatoes are bred and enhanced by allowing for manipulation (Table 9.1). This technique allows for precise changes to be made in the genome of tomatoes leading to the creation of *Solanum lycopersicum* varieties, with qualities or resistance traits, as well as an improved ability to withstand various environmental stresses.

Table 9.1 CRISPR/Cas-based genome editing for improving tomato crop characteristics

Category	Subcategory	Gene(s)	Function/modification	Reference(s)
Disease resistance/ biotic stress	Powdery mildew resistance	*SlMlo1*	Powdery mildew resistance	Nekrasov et al. (2017)
		SlPMR4	Involved in resistance against powdery mildew fungus *Oidium neolycopersi*	Huibers et al. (2013)
Viral resistance		*Ty-5 and Pelo*	Provides resistance toward tomato yellow leaf curl virus (TYLCV)	Tashkandi et al. (2018)
Fungal resistance		*ACET1a and ACET1b*	Resistance against *Botrytis cinerea* (gray mold)	Jeon et al. (2020)
		SlMYC2	Improved fruit tolerance for *Botrytis cinerea*	Min et al. (2020)
Bacterial and oomycete resistance		*SlDMR6-1*	Resistance to *Phytophthora capsici*, *Pseudomonas syringae* pv. tomato DC3000, and *Xanthomonas spp*	Ortigosa et al. (2019) and Thomazella et al. (2016)
		SlJAZ2	Resistance to *Pseudomonas syringae* pv. tomato DC3000	
		eIF4E1	Resistance against pepper mottle virus (PepMoV)	Yoon et al. (2020)
Broad-spectrum resistance		*DMR6-1*	Broad-spectrum disease resistance	Thomazella et al. (2016)
Improved yield/plant architecture/ traits	Growth habit	*SP*	Determinate growth habits for mechanized harvesting	
Fruit size		*FAS*	Increase in fruit size	
Fruit shape		*O*	Production of oval-shaped fruits	
Fruit number		*MULT*	Increase in quantity of fruits	
Fruit weight		*FW2.2*	Increase in fruit weight	
Lycopene content		*CYC B/CycB*	Increased lycopene content	

(continued)

Table 9.1 (continued)

Category	Subcategory	Gene(s)	Function/modification	Reference(s)
Color variations		*CRTISO and PSY1*	Production of orange and yellow tomatoes	Dahan-Meir et al. (2018)
Hybrid seed production		*Ms1035 and GSTAA*	Creation of male sterile plants for hybrid seed production	Liu et al. (2021) and Uluisik et al. (2016b)
Mechanical harvesting		*J2*	Enabling of mechanical harvesting	
Fruit firmness		*PL*	Improved fruit firmness by reducing pectin degradation during ripening	
Meristem size and yield		*WUS/LC and CLV3/FAS*	Improved meristem sizes, enhanced complex inflorescence, amplified locule count, elevated fruit weight, dimensions, and yield	Rodríguez-Leal et al. (2017)
Anthocyanin content		*SlANT1*	Development of purple tomatoes high in anthocyanins	Čermák et al. (2015)
Flavonoid biosynthesis		*MYB12*	Production of pink tomatoes by regulating flavonoid biosynthesis	Zsögön et al. (2017)
Enhanced nutritional content	Lycopene content	*SGR1, LCY E, Blc, LCY-B1, and LCY-B2*	Increased lycopene content	Zsögön et al. (2017)
GABA accumulation		*SlGAD2 and SlGAD3*	Increased GABA accumulation	Nonaka et al. (2017)
Carotenoid content		*SlDDB1, SlDET1, and SlCYC-B*	Enhanced carotenoid content through base editing	Hunziker et al. (2020)
Provitamins		*CRTISO and PSY1*	Production of orange and yellow tomatoes rich in provitamin A carotenoids	Dahan-Meir et al. (2018)
Anthocyanin content		*SlANT1*	Production of purple tomatoes with high anthocyanin content	Čermák et al. (2015)
Abiotic stress tolerance	Heat tolerance	*CPK28*	Improved thermotolerance by targeting APX2	Hu et al. (2021)

(continued)

Table 9.1 (continued)

Category	Subcategory	Gene(s)	Function/modification	Reference(s)
Multi-stress tolerance		*GRXS14/15/16/17*	Enhanced resistance to various abiotic stresses such as high temperatures, low temperatures, scarcity of water, toxicity caused by heavy metals, and insufficient nutrients	Kakeshpour et al. (2021)
Drought tolerance		*SlLBD40*	Enhanced drought tolerance	Liu et al. (2020b)
Herbicide resistance		*ALS1*	Enhanced tolerance to the sulfonylurea pesticide chlorsulfuron	Vu et al. (2020b)
Salinity tolerance		*HKT1;2, RAD51/54, and PR-1*	Improved salinity tolerance	Vu et al. (2020b) and Tran et al. (2021)

9.6.1 CRISPR Gene Editing for Enhancing Biotic Stress Resistance in Tomatoes

Tomatoes can be affected by diseases caused by viruses, fungi, and bacteria. Therefore, it is crucial to maintain the resistance of tomatoes against diseases. Through the use of CRISPR/Cas technology, scientists have developed tomato varieties that are resistant to harmful agents like late blight, powdery mildew, and bacterial and viral infections. The gene *SlMlo1* produces a protein that helps pathogens enter tomato plants. By modifying this gene, tomatoes can exhibit resistance to powdery mildew. The application of CRISPR/Cas technology to introduce mutations in the *SlMlo1* gene has led to the development of powdery mildew tomato plants that are transgene-free (Nekrasov et al. 2017; Santillán Martínez et al. 2020). Researchers conducted a study using genetics with the *Arabidopsis PMR4* gene as a reference point. They utilized CRISPR/Cas9 technology to target the *SlPMR4* gene and enhance resistance against the powdery mildew fungus *Oidium neolycopersi*. Their findings revealed that *SlPMR4* shares similarities with the *Arabidopsis* PMR4 gene (Santillán Martínez et al. 2020). Tashkandi and colleagues (2018) discovered that the *Ty-5* and RNA surveillance factors referring to *Pelo* genes were modified genetically to provide protection against tomato yellow leaf curl virus (TYLCV). Additionally, specific changes targeted at genes for producing coat protein (CP) and replicate (Rep) have been shown to enhance tomato plants' ability to resist TYLCV (Tashkandi et al. 2018; Yoon et al. 2020). Mutations induced by CRISPR/Cas9 at the loci of Acetylenase genes, ACET1a and ACET1b, can potentially improve resistance to the pathogen *Botrytis cinerea*, also known as gray mold. Moreover, in *SlMYC2* gene editing, a transcription factor using the CRISPR/Cas9 technique has been found to increase fruit resistance against *B. Cinerea* despite its negative effects observed in plants (Jeon et al. 2020).

Mutations in the downy mildew resistance gene, *SlDMR6,* were found to increase resistance against infections from both *Pseudomonas syringae pv.* and *Phytophthora capsici,* boosting immunity against bacterial and oomycete pathogens (Thomazella et al. 2021; Ortigosa et al. 2019). Furthermore, disabling the *SlJAZ2* gene from the jasmonate domain significantly enhanced resistance to *Pseudomonas syringae* pv. against the tomato DC3000 strain for bacterial speck disease (Ortigosa et al. 2019). Interestingly, the *elF4E1* gene mutation did not affect sensitivity to the tomato mottle virus that infects pepper (PepMoV) as reported by Yoon et al. (2020). By modifying the *DMR6 1* tomato gene ortholog, researchers observed disease resistance and a strengthened ability to combat pathogens (Thomazella et al. 2021).

9.6.2 Influence of CRISPR on Yield, Plant Structure, Flowering, and Fruit Characteristics

CRISPR/Cas technology has been widely utilized to make changes to the specific genes of tomatoes focusing on enhancing yield, plant structure, and the characteristics of flowers and fruits. This method has also brought about progress in tomato breeding. Alterations in the SELF PRUNING (*SP*) gene in tomatoes have led to the development of determinacy, a trait beneficial for harvesting. By modifying the ovate (*O*) gene, certain fruit shapes were genetically altered, while changes to the fasciated (*FAS*) gene resulted in a threefold increase in fruit size. Introducing the multiflora (*MULT*) gene led to a tenfold rise in fruit numbers, and mutations in the fruit weight gene (FW2.2) contributed to an increase in fruit weight. These combined changes led to an increase in productivity and made it easier to develop types of plants with unique fruit qualities and growth features (Zsögön et al. 2017). The fruit quality attributes were also enhanced by CRISPR/Cas. Editing of the *CYC B/ CycB* gene in tomatoes leads to an increase in lycopene content by 500%, greatly improving their nutritional value. Editing of *CRTISO* and *PSY1* genes that are involved in the carotenoid biosynthetic pathway resulted in the production of yellow and orange tomatoes. However, some nutrients, like flavonoids and carotenoids, were negatively affected as a consequence (Dahan-Meir et al. 2018).

The advancement of CRISPR/Cas9 technology has made it possible to generate plants with hybrid seeds by knocking out both *Ms1035* and *GSTAA* genes simultaneously. This breakthrough led to the development of hybrid varieties with enhanced traits. Additionally, genetic modification focusing on the *Jointless-2 (J2) gene* has greatly improved harvesting, allowing farmers to mechanically harvest and reduce time. Altering the pectate *lyase (PL)* gene leads to enhanced fruit quality by slowing down pectin breakdown during ripening. As a result, this modification extended the shelf life of fruits and increased fruit firmness (Uluisik et al. 2016a; Wang et al. 2019).

CRISPR/Cas9 has also been utilized to alter growth regulators in fruits. Changes in the *WUS/LC and CLV3/FAS* genes that control the locule number have been noted to cause meristem enlargement, leading to enhanced compound inflorescence, fruit weight, locule number, size, and yield as shown by Rodríguez-Leal et al. (2017). Additionally, pink tomatoes were developed by targeting the transcription factor, the *MYB12* gene, that regulates flavonoid synthesis (Zhu et al. 2018).

9.6.3 CRISPR-Mediated Improved Nutritional Value of Tomatoes

CRISPR/Cas genome editing has been used to enhance the value of tomatoes by increasing the levels of compounds such as lycopene, β-carotene, and other carotenoids. Scientists have focused on modifying genes to boost nutrients in tomatoes, with a primary objective being to elevate the concentration of lycopene, known for its antioxidant properties. In tomato through multiplex CRISPR/Cas various genes have been edited that are involved in carotenoid metabolic pathway like stay-green 1 (SGR1), lycopene ε-cyclase (LCY E), betalycopene cyclase (Blc), and lycopene β-cyclase 1 and 2 (LCY B1 and LCY B2) (Li et al. 2018), there was a 5.1% boost in the levels of lycopene content in tomato fruits. Another study showcased an increase in lycopene content by 500% by manipulating the *CYC B* gene using CRISPR technology (Zsögön et al. 2017), indicating significant potential for enhancing tomato nutrition through genetic modification. Furthermore, efforts have been made to enrich tomatoes with elements beyond lycopene. By altering genes for production like *SlGAD2* and *SlGAD3* that encode glutamate decarboxylase enzymes, researchers observed an increase ranging from 7% to 15% in GABA levels in tomato fruits (Nonaka et al. 2017). GABA functions as a calming neurotransmitter and a type of amino acid that is not used for building proteins but offers impacts on wellness. The amount of carotenoids has also been improved as a result of other genome editing techniques, like cytosine base editors (CBEs), for base editing. Changing the DNA bases in tomatoes through nucleotide substitutions created by CBEs led to an increase in the levels of carotenoids, lycopene, and β-carotene. This increase was linked to changes in proteins like *DDB1*, which resembles yeast DNA damage repair protein *Rad4p-like protein. DET1* is important for development and seed growth, while *CYC B* protein plays an important role in pathways for β-carotene synthesis (Nekrasov et al. 2017).

Advancements in vitamin levels have been made using CRISPR/Cas technology by altering the *CRTISO* and *PSY1* genes involved in the carotenoid biosynthesis pathway, leading to the development of orange and yellow tomatoes. These tomatoes are rich in provitamin A carotenoids (Dahan-Meir et al. 2018). Additionally, the *SlANT1* gene was altered to yield tomatoes containing an increased amount of anthocyanin, an antioxidant with potential health benefits (Vu et al. 2020b; Čermák et al. 2015). These instances demonstrate how the CRISPR/Cas methods can potentially boost the nutritional content of tomatoes by modifying genes linked to nutrient production and metabolism. The aim is to improve the health benefits of tomatoes for consumption through precise and effective technological advancements.

9.6.4 CRISPR-Based Abiotic Stress Tolerance

The CRISPR/Cas gene editing method has been extensively used to enhance the resilience of tomatoes against different abiotic stresses including heat, drought, and cold. The aim is to enhance their capacity to withstand stresses, like high temperatures, lack of water, and soil salinity. These modifications have enhanced the ability of tomato plants to thrive in conditions resulting in improved crop yields and better food quality. Rising global temperatures caused by climate fluctuations pose a substantial constraint on the cultivation of tomatoes. Research has demonstrated that the heat tolerance of tomatoes can be enhanced by modifying the *CPK28* gene, which is closely associated with the *APX2* gene responsible for the synthesis of ascorbate peroxidase 2 (Hu et al. 2021). In addition, modifying the *GRXS14/15/16/17* genes that encode CGFS-type glutaredoxin 14/15/16/17 proteins enhanced the plant's ability to withstand abiotic challenges such as severe temperatures, drought, heavy metals, and nutritional deficits (Kakeshpour et al. 2021). The ability of tomatoes to withstand dehydration was significantly improved by altering the *SlLBD40* gene, which codes for a lateral organ boundaries domain (LBD) transcription factor (Liu et al. 2020b). Moreover, modifying the tomato *ALS1* gene to control branched-chain amino acid production using CRISPR/Cas9 and cytosine base editors (CBEs) proved effective in inhibiting the sulfonylurea herbicide chlorsulfuron (Veillet et al. 2019).

Based on a study by Vu et al. (2020b), specific genetic variations like high-affinity potassium transporter 1;2 (*HKT1;2*) and *RAD51/54* have evolved to enhance salt tolerance while compromising adaptation to soil salinity. Researchers have also boosted salt stress resistance in tomatoes by targeting the pathogenesis-related protein 1 (*PR-1*) locus (Tran et al. 2021). By introducing mutations in key stress response genes of plants, scientists have successfully developed tomato plants that demonstrate improved resilience to environmental stressors. This genetic alteration has led to increased yields and enhanced resilience in tomato crops (Fig. 9.3).

9.7 Methods for Obtaining Gene-Edited Tomato Plants

To create DNA double-strand breaks (DSBs) and trigger the repair process for gene editing, the sgRNAs and Cas protein need to enter the plant cells (Fig. 9.4). This can be achieved by delivering CRISPR components or using preassembled Cas/sgRNA ribonucleoprotein (RNP) complexes (Fig. 9.5). The CRISPR–RNP complex targets and cleaves the DNA at target sites upon delivery and then degrades rapidly in those cells (Tashkandi et al. 2018; Laforest and Nadakuduti 2022; Rustgi et al. 2022; Wu et al. 2023).

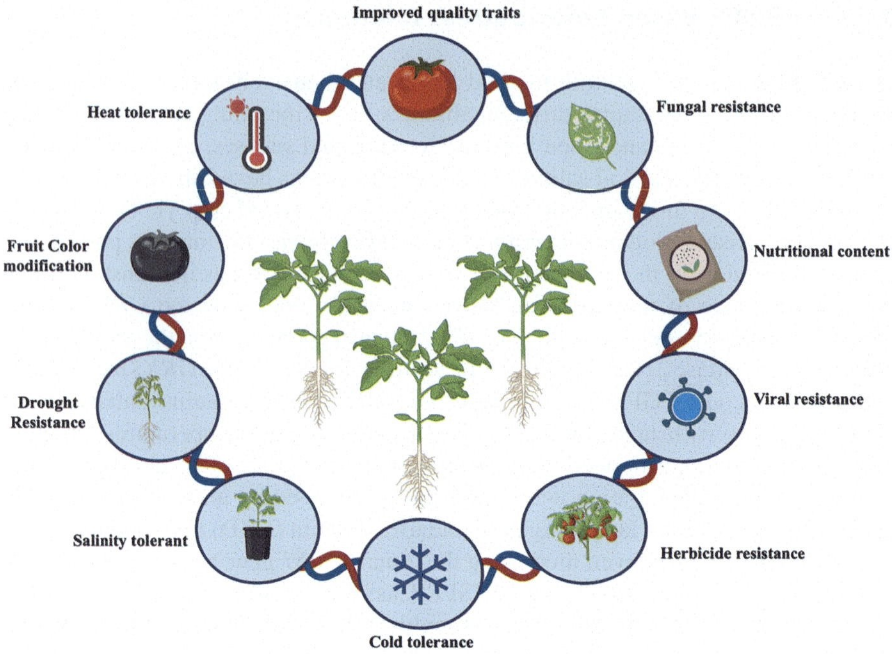

Fig. 9.3 Schematic representation of the diverse applications of the CRISPR system in tomato trait enhancement

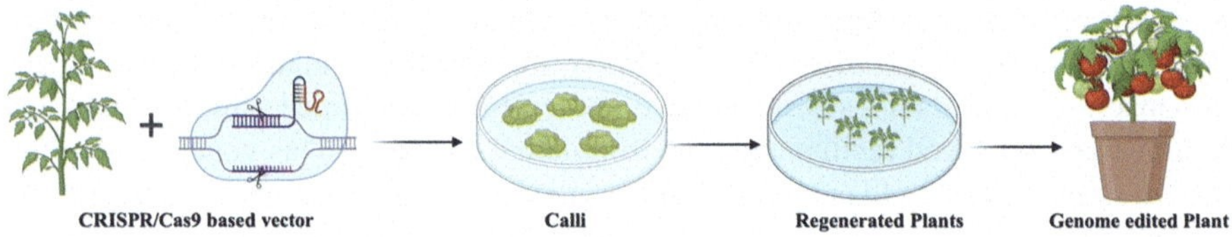

Fig. 9.4 Schematic overview of CRISPR/Cas-based genome editing precision trait enhancement in tomatoes

9.7.1 *Agrobacterium*-Mediated Transformation in Tomatoes

Agrobacterium-mediated transformation is the most commonly utilized method for CRISPR gene editing in plants including tomatoes (Yang et al. 2022; Yu et al. 2022). Additionally, methods like polyethylene glycol (PEG)–based protoplast transformation and particle bombardment have shown promising results in tomato research. When it comes to delivering the DNA into the nuclei, biological methods leverage the integration capabilities of the soil bacterium *Rhizobium* to integrate T-DNA into the plant's genomic regions (Păcurar et al. 2011). *A. Tumefaciens*, known as *Rhizobium radiobacter,* is widely utilized for introducing vectors containing CRISPR components, whereas *A. Rhizogenes* has gained considerable attention for

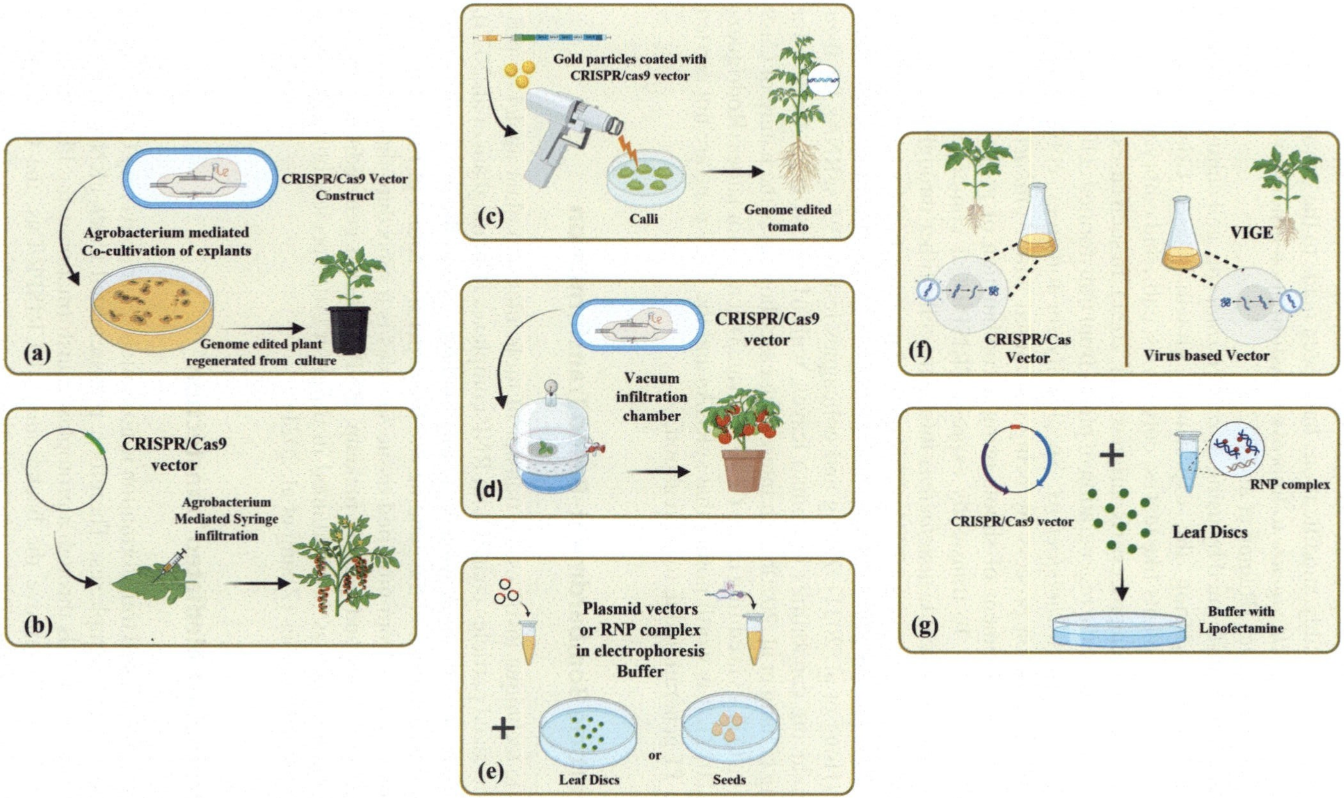

Fig. 9.5 Different methods to deliver CRISPR/Cas components into plant cells and tissues: (**a**) Agrobacterium-mediated transformation, (**b**) direct injection of the Agrobacterium culture with a plasmid vector, (**c**) particle bombardment, (**d**) vacuum infiltration, (**e**) electroporation (**f**) a virus-based plasmid vector (**g**) liposome-mediated transfection

its role in root studies (Kiryushkin et al. 2021). Various strains of *A. Tumefaciens* are accessible for transforming plant species with varying degrees of success (De Saeger et al. 2021). Notably, modifications have been made to the genomes of both *A. Tumefaciens* and *A. Rhizogenes* to improve transformation efficiency and expand their range of host species (Raman et al. 2022).

Plant tissues are transformed by using *Agrobacterium* spp. cell cultures that have the required vectors during the exponential growth phase. Coculturing of *A. Tumefaciens* with explants like leaves, cotyledons, calli, and embryos are essential for the transformation process (Rodrigues et al. 2021). In such cases, it is important to have a method for the regeneration of plants from transformed cells containing the T-DNA region of *A. tumefaciens*. Specifically for tomatoes, cotyledons can undergo a transformation process by being cocultured with *Agrobacterium*. This can lead to the development of shoots using a combination of growth hormones (Yaroshko et al. 2023). To transiently express CRISPR/Cas components in plant tissues, one can introduce *A. tumefaciens* into plant cells either through vacuum or syringe infiltration (Chincinska 2021). This method proves useful for assessing how various vectors can edit genes in the tomato genome before performing stable transformation (Hong et al. 2021). Zhang and colleagues tested 195 sgRNAs for their mutation-inducing capabilities in tomato leaves, with 61.5% proving effective for gene editing (Zhang et al. 2020b). An interesting case showcases the nature of Cas9 by modifying QTLs via editing in the *CLV3* (Clavata3) gene in tomatoes (Rodríguez-Leal et al. 2017). It is worth noting that sgRNAs and the Cas9 gene are not carried over to the T1 generation due to segregation.

9.7.2 Particle Bombardment–Based Transformation

The biolistic technique involves the transfer of molecules into plant nuclei by propelling nano-sized particles coated with RNP complexes (Cunningham et al. 2020). However, there are various drawbacks to this method, such as the integration of the DNA at various locations in the genome and the expensive nature of materials. The regeneration of plants from modified tissue samples is a procedure that also relies on the specific species and genetic makeup. RNP-mediated gene editing through particle bombardment has been attained in different crop species, but it has been never reported in tomatoes (Zhang et al. 2021).

9.7.3 Protoplast-Mediated Transfection

Agrobacterium-mediated transformation is not suitable, for delivering Cas9/ribonucleoproteins (RNPs) complexes. Therefore, preassembled Cas9–gRNA RNPs are introduced into plant cells through a protoplast transformation using PEG fusion or biolistic methods. To remove the background of CRISPR/Cas, the Cas9–gRNA RNP system is entirely DNA-free and does not involve genetic segregation (Thiruppathi 2022). PEG facilitates the incorporation of DNA or RNP complexes into plant protoplasts in the presence of cations like Ca^{2+} or Mg^{2+} at high pH, albeit with less efficiency of transformation. This method relies on successful protoplast isolation and regeneration methods. The delivery of Cas9–gRNA RNP complex offers several advantages: no need for progeny screening through selfing or backcrossing, no off-target effects, better control over CRISPR/Cas endonuclease activity, easy mutagenesis process post-transfection without lagging phase, and direct delivery of Cas9–sgRNA assembly or preassembled isolated RNPs into plant cells (Kang et al. 2024). However, challenges with the RNP system include delivery across plant cells and regeneration from cell wall. Therefore, the success of protoplast culture protocols depends on factors such as genotype, species, tissue-specificity, and cell wall characteristics; reducing the number of off-targets and investigating the potential for gene integration are factors in improving the accuracy of Cas9 endonuclease function. PEG-mediated transfection is an employed method for introducing vector DNA or RNPs into protoplasts of different plant species, such as various types of tomatoes (Slaman et al. 2023) and the wild tomato species *S. Peruvianum* (Lin et al. 2022).

9.7.4 Virus-Induced Gene Editing

The method of delivering material through *Agrobacterium* is commonly used in plants, but it is not very efficient for HDR-mediated gene editing. To overcome this issue, a technique called virus-induced genome editing (VIGE) has been developed using plant virus–based vectors to efficiently deliver sgRNAs (Butler et al. 2016) and has been used in tomatoes. VIGE utilizes viral-derived vector systems to transport CRISPR/Cas components into plant cells without the need for integrating them into the CRISPR/Cas backbone (Rössner et al. 2022), making it broader and less time-consuming. CRISPR/LbCpf1 geminiviral replicon system for HDR- or KI-based gene editing research has achieved a threefold increase in editing efficiency compared to the Cas9-based single replicon system in tomato (Vu et al. 2021). The precision of HR-based knock-in (HKI) edits in tomatoes has been enhanced by combining Cas9 with a geminivirus replicon, LbCas12a, containing multi-replicons. Additionally, enhancing the temperature of Cas12a could further improve HKI and accelerate tomato breeding efforts. Although geminivirus replicons have shown high success rates in plants, the majority of these studies have relied on selection markers with edited alleles. This implies that genome editing using geminivirus replicons remains a challenge without the use of selection markers to select edited lines. Therefore, there is a requirement for refinement in developing marker-free and selection-free mutants. Thus, CRISPR/Cas9-induced gene replacement through the HDR mechanism presents an innovative breeding approach in tomatoes. However, a key limitation of this method is that most viruses cannot infect germline or meristematic cells. As a result, regenerated plants are derived from somatic cells (Rössner et al. 2022). Different vectors derived from potato virus X and geminivirus have been successfully used to generate stable edited tomato plants through direct injection (Uranga et al. 2023). VIGE is an adaptable and reliable technology that can be utilized for the precise enhancement of tomato traits Čermák et al. (2015).

9.7.5 Other Delivery Methods

Liposomes are particles made of lipids that have proven to be effective in delivering genes and proteins. They have been successfully utilized to deliver DNA for gene editing in protoplasts (Mahmoud et al. 2022). Moreover, liposome nanoparticles have enabled the RNPs' delivery into the protoplast of tobacco, resulting in increased editing efficiencies of up to 6.0% (Liu et al. 2020c). Electroporation is a method that enhances the cell membrane permeability in plant cells (protoplast) lacking cell walls by exposing them to intense and short electric pulses. This process facilitates the introduction of the RNP complex into the cells (Ozyigit 2020). Electroporation has also been utilized in tomatoes to introduce and transiently express vectors carrying ZFNs (Hilioti 2018). However, there are no reports of instances of using CRISPR/Cas vectors or RNPs in this species.

9.8 Challenges and Limitations of Genome Editing in Tomatoes

CRISPR/Cas9 and other genome editing technologies have completely changed the field of plant biotechnology, especially with regard to tomatoes, allowing for the exact modification of features like shelf life, nutritional value, and resistance to diseases. Thanks to this invention, a robust tomato variety that can better meet the world's food demands would be developed. Nonetheless, technical limitations in offering editing tools, guaranteeing specificity, making stable genetic modifications, and managing regulatory frameworks make it harder to deploy these technologies. Additionally, it is crucial to consider the ethics of genome editing, hazards to human health, and ecological effects when having this conversation. With continuous research and development targeted at overcoming existing obstacles, the future of genome editing in tomatoes seems bright, but public acceptance and regulatory development will be critical to its success (Gan and Ling 2022). Overall, the effect on tomato agriculture might be revolutionary, resulting in more environment-friendly farming methods. However, this possibility needs to be carefully weighed against the danger and normal modification involved (Zhang et al. 2020c).

9.8.1 Off-Target Effects

The implementations of CRISPR/Cas9 for crop improvement are severely limited by the occurrence of off-target effects since even minor accidental changes can have profound implications on plant fitness and phenotype. For example, off-target mutations in tomato plants may impact characteristics like fruit size, ripening time, or disease resistance, thus negating the advantages of genome editing. Recent developments have improved the selectivity of Cas9 enzymes and improved guide RNA design to lower these dangers. CRISPR/Cas9 does not require the engineering of Cas9 to target altered genes. Usually, in the CRISPR/Cas9 technique, a 2–5-bp PAM sequence is found often in the genome; as a result, desired genes are difficult to edit with the structure constraints (Xie et al. 2014). It can be minimized by modifications of Cas9 with unalike sequences specified for PAM (Zaidi et al. 2017). One promising approach to considerably reduce off-target activity is the use of modified Cas9 variants (Kleinstiver et al. 2016). To ensure that genome editing can be used in tomatoes safely and efficiently, more studies are needed to improve these methods.

9.9 Ethical Consideration in Tomato Genome Editing

The genome editing in tomatoes raises several ethical issues that include biodiversity, environmental sustainability, and human health. The possibility of accidental mutations that could result in the emergence of unforseen consequences. Significant public health concerns are raised by these unexpected results, especially in areas where labeling and transparency laws are not strictly implemented, making it more challenging for customers to make educated decisions. Furthermore, the biodiversity of agriculture may be significantly impacted if genome-edited tomatoes are widely adopted. Genetic variety is essential for resistance against pests, diseases, and shifting environmental circumstances, but it may be lost if farmers switch more and more to genetically homogenous, modified crops. Food security may be threatened by this loss of diversity, which could render crops more vulnerable to broad epidemics. Public opinion is a major factor in determining the ethical framework surrounding genome editing. Some demographics may welcome the technology because of its possible advantages, but others are still dubious, especially because of the lack of transparent labeling. Whether or not consumers should be able to determine whether the food they eat has been genetically modified as well as the context of their control over what they eat are common topics of discussion in ethical discourse. These arguments are strongly related to more general ones concerning the duties of researchers, authorities, and businesses in guaranteeing that the advantages of genome editing are achieved without jeopardizing environmental integrity or public confidence. As emphasized by Kuzma and Kokotovich (2011), these moral questions are practical and affect how society views and regulates genome editing technology. Patent innovations enable new forms of potential ethical regulations and guidance in the field. Only two groups have achieved original CRISPR/Cas9 patents: Emmanuelle Charpentier and Jennifer Doudna from the University of California, University of Vienna, and Berkeley and for general use and Feng Zhang of the Massachusetts Institute of Technology (MIT) for use on eukaryotes, including plants and animals (Gostimskaya 2022). Legislation and regulatory framework for governing gene editing, standardizing its use, and ensuring safety, ethics, and responsibility.

9.10 Conclusion

Genome editing is a revolutionary method for enhancing tomato crops that can successfully address several agricultural issues. Scientists can improve tomatoes in several important ways by utilizing tools like CRISPR/Cas9. These include increasing the tomato's resistance to diseases, which helps reduce crop losses caused by pathogens; improving its nutritional profile, which improves health outcomes; and prolonging its shelf life, which lowers waste and eliminates the need for preservatives. By reducing the environmental impact of tomato cultivation, these advancements have the potential to greatly increase food security and support more sustainable farming methods. Putting genome editing into practice has its own set of challenges, such as managing public concerns about the moral implications of genetic alterations and negotiating regulatory frameworks. It takes a balanced approach that takes into account the wider societal implications of these technologies in addition to their scientific and commercial benefits to ensure that they are developed and used responsibly. It is anticipated that as research develops, genome editing methods will continue to advance, possibly providing even more accurate and practical answers to agricultural problems. There must be constant discussion and evaluation in these areas because the eventual success of these innovations will rely on how well they conform to societal norms and legal requirements.

References

Anzalone AV, Randolph PB, Davis JR, Sousa AA, Koblan LW, Levy JM et al (2019) Search-and-replace genome editing without double-strand breaks or donor DNA. Nature 576(7785):149–157

Anzalone AV, Koblan LW, Liu DR (2020) Genome editing with CRISPR–Cas nucleases, base editors, transposases and prime editors. Nat Biotechnol 38(7):824–844

Arimura SI, Ayabe H, Sugaya H, Okuno M, Tamura Y, Tsuruta Y,... Tsutsumi N (2020) Targeted gene disruption of ATP synthases 6-1 and 6-2 in the mitochondrial genome of Arabidopsis thaliana by mitoTALENs. Plant J 104(6):1459–1471

Barrangou R, Fremaux C, Deveau H, Richards M, Boyaval P, Moineau S et al (2007) CRISPR provides acquired resistance against viruses in prokaryotes. Science 315(5819):1709–1712

Bock R, Knoop V (eds) (2012) Genomics of chloroplasts and mitochondria (Advances in photosynthesis and respiration), vol 35 [Internet]. Dordrecht: Springer Netherlands [cited 2024 Aug 15]. Available from: http://link.springer.com/10.1007/978-94-007-2920-9

Brooks C, Nekrasov V, Lippman ZB, Van Eck J (2014) Efficient gene editing in tomato in the first generation using the clustered regularly interspaced short palindromic repeats/CRISPR-associated9 system. Plant Physiol 166(3):1292–1297

Butler NM, Baltes NJ, Voytas DF, Douches DS (2016) Geminivirus-mediated genome editing in potato (Solanum tuberosum L.) using sequence-specific nucleases. Front Plant Sci [Internet]. [cited 2024 Aug 16] 7. Available from: https://www.frontiersin.org/journals/plant-science/articles/10.3389/fpls.2016.01045/full

Cambiaso V, Gimenez MD, da Costa JHP, Vazquez DV, Picardi LA, Pratta GR et al (2019) Selected genome regions for fruit weight and shelf life in tomato RILs discernible by markers based on genomic sequence information. Breed Sci 69(3):447–454

Ceasar SA, Rajan V, Prykhozhij SV, Berman JN, Ignacimuthu S (2016) Insert, remove or replace: a highly advanced genome editing system using CRISPR/Cas9. Biochim Biophys Acta 1863(9):2333–2344

Čermák T, Baltes NJ, Čegan R, Zhang Y, Voytas DF (2015) High-frequency, precise modification of the tomato genome. Genome Biol 16(1):232
Chandrasekaran M, Boopathi T, Paramasivan M (2021) A status-quo review on CRISPR-Cas9 gene editing applications in tomato. Int J Biol Macromol 190:120–129
Chaudhuri A, Halder K, Datta A (2022) Classification of CRISPR/Cas system and its application in tomato breeding. Theor Appl Genet 135(2):367–387
Chincinska IA (2021) Leaf infiltration in plant science: old method, new possibilities. Plant Methods 17(1):83
Cunningham FJ, Demirer GS, Goh NS, Zhang H, Landry MP (2020) Nanobiolistics: an emerging genetic transformation approach. Methods Mol Biol 2124:141–159
Dahan-Meir T, Filler-Hayut S, Melamed-Bessudo C, Bocobza S, Czosnek H, Aharoni A et al (2018) Efficient in planta gene targeting in tomato using geminiviral replicons and the CRISPR/Cas9 system. Plant J 95(1):5–16
De Saeger J, Park J, Chung HS, Hernalsteens JP, Van Lijsebettens M, Inzé D et al (2021) Agrobacterium strains and strain improvement: present and outlook. Biotechnol Adv 53:107677
Eid A, Alshareef S, Mahfouz MM (2018) CRISPR base editors: genome editing without double-stranded breaks. Biochem J 475(11):1955–1964
Gan WC, Ling APK (2022) CRISPR/Cas9 in plant biotechnology: applications and challenges. BioTechnologia (Pozn) 103(1):81–93
Gao L, Gonda I, Sun H, Ma Q, Bao K, Tieman DM et al (2019) The tomato pan-genome uncovers new genes and a rare allele regulating fruit flavor. Nat Genet 51(6):1044–1051
Garneau JE, Dupuis MÈ, Villion M, Romero DA, Barrangou R, Boyaval P et al (2010) The CRISPR/Cas bacterial immune system cleaves bacteriophage and plasmid DNA. Nature 468(7320):67–71
Gostimskaya I (2022) CRISPR–Cas9: a history of its discovery and ethical considerations of its use in genome editing. Biochemistry (Mosc) 87(8):777–788
Gualberto JM, Newton KJ (2017) Plant mitochondrial genomes: dynamics and mechanisms of mutation. Annu Rev Plant Biol 68:225–252
Heler R, Marraffini LA, Bikard D (2014) Adapting to new threats: the generation of memory by CRISPR-Cas immune systems. Mol Microbiol 93(1):1–9
Hilioti Z (2018) Non-transgenic approach to deliver ZFNs in seeds for targeted genome engineering. Methods Mol Biol 1867:187–199
Hong Y, Meng J, He X, Zhang Y, Liu Y, Zhang C et al (2021) Editing miR482b and miR482c simultaneously by CRISPR/Cas9 enhanced tomato resistance to Phytophthora infestans. Phytopathology 111(6):1008–1016
Hu Z, Li J, Ding S, Cheng F, Li X, Jiang Y et al (2021) The protein kinase CPK28 phosphorylates ascorbate peroxidase and enhances thermotolerance in tomato. Plant Physiol 186(2):1302–1317
Huang TK, Puchta H (2021) Novel CRISPR/Cas applications in plants: from prime editing to chromosome engineering. Transgenic Res 30(4):529–549
Huibers RP, Loonen AEHM, Gao D, den Ackerveken GV, Visser RGF, Bai Y (2013) Powdery mildew resistance in tomato by impairment of SlPMR4 and SlDMR1. PLoS One 8(6):e67467
Hunziker J, Nishida K, Kondo A, Kishimoto S, Ariizumi T, Ezura H (2020) Multiple gene substitution by Target-AID base-editing technology in tomato. Sci Rep 10(1):20471
Jeon JE, Kim JG, Fischer CR, Mehta N, Dufour-Schroif C, Wemmer K et al (2020) A pathogen-responsive gene cluster for highly modified fatty acids in tomato. Cell 180(1):176–187.e19
Jinek M, Chylinski K, Fonfara I, Hauer M, Doudna JA, Charpentier E (2012) A programmable dual RNA-guided DNA endonuclease in adaptive bacterial immunity. Science 337(6096):816–821
Kakeshpour T, Tamang TM, Motolai G, Fleming ZW, Park JE, Wu Q et al (2021) CGFS-type glutaredoxin mutations reduce tolerance to multiple abiotic stresses in tomato. Physiol Plant 173(3):1263–1279
Kang GH, Kang BC, Han JS, Lee JM (2024) DNA-free genome editing in tomato protoplasts using CRISPR/Cas9 ribonucleoprotein delivery. Hortic Environ Biotechnol 65(1):131–142

Kawaguchi K, Takei-Hoshi R, Yoshikawa I, Nishida K, Kobayashi M, Kusano M et al (2021) Functional disruption of cell wall invertase inhibitor by genome editing increases sugar content of tomato fruit without decrease fruit weight. Sci Rep 11(1):21534

Kazama T, Okuno M, Watari Y, Yanase S, Koizuka C, Tsuruta Y et al (2019) Curing cytoplasmic male sterility via TALEN-mediated mitochondrial genome editing. Nat Plants 5(7):722–730

Kim HK, Yu G, Park J, Min S, Lee S, Yoon S et al (2021) Predicting the efficiency of prime editing guide RNAs in human cells. Nat Biotechnol 39(2):198–206

Kiryushkin AS, Ilina EL, Guseva ED, Pawlowski K, Demchenko KN (2021) Hairy CRISPR: genome editing in plants using hairy root transformation. Plants (Basel) 11(1):51

Kleinstiver BP, Pattanayak V, Prew MS, Tsai SQ, Nguyen NT, Zheng Z et al (2016) High-fidelity CRISPR-Cas9 nucleases with no detectable genome-wide off-target effects. Nature 529(7587):490–495

Kuzma J, Kokotovich A (2011) Renegotiating GM crop regulation. EMBO Rep 12(9):883–888

Laforest LC, Nadakuduti SS (2022) Advances in delivery mechanisms of CRISPR gene-editing reagents in plants. Front Genome Ed 4:830178

Larriba E, Yaroshko O, Pérez-Pérez JM (2024) Recent advances in tomato gene editing. Int J Mol Sci 25(5):2606

Li X, Wang Y, Chen S, Tian H, Fu D, Zhu B et al (2018) Lycopene is enriched in tomato fruit by CRISPR/Cas9-mediated multiplex genome editing. Front Plant Sci [Internet]. [cited 2024 Aug 16] 9. Available from: https://www.frontiersin.org/journals/plant-science/articles/10.3389/fpls.2018.00559/full

Li H, Li J, Chen J, Yan L, Xia L (2020) Precise modifications of both exogenous and endogenous genes in rice by prime editing. Mol Plant 13(5):671–674

Li N, He Q, Wang J, Wang B, Zhao J, Huang S et al (2023a) Super-pangenome analyses highlight genomic diversity and structural variation across wild and cultivated tomato species. Nat Genet 55(5):852–860

Li Y, Liang J, Deng B, Jiang Y, Zhu J, Chen L et al (2023b) Applications and prospects of CRISPR/Cas9-mediated base editing in plant breeding. Curr Issues Mol Biol 45(2):918–935

Lin Q, Zong Y, Xue C, Wang S, Jin S, Zhu Z et al (2020) Prime genome editing in rice and wheat. Nat Biotechnol 38(5):582–585

Lin CS, Hsu CT, Yuan YH, Zheng PX, Wu FH, Cheng QW et al (2022) DNA-free CRISPR-Cas9 gene editing of wild tetraploid tomato Solanum peruvianum using protoplast regeneration. Plant Physiol 188(4):1917–1930

Liu Z, Dong H, Cui Y, Cong L, Zhang D (2020a) Application of different types of CRISPR/Cas-based systems in bacteria. Microb Cell Factories 19(1):172

Liu L, Zhang J, Xu J, Li Y, Guo L, Wang Z et al (2020b) CRISPR/Cas9 targeted mutagenesis of SlLBD40, a lateral organ boundaries domain transcription factor, enhances drought tolerance in tomato. Plant Sci 301:110683

Liu W, Rudis MR, Cheplick MH, Millwood RJ, Yang JP, Ondzighi-Assoume CA et al (2020c) Lipofection-mediated genome editing using DNA-free delivery of the Cas9/gRNA ribonucleoprotein into plant cells. Plant Cell Rep 39(2):245–257

Liu J, Wang S, Wang H, Luo B, Cai Y, Li X et al (2021) Rapid generation of tomato male-sterile lines with a marker use for hybrid seed production by CRISPR/Cas9 system. Mol Breed 41(3):25

Mahmoud LM, Kaur P, Stanton D, Grosser JW, Dutt M (2022) A cationic lipid mediated CRISPR/Cas9 technique for the production of stable genome edited citrus plants. Plant Methods 18(1):33

McCourt JA, Duggleby RG (2006) Acetohydroxyacid synthase and its role in the biosynthetic pathway for branched-chain amino acids. Amino Acids 31(2):173–210

Min D, Li F, Cui X, Zhou J, Li J, Ai W et al (2020) *SlMYC2* are required for methyl jasmonate-induced tomato fruit resistance to *Botrytis cinerea*. Food Chem 310:125901

Molla KA, Sretenovic S, Bansal KC, Qi Y (2021) Precise plant genome editing using base editors and prime editors. Nat Plants 7(9):1166–1187

Nekrasov V, Wang C, Win J, Lanz C, Weigel D, Kamoun S (2017) Rapid generation of a transgene-free powdery mildew resistant tomato by genome deletion. Sci Rep 7(1):482

Nishida K, Arazoe T, Yachie N, Banno S, Kakimoto M, Tabata M et al (2016) Targeted nucleotide editing using hybrid prokaryotic and vertebrate adaptive immune systems. Science 353(6305):aaf8729

Nishimasu H, Ran FA, Hsu PD, Konermann S, Shehata SI, Dohmae N et al (2014) Crystal structure of Cas9 in complex with guide RNA and target DNA. Cell 156(5):935–949

Nonaka S, Arai C, Takayama M, Matsukura C, Ezura H (2017) Efficient increase of γ-aminobutyric acid (GABA) content in tomato fruits by targeted mutagenesis. Sci Rep 7(1):7057

Ortigosa A, Gimenez-Ibanez S, Leonhardt N, Solano R (2019) Design of a bacterial speck resistant tomato by CRISPR/Cas9-mediated editing of SlJAZ2. Plant Biotechnol J 17(3):665–673

Ozyigit II (2020) Gene transfer to plants by electroporation: methods and applications. Mol Biol Rep 47(4):3195–3210

Păcurar DI, Thordal-Christensen H, Păcurar ML, Pamfil D, Botez C, Bellini C (2011) Agrobacterium tumefaciens: from crown gall tumors to genetic transformation. Physiol Mol Plant Pathol 76:76–81

Raman V, Rojas CM, Vasudevan B, Dunning K, Kolape J, Oh S et al (2022) Agrobacterium expressing a type III secretion system delivers Pseudomonas effectors into plant cells to enhance transformation. Nat Commun 13(1):2581

Rodrigues SD, Karimi M, Impens L, Van Lerberge E, Coussens G, Aesaert S et al (2021) Efficient CRISPR-mediated base editing in Agrobacterium spp. Proc Natl Acad Sci USA 118(2):e2013338118

Rodríguez-Leal D, Lemmon ZH, Man J, Bartlett ME, Lippman ZB (2017) Engineering quantitative trait variation for crop improvement by genome editing. Cell 171(2):470–480.e8

Rössner C, Lotz D, Becker A (2022) VIGS goes viral: how VIGS transforms our understanding of plant science. Annu Rev Plant Biol 73:703–728

Rustgi S, Naveed S, Windham J, Zhang H, Demirer GS (2022) Plant biomacromolecule delivery methods in the 21st century. Front Genome Ed 4:1011934

Salava H, Thula S, Mohan V, Kumar R, Maghuly F (2021) Application of genome editing in tomato breeding: mechanisms, advances, and prospects. Int J Mol Sci 22(2):682

Santillán Martínez MI, Bracuto V, Koseoglou E, Appiano M, Jacobsen E, Visser RGF et al (2020) CRISPR/Cas9-targeted mutagenesis of the tomato susceptibility gene PMR4 for resistance against powdery mildew. BMC Plant Biol 20(1):284

Schilbert HM, Rempel A, Pucker B (2020) Comparison of read mapping and variant calling tools for the analysis of plant NGS data. Plants 9(4):439

Shao M, Xu TR, Chen CS (2016) The big bang of genome editing technology: development and application of the CRISPR/Cas9 system in disease animal models. Dongwuxue Yanjiu 37(4):191–204

Shimatani Z, Kashojiya S, Takayama M, Terada R, Arazoe T, Ishii H et al (2017 May) Targeted base editing in rice and tomato using a CRISPR-Cas9 cytidine deaminase fusion. Nat Biotechnol 35(5):441–443

Slaman E, Lammers M, Angenent GC, de Maagd RA (2023) High-throughput sgRNA testing reveals rules for Cas9 specificity and DNA repair in tomato cells. Front Genome Ed [Internet]. [cited 2024 Aug 16] 5. Available from: https://www.frontiersin.org/journals/genome-editing/articles/10.3389/fgeed.2023.1196763/full

Tashkandi M, Ali Z, Aljedaani F, Shami A, Mahfouz MM (2018) Engineering resistance against Tomato yellow leaf curl virus via the CRISPR/Cas9 system in tomato. Plant Signal Behav 13(10):e1525996

Thiruppathi D (2022) CRISPR keeps going "wild": a new protocol for DNA-free genome editing of tetraploid wild tomatoes. Plant Physiol 189(1):10–11

Thomazella DPDT, Brail Q, Dahlbeck D, Staskawicz B (2016) CRISPR-Cas9 mediated mutagenesis of a DMR6 ortholog in tomato confers broad-spectrum disease resistance [Internet]. bioRxiv [cited 2023 Apr 4]:064824. Available from: https://www.biorxiv.org/content/10.1101/064824v1

Thomazella DPDT, Seong K, Mackelprang R, Dahlbeck D, Geng Y, Gill US et al (2021) Loss of function of a DMR6 ortholog in tomato confers broad-spectrum disease resistance. Proc Natl Acad Sci USA 118(27):e2026152118

Tingting L, Jinpeng Z, Xi Y, Kejian W, Yuchun R, Chun W (2023) Development and application of prime editing in plants. Rice Sci 30(6):509–522

Tiwari JK, Singh AK, Behera TK (2023) CRISPR/Cas genome editing in tomato improvement: Advances and applications. Front Plant Sci [Internet]. [cited 2024 Jul 3] 14. Available from: https://www.frontiersin.org/journals/plant-science/articles/10.3389/fpls.2023.1121209/full

Tomato Genome Consortium (2012) The tomato genome sequence provides insights into fleshy fruit evolution. Nature 485(7400):635–641

Tran MT, Doan DTH, Kim J, Song YJ, Sung YW, Das S et al (2021) CRISPR/Cas9-based precise excision of SlHyPRP1 domain(s) to obtain salt stress-tolerant tomato. Plant Cell Rep 40(6):999–1011

Uluisik S, Chapman NH, Smith R, Poole M, Adams G, Gillis RB et al (2016a) Genetic improvement of tomato by targeted control of fruit softening. Nat Biotechnol 34(9):950–952

Uluisik S, Chapman NH, Smith R, Poole M, Adams G, Gillis RB et al (2016b) Corrigendum: genetic improvement of tomato by targeted control of fruit softening. Nat Biotechnol 34(10):1072

Uranga M, Aragonés V, García A, Mirabel S, Gianoglio S, Presa S et al (2023) RNA virus-mediated gene editing for tomato trait breeding [Internet]. bioRxiv. [cited 2024 Aug 16]:2023.10.01.560115. Available from: https://www.biorxiv.org/content/10.1101/2023.10.01.560115v1

Veillet F, Perrot L, Chauvin L, Kermarrec MP, Guyon-Debast A, Chauvin JE et al (2019) Transgene-Free genome editing in tomato and potato plants using agrobacterium-mediated delivery of a CRISPR/Cas9 cytidine base editor. Int J Mol Sci 20(2):402

Voytas DF, Gao C (2014) Precision genome engineering and agriculture: opportunities and regulatory challenges. PLoS Biol 12(6):e1001877

Vu TV, Sivankalyani V, Kim EJ, Doan DTH, Tran MT, Kim J et al (2020a) Highly efficient homology-directed repair using CRISPR/Cpf1-geminiviral replicon in tomato. Plant Biotechnol J 18(10):2133–2143

Vu TV, Das S, Tran MT, Hong JC, Kim JY (2020b) Precision genome engineering for the breeding of tomatoes: recent progress and future perspectives. Front Genome Ed 2:612137

Vu TV, Doan DTH, Tran MT, Sung YW, Song YJ, Kim JY (2021) Improvement of the LbCas12a-crRNA system for efficient gene targeting in tomato. Front Plant Sci [Internet]. [cited 2024 Aug 15] 12. Available from: https://www.frontiersin.org/journals/plant-science/articles/10.3389/fpls.2021.722552/full

Wada N, Ueta R, Osakabe Y, Osakabe K (2020) Precision genome editing in plants: state-of-the-art in CRISPR/Cas9-based genome engineering. BMC Plant Biol 20(1):234

Waltz E (2022) GABA-enriched tomato is first CRISPR-edited food to enter market. Nat Biotechnol 40(1):9–11

Wang D, Samsulrizal NH, Yan C, Allcock NS, Craigon J, Blanco-Ulate B et al (2019) Characterization of CRISPR mutants targeting genes modulating pectin degradation in ripening tomato. Plant Physiol 179(2):544–557

Wang S, Zong Y, Lin Q, Zhang H, Chai Z, Zhang D et al (2020) Precise, predictable multi-nucleotide deletions in rice and wheat using APOBEC-Cas9. Nat Biotechnol 38(12):1460–1465

Wu K, Xu C, Li T, Ma H, Gong J, Li X et al (2023) Application of nanotechnology in plant genetic engineering. Int J Mol Sci 24(19):14836

Xia X, Cheng X, Li R, Yao J, Li Z, Cheng Y (2021) Advances in application of genome editing in tomato and recent development of genome editing technology. Theor Appl Genet 134(9):2727–2747

Xiao J, Cheng H, Li X, Xiao J, Xu C, Wang S (2013) Rice WRKY13 regulates cross talk between abiotic and biotic stress signaling pathways by selective binding to different cis-elements. Plant Physiol 163(4):1868–1882

Xie K, Zhang J, Yang Y (2014) Genome-wide prediction of highly specific guide RNA spacers for CRISPR-Cas9-mediated genome editing in model plants and major crops. Mol Plant 7(5):923–926

Xu J, Hua K, Lang Z (2019) Genome editing for horticultural crop improvement. Hortic Res 6:113

Yang SH, Kim E, Park H, Koo Y (2022) Selection of the high efficient sgRNA for CRISPR-Cas9 to edit herbicide related genes, PDS, ALS, and EPSPS in tomato. Appl Biol Chem 65(1):13

Yaroshko O, Pasternak T, Larriba E, Pérez-Pérez JM (2023) Optimization of callus induction and shoot regeneration from tomato cotyledon explants. Plan Theory 12(16):2942

Yoon YJ, Venkatesh J, Lee JH, Kim J, Lee HE, Kim DS et al (2020) Genome editing of eIF4E1 in tomato confers resistance to pepper mottle virus. Front Plant Sci [Internet]. [cited 2024 Aug 16] 11. Available from: https://www.frontiersin.org/journals/plant-science/articles/10.3389/fpls.2020.01098/full

Yu Q, Han H, Vila-Aiub MM, Powles SB (2010) AHAS herbicide resistance endowing mutations: effect on AHAS functionality and plant growth. J Exp Bot 61(14):3925–3934

Yu T, Ai G, Xie Q, Wang W, Song J, Wang J et al (2022) Regulation of tomato fruit elongation by transcription factor BZR1.7 through promotion of SUN gene expression. Hortic Res 9:uhac121

Zaidi SSEA, Mahfouz MM, Mansoor S (2017) CRISPR-Cpf1: a new tool for plant genome editing. Trends Plant Sci 22(7):550–553

Zhan X, Lu Y, Zhu JK, Botella JR (2021) Genome editing for plant research and crop improvement. J Integr Plant Biol 63(1):3–33

Zhang Y, Massel K, Godwin ID, Gao C (2018) Applications and potential of genome editing in crop improvement. Genome Biol 19(1):210

Zhang T, Wu A, Yue Y, Zhao Y (2020a) uORFs: important cis-regulatory elements in plants. Int J Mol Sci 21(17):6238

Zhang N, Roberts HM, Van Eck J, Martin GB (2020b) Generation and molecular characterization of CRISPR/Cas9-induced mutations in 63 immunity-associated genes in tomato reveals specificity and a range of gene modifications. Front Plant Sci 11:10

Zhang D, Hussain A, Manghwar H, Xie K, Xie S, Zhao S et al (2020c) Genome editing with the CRISPR-Cas system: an art, ethics and global regulatory perspective. Plant Biotechnol J 18(8):1651–1669

Zhang Y, Iaffaldano B, Qi Y (2021) CRISPR ribonucleoprotein-mediated genetic engineering in plants. Plant Commun 2(2):100168

Zhou J, Deng K, Cheng Y, Zhong Z, Tian L, Tang X et al (2017) CRISPR-Cas9 based genome editing reveals new Insights into MicroRNA function and regulation in rice. Front Plant Sci [Internet]. [cited 2024 Aug 15] 8. Available from: https://www.frontiersin.org/journals/plant-science/articles/10.3389/fpls.2017.01598/full

Zhu G, Wang S, Huang Z, Zhang S, Liao Q, Zhang C et al (2018) Rewiring of the fruit metabolome in tomato breeding. Cell 172(1–2):249–261.e12

Zsögön A, Cermak T, Voytas D, Peres LEP (2017) Genome editing as a tool to achieve the crop ideotype and de novo domestication of wild relatives: case study in tomato. Plant Sci 256:120–130

CRISPR/Cas9-Mediated Genome Editing in Tomatoes: Trait Improvement and Beyond

10

Aqsa Abid, Kainat Zahra, Saba Zafar, Faiz Ahmad Joyia, Ghulam Mustafa, and Aqsa Parvaiz

Abstract

Tomatoes have a limited genetic base, which makes breeding extremely difficult. Thus, quick and efficient tomato breeding is now beneficial, thanks to the evolution of clustered regularly interspaced short palindromic repeat (CRISPR)–associated protein9 (CRISPR/Cas9) genome editing. With the use of CRISPR/Cas9, a variety of tomato traits has been functionally characterized and edited. These traits include plant structure and flower characters (such as leaf, stem, flower, male sterility, fruit, and parthenocarpy), fruit ripening, quality, nutrition (such as lycopene, carotenoid, GABA, Total Soluble Solids (TSS), anthocyanin, and shelf life), disease resistance (such as against Tomato Yellow Leaf Curl Virus (TYLCV), powdery mildew, and late blight), abiotic stress tolerance (such as heat, drought, and salinity), Carbon-Nitrogen (C-N) metabolism, and herbicide resistance. The de novo development of elite features from natural collaborators to cultivated tomatoes and vice versa has been determined by the introduction of CRISPR/Cas9. Advances in CRISPR/Cas9 empower the use of online tools for base/prime editing Target-AID (Antibiotics and Immunomodulation-Directed), homologous recombination (HR)–based gene knock-in (HKI) by geminivirus replicon, robust CRISPR/Cas constructs, and effective transformation protocols

Aqsa Abid and Kainat Zahra contributed equally with all other contributors.

A. Abid · K. Zahra · S. Zafar · A. Parvaiz (✉)
Department of Biochemistry and Biotechnology, The Women University Multan, Multan, Pakistan
e-mail: aqsa.6497@wum.edu.pk

F. A. Joyia · G. Mustafa
Centre of Agricultural Biochemistry and Biotechnology (CABB), University of Agriculture, Faisalabad, Pakistan

M. Waseem et al. (eds.), *Omics Approaches for Tomato Yield and Quality Trait Improvement*, https://doi.org/10.1007/978-981-96-3890-1_10

like *Agrobacterium* and DNA-free protoplast method for Cas9–gRNA ribonucleoprotein (RNP) complex. Due to global climate change, low biodiversity has originated hindrance for tomato breeding and agriculture. Due to the potential linkage drag of undesirable traits from their genetic backgrounds. Wild tomatoes and their relatives possess a vast, untapped reservoir of genetic diversity that remains largely unexploited in the breeding of commercial tomatoes. New plant breeding techniques (NPBTs) have made it achievable to breed tomatoes with great precision through genome editing, which distinctly uses CRISPR/Cas-based tools. Moreover, using highly precise instruments, faster introgression or de novo domestication of elite and innovative features from/to the natural tomato relatives to/from the cultivated tomatoes, respectively, has been developed.

Keywords

Solanum lycopersicum · CRISPR/Cas9 · Gene editing · Tomato breeding · Precision genome engineering

10.1 Introduction

Tomato (*Solanum lycopersicum L*) are one of the most widely cultivated and economically significant vegetable crops. Tomatoes are regarded as a healthy and nutritious food (Causse et al. 2016; Waseem and Li 2019). It is necessary to breed new tomato cultivars, which fulfill the requirements of modern consumers and industrial processing needs. This requires a huge genetic resource that is linked with the desired traits. Many varieties of tomatoes with agronomic characteristics have been promoted as a result of universal genetic variability. Moreover, it demands a lot of work, effort, and time, which can generate mutagenesis that is performed by mutagens, which are successfully used in plant breeding techniques. Phenotypic selection is also used in conventional breeding, which takes a long period of time for the whole breeding cycle. Genome editing tools like CRISPR and CRISPR-associated protein9 (Cas9) have made plant breeding much easier and more effective in overcoming the problems associated with traditional breeding methods. Modern studies have illustrated the extraordinary effectiveness of CRISPR/Cas9 gene editing in tomatoes for the engineering of useful alleles or the production of new genetic diversity with the help of precise base editing, gene disruption, and gene replacement.

It is a potent and accurate genome editing technique to promote mutants with desired traits and differentiate the functions of individual genes (Rothan et al. 2019; Chandrasekaran et al. 2021). CRISPR technology is being used to modify a different class of tomato traits. Gamma Aminobutyric Acid (GABA)-rich tomatoes with CRISPR editing have very recently knock the food market (Waltz 2022). The cloning, transformation, and regulatory aspects of sgRNA research are shown in Fig. 10.1. Nonhomologous end joining (NHEJ) repairing double-strand breaks (DSBs) in target coding regions has been the main method of gene knockout by using CRISPR/Cas9 technology. A DELLA protein was also used to have an

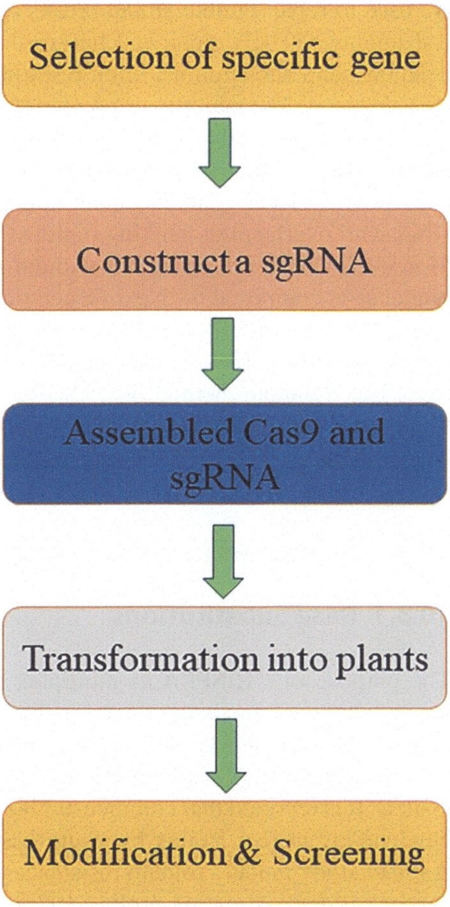

Fig. 10.1 CRISPR/Cas9 genome editing technology

effective gain of function mutation (Tomlinson et al. 2019). High numbers of heritable mutations can be obtained in the first generation (up to 100%) in tomatoes where genetic transformation is accomplished with the help of somatic embryogenesis (Brooks et al. 2014). The transgenes must be isolated in the progeny in order to make the mutation further precise and then use it for breeding. Cas9 associates with paired sgRNAs that target two ends of a particular sequence can increase the proficiency of gene disruption (Yu et al. 2017).

10.2 CRISPR/Cas9 Genome Editing Technology

The CRISPR/Cas9 protein fights invasive viruses or phages by using the rules of immunity that are present in bacteria or archaea. There are many forms and subclasses of CRISPR/Cas. Moreover, the processes called homology-directed repair (HDR), homologous recombination (HR), and nonhomologous end joining (NHEJ) are used to repair double-strand breaks (DSBs). The error-prone, non-homologous end joining (NHEJ) repair mechanism is susceptible to random small insertions, deletions, and substitutions often resulting in knockout mutations with significant consequences. The majority of point mutations or deletions induced by gene knock-in are promoted by the HDR mechanism. However, this technique has a very low progress rate at this time. Breeding becomes much simpler with CRISPR/Cas9 because it produces gene knockout mutants for desired traits. By applying template DNA with overlapping flanking regions, HDR enables the gene knock-in process. Moreover, CRISPR/Cas9 is not without its limitations. For example, the genome sequence is composed of Next Generation Genomics (NGG) protospacer adjacent motifs (PAMs). Modifying Cas9 proteins and looking for over and above Cas9 proteins in other bacteria with specific PAM sequences have therefore become a preference. It is a unique method of CRISPR/Cas9 that discusses how it is newly being used to alter tomato traits and the recent improvements in the design, cloning, transformation, and regulatory aspects of sgRNA research as shown in Fig. 10.1.

10.3 Strategies in CRISPR/Cas9 Technology

10.3.1 Base Substitutions

The purpose of CRISPR/Cas multiplex in tomato genome editing has expanded beyond targeted mutagenesis to include precise base-by-base modifications of lengthy DNA sequences. A loss function of mutation in the DELLA gene caused by direct base substitutions resulted in a phenotype with fewer numbers of serrated leaflets. It is reported that commensurable instruments were manipulated to produce ethylene-insensitive strains by using base substitutions in the ethylene receptor. Another study induces tomato resistance to herbicides by using base editing–based genome editing. Mutating proline-197 in *E. coli* to various amino acids confirmed that this residue plays a crucial role in conferring resistance to chlosulfuron. Subsequently, the complementary proline-186 in tomatoes was modified to other amino acids. Strong resistance to chlorsulfuron treatment was developed by strains with substituted bases (Veillet et al. 2019).

10.3.2 Homologous Recombination (HR)

HKI is a general precision technique that can be used in plants to modify large DNA sequences or SNPs. However, the many values of HKI, which are found in plants, are limited and not generally attainable for use due to the low adaptability of natural HR. HKI abundance has greatly increased since the development of CRISPR/Cas9 complexes as molecular scissors that are being used for creating DSBs at specific genomic locations. To produce anthocyanin over-accumulating events in tomatoes, HKI was modified by using CRISPR/Cas9 or Cas12a proteins. Herbicide-tolerant tomatoes were also successfully produced using CRISPR/Cas9-mediated HKI, which modified P186A by focusing on SlALS1 (Danilo et al. 2019).

10.3.3 CRISPR/Cas-Based Target Mutagenesis

CRISPR/Cas transformation has made it possible to utilize molecular tools for dominant precision plant breeding. Very recently, research on CRISPR/Cas-based targeted mutagenesis began to appear in 2013. These basically addressed the tool's intelligence and capacity for use in the study of gene function through knockout techniques.

10.4 Base/Prime Editing (Target-AID (Antibiotics and Immunomodulation-Directed) Technology)

Double-stranded breaks (DSBs) generated by Cas9 nuclease can lead to unwanted off-target effects. Base editing, on the other hand, offers a unique approach to modify a single base using the CRISPR-Cas 9 system without requiring donor DNA or introducing double-stranded breaks Fig. 10.2.

They promote the micro-homology-mediated end-joining technique. An unusual substitution method for medium-length DNA has been introduced. This technique is known as prime editing or precise editing. One base substitution can be altered with ease by using base or prime editing, but editing multiple bases presents challenges. Target gene sequences can be modified by using a number of CRISPR/Cas9 modifications (Vu et al. 2020b). However, they can be modified on a single base or thousands of bases because of their low persistence and difficult design (Vu et al. 2020a).

10.5 Current Status of Tomato Breeding

The development of cultivated tomato is believed to have originated in Central American, progressing through two key stages. In the first stage, Solanum lycopersicum var. cerasiform served as the direct ancestor and in the second stage, it evolved into *S. lycopersicum var. cerasiforme*. Fruit size was the development/selection parameter throughout cultivation (Lin et al. 2014; Waseem et al. 2019). Growth metrics and other fruit features are further characteristics that can be promoted to differentiate between natural cousins and cultivated tomatoes (Bai and Lindhout 2007). The genetic diversity of cultivated tomatoes has significantly decreased due to selective breading for specific geographic regions and market demands. To address this, tomato breeders including public and private institutions, individual farmers and researchers aim to develop new varieties that combine desirable traits thereby increasing genetic diversity and catering to diverse market needs and uses. Breeders of tomatoes might be assembling (public and private) or individuals, such farmers, and their aim for breeding projects can diversify (Bai and Lindhout 2007).

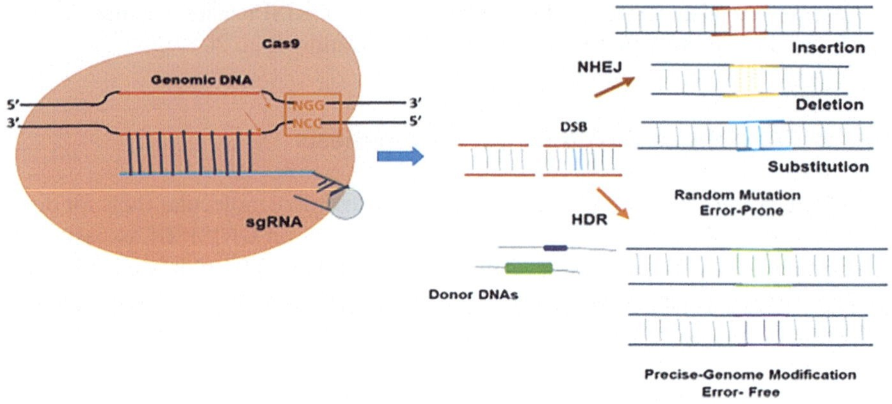

Fig. 10.2 CRISPR/Cas9 modification used for precision trait enrichment in tomatoes

Following prolonged selective breeding which reduced genetic diversity and led to inbred populations, the most effective way to restore genetic diversity and introduce new traits is through cross-pollination between distinct populations allowing for the exchange of unique characteristics. The pedigree approach guides the selection of hybrids with unusual interesting characteristics by recording the fulfillment of every progeny in multiple generations of a hybrid from genetically isolated plants. Only gene pools of parental populations can give rise to new qualities because they force careful monitoring and the selection of the best progenies across a number of generations. In the period of genomics, a process known as "marker-assisted selection" (MAS) allows genetic markers to help select specific genotypes (Collard and Mackill 2008). Traditional plant breeding has become much more efficient with the capacity to sequence a plant's entire genome at a very low cost.

10.6 Tomato Precision Genome Editing

The CRISPR/Cas9 innovation has made strong precision plant breeding achievable with the help of molecular tools. CRISPR/Cas-based mutagenesis in tomatoes concentrates on the practicability or the effective utilization of the tool and its suitability to check genome editing through knockout techniques.

10.7 Yield and Quality Improvement

High-value crops with better nutritional and resistance qualities can be produced by using CRISPR/Cas9 technology (Rasheed et al. 2021). Genes that diminish food output and negatively impact food quality can be stopped or eliminated. Here are some examples of CRISPR/Cas9 technology being used to improve crops:

- Primarily targeting the promotor sequence of the quantitative genes in tomatoes, CRISPR can be used to alter genes for fruit size, form, or architecture of the tomato plant (Rodriguez-Leal et al. 2017).
- It increases yield-related characteristics in tomatoes such as the quantity, size, and density of erect panicles by eliminating the GS3, DEP1, and Gn1a genes, which have a negative impact on yield (Li et al. 2016).
- A tomato cultivar's GDP-galactose phosphorylase 1 (GGP1) gene was altered by using CRISPR/Cas9 technology, which resulted in higher vitamin C levels (Wang et al. 2021).
- Transgenic lines intermediated by CRIPSR/Cas9 can be precisely produced in crops such as Chinese kale and Brassica campestris (Xiong et al. 2019).

10.8 Disease Resistance

To produce plants resistant to diseases, both ancestral breeding methods and transgenic approaches have been manipulated. It is understood that CRISPR/Cas9 has an extensive potential to achieve the goal in the easiest and most efficient manner. Some years ago, we could not have estimated the limit to which and the methods by which we can now decide plants and pathogens, thanks to targeted mutagenesis of a precise gene.

CRISPR/Cas9 generates disease-resistant crops by progressing the crop's natural immunity or engineering immunity into the crop, both of which will increase the crop's marketability. A remarkable improvement in CRISPR/Cas9 technology in disease resistance includes the system that has enormously expanded the potential for effective control of viral infections. Rather than healing the viral illness, Cas9 can be used to modify the DNA of the invaders to prevent them from spreading (Zaidi et al. 2016).

10.9 Abiotic Stress Tolerance

Management of abiotic stress is vital in light of the increasing influence of climate change. The process of breeding plants to become resistant to abiotic stressors has altered plant breeding. Molecular breeders discovered many genes for stress tolerance and then introduced them into crops as shown in Fig. 10.3. Researchers have utilized CRISPR-Cas9 to engineer drought-tolerant tomatoes by making targeted edits to the genome. Additionally, they have created mutant tomato cultivars with enhanced salt tolerance by modifying specific genes demonstrating the potential of gene editing for crop improvement. Using CRISPR/Cas9 to change particular sites in the genomes of tomatoes cultivars that produced mutants with an increased ratio of salt tolerance. Very recent developments in CRISPR/Cas9-based abiotic stress management include the production of drought-tolerant tomato cultivars and the editing of tomato genes to protect it against cold stress.

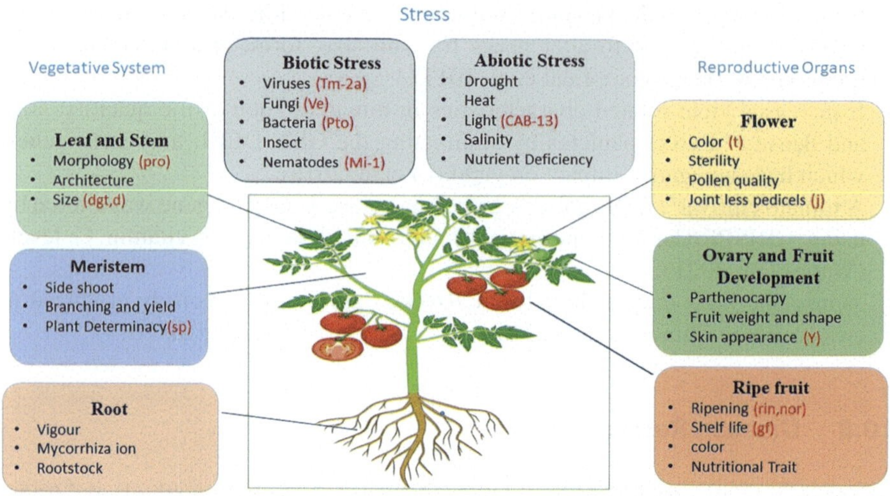

Fig. 10.3 Phenotypic traits of tomatoes (breeder discovers genes to improve the genome of tomatoes)

10.10 Nutritional Enhancement

Without compromising the useful features that are currently present in which researcher can produce stable and heritable changes. Tomato fruit characteristics are expanding, particularly for fresh uses. Moreover, some qualities of tomato fruits such as taste have vanished because of yield development and the industrial production of tomatoes at a large scale. Customizing the color of the fruit has also been achieved by controlling the accumulation of carotenoids in tomato plants.

10.11 Extended Shelf Life

Plant shelf life is a significant parameter of fruit quality and storage capability. The fruit's shelf life was increased and the ripening stage was only partially completed (Ito et al. 2015). But even in the heterozygous form, the mutants may cause tomato fruits to possess less lycopene (Herner and Sink 1973).

An intriguing example is the seedless tomato, which has applications in both fresh food and processing. The tomato AGAMOUSLIKE 6 (SlAGL6) gene was mutated by EMS (Ethyl Methane Sulfonate) to produce parthenocarpic tomato lines. This method inspired fruit development without fertilization, particularly in conditions of high heat. CRISPR/Cas9 complexes may potentially be used to accurately produce the S1AGL allele. It is interesting to note that seeds continue to develop normally even if fertilization is successful under typical circumstances, which indicates that the qualities may have breeding applications (Klap et al. 2017).

A different method of parthenocarpic tomato breeding involved particularly targeting the mutation. An SlIAA9 protein inhibits growth in the absence of fertilization. An SlIAA9 knockout mutant generated parthenocarpic fruits, but it also resulted in aberrant leaf morphology, which may have an influence on yield (Ueta et al. 2017). An intermediate metabolite of C4 plants, malic acid (also known as malate salt), is involved in root Al detoxification, fruit quality, and plant growth (Ye et al. 2017).

There are currently two possible theories that are not mutually exclusive:

- The first theory holds that glutamate, the precursor to GABA, is deficient as a result of excessive GABA buildup.
- The second theory is that GABA itself is involved in plants; therefore, an excess of it is harmful to their ability to grow. Besides genetic modification, there are alternative methods to enhance glutamate production.

A significant amount of pigments that are by-products of tomato fruit are made possible by the carotenoids generated throughout the fruit's growth and maturation. Tomato fruit contains a pigment called lycopene, which gives the fruit its color and may have health benefits (Story et al. 2010). Thus, it is interesting to breed tomatoes to increase the amount of lycopene in their fruit. A total of five genes were targeted to increase lycopene production. The main approach for employing CRISPR/Cas9 complexes is to target the genes that encode the enzymes, which are involved in the carotenoid pathway. Tomato fruit with yellow flesh was the consequence of targeted knockdown of the phytoene synthase 1, which eliminated lycopene synthesis (D'Ambrosio et al. 2018).

10.12 Conclusion

Tomato is a good candidate for genetically engineered (GE) traits, and a number of methods can currently develop better tomato varieties. However, there are significant legal challenges associated with the use of CRISPR/Cas9 technology in breeding in a number of nations. Although the United States maintains that organisms with CRISPR editing are not genetically modified organisms (GMOs), the European Union (EU) has recently made a preliminary ruling that concludes CRISPR mutagenesis qualifies as a GMO because it modifies genetic material. In actuality, any tomato variety created using GE technologies including CRISPR will have to go through a drawn-out GMO approval procedure and deal with social acceptance challenges in Europe. Thanks to advancements in Cas9, breeding can be accelerated at a lower cost by precisely editing certain genes or even multiplexing numerous genes at once. To effectively deliver Cas9, environmentally friendly techniques and a diverse CRISPR/Cas toolkit are needed to mediate the catalytic activities of CRISPR/Cas. Through base editing, prime editing, and HKI, new CRISPR/Cas tools have made editing more accurate and efficient. Targeted mutagenesis–based allele introgression, base editing, and HKI in tomatoes have all gotten simpler and require less time.

References

Bai Y, Lindhout P (2007) Domestication and breeding of tomatoes: what have we gained and what can we gain in the future? Ann Bot 100:1085–1094. https://doi.org/10.1093/aob/mcm150

Brooks C, Nekrasov V, Lippman ZB, Van Eck J (2014) Efficient gene editing in tomato in the first generation using the clustered regularly interspaced short palindromic repeats/CRISPR-associated9 system. Plant Physiol 166:1292–1297. https://doi.org/10.1104/pp.114.247577

Causse M, Giovannoni J, Bouzayen M, Zouine M (2016) The tomato genome. Springer Berlin, Heidelberg, p 259. https://doi.org/10.1007/978-3-662-53389-5

Chandrasekaran M, Boopathi T, Paramasivan M (2021) A status-quo review on CRISPR-Cas9 gene editing applications in tomato. Int J Biol Macromol 190:120–129. https://doi.org/10.1016/j.ijbiomac.2021.08.169

Collard BC, Mackill DJ (2008) Marker-assisted selection: an approach for precision plant breeding in the twenty-first century. Philos Trans R Soc Lond B Biol Sci 363:557–572. https://doi.org/10.1098/rstb.2007.2170

D'Ambrosio C, Stigliani AL, Giorio G (2018) CRISPR/Cas9 editing of carotenoid genes in tomato. Transgenic Res 27:367–378. https://doi.org/10.1007/s11248-018-0079-9

Dahan-Meir T, Filler-Hayut S, Melamed-Bessudo C, Bocobza S, Czosnek H, Aharoni A et al (2018) Efficient in planta gene targeting in tomato using Gemini viral replicons and the CRISPR/Cas9 system. Plant J 95(1):5–16. https://doi.org/10.1111/tpj.13932

Danilo B, Perrot L, Mara K, Botton E, Nogue F, Mazier M (2019) Efficient and transgene-free gene targeting using Agrobacterium-mediated delivery of the CRISPR/Cas9 system in tomato. Plant Cell Rep 38:459–462. https://doi.org/10.1007/s00299-019-02373-6

Herner RC, Sink KC (1973) Ethylene production and respiratory behavior of the rin tomato mutant. Plant Physiol 52:38–42. https://doi.org/10.1104/pp.52.1.38

Ito Y, Nishizawa-Yokoi A, Endo M, Mikami M, Toki S (2015) CRISPR/Cas9-mediated mutagenesis of the RIN locus that regulates tomato fruit ripening. Biochem Biophys Res Commun 467:76–82. https://doi.org/10.1016/j.bbrc.2015.09.117

Klap C, Yeshayahou E, Bolger AM, Arazi T, Gupta SK, Shabtai S et al (2017) Tomato facultative parthenocarpy results from SlAGAMOUS-LIKE 6 loss of function. Plant Biotechnol J 15:634–647. https://doi.org/10.1111/pbi.12662

Li M, Li X, Zhou Z, Wu P, Fang M, Pan X, Lin Q, Luo W, Wu G, Li H (2016) Reassessment of the four yield-related genes Gn1a, DEP1, GS3, and IPA1 in rice using CRISPR/Cas9 system. Front Plant Sci 7:1–13

Lin T, Zhu G, Zhang J, Xu X, Yu Q, Zheng Z et al (2014) Genomic analyses provide insights into the history of tomato breeding. Nat Genet 46:1220–1226. https://doi.org/10.1038/ng.3117

Ranc N, Munos S, Santoni S, Causse M (2008) A clarified position for Solanum lycopersicum var. cerasiforme in the evolutionary history of tomatoes (solanaceae). BMC Plant Biol 8:130. https://doi.org/10.1186/1471-2229-8-130

Rasheed A, Gill RA, Hassan MU, Mahmood A, Qari S, Zaman QU, Ilyas M, Aamer M, Batool M, Li H, Wu Z (2021) A critical review: recent advancements in the use of CRISPR/Cas9 technology to enhance crops and alleviate global food crises. Curr Issues Mol Biol 43(3):1950–1976

Rodriguez-Leal D, Lemmon Z, Barlett ME, Lippman ZB (2017) Engineering quantitative trait variation for crop improvement by genome editing. Cell 171(2):470–480

Story EN, Kopec RE, Schwartz SJ, Harris GK (2010) An update on the health effects of tomato lycopene. Annu Rev Food Sci Technol 1:189–210. https://doi.org/10.1146/annurev.food.102308.124120

Tomlinson L, Yang Y, Emenecker R, Smoker M, Taylor J, Perkins S et al (2019) Using CRISPR/Cas9 genome editing in tomato to create a gibberellinresponsive dominant dwarf DELLA allele. Plant Biotechnol J 17:132–140. https://doi.org/10.1111/pbi.12952

Ueta R, Abe C, Watanabe T, Sugano SS, Ishihara R, Ezura H et al (2017) Rapid breeding of parthenocarpic tomato plants using CRISPR/Cas9. Sci Rep 7:507. https://doi.org/10.1038/s41598-017-00501-4

Veillet F, Perrot L, Chauvin L, Kermarrec MP, Guyon-Debast A, Chauvin JE et al (2019) Transgene-free genome editing in tomato and potato plants using Agrobacterium-mediated delivery of a CRISPR/Cas9 cytidine base editor. Int J Mol Sci 20(2):402. https://doi.org/10.3390/ijms20020402

Vu TV, Das S, Tran MT, Hong JC, Kim JY (2020a) Precision genome engineering for the breeding of tomatoes: recent progress and future perspectives. Front Genome Ed 2:612137. https://doi.org/10.3389/fgeed.2020.612137

Vu TV, Sivankalyani V, Kim EJ, Doan D, Tran MT, Kim J (2020b) Highly efficient homology-directed repair using CRISPR/Cpf1-geminiviral replicon in tomato. Plant Biotechnol J 18(10):2133–2143. https://doi.org/10.1111/pbi.13373

Waltz E (2016) Gene-edited CRISPR mushroom escapes US regulation. Nature 532:293. https://doi.org/10.1038/nature.2016.19754

Waltz E (2022) GABA-enriched tomato is first CRISPR-edited food to enter market. Nat Biotechnol 40(1):9–11. https://doi.org/10.1038/d41587-021-00026-2

Wang F, Wang C, Liu P, Lei C, Hao W, Gao Y, Zhao K (2016) Enhanced rice blast resistance by CRISPR/Cas9- targeted mutagenesis of the ERF transcription factor gene OsERF922. PLoS One 11(4):e0154027

Wang T, Zhang C, Zhang H, Zhu H (2021) CRISPR/Cas9-mediated gene editing revolutionizes the improvement of horticulture food crops. J Agric Food Chem 69:13260–13269

Xia X, Cheng X, Li R, Yao J, Li Z, Cheng Y (2021) Advances in application of genome editing in tomato and recent development of genome editing technology. Theor Appl Genet 134(9):2727–2747. https://doi.org/10.1007/s00122-021-03874-3

Xiong X, Liu W, Jiang J, Xu L, Huang L, Cao J (2019) Effiecient genome editing of Brassica campestris based on the CRIPSR/Cas9 system. Mol Gen Genomics 294:1251

Ye J, Wang X, Hu T, Zhang F, Wang B, Li C et al (2017) An InDel in the promoter of Al-ACTIVATED MALATE TRANSPORTER9 selected during tomato domestication determines fruit malate contents and aluminum tolerance. Plant Cell 29:2249–2268. https://doi.org/10.1105/tpc.17.00211

Yu Q-H, Wang B, Li N, Tang Y, Yang S, Yang T et al (2017) CRISPR/Cas9 induced targeted mutagenesis and gene replacement to generate long-shelf life tomato lines. Sci Rep 7:1–9, 11874. https://doi.org/10.1038/s41598-017-12262-1

Zaidi SS-E-A, Tashkandi M, Mansoor S, Mahfouz MM (2016) Engineering plant immunity: Using CRISPR/Cas9 to generate virus resistance. Front Plant Sci 7:1673

Enhancing Abiotic Stress Tolerance in Tomato: CRISPR-Based Interventions

11

Ayesha Faiz, Aqsa Parvaiz, Ghulam Mustafa, Iqrar Ahmad Rana, Muhammad Amjad Ali, Muhammad Sarwar Khan, and Faiz Ahmad Joyia

Abstract

The discovery of CRISPR-Cas9 has revolutionized the area of genome editing as a valuable technology by virtue of its extraordinary accuracy and effectiveness in altering genetic material. The use of CRISPR-Cas9 and other cutting-edge genome editing methods to impart desired traits in tomato has been covered in detail in this chapter. Here we review the basic workings of CRISPR-Cas9, its benefits over conventional breeding techniques, and how it can accelerate the development of tomato varieties with higher yields, greater resistance to abiotic stresses, and more nutritional content. Beyond CRISPR-Cas9, we also go over the most recent progress in genome editing technologies like prime editing and base editing that provide even extra accuracy and adaptability. We provide successful cases of tomato trait augmentation, addressing both the technical difficulties and regulatory issues, using comprehensive case studies and experimental data. The present chapter offers a thorough exposition of the present situation and potential developments around genome editing in tomatoes, emphasizing the revolutionary influence it has on agricultural biotechnology and food security.

Keywords

CRISPR · Editing · Abiotic stress · Genome · Nutrients

A. Faiz · G. Mustafa · I. A. Rana · M. S. Khan · F. A. Joyia (✉)
Centre of Agricultural Biochemistry and Biotechnology (CABB), University of Agriculture, Faisalabad, Pakistan
e-mail: faizahmad1980@uaf.edu.pk

A. Parvaiz
Department of Biochemistry and Biotechnology, The Women University Multan, Multan, Pakistan

M. A. Ali
Department of Plant Pathology, University of Agriculture, Faisalabad, Pakistan

M. Waseem et al. (eds.), *Omics Approaches for Tomato Yield and Quality Trait Improvement*, https://doi.org/10.1007/978-981-96-3890-1_11

11.1 Introduction

CRISPR-Cas is a highly sophisticated combination of molecular tools found naturally in bacteria and is applied for precise genetic alterations in genomic DNA of, theoretically, all types of organisms. Clustered Regularly Interspaced Short Palindromic Repeats is abbreviated as CRISPR, while Cas stands for "CRISPR-associated proteins." Tomato yields can be maximized by its genome editing using CRISPR-Cas technology. In 2022, 186.1 million metric tons of tomatoes (*Solanum lycopersicum* L.) were produced globally, making it the most extensively farmed vegetable crop. Many nutrients and secondary metabolites that are critical to human health are abundant in it, including minerals, organic acids, phenolics, B-carotene, lycopene, flavonoids, and vitamins C and E (Fig. 11.1) (Wang et al. 2023). However, tomato production is currently threatened by increasing drought conditions, rising global temperatures, and salinity along with many biotic stresses (Kumar et al. 2022; Waseem et al. 2019). The ability of tomato plant to withstand the above-mentioned detrimental abiotic stresses has been enhanced through the application of gene-editing techniques. Moreover, it has also been applied successfully to improve lycopene and carotene levels as well as other biochemical improvements in tomato. In light of this, we provide an outline of the gene-editing technologies that are now accessible, the methods employed to produce modified plants, the outcomes obtained in tomatoes via the use of these instruments, and an overview of the advantages and disadvantages of the existing techniques in this chapter.

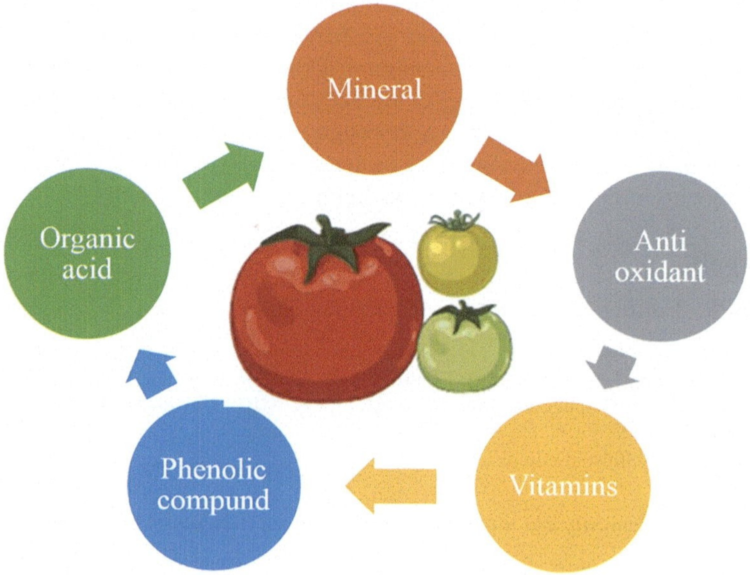

Fig. 11.1 Represents that tomatoes are an abundant supply of vital nutrients and secondary metabolites, providing a well-balanced combination of chemicals that are good for human health. Their varied nutritional composition promotes health

11.2 CRISPR/Cas9: Overview

When it comes to genetically engineered crops, CRISPR-Cas9 is the most potent technique available. Numerous environmental conditions cause epigenetic alterations to emancipate in horticultural crops, including tomato. Plant phenotypes regulated by genomic architecture are the basis for these alterations, which lead to hereditary epigenetic memory. In plants, genome editing techniques such as CRISPR-associated protein 9 (Cas9) technologies have transformed biology and predicted persistent transmission of epigenetic alterations (Qi et al. 2023). With its significant development in precise genome editing and low off-target mutation rate, the CRISPR-Cas9 approach promises improved efficiency, cost-effectiveness, adaptability, and ease of experimental design. Dicot crops, particularly tomatoes, continue to be a prime option for gene modifications based on CRISPR-Cas9, which will increase yields and productivity (Liu et al. 2022).

New Genomic Techniques are anticipated to increase in number in the upcoming years as CRISPR-Cas-based methodologies continue to advance. It has been possible to design new CRISPR-associated enzymes, such as base editors with more specific edit capabilities. To make genome editing a common technique for plant breeding, more efficient and reproducible tissue culture and transformation methods need to be established (Su et al. 2023). There are around 370,000 higher species of plants in the natural environment. However, transformation transgenesis and genome editing can only be effectively carried out by scientists in a limited number of organisms, often in the range of a few hundred.

Numerous tomato cultivars and Solanum species have been subjected to CRISPR-Cas9 gene editing during the last 10 years as well as wild species *S. lycopersicum* and *cerasiforme* (Su et al. 2023) and *Solanum peruvianum* (Lin et al. 2022). Gene editing of tomatoes was the foremost application of the CRISPR-Cas9 method. Solyc01g010970 is the tomato homolog of the ARGONAUTE7 gene. In that effort, 14 out of the 29 control plants exhibited the wiry leaf phenotype when *Agrobacterium tumefaciens*-mediated transformation of the M82 cotyledons was carried out. Similarly, Ron et al. (2014) employed the *A. rhizogenes*-mediated transformation of tomato and arabidopsis explants to deliver the CRISPR-Cas9 vector, which was intended to target the Solyc02g092370 gene coding for a short-root protein in the tomato in addition to a reporter gene (mGFP5). Later, analysis of transgenic hairy roots with a reduced root meristem and no GFP expression supports the theory that SHR (Short Root Gene) has a conserved role in both tomatoes and *Arabidopsis thaliana.* The coding region of the SHR gene exhibited minor insertions and deletions in those roots (Fig. 11.2, Tables 11.1 and 11.2).

11.3 CRISPR-Cas: Mechanism of Action

CRISPR-Cas system is an innovative tool in molecular biology and genetic engineering that has greatly enhanced the capacity for precise genome editing (Ali et al. 2023). This system serves as a natural defense against viruses that are present in

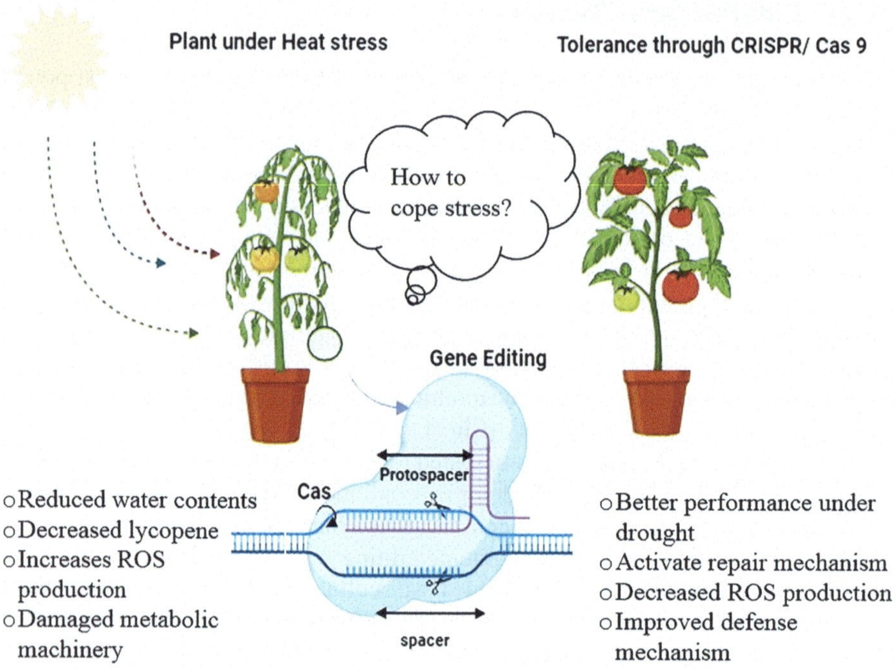

Fig. 11.2 The tolerance strategies that allow the plant to endure environmental stressors and adjust accordingly, preserving their viability and usefulness. These techniques are essential for stress management and adaptability to shifting circumstances

bacteria and archaea. This system has been modified by scientists to allow for precise gene editing in a variety of organisms, including humans. This system is composed of two components: (1) Cas9 is the most often utilized Cas protein in genome editing. It acts like a restriction enzyme to cut at a pre-decided position of DNA sequence precisely and (2) an artificial RNA molecule called guide RNA or gRNA which is designed with the purpose of directing the Cas9 protein to that a pre-decided position of a particular region of the genome. gRNA is divided into two sections: (a) a spacer region that is complementary to target DNA sequence and (b) a scaffold region that binds to and forms a complex with Cas9 protein. The gRNA-Cas complex scans the target genome and binds to the complementary sequence. Then, Cas protein causes a double-strand break at the precise spot in the DNA once the gRNA attaches to the target DNA. This step of break introduction initiates the innate DNA repair mechanisms of the cell. There are two repair mechanisms, namely, non-homologous end joining (NHEJ) and homology-directed repair (HDR). In NHEJ cut ends are directly joined together and sealed. However, it is prone to errors resulting in various mutations like insertions or deletions. On the other hand, HDR is guided by provided template to provide error-free repair (Riesenberg et al. 2022).

Table 11.1 The effect of climate change on tomato plant

Climate change effects on tomato plant		
Leaf roll	Heat exposure causes excessive leaf roll that is a result of dry conditions, from which the plant could not recover (Mirzayeva et al. 2023)	
Poor fruit set	Nights over 90 °F will stress the plant and result in fruit and blossom abortions (Pathak et al. 2023)	
Sunscald	Tomato fruits that have experienced temperature stress can develop sunscald (Sharaf-Eldin et al. 2023)	
Zippering	When a tomato undergoes cold stress, its anthers adhere to the wall of the developing fruit, resulting in zippering (Mondal et al. 2023)	
Blossom end rot	Blossom end rot is known to be induced by low humidity and high temperatures (Sethi et al. 2024)	

Table 11.2 CRISPR-based genome editing technique used in tomato cultivar in response to abiotic stress tolerance

Target protein	Target gene	Effect	Delivery techniques	References
Cas 9	SlMAPK3	Drought tolerance	Agrobacterium-mediated transformation	Wang et al. (2017)
Cas 9	*SlAGL6*	Better fruit setting when exposed to heat stress	Agrobacterium-mediated transformation	Klap et al. (2017)
Cas 9	SlCBF1	Cold tolerant	Agrobacterium-mediated tumefaciens transformation	Li et al. (2018)
Cas 9	SlNPR1	Drought tolerance	Agrobacterium-mediated transformation	Li et al. (2019)
Cas 9	SlCPK28	Heat tolerance	Agrobacterium-mediated transformation	Hu et al. (2021)
Cas 9	SlCPK28	Heat stress tolerance	Agrobacterium-mediated transformation	Hu et al. (2021)
Cas 9	SlARF4	Drought tolerance	Agrobacterium-mediated transformation	Chen et al. (2021)
Cas 9	SlHyPRP1, SlARF4	Salinity tolerant	Agrobacterium-mediated transformation	Guo et al. (2022)

11.4 Case Studies: CRISPR-Cas9 for Improving Abiotic Stress Tolerance in Tomato

11.4.1 Extreme Temperature Stress

When temperatures rise over a certain point, heat stress in plants develops and negatively impacts plant growth, development, and output. That affects the molecular, physiological, and biochemical processes of the plant. The different heat shock proteins function as tiny shock proteins that shield membranes from heat-induced damage as well as molecular chaperones HSP70, HSP90, and HSP100. HSFA1, HSFA2, and HSFA3 are additional variables that control HSPs (Pandey et al. 2024).

CRISPR-Cas9-created Slcbf1 mutants had increased amounts of malondialdehyde and increased electrolyte leakage in comparison to wild-type plants, suggesting that a SlCBF1 deletion might worsen the damage that cold stress causes to membranes. Mutants of Tomato Slcpk28 CRISPR-Cas9-produced Slcpk28 mutants showed reduced ascorbate peroxidase activity and elevated heat-stress-induced buildup of reactive oxygen species and protein oxidation level (Rahman et al. 2022). Better heat tolerance was demonstrated by SlMAPK3 deletion, which was mediated by CRISPR-Cas9 and involved lowering ROS buildup and upregulating the expression of genes encoding heat shock proteins and heat stress transcription factors. Mutations in SlAGL6 in the tomato have been shown using CRISPR-Cas9 gene knockout study to enhance tomato production under heat stress. SlIAA9 mutant plants using CRISPR-Cas9 showed fruit growth prior to fertilization, resulting in parthenocarpy. The ospin5b mutant, which was produced by CRISPR-Cas9, showed an increase in grain size, spike length, and cold tolerance (Maharajan et al. 2022). Tomato plant's ability to withstand high temperatures was improved by the slmapk3 mutant, which was created by CRISPR-Cas9 and preserved reactive oxygen species equilibrium. When two different editing techniques were used to currently knock down miR482b and miR482c, double mutants exhibited increased resistance to both fungal infections and cold stress (Li et al. 2023). Removal of the negative-response domain of SlHyPRP1 at both the germination and vegetative phases, the precise deletion of SlHyPRP1 negative-response domain greatly increased salt tolerance and resistance to cold stress (Rahman et al. 2022). The SlLBD40 gene mutation, which was induced by CRISPR-Cas9, also markedly improved tomato resilience to cold stress and drought. Tomato plants with GID1 modification, using CRISPR-Cas9 to edit GID1, having high leaf water content, were produced during a drought, hence enhancing tomato resilience to cold stress.

11.4.2 Extreme Water Stress

When soil becomes soggy, it reduces the amount of oxygen available to plant roots, a phenomenon known as waterlogging. High water table levels, inadequate drainage, or significant rains can all cause this. Abiotic stresses such as dryness, too much salt, and extreme temperatures affect almost every stage of the tomato life cycle. The SlFDH gene mutation produced by CRISPR-Cas9 showed improved resilience to waterlogging stress. The CRISPR-Cas9-mediated SlSAPK4 mutation demonstrated enhanced resilience to waterlogging stress. The SlP5CS1 mutation, which was produced by CRISPR-Cas9 (Wang et al. 2023), exhibited enhanced proline accumulation and resilience to waterlogging stress. The CRISPR-Cas9-mediated SlDREB2B mutation demonstrated improved resilience to waterlogging stress. The SlAREB1 mutation, which was produced using CRISPR-Cas9, demonstrated enhanced resilience to waterlogging stress (Bhat et al. 2024). When water availability is far less than what plants need, it is referred to as a drought, which causes water stress. Prolonged dry spells, high temps, and elevated evapotranspiration can all contribute to it. Yield loss from abiotic stress can reach 70%, contingent on the duration and plant stage of the stress. Mitogen-activated tomato protein kinase 3 (slmapk3) mutation produced by CRISPR-Cas9 demonstrated high oxidative stress and membrane damage in addition to increased resistance to drought stress (Kumar et al. 2023). Tomatoes were able to withstand drought stress better after having their ethylene-responsive factors and ERFs-like transcription factors modified using CRISPR-Cas9. Mutants with impaired osmotic stress/ABA-activated protein kinase 2 and SAPK2 were more susceptible to oxidative stress and drought than the wild type, suggesting that SAPK2 is required for drought tolerance. Members of the SnRK2 family, SnRK2 family activation in tomatoes revealed a distinctive stress regulation network that enhanced drought tolerance (Korwin Krukowski et al. 2023). Tomatoes that had AREB1 overexpression were better able to endure drought, but tomatoes that had AREB1 inactivation were more vulnerable to it.

11.4.3 Extreme Nutrient Stress

Overabundance of nutrients can induce imbalances in micronutrients; excessive nitrogen encourages foliage over fruit and excess phosphorus, potassium, and magnesium can damage plants. Increases in calcium, sulfur, iron, zinc, copper, and manganese have an impact on the absorption of vital nutrients (Kirkby 2023). Growth is impeded by too much boron, while copper equilibrium is upset by too much molybdenum. By improving nitrogen absorption and utilization, CRISPR-Cas9 editing of the NRT1.1 gene reduced the amount of surplus nitrogen in the soil. By improving phosphorus absorption and utilization, CRISPR-Cas9 editing of the PHT1 gene reduced surplus phosphorus in the soil (Muneer et al. 2024). By improving potassium absorption and utilization, CRISPR-Cas9 editing of the HAK5 gene reduced surplus potassium in the soil. By improving calcium absorption and utilization, CRISPR-Cas9 editing of the CAS1 gene reduced surplus calcium in the soil (Chongtham and Yadav 2023).

For plant growth, fruit development, and disease resistance, respectively, nitrogen, phosphorus, and potassium are essential deficits that result in yellowing, stunted growth, and weak stems. Deficits in any of these elements can cause structural problems and stunted growth. Calcium, magnesium, sulfur, iron, zinc, copper, manganese, boron, and molybdenum support a variety of physiological processes. Plant development is guaranteed when nutrient levels are kept in balance. Fruit nutrient levels were changed via CRISPR-Cas9 editing of the PSY1, MYB12, and SGR1 genes, which control fruit color (Cui et al. 2024). This led to a decrease in lycopene and an increase in naringenin chalcone. Significantly decreased levels of lycopene and chlorophyll were seen in the psy1 and sgr1 mutants, respectively. The myb12 mutant accumulated less naringenin chalcone in the peel but showed no discernible change in lycopene concentration. According to the study, it is possible to modify tomato varieties with CRISPR-Cas9 editing to add desired features without compromising other crucial agronomic variables like fruit quality and yield. It is possible that SlGRAS40 has a role in tomato abiotic and oxidative stress responses (Urhan 2024). The expression levels of the SlGRAS26 gene (Zhang et al. 2024) in wild-type tomatoes are studied in two primary circumstances that are thought to pose a danger to crop yields: dehydration and salt.

11.5 Future Prospects and Challenges

With this multidisciplinary approach, plant breeding will progress towards achieving a second Green Revolution and meeting the growing food needs of a swiftly rising worldwide population under constantly shifting climatic conditions. Genome-editing technology opens a new chapter in enhanced plant genetics by eliminating the drawbacks of existing transgenic approaches. By concentrating on the targeted genes, plant breeders and researchers may find this information helpful in their comprehensive assessment of the use of different gene-editing techniques to improve crops. It is expected that CRISPR/Cas9 technology will likely heal the rift between society and GMOs in the near future.

References

Ali A, Zafar MM, Farooq Z, Ahmed SR, Ijaz A, Anwar Z, Abbas H, Tariq MS, Tariq H, Mustafa M, Bajwa MH, Shaukat F, Razzaq A, Maozhi R (2023) Breakthrough in CRISPR/Cas system: current and future directions and challenges. Biotechnol J 18(8):2200642. https://doi.org/10.1002/biot.202200642

Bhat BA, Mir RA, Mir WR, Hamdani SS, Mir MA (2024) Transcription factors-golden keys to modulate the plant metabolism to develop salinity tolerance. Plant Stress 11:100409

Chen S, Zhang N, Zhou G, Hussain S, Ahmed S, Tian H, Wang S (2021) Knockout of the entire family of AITR genes in Arabidopsis leads to enhanced drought and salinity tolerance without fitness costs. BMC Plant Biol 21(1):137. https://doi.org/10.1186/s12870-021-02907-9

Chongtham I, Yadav P (2023) CRISPR/Cas9 genome editing and plant stress management. In: Global climate change and plant stress management, pp 335–359. https://doi.org/10.1002/9781119858553.ch23

Cui L, Zheng F, Li C, Li G, Ye J, Zhang Y, Wang T, Hong Z, Ye Z, Zhang J (2024) Defective mutations in STAY-GREEN 1, PHYTOENE SYNTHASE 1, and MYB12 genes lead to formation of green ripe fruit in tomato. J Exp Bot 75(11):3322–3336

Guo M, Wang X-S, Guo H-D, Bai S-Y, Khan A, Wang X-M, Gao Y-M, Li J-S (2022) Tomato salt tolerance mechanisms and their potential applications for fighting salinity: a review. Front Plant Sci 13:949541

Hu Z, Li J, Ding S, Cheng F, Li X, Jiang Y, Yu J, Foyer CH, Shi K (2021) The protein kinase CPK28 phosphorylates ascorbate peroxidase and enhances thermotolerance in tomato. Plant Physiol 186(2):1302–1317. https://doi.org/10.1093/plphys/kiab120

Kirkby EA (2023) Chapter 1 – introduction, definition, and classification of nutrients☆☆this chapter is a revision of the third edition chapter by E. Kirkby, pp. 3–5. https://doi.org/10.1016/B978-0-12-384905-2.00001-7. © Elsevier ltd. In: Rengel Z, Cakmak I, & White PJ (eds) Marschner's mineral nutrition of plants, 4th edn. Academic Press, pp 3–9. https://doi.org/10.1016/B978-0-12-819773-8.00016-2

Klap C, Yeshayahou E, Bolger AM, Arazi T, Gupta SK, Shabtai S, Usadel B, Salts Y, Barg R (2017) Tomato facultative parthenocarpy results from SlAGAMOUS-LIKE 6 loss of function. Plant Biotechnol J 15(5):634–647. https://doi.org/10.1111/pbi.12662

Korwin Krukowski P, Colanero S, Sutti A, Martignago D, Conti L (2023) How changes in ABA accumulation and signaling influence tomato drought responses and reproductive development. Int J Plant Biol 14(1):162–176

Kumar L, Chhogyel N, Gopalakrishnan T, Hasan MK, Jayasinghe SL, Kariyawasam CS, Kogo BK, Ratnayake S (2022) Chapter 4 – climate change and future of agri-food production. In: Bhat R (ed) Future foods. Academic Press, pp 49–79. https://doi.org/10.1016/B978-0-323-91001-9.00009-8

Kumar M, Prusty MR, Pandey MK, Singh PK, Bohra A, Guo B, Varshney RK (2023) Application of CRISPR/Cas9-mediated gene editing for abiotic stress management in crop plants [review]. Front Plant Sci 14. https://doi.org/10.3389/fpls.2023.1157678

Li X, Wang Y, Chen S, Tian H, Fu D, Zhu B, Luo Y, Zhu H (2018) Lycopene is enriched in tomato fruit by CRISPR/Cas9-mediated multiplex genome editing. Front Plant Sci 9:559

Li R, Liu C, Zhao R, Wang L, Chen L, Yu W, Zhang S, Sheng J, Shen L (2019) CRISPR/Cas9-mediated SlNPR1 mutagenesis reduces tomato plant drought tolerance. BMC Plant Biol 19(1):38. https://doi.org/10.1186/s12870-018-1627-4

Li Q, Shen H, Yuan S, Dai X, Yang C (2023) miRNAs and lncRNAs in tomato: roles in biotic and abiotic stress responses. Front Plant Sci 13:1094459

Lin C-S, Hsu C-T, Yuan Y-H, Zheng P-X, Wu F-H, Cheng Q-W, Wu Y-L, Wu T-L, Lin S, Yue J-J, Cheng Y-H, Lin S-I, Shih M-C, Sheen J, Lin Y-C (2022) DNA-free CRISPR-Cas9 gene editing of wild tetraploid tomato *Solanum peruvianum* using protoplast regeneration. Plant Physiol 188(4):1917–1930. https://doi.org/10.1093/plphys/kiac022

Liu H, Chen W, Li Y, Sun L, Chai Y, Chen H, Nie H, Huang C (2022) CRISPR/Cas9 technology and its utility for crop improvement. Int J Mol Sci 23(18):10442

Maharajan T, Krishna TA, Rakkammal K, Ceasar SA, Ramesh M (2022) Application of CRISPR/Cas system in cereal improvement for biotic and abiotic stress tolerance. Planta 256(6):106

Mirzayeva S, Huseynova I, Özmen CY, Ergül A (2023) Physiology and gene expression analysis of tomato (*Solanum lycopersicum* L.) exposed to combined-virus and drought stresses. Plant Pathol J 39(5):466

Mondal R, Das A, Bandyopadhyay A (2023) Implications of small RNAs in plant development, abiotic stress response and crop improvement in changing climate. Nucleus 66(3):321–339

Muneer MA, Afridi MS, Saddique MAB, Chen X, Zaib Un N, Yan X, Farooq I, Munir MZ, Yang W, Ji B, Zheng C, Wu L (2024) Nutrient stress signals: elucidating morphological, physiological, and molecular responses of fruit trees to macronutrients deficiency and their management strategies. Sci Hortic 329:112985. https://doi.org/10.1016/j.scienta.2024.112985

Pandey GK, Penna S, Fotopoulos V, Sanan-Mishra N, Khan A, Khan V, Pandey K, Sopory SK (2024) Frontiers Frontiers in Plant Science 2022 published: 13 May REVIEW. Abiotic stress signaling in plants: functional genomic intervention, Volume II, 94

Pathak S, Johar V, Angaria A, Vishal A (2023) Alteration in flowering and fruiting of commercial vegetable crops under protected condition: a review. International Journal of Environment and Climate Change 13(12):139–151

Qi Q, Hu B, Jiang W, Wang Y, Yan J, Ma F, Guan Q, Xu J (2023) Advances in plant epigenome editing research and its application in plants. Int J Mol Sci 24(4):3442

Rahman M-U, Zulfiqar S, Raza MA, Ahmad N, Zhang B (2022) Engineering abiotic stress tolerance in crop plants through CRISPR genome editing. Cells 11(22):3590

Riesenberg S, Helmbrecht N, Kanis P, Maricic T, Pääbo S (2022) Improved gRNA secondary structures allow editing of target sites resistant to CRISPR-Cas9 cleavage. Nat Commun 13(1):489

Ron M, Kajala K, Pauluzzi G, Wang D, Reynoso MA, Zumstein K, Garcha J, Winte S, Masson H, Inagaki S, Federici F, Sinha N, Deal RB, Bailey-Serres J, Brady SM (2014) Hairy root transformation using agrobacterium rhizogenes as a tool for exploring cell type-specific gene expression and function using tomato as a model. Plant Physiol 166(2):455–469. https://doi.org/10.1104/pp.114.239392

Sethi K, Dhaka SS, Bari VK (2024) Insights into blossom end-rot disorder in tomato (*Solanum lycopersicum*). Plant Mol Biol Report 42:1–23

Sharaf-Eldin MA, Yaseen ZM, Elmetwalli AH, Elsayed S, Scholz M, Al-Khafaji Z, Omar GF (2023) Modifying walk-in tunnels through solar energy, fogging, and evaporative cooling to mitigate heat stress on tomato. Horticulturae 9(1):77

Su W, Xu M, Radani Y, Yang L (2023) Technological development and application of plant genetic transformation. Int J Mol Sci 24(13):10646

Urhan EK (2024) Abiotic stress tolerance in plants by genome editing applications. In: Applications of genome engineering in plants. Wiley, Hoboken, pp 185–221

Wang L, Chen L, Li R, Zhao R, Yang M, Sheng J, Shen L (2017) Reduced drought tolerance by CRISPR/Cas9-mediated SlMAPK3 mutagenesis in tomato plants. J Agric Food Chem 65(39):8674–8682. https://doi.org/10.1021/acs.jafc.7b02745

Wang Y, Cao X, Zhang D, Li Y, Wang Q, Ma F, Xu X, Zhan X, Hu T (2023) SlGATA17, a tomato GATA protein, interacts with SlHY5 to modulate salinity tolerance and germination. Environ Exp Bot 206:105191

Waseem M, Rong X, Li Z (2019) Dissecting the role of a basic helix-loop-helix transcription factor, *SlbHLH22*, under salt and drought stresses in transgenic *Solanum lycopersicum* L. Front Plant Sci 10:734. https://doi.org/10.3389/fpls.2019.00734. PMID: 31231412; PMCID: PMC6558761

Zhang H, Jiang Z, Jiao P, Zhao Y, Gao B, Liu S, Guan S, Ma Y (2024) ZmGRAS46 negatively regulates flowering time in Arabidopsis thaliana. Agronomy 14(1):155

Sustainability by Design: "Omics" Applications for Environmentally Friendly Tomato Production

12

Sobia Parveen, Mazhar Ali, Muhammad Mubeen, Hafiz Muhammad Rashad Javeed, Muhammad Tahir, Shehzad Mehmood, Urooj Khan, Maria Qubtia, Abu Bakr Umer Farooq, Mazhar Saleem, and Muhammad Sabtain Abid

Abstract

Tomato is a major crop worldwide and are in tremendous demand for processed and fresh foods. Current tomato production practices are not sustainable and eco-friendly; therefore, there is a need to adopt new agricultural practices for tomato production, which improves soil fertility, uses resources sustainably, maintains biodiversity, and is eco-friendly as well, while ensuring food security and a healthy nutrition for all. This chapter elaborates various omics-based approaches that have revolutionized crop improvement and sustainable agricultural practices. The advances in omics approaches for crop improvement have made a considerable contribution to the sustainability of agriculture, and developing climate-smart cultivars is regarded as the most sustainable strategy for enhancing food production. Furthermore, to discover changes, several techniques and tools such as metabolomics, genomics, ionomics, transcriptomics, and proteomics are developed for sustainable and eco-friendly tomato production.

S. Parveen · M. Ali (✉) · M. Mubeen · M. Tahir · S. Mehmood · U. Khan · M. Qubtia · A. B. U. Farooq · M. Saleem
Department of Biotechnology, COMSATS University Islamabad, Vehari Campus, Vehari, Pakistan
e-mail: mazharali@cuivehari.edu.pk

H. M. R. Javeed
Department of Environmental Sciences, COMSATS University Islamabad, Vehari Campus, Vehari, Pakistan

M. S. Abid
Department of Biosciences, COMSATS University Islamabad, Sahiwal Campus, Sahiwal, Pakistan

M. Waseem et al. (eds.), *Omics Approaches for Tomato Yield and Quality Trait Improvement*, https://doi.org/10.1007/978-981-96-3890-1_12

Keywords
Plant growth · Bacteria · Plant growth promoting bacteria (PGPB) · Transcriptome · Proteome

12.1 Introduction

Tomatoes (*Solanum lycopersicum* L.) are widely consumed due to their nutritional, health-promoting, and cost-effective benefits. It is one of the most important vegetables in the world. The demand for tomatoes is high in any country, owing to their gastronomic versatility, and nutritional and aromatic value. The diverse range of its utility in raw, cooked, and processed food, the global demand for tomato has increased tremendously in recent years due to the increasing human population (Chaudhary et al. 2019). However, human population is projected to increase from 6.8 to 9.1 billion in 2050, necessitating a significant increase in agricultural output to meet the growing demand, despite the increased urbanization and dwindling cultivable area. However, global warming and climatic fluctuations caused by multiple abiotic and biotic stresses have hampered global crop growth and productivity, resulting in crop productivity losses in dollars (Kumar et al. 2020).

As a tropical crop, tomatoes are grown in different climatic zones, making tomato cultivation more complicated. However, to obtain higher production and better quality, an intense use of chemical-based fertilizers is essential (Chaudhary et al. 2019). Altering the environmental factors such as drought, heat, and salinity can badly affect tomato productivity. As a result, the majority of non-conventional tomato farming areas have switched to greenhouse-based agriculture to provide a year-round supply. Not only is greenhouse farming more expensive, but it also results in rapid nitrate, phosphate, and salinity accumulation in the soil, which eventually contributes to soil degradation and surface or groundwater pollution (Lin et al. 2014).

However, sensory, chemical, and biological testing are the most common conventional method used for observation of quality alteration in agricultural products throughout storage. The sensorial assessment of agricultural commodities entails an evaluation of the sensory qualities. This approach can successfully observe product quality flaws, permitting advanced problem recognition and remediation. However, a sensory assessment is restricted to judging the quality and appearance of cultivation products and relies on human education, making quantitative investigation difficult (Heymann and Lawless 2013).

Plants are unable to escape adverse environmental circumstances such as drought, salt, and cold stress. The ongoing shifts in the climate around the world which radically affect the genotype-environment connection lead to changes in the plant's basal metabolism and phenotype. Plant genetics have been purposely altered since the advent of agriculture, which occurred 8000–10,000 years ago. The domestication process of crops, known as the essential manipulation of genetics, evolved through the process of selective breeding to generate superior animals and plants, resulting in genomic alterations that were not acknowledged in the past. Plumcots, as well as other hybrids derived from plums and apricots, are still available today. Broccoflower is a widely recognized genetic modification and is reported to provide greater nutritional value than parental cauliflower or broccoli. Plant genic alterations based on radiation and chemical treatments yield well-known commodities (Van Emon 2016). Colchicine treatment of seeded watermelons produces a sterile triploid seed, resulting in the seedless. This can be useful for eating, but it eliminates the delights of eating watermelon. Mutational breeding involves exposing plants toward neutrons, γ-rays, protons, and α- and β-particles to stimulate mutations for desirable features. Chemicals like ethyl methanesulfonate and sodium azide are utilized to create mutagenic crops. Between 1930 and 2014, more than 3200 mutagenic plant varieties were officially released. Over 1000 transgenic varieties of significant staple crops have grown up around the world (Fedoroff and Brown 2004). Furthermore, massive "gamma gardens" arose in government locations in numerous nations to promote peaceful uses of nuclear energy during World War II. The Brookhaven National Laboratory used thermal neutron radiation to generate a popular red grapefruit cultivar. Plants created through mutagenic procedures or cross-breeding have random, numerous, and unspecific genetic modifications, typically resulting in numerous undesired plants before obtaining the desired features. These genetic alterations have a considerable risk of producing unanticipated and harmful compositional consequences. However, the excellent outcome of these heritable alterations has prompted researchers to look for many economical and regulated techniques to shift genetic features between plants by minimizing the danger of undesirable mutants through exactitude genetical changes (Fedoroff and Brown 2004).

12.2 Use of Plant Biostimulants for Plant Growth and Development

Plant biostimulants (PBs) could be an eco-friendly and unique tool for sustainable agriculture by lowering cropping systems and compromising the environment. PBs can be classified as non-microbial (fulvic acid, protein hydrolyzates, humic, and seaweed extracts) or microbial as arbuscular mycorrhizal fungi (AMF) and plant growth promoting bacteria (PGPB). Research suggests that PGPB and AMF can improve plant growth, nutritional efficiency, stress tolerance, quality attributes, and accessibility of nutrients in the rhizosphere and soil (Zandalinas et al. 2021). The use of PGPB-based biofertilizers for the modulation of plant microbiome is regarded as a beneficial technique of traditional breeding and hereditary manipulation for promoting plant development (Puccio et al. 2023).

PGPB, found in the *Bacillus, Pseudomonas, Kocuria, and Azospirillum* genera, improves crop yield and quality by promoting various molecular and morpho-physiological processes. These processes include secondary metabolites, volatile organic compounds (VOCs), and phytohormones production, as well as plant nutrient accessibility through Fe supply, phosphorus solubilization, and nitrogen fixation. *Kocuria rhizophila* (Gram-positive bacteria) forms circular cells in tetrads, pairs, or irregular clusters. It belongs to the *Micrococcaceae* family and the *Actinomycetales* order. It is a soil-dwelling bacteria that is often used for antimicrobial analysis and food preparation. However, research on *K. rhizophila* strains has identified genes active in the production of antifungal substances like bacilysin and cycloserine, as well as enhance the maize yield under salt conditions of stress, indicating a mechanism that preserves ion homeostasis through hormone production and absorbance of nutrients (Li et al. 2020).

Due to the growing interest in *K. rhizophila* strains as PGPB, an additional variety of the plant has recently been characterized for several traits of PGP, including the indole-3-acetic acid production, ability to withstand drought and salt stress, and phosphate solubilization. The molecular process behind the advantageous consequence of these bacteria has not been understood due to PGPB-plant interactions. A multi-omics method identifies important molecular actors triggered by *K. rhizophila* treatment in tomatoes, which play a role in plant development (Mohammadkhani et al. 2016). A multi-omics approach revealed important hub genes and metabolic pathways activated by PGPB treatment, highlighting the molecular system of *K. rhizophila's* effect on tomato. Further research is needed to understand how *K. rhizophila* treatment affects plant molecular mechanisms (Guesmi et al. 2022).

12.3 Role of Plant Growth Promoting Bacterium (*K. rhizophila*) in Tomato

Plant microbial stimulants have emerged as a potential and environmentally acceptable agricultural method for improving the yields of crops while minimizing chemical inputs to develop more sustainable agricultural plans. *K. rhizophila*, a soil-dwelling bacterium, has been classified as PGPB. However, *K. rhizophila's* PGP activity on tomatoes shows the molecular mechanisms that encourage plant development. The transcriptomic study revealed many up-regulated genes, primarily linked by amino acid and lipid secondary metabolism (Paradiso et al. 2018), organization of cell wall, and a change in DNA methylation followed by PGPB (de Andrade et al. 2023). In accordance, *K. rhizophila* accumulated a high concentration of proteins interested in cell division, photosynthesis, and the growth of plants. Furthermore, PGPB treatment had the greatest effect on the "monosaccharides," "tricarboxylic acid cycle (TCA)" and "peptides," which are referred to as metabolites. However, dopamine which is a catecholamine neurotransmitter that mediates plant evolution by SAMDC (S-adenosylmethionine decarboxylase), a gene that enhances the growth of any crop by up-regulated *K. rhizophila* in tomato (Lephatsi et al. 2021).

The PGPB-plant interaction powerfully modulates complex machinery that includes selected genes, particular plant metabolites, and proteins such as dopamine and amino acids, all of which contribute to the implementation of plant evolution. On this premise, these chosen constituents might be regarded as critical variables for increasing plant production in specialized PGPB-mediated agricultural strategies (Hayama et al. 2019). APX1 (cytosolic ascorbate peroxidase 1) is key in removing H_2O_2 and regulating ROS impact on *K. rhizophila* treatment increases the presence of important proteins in tomato leaves, including ubiquitin carboxyl-terminal hydrolase (UCH). UCH plays a crucial role in transcription and signal transduction in cells. The UCH family's de-ubiquitylation activity regulates the indole-3-acetic acid/auxin ratio in Arabidopsis. Furthermore, various phosphofructokinases over-representation in K plants is completely consistent with their roles in the metabolism of sugar, plant growth, and cell expansion.

Plant metabolites such as dopamine shown to play a specialized regulatory role in redox response to stress, photosynthetic stimulation, growth and development, and plant immunological response. The demonstration of numerous genes interconnected with abiotic stresses like salt and drought as well as those that impact plant development by increasing nitrogen and carbon metabolism, was influenced by endogenous dopamine levels (Raza et al. 2022). Notable among them is SAMDC that managed plants inoculated with *K. rhizophila*, demonstrating the cross-regulatory function of dopamine in facilitating multiple physical events and validating the supportive correspondence between metabolomics and transcriptomics data (Wi et al. 2006).

12.4 Role of Omics Approaches in Agriculture

Omics can aid in expanding agricultural research in health, chemical feedstock, food, and energy while preserving, improving, and remediating the environment. Omic technologies precisely target essential qualities of interest. Omics can improve the nutritional value of food for consumers, such as a lycopene-rich tomato, fruit with delayed ripening characteristics, and products with robust antioxidant capacities. Improving protein quantity and quality for better nutrition for humans and animals by increasing vitamin and mineral quantities to deal with nutrient inadequacies, and reducing allergens and anti-nutritional chemicals that degrade the quality of food, can all be investigated by using omics (Pazhamala et al. 2021). Another word for omics is "foodomics," which refers to the study of nutrition using interrelated omic methods. This information is helpful for better understanding how environmental interactions affect plant metabolism (Puccio et al. 2023).

12.5 Role of Plant-Microbe in Sustainable Agriculture

Plant-microbe interactions are critical to sustainable agriculture and ecosystem dynamics, and they are affected by different factors, including environmental circumstances, human activities, and host characteristics. However, the complex link between microbes and plant is critical to sustainable agriculture and ecosystem dynamics, and it is regulated by edaphic conditions, host-related aspects, human influences, and environmental dynamics. Understanding such components is critical for navigating the complexities of symbiotic partnerships. Omics methods, such as genomes, transcriptomics, genomics, and metabolomics, shed light on the molecular details of plant-microbe symbiosis (Sarim et al. 2020). Genomics reveals key genes while transcriptomics reveals gene expression. However, proteomics recognizes essential proteins, and metabolomics characterizes tiny substances, providing a holistic view. However, multi-omics approaches allow for the analysis of a wide range of interactions, particularly obligate symbiotic partnerships between arbuscular mycorrhizal plants and fungi, as well as associations with non-mycorrhizal microorganisms like biofilms (Mishra et al. 2022). These findings highlight the revolutionary potential of omics methods for enhancing sustainable agriculture by improving understanding of plant-microbe. Despite issues with ethical constraints and data integration, omics techniques hold promise for advances in resilient agriculture practices and precise intervention. Omics technologies have enormous potential for optimizing agricultural techniques and strengthening persistent plant-microbe relationships, providing the way for agricultural sustainability and environmental protection.

Transcriptomic, metabolomics, proteomic, and genomic technologies have revolutionized agriculture, food science, life science, and medicine. Plant-microbe interaction omics provides new insights into the complex interactions between microorganisms and plants, as well as its implications for crop yield, environmental resilience, and sustainability (Kumari et al. 2023). Furthermore, integration of multi-omics techniques improves interpretation, visualization, and data analysis, allowing for a better knowledge of biological processes. These advances have illuminated a diverse range of plant-associated species of bacteria and their involvement in disease control, nutrient cycle, and plant-microbe signaling system (Sindelar 2016). Furthermore, integrated plant-microbe omics contribute to the development of sustainable food production systems. However, eco-friendly agriculture and pollution reduction can be acquired using microbial alternatives instead of traditional agrochemicals. However, omics data shows microbial characteristics that improve plant development, nutrients, and stress resistance. These discoveries have led to the development of omics-driven techniques to take advantage of plant-microbe symbioses for biofertilization, plant yield, and pest control. Furthermore, these interventions promise to increase chemical reliance, agricultural sustainability, and soil fertility by improving microbial communities (Hussain et al. 2023).

12.6 Role of Biochip Technology in Detecting Modifications Related to Agriculture

One of the newest monitoring technologies is biochip technology, which allows for the quick detection of spoilage microorganisms through their proteins, cells, genes, and biological components. This technology is used to observe the transportation and storage of agricultural products. The advancement of analysis methods in sciences has opened up vast opportunities, leveraging its specificity and sensitivity in microbes' detection. Furthermore, a growing area of study involves infrared spectroscopy (IRS) techniques for the identification of alterations in agricultural quality. The development of omics technology offers fresh perspectives on how to track quality changes in agricultural products as they are being transported and stored.

Omics technology describes a set of high-throughput techniques for evaluating molecular components in biological materials. However, based on the various molecular components, these techniques are categorized into four categories: metabolomics, transcriptomics, proteomics, and genomics (Zhang et al. 2022). The creation of novel omics technologies for the study of particular substances and the integration of several omics have resulted from the research of omics technology.

Proteomics, for instance, can be used to both qualitatively and quantitatively proteins investigation of vegetables and fruits after harvest, and their roles in the development of fruit and vegetables, which conduct extensive research on the regulating mechanism of exogenous fruit development, stress resistance, and other related areas. Omics technology has become a popular field of study due to its real-time dynamic monitoring capabilities, high sensitivity, short processing cycles, high throughput, and exceptional accuracy (Lee et al. 2017).

Genomics, metabolomics, transcriptomics, epigenomics, proteomics, bioinformatics, ionomics, proteogenomics, interactomics, and phenomics are examples of multi-omics approaches that provide much information about physiological mechanisms in vegetation under stressful circumstances and arduous tasks of combating the harmful effects of those stresses (Beyer et al. 2002). However, using merely a mono-omics approach fails to yield enough information to comprehend the complexities of plant responses during stress situations. As a result, multi-omics techniques must be applied or integrated to produce such outcomes (Mohammadkhani et al. 2016). Innovative methodologies incorporating data from various omics sections, such as genomes, proteomics, transcriptomics, and metabolomics, show significant promise for enhancing crop growth strategies (Valdés et al. 2013).

12.7 Genome Selection Techniques and Their Role in Tomato Yield

Marker-based techniques like genetic fingerprinting, quantitative trait loci (QTL) mapping, and linkage maps require a large amount of genotype data. QTLs have been discovered by association mapping for finding marker-trait relationships. In contrast, a genome-wide association study (GWAS) technique has an effective response over the linkage mapping since it investigates the genetic diversity found in germplasm and provides the enhanced mapping resolution (Uffelmann et al. 2021).

As a result, GWAS has been widely used to identify SNPs (single nucleotide polymorphism) for agronomical attributes in a worldwide tomato resource. However, GWAS in tomato was conducted by markers (182SSR) to find the chromosomal domain which linked 28 distinct inconstant elements that specify the flavor of tomato. With the lowering cost of SNP tests, genotyping an extensive amount of experimental groups is now feasible, allowing GS proficiency to be implemented in crop breeding projects (Zhang et al. 2015).

The GS technique is effective at tracking all loci that contribute to trait development at the same time, regardless of their size. The issue of evolution by QTL mapping is addressed by the GS approach. The majority of economically significant traits are complex, and epistatic interactions produce unanticipated trait expression. As a result, GS is a highly effective strategy for analyzing the genetic outcomes for selection since it combines every possible genetic marker with the phenotypic variables from an experimental population (Deshmukh et al. 2014).

12.8 Use of Genomics for Agricultural Product Transportation

Genomics is the foundation of omics. Post-genomics encompasses transcriptomics and proteomics. The basic use of the genomics technique is to recognize the target gene by sequencing methods. Whole-genome sequencing is one of the most commonly used approaches. Genomic applications in agricultural commodity storage and transportation have reached an advanced level of development. The first is to employ gene sequencing technologies to measure the genotypes of bacteria and fungi to gain a thorough understanding of these microbes that influence agricultural product transportation and storage at the genetic level as well as changes in quality that occur during the storage processes. Genomic research can increase agricultural product variability and storage life by altering genes (Yang et al. 2021).

12.9 Genomics

Genomics is an excellent starting point for the subject of omics, especially since the term "genomics" was coined. A plant's genome contains information about its domestication and breeding history. GenBank is an annotated library of easily available DNA sequences and is constantly revised to deliver the most recent information. Genomics offers knowledge-based strategies to agricultural plant biotechnology, allowing for accurate and controllable breeding methods and the selection of markers, hence expediting the production of fresh varieties of crops (Shabbir et al. 2022). Gene expression analyses reveal functional genes that influence the phenotype, offering helpful details for plant enhancement (Fig. 12.1). Traditional breeding of plants involves introducing a specific gene into a plant's genome or shutting down a particular gene with RNA, which can produce the desired phenotype faster. Genomics is the study of genes and genomes, including function, mapping, evolution, structure, epigenomic, genome mapping, and mutagenomic (Muthamilarasan et al. 2019). Genomics can help to elucidate genetic variance by increasing the efficiency of crop breeding and genetic enhancement. Structural genomics involves the sequence of chromosomal as well as the creation of genetic and physical maps to determine the features of plant biology.

Functional genetics also sheds light on how genes regulate traits of interest. Epigenomics is the term used to describe epigenetic changes at the genomic level, such as modifications of histone and small RNA or DNA methylations. Functional genomics examines data collected from whole or partial genome sequences to identify gene functions, and it uses two methodologies (forward and reverse genetics) to

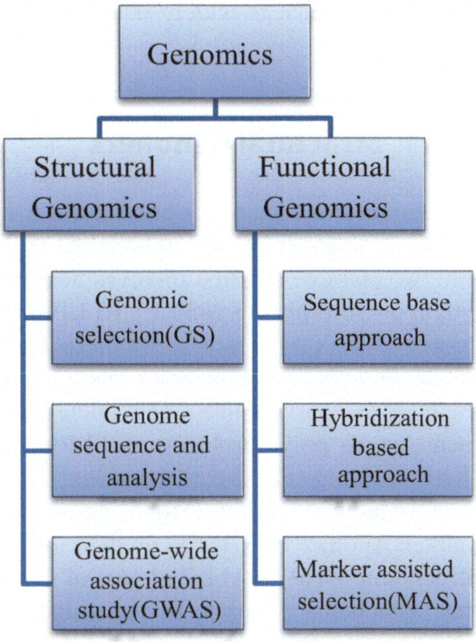

Fig. 12.1 Approaches used in genomics to study the gene of interest

determine individual genes. The forward technique studies a haphazardly selected mutant with a fascinating phenotype and determines its relevant genes. In contrast, the reverse genetic technique involves analyzing the phenotype of any organism by disrupting the best-known gene.

Functional genomics is a technology that aids in the investigation of gene connections and regulatory networks. Mutagenomics focuses on the mutations that induce genetic changes in mutant attributes. Pangenomics can be defined as the integration of a core genome possessed by all individuals and a redundant genome that is either partially identical or individually distinct. Pangenomics and metagenomics are contemporary omics techniques in agricultural sciences that focus on pangenome and mutagenesis (Muthamilarasan et al. 2019).

12.10 Sequencing of Whole Genome

Sequencing the complete genome is the most effective way to accelerate molecular studies. In crops, Arabidopsis was first genome mapped by a multinational collaborative. Genome sequencing contributed to a better knowledge of genome organization, and evolution.

Furthermore, development of next-generation sequencing (NGS) technologies that allow for the simultaneous sequencing of millions of molecules, whole-genome sequencing has become substantially less expensive and faster than traditional methods. However, numerous crop genomes like tomatoes, are sequenced by NGS and Sanger's technologies. The high-quality, well-annotated tomato reference genome sequence is widely used in genomics and transcriptome research (Aoki et al. 2013).

12.11 Structural Genomics

Structural genomics relies on molecular markers to map and tag genes of interest before these are employed in breeding crops. However, marker methods are divided into two classes. The first involves non-PCR techniques called restriction fragment length polymorphisms (RFLP). RFLP studies DNA polymorphism by integrating a tagged DNA probe into a Southern blot treated with restriction enzymes. This results in a changed pattern of DNA fragments. The second method is PCR-based methods for markers like single nucleotide polymorphisms (SNPs), amplified fragment length polymorphisms (AFLP) and random amplified polymorphic DNA (RAPD) (Yang et al. 2021). However, using a single primer with an arbitrary nucleotide sequence, random DNA segments are amplified by PCR to create an RAPD marker. Another PCR-based method that selectively amplifies restriction fragments of genomic DNA is AFLP (Agarwal et al. 2008). Appleby et al. (2009) suggested that sequencing genomic PCR results from different individuals can be done by SNP analysis. Diversity array technology (DArT) is a large-scale approach that uses microarray hybridization to genotype many polymorphic sites across the genome. The introduction of NGS enabled the detection and use of SNPs (Fig. 12.1).

12.12 Approaches Used in Multiple Traits of Crops

GWAS and QTL mapping approaches are used to analyze crops' numerous attributes (McCarthy et al. 2008). QTL mapping is a best scheme for integrating the types of data: complex genotypes and phenotype. However, molecular markers like AFLPs and SNPs are routinely exploited to map the QTLs and connect them with reported phenotypic data. GWAS use SNPs in sequencing data to detect association between genic variants and traits of any organism (Challa and Neelapu 2018).

DNA markers are used for various applications in genomics, molecular biology, and plant breeding, especially in tomatoes. However, in plant breeding, molecular markers are frequently used for QTL and gene mapping, population characterization, marker-assisted breeding, germplasm evaluation, diversification research, and genomic fingerprinting. Marker-assisted selection (MAS) for each trait during breeding demands exact knowledge about molecular markers and map position (Fig. 12.1). The generation of markers for MAS includes high-resolution QTL mapping, and confirmation of linked markers. MAS is often used to produce disease-resistant tomato varieties. MAS is expeditious by the genotyping approaches based on SNPs (single nucleotide polymorphism). High-throughput marker genotyping techniques are widely available, and having well-distributed markers allows for more efficient MAS. Whole-genome sequencing of 84 tomato varieties revealed millions of SNPs spread across the genome. Despite the high economic and commercial value of tomatoes, the development of climate-smart cultivars with high yields under severe climate circumstances requires integrating multidisciplinary knowledge (Foolad 2007).

12.13 Role of GWAS in Genomics

Using GWAS in the genomics method is critical for improving crop tolerance to abiotic stress. Furthermore, structural variations (SVs) are crucial in the genetic regulation of agronomically key traits in crops. Breeders can enhance hybrid propagation by combining genotype-by-sequencing (GBS) with MAS to modify the crop yield and quality (Figure1). Multiparent mapping, including multiparent advanced generation intercrosses (MAGIC) and nested association mapping (NAM) in model plants and crops, reveals significant variation in phenotype, it can be accomplished via testing. Examination of genotype-phenotype interactions can determine QTLs, which can be verified using useful genomics techniques (Kover et al. 2009).

12.14 Agricultural Sustainability and Genomics

One use of omics data is to find functional genes related to sustainable agriculture. These genes are then modified to create new varieties of crops for agricultural sustainability. The knowledge gathered from omics data, when combined with modern technologies such as targeted gene editing, could help in the development of novel types of crops. Many plants, particularly those at mid-to-high latitudes, face a high survival risk due to changes in the climate worldwide (Rivero et al. 2022). As a result, climate-resistant cultivars and crops with consistent and high output need to be produced. Until now, crop genetic development has relied on inefficient artificial selection of existing natural variants. However, growing crops in well-controlled urban areas for producing climate-resistant varieties of crops, can contribute to sustainable agriculture (Li and Yan 2020).

12.15 Transcriptomics

The transcriptome, or complete collection of RNA, is altered and transformed into messenger RNA, which carries instructions toward the ribosome, where it gets converted into protein. A specific cell line's transcriptome shows a dynamic nature, representing the genes that can be successfully transcribed at any one point under various circumstances, in contrast to the genome, which has largely fixed barring mutations (Fig. 12.2). A useful technique for researching biological processes is transcriptomics, can help us better comprehend the genome's functional elements. RNA sequencing (RNA-seq) has emerged as the most efficient, productive, and high-yield transcriptomic method due to the quick advancement of NGS. In contrast to microarray, the RNA-sequencing method helps discover new genes and spliced

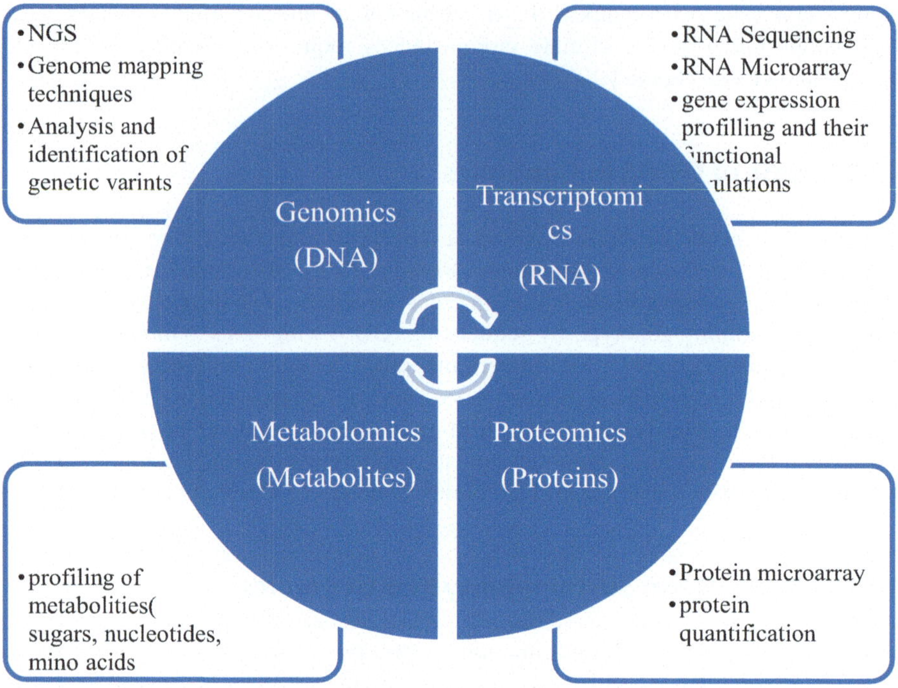

Fig. 12.2 Relationship among different omics approaches

variants, particularly in non-model plants. It is however limited to comparing transcript levels (Muthamilarasan et al. 2019).

12.16 Proteomics

Proteins are found throughout plants and are essential for various cellular processes. Proteomics can be used to assess mRNA expression in protein synthesis, which can help to understand gene function (Al-Amrani et al. 2021). Plants contain thousands of proteins, which perform important functional roles in almost all food items' texture, flavor, yield, and nutritional value. This technique aids in the comprehension of the roles played by different proteins. Proteomics may also improve our knowledge of pesticide biodegradation, mode of action, and resistance mechanisms. Translational proteomics can be used to apply proteomics findings for food authenticity, energy sustainability, great economic value, human health, and environmental stewardship (Muthurajan et al. 2011). In proteomics, researchers often use mass spectrometry (MS), ELISA, two-dimensional (2-D) gel electrophoresis, and western blots in combination with advanced bioinformatic tools for protein identification. However, MS-based approaches, when combined with computational tools, may process numerous peptide transitions at once while maintaining high predictability, high repeatability, and accuracy. Microsatellites, also known as simple sequence repeats (SSRs), are considered the most promising marker systems for laboratories where SNPs (single nucleotide polymorphism) genotyping is not possible. The genome-wide analysis of microsatellites and development of subsequent marker provides a valuable resource for breeding processes. Among the various SNP-based genotyping technologies, GBS methodology is a multiplexed platform for reduced representation libraries (RRL) construction, genotype improvement of crops, and molecular marker identification. For example, an NGS study of tomatoes using GBS discovered 8784 SNPs (Rayaprolu et al. 2022).

12.17 Metabolomics

Metabolomics is an analysis of chemical processes that link phenotypes and genotypes and are affected by both external and internal factors. Proteomics detects gene products, whereas metabolomic investigations assess whether proteins produced by genes are metabolically active or not, as well as assess the function of different metabolites. The ability to observe metabolic pathways simultaneously facilitates the identification of changes brought about by biotic/abiotic stress, and this may help develop better crop types and provide a basic understanding of systems physiology (Pandey et al. 2017). However, by identifying the essential chemicals responsible for imparting taste and aroma, a metabolic profile provides a real-time picture of what is happening within the cell during fruit ripening. Improvement of plant health can result from observing changes in metabolite levels. However, modifications to the metabolome can aid in determining the modes of action of pesticides, which is crucial for novel pesticide discoveries (Helms et al. 2015).

Metabolomics can help identify the compositional or equivalency similarity between changed and conventional plants, as well as whether undesired alterations have taken place in the overall makeup of metabolites. Cultured tomato cells are utilized for the metabolic investigation of low oxygen levels. Nevertheless, the findings showed that the low stress of oxygen changes the tomato cells' metabolic profile by causing them to accumulate glycolysis intermediates along with higher levels of sugar, lactate, and alcohols. The combination of metabolomics, GWAS based on metabolome variation, and linkage mapping offers a thorough understanding of the amount of natural variation in tomato metabolism and the biochemical and genetic factors governing it. Identification of metabolites by using nuclear magnetic resonance and MS analysis helps to determine metabolic reactions to herbicides, as well as to examine the metabolic regulation and alterations brought on by environmental factors such as soil type, light, salt, pests, temperature, genetic disturbances, fertilizers, and humidity (Raza et al. 2021). In-depth analysis of metabolites can be useful for generating novel insecticides, reducing pesticide use, enhancing nutritional values, or contributing to other crucial features (Jorge et al. 2016).

12.18 Proteomics and Metabolomics

Proteomics and metabolomics, which allow for examining changes and interactions between proteomics and metabolites downstream of plant gene expression, can be usefully integrated with small RNA, genome, and transcriptome sequencing technologies. However, such a method can improve our understanding of plant response mechanisms to pathogens/pests and other related microbes. Furthermore, metagenomic techniques allow for a better understanding of plant-associated microorganisms, providing a new possibility to sustain and monitor agricultural processes based on microbiomes (Yang et al. 2014).

12.19 Ionomics

Ionomics is a study of non-essential and essential components. The ionomics profile is affected by several factors, including environment, variety, plant species, and organs. Ionomics techniques use ICP-AES (Inductively Coupled Plasma-Atomic Emission Spectrometry) and ICP-MS (Inductively Coupled Plasma-Mass Spectrometry) to study the role of mineral components in plants. Ionomics is emerging as a promising technique for sustainable crop development because it has led to the discovery of key genes and processes that regulate the concentration of trace elements and mineral nutrients in plants (Zargar et al. 2016).

12.20 Bioinformatics

Bioinformatics is a broad integrative area that includes both practical and theoretical approaches for understanding, generating, analyzing, and disseminating biologic information. Bioinformatics shows a connection between biological input and computational technologies. Integrative Omics-Metabolic Analysis (IOMA) is performed to integrate proteomic facts with cell-related metabolic knowledge. Bioinformatics has grown throughout time and now serves as the foundation for dynamic "omics" technologies (Roychowdhury et al. 2023). However, bioinformatics offers the framework for combining omics techniques by developing an increasing number of new technologies to offer a comprehensive understanding of the complicated data set necessary for both individual studies and comparison analyses. Clustered regularly interspaced palindromic repeats (CRISPR/Cas9) is a gene-editing method that requires two basic components: a guide RNA that matches a desired target gene and CRISPR-associated protein 9 (cas9), an enzyme known as an endonuclease that induces a DNA double-strand break and enables genome alterations. For instance, CRISPR-Cas9 is used to study different situations of crop plants; nevertheless, it necessitates the creation of pipelines in bioinformatics and genomics that provide more comprehensive guidance on how to achieve this. Several software tools, including E-CRISP, CHOPCHOP, CCTop, and TIDE have been created to visualize and choose these for genome editing. The gene ontology (GO) analysis toolbox for agricultural communities, formerly referred to as AgriGO, has been updated with the program AgriGO 2.0. Further bioinformatics investigations included in this toolkit are transferable IDs employing BLAST4ID, parametric analysis of gene set enrichment (PAGE), and singular enrichment analysis (SEA). However, by combining multi-omics data and utilizing new web-based methods, researchers can acquire full insights into the genomic, proteomic, transcriptional, and metabolic landscapes of plant-micro-organisms interactions (Du et al. 2010).

12.21 Conclusion

The use of native genetic material by utilizing proven omics-based techniques is essential for agricultural production and its sustainability. As a result, by employing omics-based methods, a desired agricultural trait can be transmitted to a crop to enhance its performance. Genomics-assisted breeding is a critical frontier for improving the balance of environmental preservation, food security, and sustainable agriculture. Its efficiency and precision have the potential to develop resilient crops and biodiversity protection, resulting in secure and sustainable food production. However, when combined with improved computational techniques, omics data enables microbial consortium design, predictive modeling, and agriculture management; hence increasing resilience and productivity. Furthermore, to discover changes, several tools and techniques such as genomics, transcriptomics, ionomics, phenomics, and metabolomics have been developed that enable an in-depth understanding of signaling cascades, genetic makeup, and adaptability under stress. A multi-omics approach has identified the important hub genes and metabolic processes triggered after PGPB treatment, highlighting the molecular mechanisms involved in *K. rhizophila's* effect on tomato plants. The PGPB-plant interaction powerfully modulates complex machinery that involves selected proteins, genes, and plant metabolites such as energy regulators, amino acids, sugars, and other substances such as dopamine, all of which contribute to the implementation of plant growth. On this premise, these selected elements might be regarded as critical variables for increasing plant production in specialized PGPB-mediated sustainable and eco-friendly agricultural approaches.

References

Agarwal M, Shrivastava N, Padh H (2008) Advances in molecular marker techniques and their applications in plant sciences. Plant Cell Rep 27:617–631

Al-Amrani S, Al-Jabri Z, Al-Zaabi A, Alshekaili J, Al-Khabori M (2021) Proteomics: concepts and applications in human medicine. World J Biol Chem 12(5):57–69. https://doi.org/10.4331/wjbc.v12.i5.57. PMID: 34630910; PMCID: PMC8473418

Aoki K, Ogata Y, Igarashi K, Yano K, Nagasaki H, Kaminuma E, Toyoda A (2013) Functional genomics of tomato in a post-genome-sequencing phase. Breed Sci 63(1):14–20

Appleby N, Edwards D, Batley J (2009) New technologies for ultra-high throughput genotyping in plants. In: Plant genomics: methods and protocols. Humana Press, pp 19–39

Beyer P, Al-Babili S, Ye X, Lucca P, Schaub P, Welsch R, Potrykus I (2002) Golden rice: introducing the β-carotene biosynthesis pathway into rice endosperm by genetic engineering to defeat vitamin A deficiency. J Nutr 132(3):506S–510S

Challa S, Neelapu NR (2018) Genome-wide association studies (GWAS) for abiotic stress tolerance in plants. In: Biochemical, physiological and molecular avenues for combating abiotic stress tolerance in plants. Academic press, pp 135–150

Chaudhary J, Alisha A, Bhatt V, Chandanshive S, Kumar N, Mir Z, Kumar A, Yadav SK, Shivaraj SM, Sonah H et al (2019) Mutation breeding in tomato: advances, applicability and challenges. Plants 8(5):128. https://doi.org/10.3390/plants8050128

de Andrade LA, Santos CHB, Frezarin ET, Sales LR, Rigobelo EC (2023) Plant growth-promoting rhizobacteria for sustainable agricultural production. Microorganisms 11(4):1088

Deshmukh R, Sonah H, Patil G, Chen W, Prince S, Mutava R et al (2014) Integrating omic approaches for abiotic stress tolerance in soybean. Front Plant Sci 5:244

Du Z, Zhou X, Ling Y, Zhang Z, Su Z (2010) agriGO: a GO analysis toolkit for the agricultural community. Nucleic Acids Res 38(suppl_2):W64–W70

Fedoroff N, Brown NM (2004) Mendel in the kitchen–a scientist's view of genetically modified foods. Joseph Henry Press, Washington

Foolad MR (2007) Current status of breeding tomatoes for salt and drought tolerance. In: Advances in molecular breeding toward drought and salt tolerant crops. Springer, Dordrecht, pp 669–700

Guesmi S, Mahjoubi M, Pujic P, Cherif A, Normand P, Sghaier H, Boubakri H (2022) Biotechnological potential of Kocuria rhizophila PT10 isolated from roots of Panicum turgidum. Int J Environ Sci Technol 19:1–14

Hayama R, Yang P, Valverde F, Mizoguchi T, Furutani-Hayama I, Vierstra RD, Coupland G (2019) Ubiquitin carboxyl-terminal hydrolases are required for period maintenance of the circadian clock at high temperature in Arabidopsis. Sci Rep 9(1):17030

Helms JB, Kaloyanova DV, Strating JR, van Hellemond JJ, van der Schaar HM, Tielens AG et al (2015) Targeting of the hydrophobic metabolome by pathogens. Traffic 16(5):439–460

Heymann H, Lawless HT (2013) Sensory evaluation of food: principles and practices. Springer Science & Business Media

Hussain M, Zahra N, Lang T, Zain M, Raza M, Shakoor N et al (2023) Integrating nanotechnology with plant microbiome for next-generation crop health. Plant Physiol Biochem 196:703–711. https://doi.org/10.1016/j.plaphy.2023.02.022

Jorge TF, Rodrigues JA, Caldana C, Schmidt R, van Dongen JT, Thomas-Oates J, António C (2016) Mass spectrometry-based plant metabolomics: metabolite responses to abiotic stress. Mass Spectrom Rev 35(5):620–649

Kover PX, Valdar W, Trakalo J, Scarcelli N, Ehrenreich IM, Purugganan MD et al (2009) A multiparent advanced generation inter-cross to fine-map quantitative traits in Arabidopsis thaliana. PLoS Genet 5(7):e1000551

Kumar A, Kumar R, Sengupta D, Das SN, Pandey MK, Bohra A et al (2020) Deployment of genetic and genomic tools toward gaining a better understanding of rice-Xanthomonas oryzae pv oryzae interactions for development of durable bacterial blight resistant rice. Front Plant Sci 11:1152

Kumari A, Kumari A, Sharma H, Sharma P, Bhattacharya S, Mishra T et al (2023) Modern approaches in studying the role of plant-microbial interactions: a way towards the development of sustainable agriculture. In: New frontiers in plant-environment interactions: innovative technologies and developments. Springer Nature Switzerland, Cham, pp 69–91

Lee S, Noh TG, Choi JH, Han J, Ha JY, Lee JY, Park Y (2017, May) NIR spectroscopic sensing for point-of-need freshness assessment of meat, fish, vegetables and fruits. In: Sensing for agriculture and food quality and safety IX, vol 10217. SPIE, pp 51–57

Lephatsi MM, Meyer V, Piater LA, Dubery IA, Tugizimana F (2021) Plant responses to abiotic stresses and rhizobacterial biostimulants: metabolomics and epigenetics perspectives. Meta 11(7):457

Li Q, Yan J (2020) Sustainable agriculture in the era of omics: knowledge-driven crop breeding. Genome Biol 21:1–5

Li H, Liu H, Wang Y, Teng RM, Liu J, Lin S, Zhuang J (2020) Cytosolic ascorbate peroxidase 1 modulates ascorbic acid metabolism through cooperating with nitrogen regulatory protein P-II in tea plant under nitrogen deficiency stress. Genomics 112(5):3497–3503

Lin T, Zhu G, Zhang J, Xu X, Yu Q, Zheng Z et al (2014) Genomic analyses provide insights into the history of tomato breeding. Nat Genet 46(11):1220–1226

McCarthy MI, Abecasis GR, Cardon LR, Goldstein DB, Little J, Ioannidis JP, Hirschhorn JN (2008) Genome-wide association studies for complex traits: consensus, uncertainty and challenges. Nat Rev Genet 9(5):356–369

Mishra AK, Sudalaimuthuasari N, Hazzouri KM, Saeed EE, Shah I, Amiri KMA et al (2022) Tapping into plant-microbiome interactions through the lens of multi-omics techniques. Cells 11:3254. https://doi.org/10.3390/cells11203254

Muthamilarasan M, Singh NK, Prasad M (2019) Multi-omics approaches for strategic improvement of stress tolerance in underutilized crop species: a climate change perspective. Adv Genet 103:1–38

Muthurajan R, Shobbar ZS, Jagadish SVK, Bruskiewich R, Ismail A, Leung H, Bennett J (2011) Physiological and proteomic responses of rice peduncles to drought stress. Mol Biotechnol 48:173–182

Pandey P, Irulappan V, Bagavathiannan MV, Senthil-Kumar M (2017) Impact of combined abiotic and biotic stresses on plant growth and avenues for crop improvement by exploiting physio-morphological traits. Front Plant Sci 8:237767

Paradiso VM, Pasqualone A, Summo C, Caponio F (2018) An "omics" approach for lipid oxidation in foods: the case of free fatty acids in bulk purified olive oil. Eur J Lipid Sci Technol 120(7):1800102

Pazhamala LT, Kudapa H, Weckwerth W, Millar AH, Varshney RK (2021) Systems biology for crop improvement. Plant Genome 14(2):e20098

Puccio G, Ingraffia R, Mercati F, Amato G, Giambalvo D, Martinelli F et al (2023) Transcriptome changes induced by Arbuscular mycorrhizal symbiosis in leaves of durum wheat (Triticum durum Desf.) promote higher salt tolerance. Sci Rep 13(1):116

Rayaprolu L, Deshpande SP, Gupta R (2022) Genotyping-by-sequencing (GBS) method for accelerating marker-assisted selection (MAS) program. In: Genomics of cereal crops. Springer US, New York, pp 245–257

Raza A, Tabassum J, Kudapa H, Varshney RK (2021) Can omics deliver temperature resilient ready-to-grow crops? Crit Rev Biotechnol 41(8):1209–1232

Raza A, Salehi H, Rahman MA, Zahid Z, Madadkar Haghjou M, Najafi-Kakavand S et al (2022) Plant hormones and neurotransmitter interactions mediate antioxidant defenses under induced oxidative stress in plants. Front Plant Sci 13:961872

Rivero RM, Mittler R, Blumwald E, Zandalinas SI (2022) Developing climate-resilient crops: improving plant tolerance to stress combination. Plant J 109(2):373–389

Roychowdhury R, Das SP, Gupta A, Parihar P, Chandrasekhar K, Sarker U et al (2023) Multi-omics pipeline and omics-integration approach to decipher plant's abiotic stress tolerance responses. Genes 14(6):1281

Sarim KM, Srivastava R, Ramteke PW (2020) Chapter 9 – Next-generation omics technologies for exploring complex metabolic regulation during plant-microbe interaction. In: Singh JS, Vimal SR (eds) Microbial services in restoration ecology. Elsevier, pp 123–138. https://doi.org/10.1016/B978-0-12-819978-7.00009-9

Shabbir R, Singhal RK, Mishra UN, Chauhan J, Javed T, Hussain S et al (2022) Combined abiotic stresses: challenges and potential for crop improvement. Agronomy 12(11):2795

Sharifi P, Mohammadkhani N (2016) Effects of drought stress on photosynthesis factors in wheat genotypes during anthesis. Cereal Res Commun 44(2):229–239

Sindelar RD (2016) Genomics, other "omics" technologies, personalized medicine, and additional biotechnology-related techniques. In: Pharmaceutical biotechnology. CRC Press, pp 149–190

Uffelmann E, Huang QQ, Munung NS, De Vries J, Okada Y, Martin AR et al (2021) Genome-wide association studies. Nat Rev Methods Primers 1(1):59

Valdés A, Ibáñez C, Simó C, García-Cañas V (2013) Recent transcriptomics advances and emerging applications in food science. TrAC Trends Anal Chem 52:142–154

Van Emon JM (2016) The omics revolution in agricultural research. J Agric Food Chem 64(1):36–44. https://doi.org/10.1021/acs.jafc.5b04515. Epub 2015 Nov 17. PMID: 26468989; PMCID: PMC4714296

Wi SJ, Kim WT, Park KY (2006) Overexpression of carnation S-adenosylmethionine decarboxylase gene generates a broad-spectrum tolerance to abiotic stresses in transgenic tobacco plants. Plant Cell Rep 25:1111–1121

Yang D, Du X, Yang Z, Liang Z, Guo Z, Liu Y (2014) Transcriptomics, proteomics, and metabolomics to reveal mechanisms underlying plant secondary metabolism. Eng Life Sci 14(5):456–466

Yang Y, Saand MA, Huang L, Abdelaal WB, Zhang J, Wu Y et al (2021) Applications of multi-omics technologies for crop improvement. Front Plant Sci 12:563953

Zandalinas SI, Sengupta S, Fritschi FB, Azad RK, Nechushtai R, Mittler R (2021) The impact of multifactorial stress combination on plant growth and survival. New Phytol 230(3):1034–1048

Zargar SM, Gupta N, Nazir M, Mir RA, Gupta SK, Agrawal GK, Rakwal R (2016) Omics–a new approach to sustainable production. In: Breeding oilseed crops for sustainable production. Academic Press, pp 317–344

Zhang J, Zhao J, Xu Y, Liang J, Chang P, Yan F et al (2015) Genome-wide association mapping for tomato volatiles positively contributing to tomato flavor. Front Plant Sci 6:1042

Zhang Y, Hu X, Wang Q, Zhang Y (2022) Recent advances in microchip-based methods for the detection of pathogenic bacteria. Chin Chem Lett 33(6):2817–2831

Marker-Assisted Selection (MAS) Revolutionized: Integrating 'Omics' Data into Tomato Breeding

13

Vidadala Rajendra

Abstract

Tomato breeding has witnessed a transformative shift with the integration of 'omics' data into marker-assisted selection (MAS). This chapter delves into the revolutionary potential of this synergy, exploring how 'omics' technologies like genomics, transcriptomics, proteomics, and metabolomics are empowering MAS to accelerate the development of superior tomato varieties. We discuss the principles of integrating 'omics' data with MAS, highlighting the crucial role of genetic markers in identifying and utilizing desirable traits. The chapter outlines the integrated steps in marker discovery and 'omics' integration, including phenotyping, genotyping, quantitative trait loci (QTL) mapping, and validation. We delve into the comprehensive marker-trait relationships revealed through 'omics' data, providing a deeper understanding of the genetic and biochemical mechanisms underlying complex traits. The chapter explores the specific applications of 'omics' technologies in tomato breeding, showcasing how genomics, transcriptomics, proteomics, and metabolomics contribute to enhancing disease resistance, improving fruit quality, and promoting tolerance to abiotic stresses. We also discuss the role of advanced technologies like CRISPR/Cas9 and single-cell RNA sequencing in further refining MAS. Finally, we examine the emerging field of 'omics'-assisted breeding and big data analytics, highlighting the role of machine learning, cloud computing, and enviromics in shaping the future of tomato breeding. The chapter concludes by addressing the challenges and prospects of integrating 'omics' data with MAS, emphasizing the transformative potential of this approach in ensuring global food security and agricultural sustainability.

V. Rajendra (✉)
M.S.Swaminathan School of Agriculture, Centurion University, R.Sitapur, Odisha, India

M. Waseem et al. (eds.), *Omics Approaches for Tomato Yield and Quality Trait Improvement*, https://doi.org/10.1007/978-981-96-3890-1_13

Keywords

Marker-assisted selection (MAS) · Tomato breeding · Omics technologies · Genomics · Transcriptomics · Proteomics · Metabolomics

13.1 Introduction

Marker-assisted selection (MAS) is a revolutionary method in plant breeding that uses molecular markers to select desirable traits, significantly improving the precision and efficiency of traditional breeding techniques. By enabling breeders to make informed decisions based on genetic information, MAS accelerates the development of new crop varieties, reducing breeding cycles and increasing the accuracy of selecting plants with beneficial traits (Collard and Mackill 2008). Historically, conventional breeding relied on phenotypic selection, which could be slow and imprecise due to environmental influences. The advent of genetic markers in the 1980s allowed for the selection of plants based on their DNA, independent of environmental conditions, thus transforming the breeding process (Tanksley et al. 1989). Over time, marker systems evolved from Restriction Fragment Length Polymorphisms (RFLPs) to more advanced markers like simple sequence repeats (SSRs) and single nucleotide polymorphisms (SNPs), enabling more precise and efficient selection processes (Collard et al. 2005). This shift towards genomics-focused approaches has led to significant improvements in various crops, including tomatoes, by allowing the tracking and introgression of multiple traits simultaneously (Collard and Mackill 2008). Integrating 'omics' data with MAS, however, presents challenges such as the complexity and heterogeneity of 'omics' data, high dimensionality, and the need for advanced computational tools and standardized protocols to ensure data quality and consistency (López de Maturana et al. 2019). Despite these challenges, the benefits of integrating 'omics' data with MAS—enhanced precision in selecting desirable traits and a deeper understanding of genetic and biochemical mechanisms—make it a worthwhile endeavor. Innovations in high-throughput genotyping, sequencing technologies, machine learning (ML), artificial intelligence (AI)-driven predictive modeling, and cloud-based big data analytics are further refining the selection process, enabling breeders to predict trait outcomes with unprecedented accuracy and tailor breeding strategies to specific environmental conditions (Cao and Gao 2022). The adoption of enviromics is also crucial for developing climate-resilient crops ((Krzemińska et al. 2023)). Researchers, breeders, and policymakers must collaborate to develop standardized protocols, share data, and promote the adoption of integrated MAS and 'omics' approaches, ultimately contributing to global food security and agricultural sustainability (Collard and Mackill 2008; Varshney et al. 2014).

13.2 Principles of Integrating 'Omics' Data with Marker-Assisted Selection (MAS)

13.2.1 The Role of Genetic Markers in MAS for Tomato Breeding

Marker-assisted selection (MAS) and omics technologies have revolutionized tomato breeding by enabling the identification and utilization of genetic markers associated with desirable traits. These advancements have significantly improved the efficiency and precision of breeding programs aimed at enhancing disease resistance, stress tolerance, and other economically important traits in tomatoes. Markers have been extensively used for genetic mapping and identification of genes and QTLs for traits such as disease resistance, abiotic stress tolerance, and fruit characteristics. The advent of next-generation sequencing (NGS) technologies, including genotyping-by-sequencing (GBS), has facilitated the discovery and genotyping of SNPs, accelerating the use of MAS in tomato breeding. The high-quality tomato genome sequence released in 2012 has enabled the development of numerous DNA markers and promoted new genotyping methods (Shirasawa and Hirakawa 2013). Despite the availability of many markers, not all are validated across different tomato genotypes, reducing their utility in breeding programs. Omics approaches, such as transcriptomics, metabolomics, and proteomics, have been used to identify genetic variations, gene expression profiles, and metabolites, providing insights into the mechanisms underlying complex traits. Omics approaches, such as transcriptomics, metabolomics, and proteomics, have been used to identify genetic variations, gene expression profiles, and metabolites, providing insights into the mechanisms underlying complex traits. These are some of the important markers used in MAS.

- Simple Sequence Repeats (SSRs): SSRs, or microsatellites, are highly polymorphic, repetitive DNA sequences found throughout the genome. Their high level of polymorphism and broad genome coverage make them particularly useful in tomato breeding for traits like disease resistance and fruit quality (Powell et al. 1996).
- Single Nucleotide Polymorphisms (SNPs): SNPs are the most abundant type of genetic variation in genomes, representing single base pair changes. They provide extensive genome coverage and are suitable for high-throughput genotyping. Their stability and abundance make them ideal for detailed genetic mapping and association studies in tomatoes (Collard et al. 2005).
- Insertion-Deletion Polymorphisms (INDELs): INDELs are variations caused by the insertion or deletion of small DNA fragments. They are co-dominant and can be detected using PCR and sequencing. INDELs provide valuable genetic information and are useful for various genetic studies in tomato breeding (Yang et al. 2014).
- Diversity Arrays Technology (DArT): DArT offers a powerful microarray-based marker system that detects DNA polymorphism at multiple loci across the genome. This technology is particularly useful for genetic mapping and diversity

studies, allowing breeders to assess genetic variation and structure within tomato populations more effectively. DArT has been instrumental in advancing our understanding of the genetic basis of important traits and in facilitating the selection of superior varieties (Schalkwyk et al. 2012).

13.2.2 Integrated Steps in Marker Discovery and 'Omics' Integration for Tomato Breeding

Phenotyping: Phenotyping involves the accurate measurement of observable traits in tomato plants, such as fruit size, color, disease resistance, and stress tolerance. This step forms the basis for linking genetic markers to specific traits. Advanced imaging techniques and high-throughput phenotyping platforms enable the rapid and accurate capture of a wide range of phenotypic data. For instance, automated phenotyping systems can measure fruit characteristics in real time, providing a robust dataset for subsequent analyses (Yang et al. 2020). Integrating 'omics' data into this step involves using whole-genome sequencing to identify genetic variants associated with phenotypic traits, RNA-Seq data to correlate gene expression patterns with observed traits, and proteomics and metabolomics to link proteins and metabolites to the phenotypic traits, providing a deeper understanding of the underlying biological pathways.

Genotyping: Genotyping involves identifying genetic variations, such as SNPs, SSRs, and INDELs, within the tomato genome. High-throughput sequencing technologies like genotyping-by-sequencing (GBS) facilitate the rapid and cost-effective discovery of thousands of genetic markers across the genome, which are then used to create genetic maps linking genetic variation to phenotypic traits (Rodriguez et al. 2020). Integration of 'omics' data in this step includes using genomics to provide comprehensive data on genetic variants for constructing high-density genetic maps, transcriptomics to identify expression quantitative trait loci (eQTLs) that link gene expression levels to genetic markers, and proteomics and metabolomics to identify biomarkers and metabolic pathways associated with genetic markers, thereby enhancing the functional annotation of genetic variants.

QTL Mapping: Quantitative trait loci (QTL) mapping involves identifying genomic regions associated with quantitative traits by combining phenotypic and genotypic data to locate QTLs on the genetic map. Advanced statistical methods and software analyze the data, identifying regions of the genome contributing to trait variation (Adhikari et al. 2020). 'Omics' integration here includes using high-density genetic maps from whole-genome sequencing for detailed QTL mapping, RNA-Seq data to identify eQTLs linking gene expression differences to genetic loci, and proteomics and metabolomics to map protein QTLs (pQTLs) and metabolite QTLs (mQTLs), thus linking genetic loci to functional outcomes.

Validation: Validation confirms the identified genetic markers' association with desired traits through methods such as marker-assisted backcrossing, functional genomics, and field trials, ensuring the markers are reliable for breeding programs. Integrating 'omics' data in this step involves using genomics for functional validation of candidate genes within QTL regions via CRISPR-Cas9 or other gene-editing technologies, transcriptomics to validate gene expression patterns under different conditions, and proteomics and metabolomics for functional assays to confirm the association of proteins and metabolites with specific traits (Celik 2023).

Through these integrated steps, the identification of genetic markers is enhanced, providing a comprehensive approach to improving tomato breeding programs by leveraging the full potential of 'omics' technologies (Fig. 13.1).

13.2.3 Integration of 'Omics' Data for Comprehensive Marker-Trait Relationships

The integration of 'omics' data—genomics, transcriptomics, proteomics, and metabolomics—provides a comprehensive understanding of marker-trait relationships, facilitating more informed marker selection in tomato breeding.

Genomics: Whole-genome sequencing (WGS) and genotyping-by-sequencing (GBS) provide a comprehensive catalog of genetic variants across the tomato genome. These datasets enable the identification of SNPs, SSRs, and INDELs linked to important traits such as disease resistance, fruit quality, and abiotic ability for stress tolerance. High-density genetic maps constructed from these variants enhance the precision of QTL mapping, allowing breeders to pinpoint genomic regions associated with desirable traits (Tomato Genome Consortium 2012).

Transcriptomics: RNA-Seq data provides insights into gene expression patterns under different developmental stages and environmental conditions. This information is crucial for identifying expression QTLs (eQTLs) that link genetic markers to gene expression changes. For example, QTL mapping in tomatoes can identify genes associated with fruit ripening and quality traits, such as flavor, texture, and color. Transcriptomics also helps in understanding the molecular mechanisms underlying complex traits, enabling breeders to target specific genes for improvement (Liu et al. 2020).

Proteomics: Proteomic analyses identify and quantify proteins and their modifications, providing a direct link between genetic variation and phenotypic traits. By integrating proteomic data with genomics and transcriptomics, researchers can identify protein QTLs (pQTLs) that associate protein abundance with genetic loci. In tomato breeding, proteomics can help identify biomarkers for disease resistance and fruit quality, enhancing the selection of robust genetic markers (Momo et al. 2022).

Metabolomics: Metabolomics focuses on the comprehensive analysis of metabolites, the small molecules involved in metabolic pathways. Metabolomic profil-

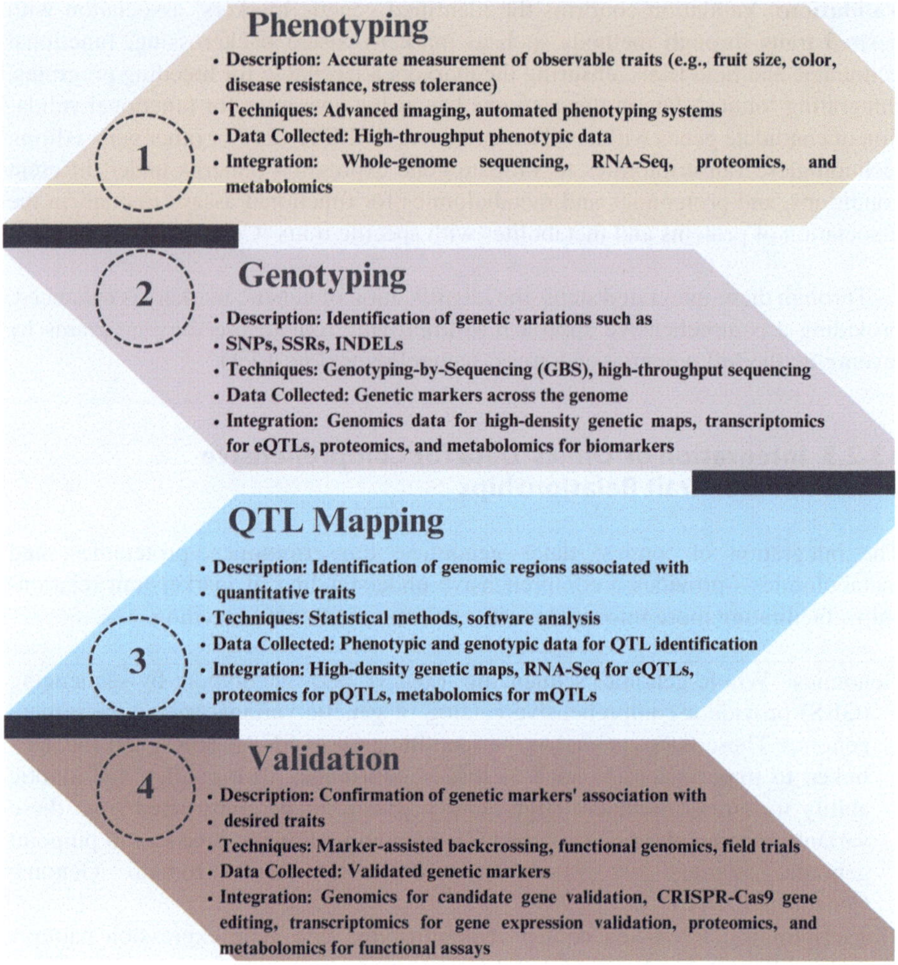

Fig. 13.1 Illustration depicting integrated steps in marker discovery pipeline and 'Omics' integration for tomato breeding

ing provides insights into the biochemical processes underlying phenotypic traits. In tomato breeding, metabolomics can identify metabolic QTLs (mQTLs) that link metabolites to genetic loci, offering valuable information for selecting markers associated with traits such as flavor, nutritional content, and stress tolerance. Integrating metabolomic data with other 'omics' data provides a holistic view of the genetic basis of complex traits.

The integration of 'omics' data into MAS for tomato breeding provides a comprehensive understanding of marker-trait relationships, facilitating the identification and validation of robust genetic markers. By combining genomics, transcriptomics, proteomics, and metabolomics, breeders can make informed decisions and develop superior tomato varieties with enhanced traits. This holistic approach represents a significant advancement in tomato breeding, enabling more efficient and precise selection processes.

13.2.4 Synergizing MAS and 'Omics' for Informed Breeding Decisions

Integrating MAS and 'omics' data is a transformative advancement in tomato breeding, enabling a more informed and precise selection of parental and donor lines, accelerating the breeding process, and enhancing the precision of trait introgression. By leveraging the synergistic potential of genetic markers and comprehensive 'omics' information, breeders can develop tomato varieties with superior disease resistance, improved fruit quality, and greater tolerance to abiotic stresses. The success of this integrated approach in various studies underscores its potential to revolutionize plant breeding practices and deliver significant improvements in crop performance.

A. Enabling Informed Parental/Donor Selection: The integration of marker-assisted selection (MAS) with 'omics' data—including transcriptomics, proteomics, and metabolomics—revolutionizes the selection of parental and donor lines in tomato breeding. This approach provides a multifaceted view of the genetic, transcriptional, and metabolic profiles of potential breeding lines, offering a holistic perspective on their phenotypic potential. For instance, transcriptomic data reveal differential gene expression patterns, which can be correlated with traits of interest, such as disease resistance or fruit quality. This comprehensive profiling ensures that breeders can identify and select individuals with the most advantageous genetic compositions, thereby optimizing the genetic gains in progeny. For example, integrating transcriptomic data with MAS has elucidated gene expression networks associated with enhanced fruit firmness and shelf life, facilitating the targeted selection of superior breeding lines (Tiwari et al. 2022).

B. Accelerating the Breeding Process: The utilization of 'omics' data significantly accelerates the breeding process by refining the identification and validation of genetic markers linked to complex traits. This integration enhances the predictive power of MAS, allowing for the early selection of desirable traits and reducing the breeding cycle duration. For example, proteomic data can identify key proteins and their post-translational modifications involved in stress responses, while metabolomic data can reveal the biochemical pathways activated under stress conditions. Such insights enable breeders to predict trait performance more accurately. Studies have shown that combining proteomic and metabolomic data with MAS enhances the prediction accuracy for traits such as drought tolerance and pathogen resistance, thereby expediting the development of robust tomato varieties (Osei et al. 2018).

C. Enhancing Precision of Trait Introgression: The precision of trait introgression is markedly enhanced by integrating 'omics' data with MAS. This synergy enables breeders to dissect the genetic and molecular bases of complex traits and to pinpoint specific alleles that confer beneficial attributes. Genetic markers can highlight regions of interest within the genome, while 'omics' data provide detailed insights into the underlying biological processes and pathways. For instance, metabolomic profiling can uncover metabolic signatures associated with improved fruit quality traits, such as higher levels of sugars and antioxidants. This detailed molecular information aids in the precise introgression of target traits, minimizing linkage drag and ensuring that only the desired alleles are incorporated into the new varieties. Utilizing genomic selection (GS) models that incorporate transcriptomic and metabolomic data has been shown to significantly improve the accuracy of predicting and introgressing traits like fruit quality and abiotic stress tolerance (Foolad and Panthee 2012).

13.3 'Omics' Technologies Empowering MAS in Tomato Breeding

13.3.1 Genomics: Expanding the Repertoire of MAS through Genomic Information

Genomics, the study of an organism's complete set of DNA, has significantly advanced marker-assisted selection (MAS) in tomato breeding by providing detailed genetic information. Genomics is critical for understanding the genetic basis of traits, enabling the identification and utilization of molecular markers linked to desirable characteristics such as disease resistance, stress tolerance, and improved fruit quality. The integration of genomics into MAS has revolutionized tomato breeding by providing powerful tools for identifying and utilizing genetic markers associated with desirable traits. Advances in DNA sequencing, bioinformatics, and genome editing technologies continue to enhance the precision and efficiency of breeding programs, enabling the development of tomato varieties with improved traits. The ongoing innovations in genomic technologies promise to accelerate progress in tomato breeding further, offering new possibilities for crop improvement.

13.3.1.1 Utilization of Genomic Information in MAS

Genomic information in MAS is primarily utilized through technologies such as DNA sequencing and bioinformatics tools. These technologies facilitate the identification of genetic markers and the construction of dense genetic maps, which are essential for locating quantitative trait loci (QTLs) associated with important agronomic traits.

(i) *DNA Sequencing Technologies*:
   (a) *Next-Generation Sequencing (NGS)*: NGS platforms like Illumina and PacBio have revolutionized plant genomics by enabling high-throughput

sequencing of entire genomes at a reduced cost. This technology supports the identification of single nucleotide polymorphisms (SNPs), which are widely used in MAS (Simko et al. 2021).

(b) *Long-Read Sequencing*: Technologies such as PacBio and Oxford Nanopore provide long-read sequencing capabilities, which are beneficial for assembling complex genomes and identifying structural variations that short-read sequencing might miss (Salava et al. 2021).

(ii) *Bioinformatics Tools*:

(a) *Genome Assembly and Annotation*: Tools like SPAdes and CLC Genomics Workbench are used for assembling genomic sequences and annotating genes, providing a comprehensive understanding of the genetic architecture (Valsala Sudarsanan and Sreekumar 2024).

(b) *Genomic Databases*: Databases such as GenBank and the Tomato Genome Database (TGRD) are crucial for storing and accessing genomic data and facilitating research and breeding efforts (Aglawe et al. 2020).

13.3.1.2 Linkage Mapping and QTL Mapping

(i) *Linkage Mapping*: Linkage mapping involves creating genetic maps based on the recombination frequency between markers. These maps are used to identify the location of genes associated with specific traits. The process includes crossing plants with contrasting traits and analyzing the inheritance patterns of markers in the progeny.

(ii) *QTL Mapping*: QTL mapping is used to identify genomic regions associated with quantitative traits. This method involves statistical analysis to link phenotypic data with genotypic data, identifying regions of the genome that control traits like yield, disease resistance, and stress tolerance (Collard and Mackill 2008).

13.3.1.3 Genome-Wide Association Studies (GWAS)

GWAS involves scanning the entire genome to find associations between genetic variants and traits. This approach uses a diverse set of genotypes to identify SNPs linked to complex traits. GWAS is particularly powerful for identifying markers in crops with high genetic diversity and complex traits.

Applications in Tomato Breeding: GWAS has been used to identify markers associated with fruit quality, disease resistance, and abiotic stress tolerance in tomatoes. For example, studies have identified loci linked to fruit shape, size, and color, aiding the development of varieties with improved marketability (Scheben et al. 2017).

13.3.1.4 Genomic Selection (GS)

Genomic selection (GS) uses genome-wide marker data to predict the breeding value of plants. This method combines phenotypic and genotypic data to develop prediction models, which are then used to select superior individuals for breeding.

Advantages of GS

(i) *Increased Accuracy*: GS provides a more accurate selection by considering the effects of all markers across the genome.
(ii) *Reduced Breeding Cycle*: GS accelerates the breeding process by allowing early selection of desirable traits, reducing the need for extensive phenotyping in later generations.

13.3.2 Transcriptomics: Assessing Gene Expression for Informed Marker Selection in Tomato Breeding

Transcriptomics: Assessing Gene Expression for Informed Marker Selection in Tomato Breeding In tomato breeding, transcriptomics provides insights into the molecular mechanisms underlying important traits. By analyzing the transcriptome, researchers can identify genes that are differentially expressed under different conditions, aiding in the selection of markers linked to desirable traits. This information is instrumental in marker-assisted selection (MAS), where genetic markers associated with beneficial traits are used to guide breeding decisions, ultimately improving crop quality and yield.

13.3.2.1 RNA Sequencing (RNA-Seq) and Its Applications in Tomato Breeding

RNA-Seq has revolutionized transcriptome analysis by enabling the precise quantification of RNA levels across the genome. This technique involves sequencing cDNA derived from RNA and mapping the reads back to a reference genome or transcriptome, allowing for comprehensive profiling of gene expression.

Applications in Tomato Breeding

(i) *Gene Expression Quantification*: RNA-Seq allows for the accurate measurement of gene expression levels, helping to identify up or down-regulated genes in response to various stimuli or developmental stages (Chen et al. 2019).
(ii) *Identification of Differentially Expressed Genes (DEGs)*: By comparing gene expression profiles between different conditions, RNA-Seq can identify DEGs that are critical for traits such as disease resistance, fruit quality, and abiotic stress tolerance (Costa-Silva et al. 2017).provides a more accurate selection by consider
(iii) *Discovery of Novel Transcripts and Splice Variants*: RNA-Seq can detect previously unknown transcripts and alternative splicing events, providing a more detailed view of the transcriptome (Van Den Berge et al. 2018).

13.3.2.2 Gene Expression Analysis, eQTL Identification, and Functional Genomics in Tomato Breeding

(i) *Gene Expression Analysis*: Gene expression analysis using RNA-Seq involves several steps, including quality control, read mapping, and quantification. Various computational tools and statistical methods are used to ensure accurate measurement of gene expression levels and to identify DEGs.

(ii) *Expression Quantitative Trait Loci (eQTL) Identification*: eQTL mapping links variations in gene expression to genetic loci, providing insights into the genetic architecture of complex traits. RNA-Seq enhances eQTL studies by offering high-resolution data on allele-specific expression and isoform-level expression, which are not accessible through traditional microarrays (Zhu et al. 2022a).

(iii) *Functional Genomics*: Functional genomics aims to understand the roles of genes and regulatory elements in complex biological systems. In tomato breeding, RNA-Seq data can be integrated with other 'omics' data (such as proteomics and metabolomics) to construct gene regulatory networks, identify key regulatory genes, and understand the molecular basis of important traits.

13.3.2.3 Modern Transcriptomic Technologies: Single-Cell RNA Sequencing and Spatial Transcriptomics

(i) *Single-Cell RNA Sequencing (scRNA-Seq)*: scRNA-Seq enables the analysis of gene expression at the single-cell level, providing unprecedented resolution in transcriptomic studies. This technology allows researchers to dissect the cellular heterogeneity within tissues, identify rare cell types, and understand cell-type-specific responses to environmental stimuli (Maria et al. 2022).

Applications in Tomato Breeding

(a) *Cell Type Identification*: scRNA-Seq can identify distinct cell types within tomato tissues, aiding in the study of development and differentiation processes.

(b) *Gene Regulatory Networks*: By analyzing gene expression patterns in individual cells, scRNA-Seq helps to reconstruct gene regulatory networks at a finer resolution.

(ii) *Spatial Transcriptomics*: Spatial transcriptomics combines RNA-Seq with spatial information, allowing researchers to map gene expression within the context of tissue architecture. This technology is valuable for understanding the spatial organization of gene expression and its role in tissue function and development.

Applications in Tomato Breeding

(a) *Developmental Biology*: Spatial transcriptomics can reveal spatial patterns of gene expression during tomato fruit development, providing insights into the molecular mechanisms of tissue differentiation.
(b) *Stress Responses*: By mapping gene expression in response to biotic and abiotic stresses, spatial transcriptomics helps identify regions within tissues that are critical for stress tolerance (Chaudhary et al. 2019).

13.3.2.4 Integration of Transcriptomics Data with MAS

The integration of transcriptomics data with MAS enhances the precision and efficiency of tomato breeding programs. By combining gene expression data with genetic markers, breeders can make more informed decisions and accelerate the development of improved varieties.

Key Benefits

(i) *Enhanced Marker Discovery*: Transcriptomics data can identify novel markers linked to important traits, improving the accuracy of MAS (Simko et al. 2021).
(ii) *Trait Dissection*: Integrating eQTL data with transcriptomics helps dissect complex traits into their genetic and regulatory components, facilitating targeted breeding strategies.
(iii) *Functional Validation*: Transcriptomics data can validate the functional relevance of identified markers, ensuring that they are truly associated with the traits of interest (Maria et al. 2022).

13.3.3 Proteomics and Metabolomics: Translating Complex Data into Breeding Strategies for Tomatoes

Proteomics is the large-scale study of proteins, including their structures and functions. Proteins are the primary functional molecules in biological systems, driving virtually all cellular processes. Proteomics aims to identify and quantify the entire protein complement of a cell, tissue, or organism, providing insights into the mechanisms that control phenotypic traits. Metabolomics, on the other hand, is the study of the complete set of small-molecule metabolites found within a biological sample. These metabolites, which include sugars, lipids, amino acids, and nucleotides, reflect the physiological state of an organism and its metabolic pathways. Metabolomics provides a direct readout of cellular activity and physiological responses to genetic and environmental changes. In tomato breeding, both proteomics and metabolomics are invaluable. They offer a deeper understanding of the biochemical pathways and molecular mechanisms underlying traits such as disease resistance, fruit quality, and stress tolerance. By integrating proteomic and metabolomic data with marker-assisted selection (MAS), breeders can develop more targeted and effective breeding strategies.

13.3.3.1 Overview of Proteomic Techniques and Applications in Identifying Proteins Associated with Desirable Traits in Tomatoes

Proteomics involves several key techniques, including mass spectrometry (MS), which is crucial for identifying and quantifying proteins. MS-based proteomics can profile the protein content of tomato tissues under various conditions, helping to identify proteins linked to desirable traits.

Key Proteomic Techniques

(i) *Mass Spectrometry (MS)*: MS is the cornerstone of proteomics, enabling the identification and quantification of proteins in complex mixtures. High-throughput MS techniques such as data-independent acquisition (DIA) and tandem mass spectrometry (MS/MS) are widely used.

(ii) *Two-Dimensional Gel Electrophoresis (2-DE)*: This technique separates proteins based on their isoelectric points and molecular weights, allowing for the resolution of complex protein mixtures. It is often used in combination with MS for protein identification.

(iii) *Protein Microarrays*: These are high-throughput platforms used to detect and quantify protein interactions, modifications, and functions.

Applications in Tomato Breeding

(a) *Disease Resistance*: Proteomics can identify proteins involved in plant immune responses, such as pathogenesis-related (PR) proteins, which are critical for developing disease-resistant tomato varieties. Proteomic analyses reveal that both resistant and susceptible tomato plants show varied protein expression when inoculated with pathogens. Resistant plants tend to elevate pathogenesis-related and other defense-related proteins, while susceptible plants show changes in cell wall metabolic proteins and stress-related proteins (Jain et al. 2020).

(b) *Fruit Quality*: Proteomic studies have identified proteins associated with fruit ripening, flavor, texture, and nutritional content. For instance, enzymes involved in cell wall modification and ethylene signaling pathways play vital roles in these traits (Xie et al. 2021).

(c) *Stress Tolerance*: Stress-responsive proteins, including heat shock proteins and antioxidants, are crucial for enhancing tolerance to abiotic stresses like drought and salinity.

13.3.3.2 Overview of Metabolomic Techniques and Applications in Identifying Metabolic Pathways Linked to Trait Development in Tomatoes

Metabolomics employs various analytical techniques to profile the metabolites within a biological sample. These techniques provide comprehensive data on the metabolic pathways active in tomato plants, linking metabolic profiles to phenotypic traits.

Key Metabolomic Techniques

(i) *Gas Chromatography-Mass Spectrometry (GC-MS)*: GC-MS is widely used for volatile and semi-volatile compound analysis, making it ideal for studying metabolites like organic acids, fatty acids, and amino acids in tomatoes.
(ii) *Liquid Chromatography-Mass Spectrometry (LC-MS)*: LC-MS is suitable for a broad range of metabolites, including those with high polarity and molecular weight. It is commonly used for profiling secondary metabolites like flavonoids and alkaloids.
(iii) *Nuclear Magnetic Resonance (NMR) Spectroscopy*: NMR provides detailed information on the structure and quantity of metabolites in a sample. It is highly reproducible and non-destructive, making it a valuable tool for metabolomics (Abreu and Fernández 2020; Jiang et al. 2023).

Applications in Tomato Breeding

(a) *Nutritional Quality*: Metabolomics can identify key metabolites associated with nutritional quality, such as vitamins, minerals, and antioxidants, which are essential for developing healthier tomato varieties (Zhu et al. 2022b).
(b) *Flavor and Aroma*: Volatile organic compounds (VOCs) are critical for tomato flavor and aroma. Metabolomic profiling helps identify and quantify these VOCs, guiding the breeding of tomatoes with superior taste (Tang et al. 2020).
(c) *Stress Response*: Metabolites involved in stress responses, such as osmoprotectants and antioxidants, can be profiled to enhance tomato resilience to environmental stresses (Zeiss et al. 2018).

13.3.3.3 Integration of Proteomic and Metabolomic Data with MAS

Integrating proteomic and metabolomic data with MAS enhances the precision and effectiveness of breeding programs by linking molecular data to phenotypic traits. This multi-omics approach allows breeders to understand the underlying biochemical pathways and identify key biomarkers for selection.

Benefits of Integration

(i) *Comprehensive Trait Analysis*: By combining proteomic and metabolomic data, breeders can obtain a holistic view of the biological processes influencing trait development. This comprehensive analysis enables more accurate identification of biomarkers linked to desirable traits (Di Giacomo et al. 2023).
(ii) *Improved Marker Discovery*: Multi-omics data can reveal novel markers that are not detectable through single-omics approaches. These markers can be

used to track complex traits more effectively in breeding programs (Razzaq et al. 2022).

(iii) *Functional Validation*: Integrating proteomic and metabolomic data with genomic information provides functional validation of candidate genes and markers, ensuring their relevance to the traits of interest (Fig. 13.2).

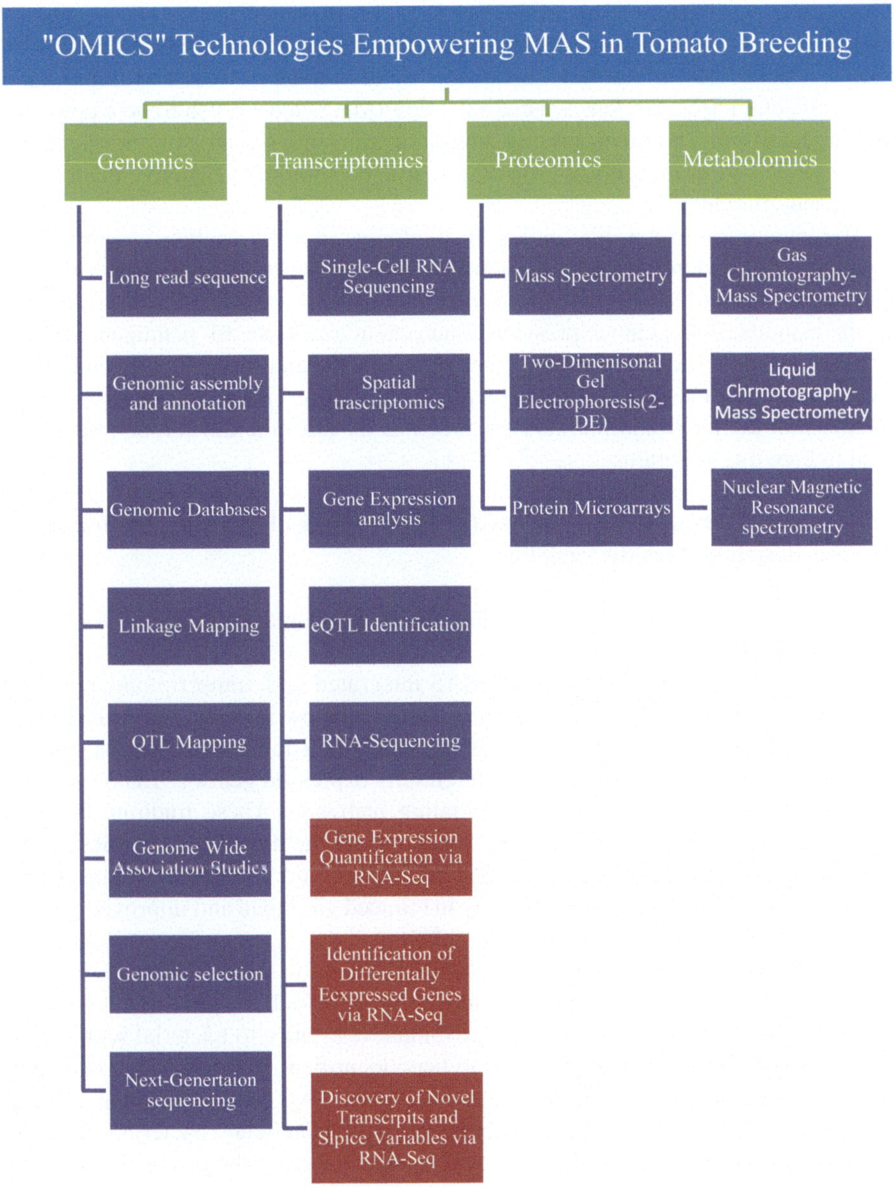

Fig. 13.2 Mind map illustrating the 'Omics' technologies empowering MAS in tomato breeding

13.4 MAS in Enhancing Disease Resistance and Tolerance to Abiotic Stresses

13.4.1 Enhancing Disease and Pest Resistance via Robust MAS Strategies

Disease and pest resistance are critical goals in tomato breeding due to their significant impact on yield and quality. Breeding for resistance helps reduce reliance on chemical pesticides, promoting sustainable agricultural practices and minimizing environmental harm. Marker-assisted selection (MAS) has proven to be a powerful tool in identifying and incorporating resistance traits into tomato varieties, leveraging genomics, transcriptomics, proteomics, and metabolomics data to enhance precision and efficiency.

The integration of various 'omics' technologies in MAS enables a comprehensive approach to identifying and validating resistance traits. Genomics provides the foundation by identifying genetic markers associated with resistance. Transcriptomics offers insights into gene expression changes in response to pathogen attacks. Proteomics helps identify resistance-related proteins, and metabolomics uncovers the metabolic pathways involved in defense mechanisms. This multi-omics approach ensures that the identified markers are functionally relevant and can be effectively used in breeding programs.

Case Studies Demonstrating Successful Applications of MAS in Developing Disease-Resistant Tomato Varieties

A. Developing Tomato Varieties Resistant to Tomato Yellow Leaf Curl Virus (TYLCV)

   A 2022 study in China utilized MAS integrated with transcriptome profiling to develop tomato varieties resistant to tomato yellow leaf curl virus (TYLCV). Researchers conducted RNA-seq on a TYLCV-resistant cultivar and a susceptible line, identifying over 1000 differentially expressed genes (DEGs) linked to phytohormone biosynthesis and signaling pathways. These findings revealed key genes involved in the resistance mechanism. The developed varieties exhibited significantly lower lipid peroxidation levels and maintained higher ROS-scavenging enzyme activity, resulting in reduced viral load and improved growth under TYLCV infection (Song et al. 2022).

B. Enhancing Resistance to Bacterial Wilt

   A 2023 study in China used a multi-omics approach integrating microbiome and metabolome analyses to enhance tomato resistance to bacterial wilt caused by Ralstonia solanacearum. Researchers identified key microbial taxa and metabolites associated with disease resistance. They discovered that healthy tomato plants had a higher abundance of beneficial bacteria, such as Sphingomonas and Pseudomonas, which were linked to the production of antibacterial compounds. The resistant varieties showed improved growth and reduced bacterial wilt symptoms under field conditions (Wei et al. 2023).

C. Combating Fusarium Wilt in Brazil Using Multi-Omics Approaches
   A recent study in Brazil employed multi-omics approaches to tackle Fusarium wilt in tomatoes, a critical disease affecting yield. Researchers collected samples from tomato-producing areas, identified Fusarium oxysporum and Fusarium solani species, and used molecular techniques to classify isolates. This comprehensive approach revealed novel pathogens like Fusarium falciform and Fusarium suttonianum, demonstrating the complexity of Fusarium-induced diseases and highlighting the importance of advanced molecular tools in accurate pathogen identification and disease management (Severo et al. 2024).
D. Utilizing CRISPR/Cas9 and MAS for Late Blight Resistance in tomatoes in the United States
   A 2022 study applied CRISPR/Cas9 and marker-assisted selection (MAS) to enhance late blight resistance in tomatoes. Researchers targeted the SlPMR4 gene using CRISPR/Cas9, introducing precise mutations that reduced susceptibility to Phytophthora infestans. This combined approach enabled the development of tomato cultivars with significantly lower disease symptoms, showcasing the potential of integrating advanced genomic tools with traditional breeding methods (Li et al. 2022).
E. Addressing Tomato Spotted Wilt Virus (TSWV) Resistance
   A 2018 study integrated advanced genomics and molecular techniques, including TaqMan™ SNP Genotyping and RNA-Seq, to enhance TSWV resistance in tomatoes. Researchers identified key single nucleotide polymorphisms (SNPs) associated with resistance by genotyping diverse tomato lines. Concurrently, RNA-Seq revealed differentially expressed genes linked to antiviral responses, while proteomic analysis highlighted upregulated defense proteins. This comprehensive approach allowed for the precise selection of resistant varieties, resulting in tomato lines that exhibited significantly improved resistance and yield under TSWV pressure (di Rienzo et al. 2018).

Integrating genomics, transcriptomics, proteomics, and metabolomics in MAS has revolutionized the development of disease-resistant tomato varieties. These case studies highlight the effectiveness of multi-omics approaches in identifying and validating resistance traits, leading to the successful breeding of robust tomato lines capable of withstanding various pathogens. Continued advancements in 'omics' technologies and their integration into MAS will further enhance the precision and efficiency of tomato breeding programs, contributing to sustainable agricultural practices and improved crop resilience.

13.4.2 Improving Fruit Quality Traits Through the Transfer of Favorable Alleles

Fruit quality traits such as flavor, texture, color, and nutritional content are paramount in tomato breeding as they directly influence consumer preference, market value, and nutritional benefits. Improving these traits through breeding requires precise identification and incorporation of favorable alleles, which can be efficiently achieved by integrating various omics approaches and marker-assisted selection (MAS).

Utilization of Omics Approaches to Identify and Transfer Favorable Alleles for Fruit Quality Improvement

Omics technologies, including genomics, transcriptomics, proteomics, and metabolomics, provide comprehensive insights into the genetic and biochemical basis of fruit quality traits. These approaches enable the identification of genetic markers associated with desirable traits, facilitating their incorporation into breeding programs through MAS. Integrating multi-omics data ensures a thorough understanding of the molecular mechanisms driving fruit quality, leading to the development of superior tomato varieties.

Case Studies Illustrating the Enhancement of Fruit Quality Traits Through Integrated MAS and Omics Approaches

A. Enhancing Flavor and Nutritional Content in Tomatoes

   A recent study combined transcriptomics and metabolomics to investigate the impact of low-temperature storage on tomato fruit quality. The findings revealed significant differential expression of genes involved in ripening, cell wall metabolism, and the biosynthesis of flavor and aroma compounds under low-temperature conditions. Metabolomic analysis corroborated these transcriptomic changes, showing reduced sugars, organic acids, and health-promoting compounds like carotenoids. This integrative 'omics' approach provides insights into the molecular mechanisms underlying the suppression of desirable flavor and nutritional attributes during low-temperature storage, informing strategies to develop tomato varieties with enhanced postharvest performance (Bai et al. 2023).

B. Improving Texture and Shelf Life in Tomatoes in Spain

   A 2023 study integrated transcriptomics and proteomics to enhance tomatoes' shelf life and texture. Researchers analyzed two near-isogenic lines (NIL115 and NIL080) with Solanum pimpinellifolium introgressions, revealing key genes involved in fruit ripening. NIL115, in particular, showed improved firmness and extended shelf life attributed to differentially expressed genes like AP2/ERF and proteins such as pectate lyase and ACO3. This comprehensive approach provided valuable insights for breeding tomatoes with superior texture and longevity (Di Giacomo et al. 2023).

C. Developing High-Lycopene Tomatoes in Italy

   An Italian study in 2022 used metabolomics and genomics to breed tomatoes with high-lycopene content. Metabolomic analysis identified key metabolic pathways associated with lycopene biosynthesis. Genomic data revealed SNP markers linked to these pathways, which were used in MAS to develop high-lycopene tomato varieties. These tomatoes exhibited significantly higher lycopene content, enhancing their nutritional value (Tripodi and Francese 2023).

D. Leveraging Metabolomic Selection for Enhanced Flavor in Tomatoes

At the University of Florida, a 2022 study by Colantonio et al. leveraged metabolomic selection to enhance tomato flavor. Researchers utilized targeted metabolomics to quantify sugars, acids, and volatile organic compounds (VOCs) across diverse tomato accessions. Key metabolites, including glucose, fructose, and VOCs such as 2-phenyl ethanol and E-2-personal, were identified as significant contributors to flavor. Machine learning models, including XGBoost and gradient boosting machines, were employed to predict consumer sensory perceptions. This integrative approach facilitated the selection of tomato varieties with improved sweetness, umami, and overall flavor intensity, advancing the precision of flavor phenotyping in breeding programs.

E. Enhancing Sugar Content in Tomatoes Using CRISPR/Cas9 Gene Editing

In a study conducted at Xinjiang Academy of Agricultural Sciences, researchers employed CRISPR/Cas9-mediated gene editing to enhance the sugar content in tomatoes. By targeting and knocking out the SlINVINH1 and SlVPE5 genes, which inhibit sugar accumulation, they developed mutant lines with significantly increased glucose levels, fructose, and total soluble solids (TSS). The synergistic effect of knocking out both genes resulted in a substantial improvement in the sweetness and quality of the tomatoes, providing a promising approach for future tomato breeding efforts (Wang et al. 2021).

Integrating omics approaches with MAS has significantly improved fruit quality traits in tomato breeding. These case studies demonstrate the successful application of multi-omics data in identifying and transferring favorable alleles, leading to the development of tomato varieties with enhanced flavor, texture, color, and nutritional content. Continued advancements in omics technologies and their integration into breeding programs will further enhance the precision and efficiency of tomato breeding, ultimately benefiting both producers and consumers.

13.4.3 Tolerance to Abiotic Stresses Using Integrated Omics Approaches

Abiotic stresses such as drought, salinity, heat, and cold significantly impact tomato yield and quality, posing challenges to sustainable agriculture. Developing stress-tolerant tomato varieties is crucial for ensuring food security and agricultural resilience. Integrating various omics approaches—genomics, transcriptomics, proteomics, and metabolomics—into breeding programs enhances identifying and validating traits associated with abiotic stress tolerance. This comprehensive approach enables the development of robust tomato varieties that can withstand adverse environmental conditions.

Integration of Omics Data to Identify and Validate Traits Associated with Tolerance to Abiotic Stresses

Omics technologies provide detailed insights into the genetic, transcriptional, protein, and metabolic changes that occur in response to abiotic stresses. Genomics identifies genetic markers linked to stress tolerance, transcriptomics reveals changes in gene expression, proteomics uncovers stress-responsive proteins, and metabolomics profiles metabolic adjustments. Integrating these data layers through marker-assisted selection (MAS) allows breeders to efficiently develop varieties with enhanced stress tolerance.

Case Studies Illustrating Successful MAS Applications for Abiotic Stress Tolerance

A. Drought Tolerance in Tomatoes Using Genomics and Transcriptomics

   A 2021 study analyzed the effects of drought stress on tomato anthers at different developmental stages using transcriptomics and morpho-physiological methods. Researchers found that drought stress caused male sterility by inducing abnormal anther development, including pollen abortion and inadequate pollen starch accumulation. RNA-Seq analysis revealed over 15,000 differentially expressed genes involved in various pathways. Key findings suggested that tapetum development and ABA homeostasis genes were induced, while sugar utilization and IAA metabolic genes were repressed, providing insights into the molecular mechanisms underlying drought tolerance in tomatoes (Lamin-Samu et al. 2021).

B. Unveiling Heat Stress Tolerance Mechanisms in Tomatoes through Integrated Omics Analysis

   In an intriguing study, integrating transcriptome and proteome analyses has unveiled critical insights into heat stress tolerance in tomatoes (Solanum lycopersicum). Researchers identified 3531 differentially expressed genes (DEGs) and 268 differentially expressed proteins (DEPs) under high-temperature stress (42 °C for 4 h). Notably, the enrichment of the "protein processing in the endoplasmic reticulum" pathway suggests a pivotal role of endoplasmic reticulum stress in heat tolerance. Among the significant findings, the gene SlBAG5 emerged as a key player, with its ectopic overexpression in Arabidopsis leading to a thermosensitive phenotype. This integrative omics approach deepens our understanding of the molecular mechanisms underpinning thermotolerance and holds promise for developing heat-resilient tomato cultivars (Ding et al. 2020).

C. Enhancing Cold Stress Tolerance in Tomatoes Through Integrative Genomics and Metabolomics Approaches

   A 2024 study focused on enhancing cold stress tolerance in tomatoes through genomics and metabolomics. Researchers used genotyping-by-sequencing on 50 diverse genotypes, identifying 3854 significant SNPs linked to cold tolerance traits via GWAS. Key traits, such as proline content, were associated with stress-responsive genes, leading to improved cold tolerance and yield in selected lines. This integrative approach demonstrates the potential of combining genomic and metabolomic data to breed tomatoes resilient to cold stress (Shah et al. 2024).

Integrating various omics approaches into MAS has significantly advanced the development of abiotic stress-tolerant tomato varieties. These case studies demonstrate the effectiveness of multi-omics data in identifying and validating traits associated with stress tolerance, leading to the successful breeding of robust tomato lines capable of withstanding adverse environmental conditions. Continued advancements in omics technologies and their integration into breeding programs will further enhance the precision and efficiency of developing stress-tolerant crops, contributing to sustainable agriculture and food security.

13.5 Technological Advances and Modern Omics Approaches in MAS

13.5.1 Modern and Futuristic Genomic Technologies

A. CRISPR/Cas9 and Genome Editing: CRISPR/Cas9, a revolutionary genome editing tool, allows for precise, targeted modifications to specific genes. It uses a guide RNA to direct the Cas9 nuclease to a specific location in the genome, where it creates double-strand breaks. These breaks can be repaired by either knocking out genes or introducing new genetic material, enabling a wide range of applications from basic research to agricultural improvements (Wan et al. 2021). CRISPR/Cas9 has been instrumental in improving tomato traits. For instance, genes related to disease resistance, such as the mildew resistance locus (MLO1), have been edited to produce mildew-resistant tomatoes. Shelf life has been extended by modifying genes like RIN (ripening inhibitor), resulting in delayed ripening. Nutritional content has been enhanced by increasing lycopene production by modifying the LCY-E (lycopene epsilon-cyclase) gene (Tiwari et al. 2023).
B. Long-Read Sequencing Technologies: Technologies like PacBio and Oxford Nanopore offer long-read sequencing capabilities, crucial for resolving complex genomic regions and providing more complete and accurate genomic information. These technologies can sequence entire genes and even whole genomes without the fragmentation issues associated with short-read sequencing (Chaudhuri et al. 2022). In tomato genomics, long-read sequencing has facilitated the creation of more accurate reference genomes. This comprehensive genomic information aids in identifying genes and markers associated with important traits like disease resistance, fruit quality, and yield, significantly advancing tomato breeding programs (Chaudhuri et al. 2022).
C. Synthetic Biology: Synthetic biology combines biology and engineering to design and construct new biological parts, devices, and systems. This approach goes beyond traditional breeding by enabling the design of entirely new traits that do not exist in nature, potentially transforming tomato breeding. Synthetic biology has been used to enhance the nutritional content of tomatoes, such as increasing anthocyanin levels for better health benefits. Additionally, stress resistance has been improved by engineering pathways that enhance tolerance to drought and salinity. Yield can also be increased by optimizing photosynthetic efficiency and carbon assimilation pathways.
D. Genome-wide association Studies (GWAS) and Genomic Selection (GS)

GWAS identifies genetic loci associated with traits by scanning the genomes of many individuals. Genomic selection (GS) uses these loci to predict the performance of breeding lines, accelerating the selection process by focusing on genetic potential rather than phenotypic traits alone. GWAS has identified loci linked to important tomato traits such as yield, flavor, and shelf life. For example, the FAS gene associated with fruit size and the MULT gene related to fruit number have been identified through GWAS. GS has been used to predict the performance of new breeding lines, making the selection process faster and more efficient (Causse et al. 2022).

13.5.2 Modern and Futuristic Transcriptomic Technologies

A. Single-Cell RNA Sequencing: Single-cell RNA sequencing (scRNA-seq) provides high-resolution gene expression profiles at the individual cell level, revealing cell-specific responses and heterogeneity that bulk RNA sequencing cannot capture. In tomatoes, scRNA-seq has been used to understand cell-specific responses to environmental stresses and pathogen attacks. This technology provides insights into the molecular mechanisms underlying trait expression, such as identifying specific cell types involved in disease resistance or stress responses.
B. Spatial Transcriptomics: Spatial transcriptomics maps gene expression in the spatial context of tissues, revealing the location and distribution of gene activity. This method integrates spatial information with transcriptomic data to understand tissue-specific functions. Spatial transcriptomics has been applied to study tissue-specific gene expression during tomato fruit development and ripening. This approach aids in identifying genes involved in processes like fruit quality and flavor development by mapping where these genes are active within the fruit (Shinozaki et al. 2018).
C. Advanced RNA Sequencing (RNA-Seq) Technologies: RNA-Seq technologies identify genes involved in specific traits and analyze gene expression changes in response to various stimuli. These technologies provide comprehensive gene expression profiles crucial for understanding trait development and responses to environmental changes. In tomato research, RNA extraction methods must preserve the quality and integrity of RNA. Library preparation involves converting RNA into a format suitable for sequencing, and sequencing technologies must be chosen based on the study's specific needs, ensuring accurate and reliable data.

13.5.3 Modern and Futuristic Proteomic Technologies

A. High-Throughput Mass Spectrometry

   High-throughput mass spectrometry enables large-scale proteomic analyses, identifying and quantifying a wide range of proteins. This technology is essential for understanding the functional roles of proteins in trait expression. In tomatoes, high-throughput mass spectrometry has been used to identify protein profiles associated with desired traits, such as disease resistance or enhanced nutritional content. This information helps us understand how proteins contribute to these traits and guide breeding decisions.

B. Protein–Protein Interaction (PPI) Networks: Studying PPIs is crucial for understanding how proteins function in complex networks to carry out specific cellular processes. PPIs provide insights into the molecular mechanisms underlying trait expression. Techniques such as yeast two-hybrid screening and co-immunoprecipitation are used to identify and analyze PPIs in tomato breeding. These methods help elucidate the interactions between proteins involved in important traits like stress response and fruit development (Krzemińska et al. 2023).

C. Proteomic Imaging Techniques: Proteomic imaging techniques, such as MALDI imaging mass spectrometry, allow for the visualization of the spatial distribution of proteins within tissues. This technique provides a deeper understanding of protein dynamics in situ (Lu et al. 2021). Proteomic imaging has been used to visualize protein distribution during tomato fruit ripening and stress responses. This approach helps identify where specific proteins are active within the fruit, providing insights into their roles in trait expression (Xie et al. 2021).

13.5.4 Modern and Futuristic Metabolomic Technologies

A. Metabolite Imaging: Metabolite imaging techniques, such as MALDI imaging mass spectrometry, visualize the spatial distribution of metabolites within tissues. This method is crucial for understanding metabolic processes and pathways. In tomatoes, metabolite imaging has been used to study metabolic changes during fruit ripening and stress responses. This technique helps identify the location and abundance of metabolites related to important traits like flavor and nutritional content (Almeida et al. 2021).

B. Comprehensive Metabolite Profiling: Metabolomic techniques such as gas chromatography-mass spectrometry (GC-MS), liquid chromatography-mass spectrometry (LC-MS), and nuclear magnetic resonance (NMR) spectroscopy provide comprehensive profiles of metabolites. These techniques are essential for linking metabolic pathways to specific traits. Metabolomics has been used to identify metabolic pathways linked to traits such as flavor, nutritional content, and stress resistance in tomatoes. This information guides breeding programs to select desirable metabolic profiles. For example, the flavor of tomatoes has been improved by targeting metabolites involved in the synthesis of sugars and acids. Nutritional content has been enhanced by increasing metabolites like vitamins and antioxidants. Stress resistance has been bolstered by identifying and selecting metabolites involved in stress response pathways (Tang et al. 2020).

C. Integration of Metabolomic Data: Integrating metabolomic data with genomic, transcriptomic, and proteomic data creates a multi-omics approach that provides a more comprehensive understanding of biological processes. This integration allows for the correlation of genetic variations, gene expression changes, protein functions, and metabolic outcomes. The benefits of multi-omics integration include enhanced accuracy in trait selection and the development of holistic breeding strategies. Breeders can identify key biomarkers and pathways that contribute to desired traits by combining data from different omics layers. This approach leads to more efficient and effective breeding programs, ultimately resulting in improved tomato varieties with optimal characteristics (Nakamura et al. 2017).

13.6 Omics-Assisted Breeding and Big Data Analytics

13.6.1 Machine Learning and Predictive Modeling

The Role of Machine Learning Algorithms in Analyzing Complex Omics Data Machine learning (ML) algorithms have become essential tools in analyzing complex omics data, revolutionizing the field of genomics-assisted breeding. These algorithms can process large, high-dimensional datasets, uncover patterns, and make predictions that traditional statistical methods cannot handle. By integrating genomics, transcriptomics, proteomics, and metabolomics data, ML can identify associations between genetic markers and phenotypic traits, facilitating more efficient and precise breeding decisions (Reel et al. 2021).

Artificial Intelligence (AI) for Data Analysis: AI and machine learning algorithms are increasingly being applied to analyze large-scale proteomic and metabolomic data. These technologies can identify complex patterns and relationships within the data, aiding in discovering biomarkers and developing predictive models (Chao et al. 2023).

Applications

A. Biomarker Discovery: AI can analyze proteomic and metabolomic data to identify potential biomarkers for disease resistance, fruit quality, and ability for stress tolerance.
B. Predictive Modeling: Machine learning models can predict trait outcomes based on multi-omics data, helping breeders to make informed selection decisions.

13.6.2 Techniques for Integrating Data from Genomics, Transcriptomics, Proteomics, and Metabolomics

A. Multi-Omics Data Integration Techniques

Integrating data from multiple omics platforms is crucial for comprehensively understanding biological systems. Several techniques have been developed to combine these diverse datasets effectively:

(i) Dimensionality Reduction: Methods such as principal component analysis (PCA) and t-distributed stochastic neighbor embedding (t-SNE) reduce the complexity of multi-omics data, making it more manageable for analysis (Reel et al. 2021).
(ii) Network-Based Integration: These methods construct networks representing the relationships between different types of omics data. For example, protein–protein interaction networks can be integrated with gene expression data to uncover the molecular pathways involved in trait expression (Krzemińska et al. 2023).
(iii) Machine Learning Models: Advanced ML techniques, including deep and ensemble learning, can integrate multi-omics data by learning complex patterns and interactions within the datasets. These models are particularly effective for predicting phenotypic outcomes from genotypic data.

13.6.3 Benefits of a Systems-Level Approach in Understanding Complex Trait Expression and Interactions

A systems-level approach to integrating multi-omics data offers several advantages:

A. Holistic View: This approach provides a comprehensive understanding of the biological processes underlying complex traits, considering all levels of molecular data.
B. Identification of Key Regulators: By examining data from multiple omics layers, researchers can identify key regulatory genes, proteins, and metabolites that control important traits.
C. Improved Trait Prediction: Multi-omics integration enhances the accuracy of predictive models, leading to the more reliable selection of desirable traits in breeding programs.

13.6.4 Cloud Computing and Big Data Analytics

The Role of Cloud Computing in Managing and Analyzing Large-Scale Omics Data

Cloud computing has revolutionized managing and analyzing large-scale omics data by providing scalable, flexible, and cost-effective resources. It enables researchers to store vast amounts of data and perform complex analyses without extensive local infrastructure.

A. Scalability: Cloud platforms offer scalable storage and computing resources, allowing researchers to handle large datasets efficiently.

B. Collaboration: Cloud computing facilitates collaboration among researchers by providing a centralized platform for data sharing and joint analysis.
C. Cost-Effectiveness: Cloud services reduce the need for expensive hardware and maintenance, making high-throughput data analysis more accessible (Huang et al. 2023).

13.6.5 Environmental Genomics (Enviromics)

Enviromics involves studying how environmental factors interact with genetic traits to influence phenotypic outcomes. This field integrates environmental data, such as climate, soil conditions, and agricultural practices, with genomic data to understand how different genotypes perform under various conditions.

Enviromics Can Be Integrated with MAS to Develop Climate-Resilient Tomato Varieties

A. Drought Resilience: By combining environmental data on drought conditions with genomic data, researchers can identify genetic markers associated with drought resilience and use MAS to develop varieties that thrive in water-limited environments (Xu 2016).
B. Heat Tolerance: Enviromics can integrate temperature data with genomic and transcriptomic information to identify heat-tolerant genotypes. These markers can then be incorporated into breeding programs to produce tomato varieties that maintain high yield and quality under high temperatures (Raza et al. 2023).

13.6.6 Molecular Breeding Decision Support Systems

Decision support systems (DSS) in molecular breeding integrate omics data, predictive models, and breeding objectives to assist breeders in making informed decisions. These systems provide recommendations on parent selection, cross combinations, and trait selection based on comprehensive data analysis. These Systems Can Aid Breeders in Making Informed Decisions Based on Integrated Omics Data

A. Breeding Program Optimization: DSS can analyze multi-omics data to identify optimal breeding strategies, reducing the time and cost of developing new tomato varieties.
B. Trait Prioritization: By integrating data on multiple traits, DSS can help breeders prioritize traits that align with market demands and environmental conditions, ensuring the development of commercially viable and resilient varieties (Reel et al. 2021).

13.7 Overcoming Challenges and Embracing Future Prospects

13.7.1 Challenges of Integrating 'Omics' with MAS

Integrating 'omics' data with marker-assisted selection (MAS) in tomato breeding presents several significant challenges. One of the primary difficulties lies in the complexity and heterogeneity of 'omics' data. Different 'omics' technologies, such as genomics, transcriptomics, proteomics, and metabolomics, generate data that vary in structure, scale, and type. This diversity complicates data integration, as each 'omics' layer requires specific preprocessing, normalization, and transformation methods to be combined effectively (Ghag et al. 2021). Moreover, the high dimensionality of 'omics' datasets, where the number of variables (genes, proteins, metabolites) far exceeds the number of samples, poses statistical challenges. Techniques such as dimensionality reduction and network-based integration are often employed to manage this complexity, but they are only sometimes straightforward to implement and can sometimes lead to the loss of critical information (Simko et al. 2021).

Additionally, ensuring the quality and consistency of data across different 'omics' platforms is crucial. Variability in experimental conditions, sample handling, and technological platforms can introduce biases and systematic errors, making it challenging to draw reliable conclusions from integrated datasets (Mahmood et al. 2022). Another major challenge is interpreting integrated 'omics' data to derive biologically meaningful insights. While the integration aims to provide a holistic view of the biological systems and improve the accuracy of marker selection, the resulting data complexity can be overwhelming. Advanced bioinformatics tools and machine learning algorithms are required to uncover patterns and associations within multi-layered 'omics' data (Tiwari et al. 2022). However, these methods often necessitate high computational power and sophisticated analytical skills, which may only be readily available in some breeding programs.

Moreover, translating multi-omics data into actionable markers for MAS involves validating the biological relevance of identified markers in diverse genetic backgrounds and environmental conditions. This validation process is resource-intensive and time-consuming. There is also a need for standardized protocols and frameworks to guide the integration and interpretation processes, ensuring reproducibility and consistency across studies. Despite these challenges, the potential benefits of integrating 'omics' data with MAS—such as enhanced precision in selecting desirable traits and a deeper understanding of these traits' genetic and biochemical basis—make it a worthwhile endeavor. Continued advancements in 'omics' technologies, data integration methodologies, and computational tools are essential to overcoming these challenges and fully realizing the potential of integrated 'omics' and MAS in tomato breeding (Yang et al. 2021).

13.7.2 Innovations and Future Directions

The future of marker-assisted selection (MAS) in tomato breeding is poised for significant transformation, driven by innovations in 'omics' technologies and big data analytics. One of the promising directions is the enhanced integration of multi-omics data using advanced machine learning and artificial intelligence (AI) techniques. Machine learning algorithms, such as deep learning and ensemble learning, can process large and complex 'omics' datasets to identify patterns and accurately predict trait outcomes. These algorithms can handle the high dimensionality and heterogeneity of 'omics' data, making them ideal for integrating genomics, transcriptomics, proteomics, and metabolomics information. For instance, deep generative models have been employed to create joint embeddings of single-cell multi-omics data, providing comprehensive insights into the molecular mechanisms underlying complex traits. These models can help predict how different genetic and environmental factors influence trait expression, facilitating more precise and efficient selection in breeding programs (Ghag et al. 2021).

Another critical innovation is the application of cloud computing and big data analytics in managing and analyzing large-scale 'omics' data. Cloud platforms offer scalable storage and computational resources, enabling researchers to handle the vast amounts of data generated by high-throughput 'omics' technologies (Amoroso et al. 2022). This scalability is essential for conducting comprehensive multi-omics studies integrating data from multiple sources. Additionally, big data analytics tools can process and analyze these datasets efficiently, uncovering complex biological interactions and identifying key genetic markers for MAS. For example, cloud-based platforms can integrate environmental data with 'omics' data to study the interactions between tomato genotypes and their growing conditions, enhancing the development of climate-resilient varieties. Integrating environmental genomics, or enviromics, with MAS, can help breeders develop tomato varieties that are not only high-yielding but also capable of thriving under diverse environmental stresses. This approach combines genomic data with environmental information, providing a holistic view of how different factors influence plant performance (Mahmood et al. 2022). Furthermore, the development of molecular breeding decision support systems, which integrate multi-omics data and predictive modeling, can assist breeders in making informed decisions. These systems use advanced algorithms to analyze complex datasets and provide recommendations on the best breeding strategies, improving breeding programs' efficiency and success rate (Yang et al. 2021).

By incorporating these recent advancements and leveraging cutting-edge technologies, the integration of 'omics' data with MAS in tomato breeding holds great promise for revolutionizing the field and achieving significant improvements in crop quality and resilience (Tiwari et al. 2022).

13.8 Conclusion: The Next Frontier of MAS in the 'Omics' Era

Integrating marker-assisted selection (MAS) with advanced 'omics' technologies marks a transformative frontier in tomato breeding, offering numerous benefits and profound impacts on crop improvement. This synergy enhances the precision and efficiency of breeding programs by leveraging comprehensive datasets from genomics, transcriptomics, proteomics, and metabolomics, providing a multi-dimensional understanding of trait development and expression. The benefits include accelerated breeding cycles, increased accuracy in selecting desirable traits, and the ability to dissect complex genetic traits that were previously challenging to address. These integrated approaches facilitate the identification of robust genetic markers and validate their functional relevance, ensuring the development of superior tomato varieties with enhanced yield, quality, and resilience to biotic and abiotic stresses. As we move into the era of precision breeding, the future trajectory of MAS will increasingly rely on cutting-edge innovations such as machine learning, AI-driven predictive modeling, and cloud-based big data analytics. These technologies will refine the selection process, enabling breeders to predict trait outcomes with unprecedented accuracy and tailor breeding strategies to specific environmental conditions. The adoption of enviromics will also play a crucial role in developing climate-resilient crops, and addressing the challenges of global climate change. A collaborative effort is essential to harness the potential of these advancements fully. Researchers, breeders, and policymakers must work together to develop standardized protocols, share data and insights, and promote the adoption of integrated MAS and 'omics' approaches in breeding programs worldwide. This collaboration will not only enhance the sustainability and efficiency of crop improvement practices but also ensure that the benefits of these technological advancements are accessible to all, ultimately contributing to global food security and agricultural sustainability. This call to action underscores the need for continued investment in research and development, fostering an environment where innovation can thrive and drive the next generation of breakthroughs in tomato breeding and beyond.

References

Abreu AC, Fernández I (2020) NMR metabolomics applied on the discrimination of variables influencing tomato (*Solanum lycopersicum*). Molecules (Basel, Switzerland) 25(16):3738. https://doi.org/10.3390/molecules25163738

Adhikari P, McNellie J, Panthee D (2020) Detection of quantitative trait loci (QTL) associated with the fruit morphology of tomato. Genes 11:1117. https://doi.org/10.3390/genes11101117

Aglawe SB, Verma AK, Upadhyay AK (2020) Bioinformatics tools and databases for genomics-assisted breeding and population genetics of plants: a review. Curr Bioinforma 16(6):766–773. https://doi.org/10.2174/1574893615999200831144028

Almeida J, Perez-Fons L, Fraser PD (2021) A transcriptomic, metabolomic, and cellular approach to the physiological adaptation of tomato fruit to high temperature. Plant Cell Environ 44(7):2211–2229. https://doi.org/10.1111/pce.13854

Amoroso CG, Panthee DR, Andolfo G, Ramìrez FP, Ercolano MR (2022) Genomic tools for improving tomato to biotic stress resistance. In: Kole C (ed) Genomic designing for biotic stress resistant vegetable crops. Springer, Cham. https://doi.org/10.1007/978-3-030-97785-6_1
Bai C, Wu C, Ma L, Fu A, Zheng Y, Han J, Li C, Yuan S, Zheng S, Gao L, Zhang X, Wang Q, Meng D, Zuo J (2023) Transcriptomics and metabolomics analyses provide insights into postharvest ripening and senescence of tomato fruit under low temperature. Hortic Plant J 9(1):109–121. https://doi.org/10.1016/j.hpj.2021.09.001
Cao ZJ, Gao G (2022) Multi-omics single-cell data integration and regulatory inference with graph-linked embedding. Nat Biotechnol 40(11):1458–1466. https://doi.org/10.1038/s41587-022-01284-4
Causse M, Bénéjam J, Bineau E, Bitton F, Brault M, Carretero Y, Desaint H, Hereil A, Pellegrino K, Pelpoir E, Zhao J (2022) Genetic control of tomato fruit quality: from QTL mapping to genome-wide association studies and breeding. C R Biol 345(4):3–13. https://doi.org/10.5802/crbiol.99
Celik I (2023) In Silico integrated analysis of genomic, transcriptomic, and proteomic data reveals QTL-specific genes for bacterial canker resistance in tomato (Solanum lycopersicum L.). Curr Issues Mol Biol 45:1387–1395. https://doi.org/10.3390/cimb45020090
Chao H, Zhang S, Hu Y, Ni Q, Xin S, Zhao L, Ivanisenko V, Orlov Y, Chen M (2023) Integrating omics databases for enhanced crop breeding. J Integr Bioinform 20(4):20230012. https://doi.org/10.1515/jib-2023-0012
Chaudhary J, Khatri P, Singla P, Kumawat S, Kumari A, Vikram A, Jindal SK, Kardile H, Kumar R, Sonah H, Deshmukh R (2019) Advances in omics approaches for abiotic stress tolerance in tomato. Biology 8(4). https://doi.org/10.3390/biology8040090
Chaudhuri A, Halder K, Datta A (2022) Classification of CRISPR/Cas system and its application in tomato breeding. Theor Appl Genet 135(2):367–387. https://doi.org/10.1007/s00122-021-03984-y
Chen G, Ning B, Shi T (2019) Single-cell RNA-Seq technologies and related computational data analysis. Front Genet 10:317. https://doi.org/10.3389/fgene.2019.00317
Colantonio V, Rajagopalan N, Williams MC, Brock T, Goulet KM, Jones AD, Tieman DM (2022) Metabolomic selection for enhanced fruit flavor. Plant J 110(4):1152–1165. https://doi.org/10.1111/tpj.15654
Collard BC, Mackill DJ (2008) Marker-assisted selection: an approach for precision plant breeding in the twenty-first century. Philos Trans R Soc Lond Ser B Biol Sci 363(1491):557–572. https://doi.org/10.1098/rstb.2007.2170
Collard BCY, Jahufer MZZ, Brouwer JB et al (2005) An introduction to markers, quantitative trait loci (QTL) mapping and marker-assisted selection for crop improvement: the basic concepts. Euphytica 142(1):169–196. https://doi.org/10.1007/s10681-005-1681-5
Costa-Silva J, Domingues D, Lopes FM (2017) RNA-Seq differential expression analysis: an extended review and a software tool. PLoS One 12(12):e0190152. https://doi.org/10.1371/journal.pone.0190152
Di Giacomo M, Vega TA, Cambiaso V, Picardi LA, Rodríguez GR, Pereira da Costa JH (2023) An integrative transcriptomics and proteomics approach to identify putative genes underlying fruit ripening in tomato near isogenic lines with long shelf life. Plants (Basel, Switzerland) 12(15):2812. https://doi.org/10.3390/plants12152812
di Rienzo V, Bubici G, Montemurro C, Cillo F (2018) Rapid identification of tomato Sw-5 resistance-breaking isolates of tomato spotted wilt virus using high-resolution melting and TaqMan SNP genotyping assays as allelic discrimination techniques. PLoS One 13(4):e0196738. https://doi.org/10.1371/journal.pone.0196738
Ding H, Mo S, Qian Y, Yuan G, Wu X, Ge C (2020) Integrated proteome and transcriptome analyses revealed key factors involved in tomato (Solanum lycopersicum) under high temperature stress. Food Energy Secur 9(4):e239. https://doi.org/10.1002/fes3.239
Foolad M, Panthee D (2012) Marker-assisted selection in tomato breeding. Crit Rev Plant Sci 31:123–193. https://doi.org/10.1080/07352689.2011.616057

Ghag SB, Ganapathi T, Jain SM, Penna S (2021) Omics technologies and breeding of horticultural crops. In: Omics in horticultural crops. Academic Press, pp 75–90. https://doi.org/10.1016/B978-0-323-89905-5.00024-0

Huang L, Song M, Shen H, Hong H, Gong P, Deng H-W, Zhang C (2023) Deep learning methods for omics data imputation. Biology 12(10):1313. https://doi.org/10.3390/biology12101313

Jain A, Singh H, Das S (2020) Deciphering plant-microbe crosstalk through proteomics studies. Microbiol Res 242:126590. https://doi.org/10.1016/j.micres.2020.126590

Jiang H, Waseem M, Wang Y, Basharat S, Zhang X, Li Y, Liu P (2023) Development of simple sequence repeat markers for sugarcane from data mining of expressed sequence tags. Front Plant Sci 14:1199210. https://doi.org/10.3389/fpls.2023.1199210

Krzemińska A, Miller T, Nowakowska A, Adamski C, Radaczyńska M, Mikiciuk G, Mikiciuk M (2023) An integrated multi-omics and artificial intelligence framework for advanced plant phenotyping in horticulture. Biology 12(10):1298. https://doi.org/10.3390/biology12101298

Lamin-Samu AT, Farghal M, Ali M, Lu G (2021) Morpho-physiological and transcriptome changes in tomato anthers of different developmental stages under drought stress. Cells 10(7):1809. https://doi.org/10.3390/cells10071809

Li R, Maioli A, Yan Z, Bai Y, Valentino D, Milani AM, Pompili V, Comino C, Lanteri S, Moglia A, Acquadro A (2022) CRISPR/Cas9-based knock-out of the PMR4 gene reduces susceptibility to late blight in two tomato cultivars. Int J Mol Sci 23(23):14542. https://doi.org/10.3390/ijms232314542

Liu D, Yang L, Zhang J, Zhu G, Lü H, Lü Y, Wang Y, Cao X, Sun T, Huang S, Wu Y (2020) Domestication and breeding changed tomato fruit transcriptome. J Integr Agric 19:120–132. https://doi.org/10.1016/s2095-3119(19)62824-8

López de Maturana E, Alonso L, Alarcón P, Martín-Antoniano IA, Pineda S, Piorno L, Calle ML, Malats N (2019) Challenges in the integration of omics and non-omics data. Genes 10(3):238. https://doi.org/10.3390/genes10030238

Mahmood U, Li X, Fan Y, Chang W, Niu Y, Li J, Qu C, Lu K (2022) Multi-omics revolution to promote plant breeding efficiency. Front Plant Sci 13. https://doi.org/10.3389/fpls.2022.1062952

Maria M, Pouyanfar N, Örd T, Kaikkonen MU (2022) The power of single-cell RNA sequencing in eQTL discovery. Genes 13(3):502. https://doi.org/10.3390/genes13030502

Momo J, Rawoof A, Kumar A, Islam K, Ahmad I, Ramchiary N (2022) Proteomics of reproductive development, fruit ripening, and stress responses in tomato. J Agric Food Chem. https://doi.org/10.1021/acs.jafc.2c06564

Nakamura J, Morikawa-Ichinose T, Fujimura Y et al (2017) Spatially resolved metabolic distribution for unraveling the physiological change and responses in tomato fruit using matrix-assisted laser desorption/ionization–mass spectrometry imaging (MALDI–MSI). Anal Bioanal Chem 409(7):1697–1706. https://doi.org/10.1007/s00216-016-0118-4

Osei M, Prempeh R, Adjebeng-Danquah J, Opoku J, Danquah A, Danquah E, Blay E, Adu-Dapaah H (2018) Marker-assisted selection (MAS): a fast-track tool in tomato breeding. In: Recent advances in tomato breeding and production. https://doi.org/10.5772/INTECHOPEN.76007

Powell W, Morgante M, Andre C et al (1996) The comparison of RFLP, RAPD, AFLP, and SSR (microsatellite) markers for germplasm analysis. Mol Breed 2:225–238. https://doi.org/10.1007/BF00564200

Raza A, Bashir S, Khare T, Karikari B, Copeland RG, Jamla M, Abbas S, Charagh S, Nayak SN, Djalovic I, Rivero RM, Siddique KH, Varshney RK (2023) Temperature-smart plants: a new horizon with omics-driven plant breeding. Physiol Plant 176(1):e14188. https://doi.org/10.1111/ppl.14188

Razzaq A, Wishart DS, Wani SH, Hameed MK, Mubin M, Saleem F (2022) Advances in metabolomics-driven diagnostic breeding and crop improvement. Meta 12(6):511. https://doi.org/10.3390/metabo12060511

Reel PS, Reel S, Pearson E, Trucco E, Jefferson E (2021) Using machine learning approaches for multi-omics data analysis: a review. Biotechnol Adv 49:107739. https://doi.org/10.1016/j.biotechadv.2021.107739

Rodriguez M, Scintu A, Posadinu C, Xu Y, Nguyen C, Sun H, Bitocchi E, Bellucci E, Papa R, Fei Z, Giovannoni J, Rau D, Attene G (2020) GWAS based on RNA-Seq SNPs and high-throughput phenotyping combined with climatic data highlights the reservoir of valuable genetic diversity in regional tomato landraces. Genes 11:1387. https://doi.org/10.3390/genes11111387

Salava H, Thula S, Mohan V, Kumar R, Maghuly F (2021) Application of genome editing in tomato breeding: mechanisms, advances, and prospects. Int J Mol Sci 22(2):682. https://doi.org/10.3390/ijms22020682

Schalkwyk AV, Wenzl P, Smit S, Lopez-Cobollo R, Kilian A, Bishop G, Hefer C, Berger DK (2012) Bin mapping of tomato diversity array (DArT) markers to genomic regions of Solanum lycopersicum × Solanum pennellii introgression lines. Theor Appl Genet 124(5):947–956. https://doi.org/10.1007/s00122-011-1759-5

Scheben A, Batley J, Edwards D (2017) Genotyping-by-sequencing approaches to characterize crop genomes: choosing the right tool for the right application. Plant Biotechnol J 15(2):149–161. https://doi.org/10.1111/pbi.12645

Severo R, Shibutani LJS, Silva GF, Guimarães SSC, Neto FDA, Beserra JEA Jr, Souza JFF, Melo MP (2024) Fusarium species causing root rot and wilt in tomato in Brazil. J Phytopathol 172:e13261. https://doi.org/10.1111/jph.13261

Shah LR, Ahmed N, Hussain K, Mansoor S, Khan T, Khan I, Narayan S, Afroza B, Murtaza I, Shikari AB, Bhat B, Masoodi KZ (2024) Mapping phenotypic performance and novel SNPs for cold tolerance in tomato (Solanum lycopersicum) genotypes through GWAS and population genetics. BMC Genom Data 25(1):9. https://doi.org/10.1186/s12863-024-01190-5

Shinozaki Y, Nicolas P, Fernandez-Pozo N, Ma Q, Evanich DJ, Shi Y, Xu Y, Zheng Y, Snyder SI, Martin LBB, Ruiz-May E, Thannhauser TW, Chen K, Domozych DS, Catalá C, Fei Z, Mueller LA, Giovannoni JJ, Rose JKC (2018) High-resolution spatiotemporal transcriptome mapping of tomato fruit development and ripening. Nat Commun 9(1):364. https://doi.org/10.1038/s41467-017-02782-9

Shirasawa K, Hirakawa H (2013) DNA marker applications to molecular genetics and genomics in tomato. Breed Sci 63:21–30. https://doi.org/10.1270/jsbbs.63.21

Simko I, Jia M, Venkatesh J, Kang BC, Weng Y, Barcaccia G et al (2021) Genomics and marker-assisted improvement of vegetable crops. Crit Rev Plant Sci 40(4):303–365. https://doi.org/10.1080/07352689.2021.1941605

Song L, Wang Y, Zhao L, Zhao T (2022) Transcriptome profiling unravels the involvement of phytohormones in tomato resistance to the Tomato Yellow Leaf Curl Virus (TYLCV). Horticulturae 8(2):143. https://doi.org/10.3390/horticulturae8020143

Tang H, Zhang X, Gong B, Yan Y, Shi Q (2020) Proteomics and metabolomics analysis of tomato fruit at different maturity stages and under salt treatment. Food Chem 311:126009. https://doi.org/10.1016/j.foodchem.2019.126009

Tanksley SD, Young ND, Paterson AH, Bonierbale MW (1989) RFLP mapping in plant breeding: new tools for an old science. Bio/Technology 7(3):257–264. https://doi.org/10.1038/nbt0389-257

Tiwari J, Yerasu S, Rai N, Singh D, Singh A, Karkute S, Singh P, Behera T (2022) Progress in marker-assisted selection to genomics-assisted breeding in tomato. Crit Rev Plant Sci 41:321–350. https://doi.org/10.1080/07352689.2022.2130361

Tiwari JK, Singh AK, Behera TK (2023) CRISPR/Cas genome editing in tomato improvement: advances and applications. Front Plant Sci 11:1121209. https://doi.org/10.3389/fpls.2023.1121209

Tomato Genome Consortium (2012) The tomato genome sequence provides insights into fleshy fruit evolution. Nature 485(7400):635–641. https://doi.org/10.1038/nature11119

Tripodi P, Francese G (2023) An integrated genomic and biochemical approach to investigate the potentiality of heirloom tomatoes: breeding resources for food quality and sustainable agriculture. Front Plant Sci 13:1031776. https://doi.org/10.3389/fpls.2022.1031776

Valsala Sudarsanan SK, Sreekumar N (2024) Bioinformatic resources for plant genomic research. Curr Bioinforma 19(6):513–529. https://doi.org/10.2174/1574893618666230725123211

Van Den Berge K, Hembach K, Soneson C, Tiberi S, Clement L, Love MI, Patro R, Robinson M (2018) RNA sequencing data: hitchhiker's guide to expression analysis. PeerJ Preprints 6:e27283v2. https://doi.org/10.7287/peerj.preprints.27283v2

Varshney RK, Terauchi R, McCouch SR (2014) Harvesting the promising fruits of genomics: applying genome sequencing technologies to crop breeding. PLoS Biol 12(6):e1001883. https://doi.org/10.1371/journal.pbio.1001883

Wan L, Wang Z, Tang M, Hong D, Sun Y, Ren J, Zhang N, Zeng H (2021) CRISPR-Cas9 gene editing for fruit and vegetable crops: strategies and prospects. Horticulturae 7(7):193. https://doi.org/10.3390/horticulturae7070193

Wang B, Li N, Huang S, Hu J, Wang Q, Tang Y, Yang T, Asmutola P, Wang J, Yu Q (2021) Enhanced soluble sugar content in tomato fruit using CRISPR/Cas9-mediated SlINVINH1 and SlVPE5 gene editing. PeerJ 9:e12478. https://doi.org/10.7717/peerj.12478

Wei C, Liang J, Wang R, Chi L, Wang W, Tan J, Shi H, Song X, Cui Z, Xie Q, Cheng D, Wang X (2023) Response of bacterial community metabolites to bacterial wilt caused by *Ralstonia solanacearum*: a multi-omics analysis. Front Plant Sci 14:1339478. https://doi.org/10.3389/fpls.2023.1339478

Xie Q, Tian Y, Hu Z, Zhang L, Tang B, Wang Y (2021) Roles and phosphorylation modification regulation mechanisms of tomato (Solanum lycopersicum) fruit ripening revealed by integrative proteomics and phosphoproteomics. Int J Mol Sci 22(21):11782. https://doi.org/10.3390/ijms222111782

Xu Y (2016) Envirotyping for deciphering environmental impacts on crop plants. Theor Appl Genet 129:653–673. https://doi.org/10.1007/s00122-016-2691-5

Yang J, Wang Y, Shen H, Yang W (2014) In silico identification and experimental validation of insertion–deletion polymorphisms in tomato genome. DNA Res 21:429–438. https://doi.org/10.1093/dnares/dsu008

Yang W, Feng H, Zhang X, Zhang J, Doonan J, Batchelor W, Xiong L, Yan J (2020) Crop phenomics and high-throughput phenotyping: past decades, current challenges and future perspectives. Mol Plant 13(2):187–214. https://doi.org/10.1016/j.molp.2020.01.008

Yang Y, Saand MA, Huang L, Abdelaal WB, Zhang J, Wu Y, Li J, Sirohi MH, Wang F (2021) Applications of multi-omics technologies for crop improvement. Front Plant Sci 12:563953. https://doi.org/10.3389/fpls.2021.563953

Zeiss DR, Mhlongo MI, Tugizimana F, Steenkamp PA, Dubery IA (2018) Comparative metabolic phenotyping of tomato (*Solanum lycopersicum*) for the identification of metabolic signatures in cultivars differing in resistance to *Ralstonia solanacearum*. Int J Mol Sci 19(9):2558. https://doi.org/10.3390/ijms19092558

Zhu F, Jadhav SS, Tohge T, Salem MA, Lee JM, Giovannoni JJ, Cheng Y, Alseekh S, Fernie AR (2022a) A comparative transcriptomics and eQTL approach identifies SlWD40 as a tomato fruit ripening regulator. Plant Physiol 190(1):250–266. https://doi.org/10.1093/plphys/kiac200

Zhu F, Wen W, Cheng Y et al (2022b) The metabolic changes that affect fruit quality during tomato fruit ripening. Mol Hortic 2(2). https://doi.org/10.1186/s43897-022-00024-1

Metabolomics for Flavor and Nutrition: Enhancing Tomato Quality Through "Omics"

14

Muhammad Saeed, Ghulam Mustafa, Faiz Ahmad Joyia, Aneela Shadab, Mishal Subhan, Kousar Parveen, and Aqsa Parvaiz

Abstract

In this chapter, we delve into the burgeoning field of metabolomics applied to enhance the flavor and nutritional quality of tomatoes. Metabolomics offers a comprehensive approach to understanding the biochemical composition of tomatoes, encompassing the intricate interplay between genes, proteins, and metabolites. By employing cutting-edge "OMICS" technologies, including genomics, transcriptomics, proteomics, and metabolomics, we can gain insights into the complex molecular networks governing tomato flavor and nutritional attributes. Our discussion centers on how metabolomics can identify key metabolites associated with desirable flavor profiles and nutritional benefits in tomatoes. We explore the impact of genetic factors, environmental conditions, and agronomic practices on tomato metabolite composition. Furthermore, we highlight innovative strategies for optimizing tomato quality through targeted metabolic engineering and breeding approaches. It also underscores the significance of metabolomics in unraveling the biochemical pathways that contribute to specific flavor compounds and health-promoting constituents in tomatoes. By integrating metabolomics data with other "OMICS" data sets, we aim to advance our understanding of tomato metabolism and facilitate the development of superior tomato varieties with enhanced flavor and nutritional attributes. Overall, this chapter provides a comprehensive overview of the application of metabolomics in the context of tomato quality enhancement, emphasizing its potential to revolution-

M. Saeed · G. Mustafa · F. A. Joyia · A. Shadab · K. Parveen
Centre of Agricultural Biochemistry and Biotechnology (CABB), University of Agriculture, Faisalabad, Pakistan

M. Subhan · A. Parvaiz (✉)
Department of Biochemistry and Biotechnology, The Women University Multan, Multan, Pakistan
e-mail: aqsa.6497@wum.edu.pk

M. Waseem et al. (eds.), *Omics Approaches for Tomato Yield and Quality Trait Improvement*, https://doi.org/10.1007/978-981-96-3890-1_14

ize tomato breeding and cultivation practices for improved consumer satisfaction and health outcomes.

Keywords

Omics · Tomatoes · Flavor · Metabolites · Breeding

14.1 Introduction to Metabolomics and Tomato Quality

14.1.1 Overview of Metabolomics

Metabolomics can be described as the capability to analyze various small molecules, or metabolites, found in a cell, tissue, or organism. This field entails the identification of small molecules in biological samples and typically interfaces with techniques such as GC-MS, LC-MS, and nuclear magnetic resonance (NMR) spectroscopy. Chemometrics when combined with these sophisticated analytical tools has greatly improved the comprehension of the global metabolome and its related pathway systems.

In metabolomics, hypothesis can be either untargeted or targeted based on the questions under experimentation and goals set. Untargeted studies' objectives are less selective and are known as metabolic profiling or metabolic profiling; this way, it is possible to simultaneously assess the global metabolome and various chemical classes of metabolites related to different pathways. On the other hand, hypothesis-driven analysis is concerned with specific metabolite or biological pathways for which an investigator has developed a hypothesis. In some cases, both the approaches are required to get the complete information (Hopfer et al. 2015).

Over the last few decades, developments in the technologies of analytical devices and data processing tools have furthered the utilization of metabolomics across different disciplines. However, metabolomics is still considered as an emergent discipline, and already it has gained a lot of attention in the agri-food domain for uses in food transformation, assessing food quality, crop improvement, and innovation of new products. However, to date, metabolomics has been rarely implemented intentionally by the regulatory agencies in the assessment of the quality of foods, to which these methods have been proven to be superior to the ones traditionally used in some cases. For example, metabolomics has extended its application to illustrate the identification of the aroma in both fresh and processed plant-based products (PBPs). Most of the current food metabolomic studies are related to the quality evaluation of processed or semi-processed foods and only a few are concerned with fresh produce; even in the case of fresh produce, most of the focus is centered on economically important PBPs or crops with special uses (Botella-Pavía et al. 2010).

14.2 Metabolomics for Evaluating Flavor-Associated Metabolites in Fresh PBPs

Concerning the metabolomics application in the agri-food sector, these strategies are used in different fields for numerous purposes. Specifically for one application the metabolites associated with the flavor, which with regard to fresh PBPs can be affected by biochemical properties. Taste plays an important role in influencing consumer's choices and patterns of consumption, which gives the accent to the search for flavor-related metabolites in PBPs. Flavor or taste sensation clearly involves both the volatile aroma metabolites and the non-aroma or taste metabolites which are in different classes (Coyago-Cruz et al. 2019).

In fresh plant-based products (PBPs), a large number of volatile chemical compounds participate in their natural odor, thus increasing the difficulty of these aroma-related metabolites. This is further compounded by the fact that various (Library of Identified Volatile Compounds) LIVCs lose or gain aromas when they interact with other LIVCs, creating library individuality that cannot be deduced merely by the addition of all volatile compounds displayed in the LIVC library. Currently, over seven thousand volatile compounds have been described in foods; however, it has turned out that only 300–400 separate components and their concentrations in the composition are mainly responsible for the odor of a particular product.

Fresh PBPs have various volatile aroma metabolites belonging to esters, alcohols, aldehydes, ketones, lactones, terpenoids, and apocarotenoids. Among these, amino acid derivatives, lipids, phenolic acids, and sesquiterpenes have been times a relevant role to play in the aroma of PBPs (Liu et al. 2018). However, in some of the PBPs particularly fruits and vegetables, sulfur-containing compounds and their conjugates also participate in their flavors. Most of the flavor-active compounds that are liberated in the fresh PBPs can be sourced from phytochemicals that fall under fatty acids, amino acids, carotenes, and terpenes and by going through not more than four major biosynthetic pathways. Most of these pathways are more or less aligned in their role of creating the skeleton of the volatile compounds, although there are additional steps in the pathways that introduce more complexity to the chains and related transformations. The fatty acid-derived volatiles including the alcohols, esters, ketones, acids, and lactones are supposed to be the character impact compounds and take part in the contribution of the fresh fruits flavor, which is generated through α oxidation of fatty acids, β oxidation of fatty acids and the lipoxygenase pathway (Chong et al. 2018).

14.3 Framework for Integrating Metabolomics with Current State-of-the-Art Technologies

Metabolomics with some of today's developed approaches to quality assessment is a process executed in several stages, where the implementation of several technological tools. To optimize metabolomics, the data gathered from different platforms must be compiled and the result must be properly understood. Despite the fact that this integration can prove to be quite helpful in most categories of PBPs and the types of information that necessitate processing, it can pose a number of technical difficulties. Some of the difficulties are as follows: (i) The access to suitable instrumental platforms or extraction is under investigation (Eichelmann et al. 2016).

The continuous development of extraction and powerful analytical techniques have offered several possibilities in addressing the chemically diverse metabolites. The first technical obstacle that is usually encountered can be resolved by way of rounds of tests and adjustments. Nonetheless, the second and the third challenges—the absence of reference databases and the limited and sparse information on metabolome, are considerably more rigorous and impact not only the agri-food sector of study but also other scientific disciplines. Due to this shortage of information, data analysis and the achievement of insights turn out to be challenging (Mata et al. 2021).

Few years back, several focused databases have been constructed for various phytochemicals and bioactive metabolites in plant sciences to support researchers. The plant metabolome's high diversity, encompassing thousands of metabolites, makes annotating each metabolite a labor-intensive and technically challenging task. To address this, efforts can be focused on identifying key candidate metabolites from various types.

Nevertheless, this realization is effective when used for preliminary selection, detailed investigations are suggested since their outcomes are numerically based (Gerbens-Leenes et al. 2010).

Apart from these socio-technical factors, there is a need for a common approach on how these platforms' data can be integrated and summarized in order to derive maximum benefits common in the sensory evaluation of fresh PBPs from multi-platform approaches. In the last couple of years, there has been a lot of development in the aspect of data integration and chemometrics pipeline which enabled easy production of biological outputs from different platforms. A large number of tools have been created in popular languages like Python, R, and Matlab® to help with the metabolomics data curation. In the same regard, other open-access structures such as GitHub and Stack Overflow offer grounds through which scripts and workflows can be shared. Self-service and cognitive data integration that leverages artificial intelligence (AI)/machine learning (ML) is also being established.

Data integration platforms with e-noses and e-tongues along with high-resolution MSs, and other analytical instruments can supplement the current sensory tests by humans and metabolic estimations. Despite the fact that these artificial sensory techniques are unable to mimic the real human taste and smell, they are capable of providing data in large amounts with rather high accuracy. If one has the required computation power one can amalgamate discrete data from these artificial sensors and come up with something fruitful (Goff and Klee 2006).

This integration is more helpful where there is novation and creation of new products for instance with the changing world's palate as concerns foods. Also, connecting these platforms with state-of-the-art artificial intelligence makes them become smart sensing system that can forecast new opportunities in food safety risks including adulterants and pathogens. Realizing this will therefore entail

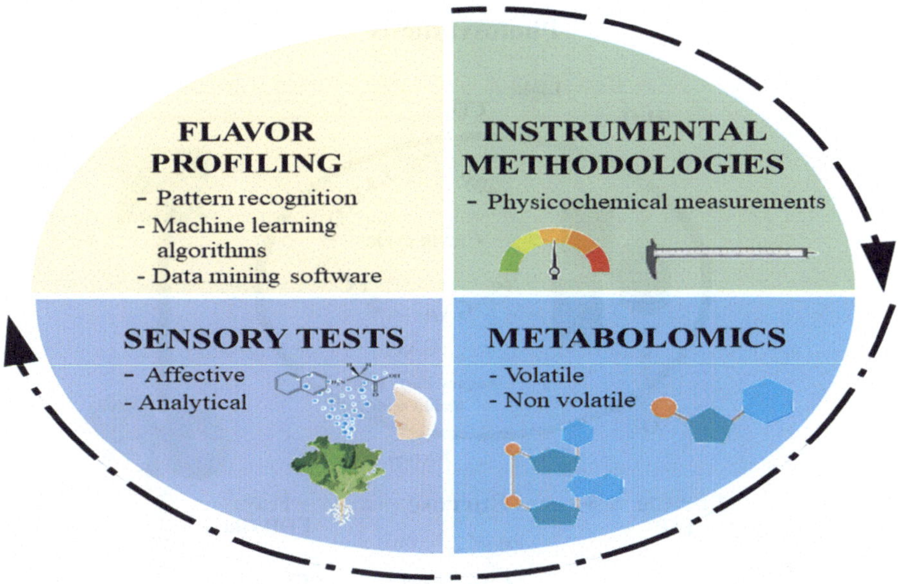

Fig. 14.1 Strategy for the incorporation of metabolomics in the organoleptic analysis of new fresh PBPs with the use of sophisticated technologies. While the physicochemical tests give a general approximation, the metabolomics, and sensory-based tests are more precise for the analysis of fresh PBPs' flavor for a correct classification. Data integration platforms are crucial as they allow the efficient aggregation of data in order to draw useful conclusions. Precisely, it explains the intensity of nutrient and flavor build up in influencing tomato fruit consumption

collective action and integration of remedies and approaches from multiple players in this fast-growing area (Fig. 14.1).

14.4 Importance of Flavor and Nutritional Quality in Tomatoes

Tomato flavor is derived from a complex combination of taste metabolites and volatile compounds (Wishart et al. 2018). Carotenoids, polyphenols, and vitamin C for instance are antioxidants that maintain the balance of biochemical reactions in the body and also reduce the incidence of chronic diseases (Tilahun et al. 2017). Progress in the field has enhanced comprehension of the biosynthesis, and the regulatory elements of these factors in tomatoes. Some of them include; SlNAC1, SlCCD, SlbHLH22, and SlHY5 which have been proven to play the role of promoting the levels of nutrient accumulation (Waseem et al. 2019). However, it must be noted that the regulation of nutrient metabolism during ripening is quite intricate, and the factors defining the content of fruits are still rather vague and warrant further research (McMacken and Shah 2017). Besides nutritional attributes, customers also pay much attention to the taste qualities of tomatoes. However, consumers have

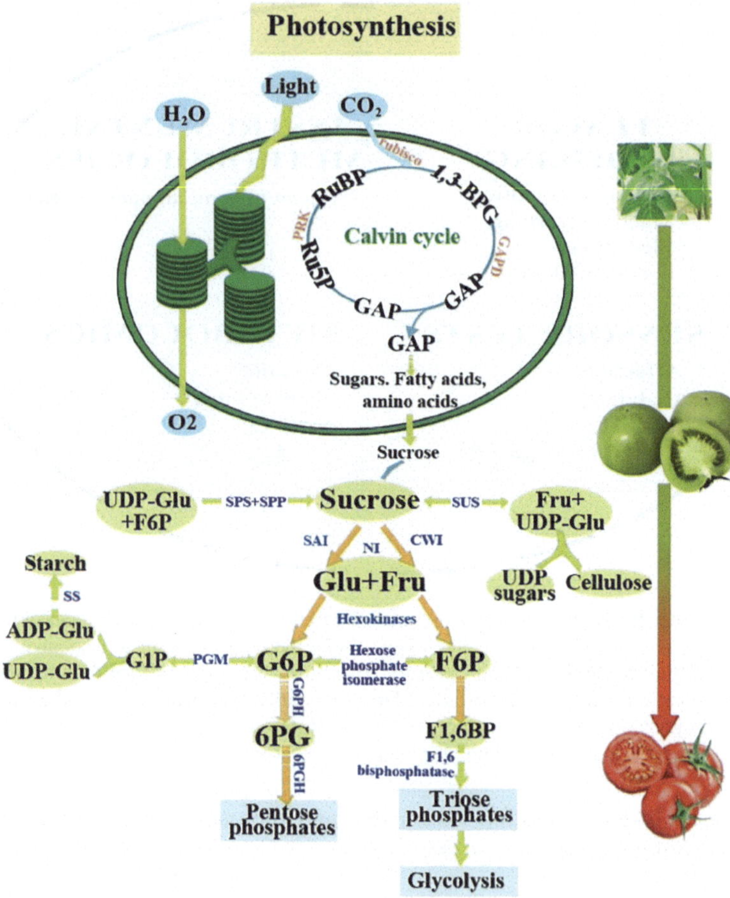

Fig. 14.2 Plant sugar metabolism

noticed a decline in nutrient content and flavor in many tomato varieties due to the focus on fruit yield and size. Consequently, improving nutritional and flavor quality is a primary goal of many breeding programs. To effectively enhance these qualities in tomato (Manach et al. 2004) (Figs. 14.2 and 14.3).

Among these, tomato has received the most extensive research focus. This summary highlights recent advancements in tomato flavor research, particularly emphasizing the integration of machine learning techniques to enhance these studies. It also underscores the benefits of integrating multi-omics strategies to elucidate the genetic foundations of tomato flavor. Finally, it discusses modern molecular marker breeding techniques based on metabolomics, coupled with machine learning approaches, to advance the genetic enhancement of crop nutrition and flavor quality (Mathieu et al. 2009).

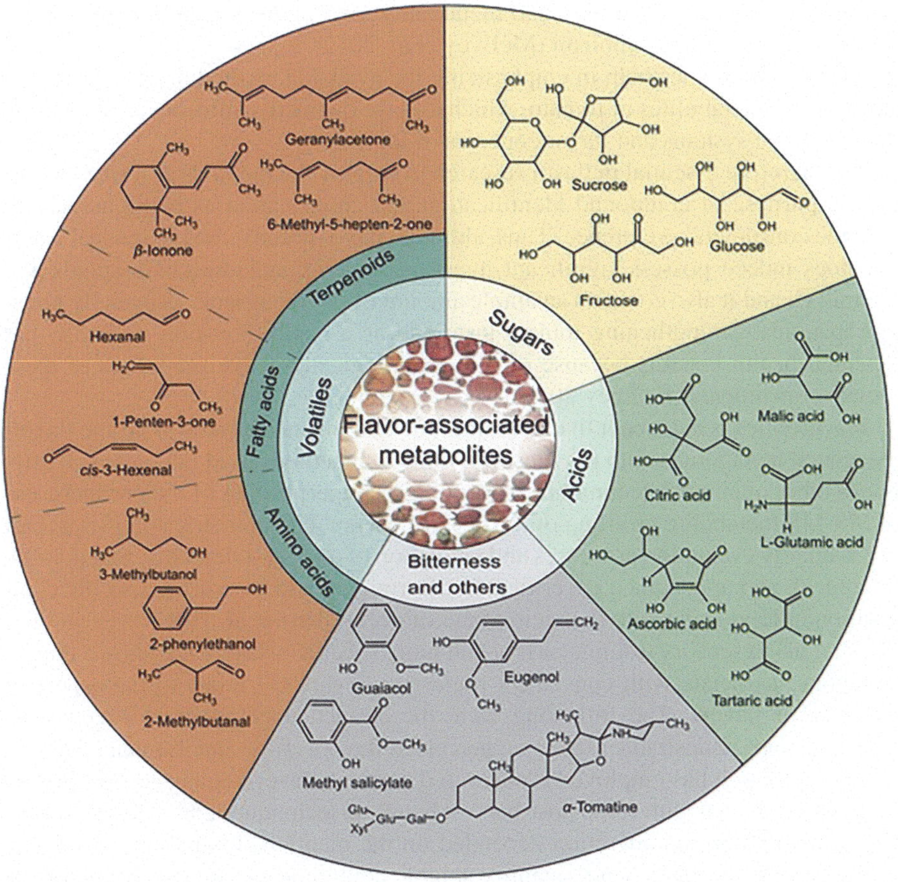

Fig. 14.3 Categorization of key metabolites linked to tomato flavor includes representative compounds. (Adopted from Wang et al. 2023)

14.5 Characterization of Flavor-Related Metabolites Based on Metabolomic Technology

The composition of metabolites in the tomato is responsible for the nutritional value, flavor profile, and color of the tomatoes, therefore, the detection and identification of metabolites are important in developing the flavor of tomatoes. GC-MS, LC-MS, NMR and other improved detection technologies have greatly improved the capabilities of the metabolomic analysis of tomatoes (Castada et al. 2019).

LC-MS and GC-MS are used in the determination of various kinds of metabolites. LC-MS is usually applied in semipolar metabolites' identification, for instance, SA and flavonoids. A current work used high-throughput LC-MS for positional and developmental profiling of the metabolome of 20 major tissues and growth phases of tomato, and recognized over 540 metabolites, as well as helped to explicate the dynamics of tomato metabolism (McEvoy et al. 2012).

These technologies help in emphasizing the available methodologies for understanding the metabolites of tomato, which all give distinctive information regarding the metabolic systems and flavor concepts of tomato.

It is therefore essential in flavor research to employ the methods described above for the purpose of compound identification and confirmation as being associated with the consumers' preference. Thus, although flavor is consumed in sensual ways, tomatoes indeed possess significant nutritional values including carotenoids and vitamin C, and it also contains an ample amount of phytonutrients, especially phenylpropanoids. Strengthening tomato taste and as a result, boosting its intake has potential health benefits because of the physiologically active properties of these tomato constituents (Kearney 2010).

Flavoromics is a concept that combines metabolomics analyses of all the flavor-associated metabolites with the sensory evaluation methods used in humans. It integrates other qualitative techniques such as flavor fingerprinting of the products, use of MS-MS metabolite profiling of flavors and flavors' chemometric analysis as part of the quality evaluation processes and validation of quality statements. Smell, taste, mouthfeel, and appearance are critical in determining and measuring specific composition in foods related to the view of the subject (Glanz et al. 1998).

After using sensory testing, current and more specific research has shown that 33 metabolite associate with consumers preferences while 37 relate to the aspect of intensity of flavors. The following were the extractant compounds which were found to cause undesirable effects on perceived flavors (Klee and Tieman 2013).

Also, research has employed e-noses to determine flavor-associated metabolites in tomatoes treated with some post-harvest handling techniques and chilling storage and found that aroma alteration depended on the mentioned handling treatments. E-tongues, on the other hand, quantify soluble compounds in tomato fruits for the objective evaluation of the fruits' composition and therefore help in tracking the genuineness of processed products based on their sensory profiles (Liu et al. 2016).

14.6 Technical Approaches to Assess the Metabolome

Metabolomics aims to comprehensively quantify the metabolites present in organisms, yet current methodologies fall short given the estimated vast diversity of over 200,000 metabolites in plants. These include primary metabolites like sugars and amino acids essential for basic cellular functions, and a plethora of secondary metabolites produced to ward off stresses or attract pollinators. To study tomato metabolomics, three major technologies are predominant (Bartel et al. 2013).

GC-MS is widely employed in plant metabolomics due to its ability to analyze polar metabolites by converting them into volatile derivatives for separation and detection. Time-of-flight (TOF)-MS is favored for its rapid scan times and high mass accuracy, facilitating efficient deconvolution and analysis of complex mixtures. The method's robustness allows for standardized protocols and shared libraries of retention time and mass spectra data, enhancing reproducibility across laboratories. However, GC-MS is limited to thermally stable volatile compounds, thus excluding high molecular weight substances (>1 kDa) from analysis (Popkin et al. 2002).

In contrast, LC-MS does not need derivatization and enables the separation of metabolites in accordance with the type of the used column regarding chemistry of the metabolite. Compared to HPLC, UPLC offers increased separation and increased sensitivity while different MS types such as QTOF and Orbitrap offer different levels of sensitivity and resolution. However, the genericity of the method requires each research unit to develop its own mass spectral libraries, which presents some difficulties in the process of unification of investigations.

In general, GC-MS is suitable for primary metabolites such as sugars, amino acids and organic acids while LC-MS is suitable for a wide range including diverse secondary metabolites. Although, both technologies play a vital role in enhancing the knowledge of the metabolism of tomatoes and its correlation with biological findings which could be used in further developments in the related field as pointed out by Aleksandrowicz et al. (2016), Paulionis (2008), Wang et al. (2020), and Harker et al. (2003).

These innovative techniques are poised to significantly enhance our understanding of the tomato metabolome, offering new avenues for comprehensive metabolic profiling and flux analysis in future research.

14.7 Integrative Omics Analysis Can Reveal the Genetic Basis for Tomato Flavor Formation

This progress has led to the construction of high-density physical and genetic linkage maps, which are crucial for precise genetic variation studies (Blaney and Diehl 2009).

The integration of comprehensive transcriptomic and metabolomic analyses has proven to be a powerful approach for identifying genes involved in metabolic pathways and understanding their regulatory mechanisms. For instance, comparative co-expression analyses between tomato and potato have elucidated the glycoalkaloid metabolism (GAME) genes responsible for the biosynthesis of steroidal glycoalkaloids (SGAs). These studies identified genes like GAME7, GAME8a, and GAME8b, which are necessary for the synthesis of cytochrome P450 enzymes of SGA. More recent metabolomic studies carried out in other species of tomato added to this knowledge of the relevant pathways. Moreover, previously, the transcription factor GAME9, belonging to the AP2/ERF family, was shown to have an influence on the synthesis of steroidal alkaloids through the regulation of SGA biosynthesis and proto genes of the upstream mevalonate/cholesterol synthetic pathways (Li et al. 2021).

The major strategies that provided important insights into the genetics of natural variation in the tomato metabolome with regard to the factors associated with fruit flavors are linkage mapping and GWAS. These approaches have noted several genes or loci that influence or are related to the various flavor parameters. For instance, GWAS predicted the sugar transporter gene, namely STP11 (Solyc06g054270) involved in determining the SSC of the tomato fruits through allelic differences [93]. The graphics of variation in citric acid and malic acid contents have been associated with natural alleles of the genes encoding citrate synthase (Solyc07g055840) and malic enzyme (Solyc12g008430), respectively. Also, the genes are puttingatively involved in fatty acid-derived metabolite synthesis such as LIP8 (Solyc09g091050) (Kushad et al. 1999).

In summary, these integrated approaches added to the knowledge about the genetics of tomato flavor and are beginning to point to the way for how breeding strategy based on genetic information could improve the tomato flavor and nutritional value (Table 14.1).

This analysis suggests that while sugar content has not been rigorously selected during tomato domestication and improvement, modern cultivars have reduced allelic diversity compared to ancestral varieties, impacting sugar, acid, and volatile content regulation. However, the accumulation of multiple elite alleles shows promise for breeding new tomato varieties with enhanced sweetness (Fantini et al. 2013).

Furthermore, genomic studies focusing on malate content identified a genomic region on chromosome 6 containing multiple SNPs positively associated with malic acid levels. Within this region, two candidate genes, Solyc06g072910 and Solyc06g072920, correspond to the Sl-ALMT9 gene, which functions as a malate efflux transporter crucial for regulating fruit malate accumulation in tomato roots.

To understand why modern cultivated tomato fruits have lost their flavor, recent research examined 398 tomato germplasm resources, encompassing modern, heirloom, and wild accessions, using metabolomic methods. Through consumer sensory evaluations, 101 distinct accessions were identified, revealing 33 and 37 chemicals significantly correlated with consumer liking and flavor intensity, respectively, with 28 compounds associated with both metrics. Compared to their wild relatives, modern commercial varieties were found to contain significantly lower levels of crucial flavor-related compounds (Fenwick et al. 1983). A metabolite-based GWAS analyzed 27 volatiles and key nutrients in ripe fruits, identifying 251 associated signals across 20 traits, including both non-volatile and volatile flavor chemicals. These results suggest a negative relationship between tomato fruit weight and sugar content, likely linked to the loss and enhancement of high-sugar alleles during domestication when larger fruits were favored. Recent attention has focused on metabolic changes during domestication, showing lower levels of antinutritional steroidal glycoalkaloids in ripe tomato fruits over generations. Metabolomic analyses across tomato subpopulations indicated substantial negative selection against bitter and toxic compounds, revealing a gene cluster on chromosome 10 potentially involved in their biosynthesis (Florkowski et al. 2014). Despite inadvertent selection for traits like yield and disease resistance, breeders may have inadvertently reduced elite alleles responsible for tomato nutrition and flavor. Nevertheless, significant flavor differences persist among tomato varieties, underscoring the need for further research to fully grasp the complexities of flavor-related metabolite compositions (Bartłomiej et al. 2012; Ariel et al. 2009) (Fig. 14.4).

Table 14.1 Genes related to flavor and their linkage analysis

Primary pathway	Metabolites	Related genes
Sugars	Fructose, glucose	*Lin5, STP11, SWEET*
	Citric acid	Citrate synthase
Acids	Malic acid	*SlALMT, TFM6,* Malic enzyme
Alkaloids	Steroidal alkaloids	*GAME1, GAME2, GAME5–9, GAME11, GAME12, GAME17, GAME18, GAME25, GAM*
	Proline	Serine incorporator
	2-Isobutylthiazole, 3-methylbutyraldehyde oxime, 1-nitro-3-methylbutane, 3-methylbutanenitrile	*SlTNH1*
	Asparagine, GABA, glutamine, threonine	*GDSL* esterase/lipase
Amino acids	2-Methyl-1-butanol, 3-methyl-1-butanol	*BCKDC, SlBCAT1, SlBCAT2, AAT1*
	2-Phenylethanol	*LeAADC1A, LeAADC1B, LeAADC2, PPEAT, FLORAL4*
	Phenylacetaldehyde	*LeAADC1A, LeAADC1B, LeAADC2, GT*
	Guaiacol, methyl salicylate	*E8*
Phenylalanines	Methyl salicylate	*SlSAMT*
Carotenoids	6-Methyl-5-hepten-2-one, geranylacetone	Phytoene synthase
	Geraniol	*GES*
Monoterpenes	β-Myrcene, limonene	*TPS7*
Sesquiterpenes	Linalool, 1,8-cineole,β-phellandrene, nerolidol	*TPS3, TPS5, TPS20, TPS37, TPS39*
Terpenoids	Geranylacetone, pseudoionone, β-ionone	*LeCCD1A, LeCCD1B*
	1-Octen-3-ol	*SlFAD7*
	1-Pentanol	*SlFAD7, Sl-LIP8*
	1-Penten-3-ol,*E*-2-hexen-1-ol,hexyl alcohol,*Z*-3-hexen-1-ol,	*Sl-LIP8*
	C5 or C6 volatiles	*LOXH1, LoxF, TomloxC*
	Alcohols, acetic acid	*SlCXE1, SlCXE5*
Fatty acids	Hexanol, *Z*-3-hexanol	*ADH2*

Adopted from Wang et al. (2023)

14.8 Multi-omics-Based Sensory Evaluation and Machine Learning for Delicious Tomato Breeding

Advancements in metabolomics, particularly through LC-MS and GC-MS techniques, have revolutionized flavor analysis by enabling cost-effective, high-throughput profiling with exceptional sensitivity across a wide range of metabolites. These methods are pivotal in identifying subtle yet significant flavor compounds

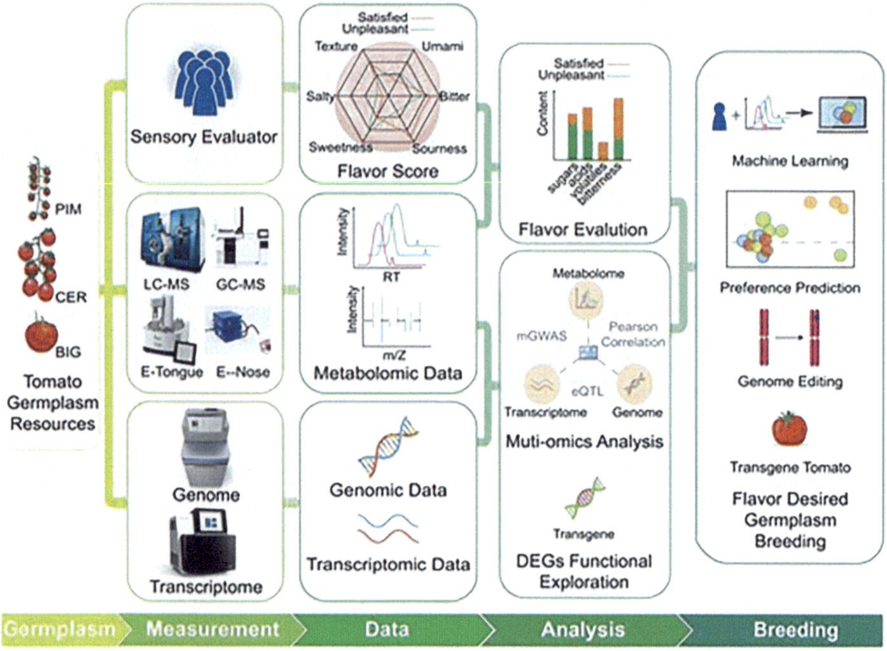

Fig. 14.4 The strategy for exploring tomato flavor based on multi-omics analysis involves a structured approach divided into five key stages: collection of tomato germplasm resources, measurement of samples and sensory evaluation, data acquisition, comprehensive analysis of multi-omics data, and practical application. (Adopted from Wang et al. 2023)

such as sugars, acids, amino acids, lipids, and volatiles, whose profiles are influenced by both genetic variations and environmental conditions.

Metabolomics applications have facilitated extensive analyses of diverse tomato populations, including wild varieties and heirlooms, providing insights into flavor profiles early in breeding cycles (Kader 2002). This capability helps prevent the inadvertent loss of superior flavor genotypes during breeding efforts, thereby supporting the development of tomato varieties with enhanced flavor traits (Cedillo-Jimenez et al. 2020).

As detection technologies advance and metabolomics data accumulate, researchers are increasingly turning to artificial intelligence (AI) for analyzing metabolomic big data, particularly in the context of tomato fruit flavor research. However, there has been limited progress in applying in-depth algorithms and novel machine learning models at the metabolic level for studying tomato fruit flavors. To enhance the efficiency and throughput of flavor phenotype identification and analysis, statistical and machine learning models are being employed to predict consumer sensory panel ratings based on fruit metabolite compositions.

It starts with determining the possibility of identifying the different metabolites that exist in fruits as well as the quantity and concentration of these metabolites, and at the same time, identifying the corresponding perceived consumer scores of the fruits. These are then used to model future metabolite consumer taste preference patterns from the accumulated data. There is a problem in defining relations between the metabolites and specific sensory attributes; there is a complex interaction between different metabolites; large sensory panels are needed to establish and validate the prediction models (Dorais et al. 2008).

When searching for the metabolites that have a strong impact on tomato flavor, a model was built based on data on the metabolome of 209 tomatoes and their assessment by the organoleptic panel. The modeling was carried out using such modeling techniques as identification of principal metabolites. For example, glucose and fructose were identified to be significant factors hence enhancing perception of sweetness; gradient boosting machine learning showing relationships between 1-penten-3-one and 2-phenylethanol and sweetness perception as well as the Bayes A model in making conclusions between E-2-pentenal and 4-carene and sweetness. On the same note, a new volatile compound, E-2-pentenal was also deemed to play a major role in overall flavors and sensation including umami in tomatoes.

As an applied AI approach, integrating genomics with artificial intelligence could help identify various molecular regulations that go into the production of tomato flavor, and use large datasets to understand the flavor conclusively (Krauss et al. 2006; Bravo 1998).

14.9 Conclusions and Future Directions

Tomato breeding has evolved beyond traditional agronomic traits like yield and disease resistance to encompass nutritional enhancement and flavor enhancement. The rapid advancements in genomics, transcriptomics, and metabolomics have enabled detailed analyses of metabolic pathways responsible for tomato flavor compounds. The current coverage of metabolite profiling, encompassing around 350 metabolites, represents a significant advance from earlier studies and underscores metabolomics' pivotal role in both fundamental research and quality testing of food products.

In tomato fruit research, metabolomics is crucial for understanding metabolic changes during development and unraveling the genetic basis of metabolite accumulation. Integrative approaches combining metabolomics with other post-genomic profiling methods promise broader insights into complex biological questions. Moving forward, enhancing the comprehensiveness of metabolomics, developing methods for assessing subcellular metabolism, and advancing isotope-based flux profiling techniques will be critical for further advancing our understanding of tomato metabolism and its implications for plant biology.

References

Aleksandrowicz L, Green R, Joy EJM, Smith P, Haines A (2016) The impacts of dietary change on greenhouse gas emissions, land use, water use, and health: a systematic review. PLoS One 11:e0165797

Ariel RV, George AM, Gabriel OS, Carlos HC (2009) Chapter 5—Nutritional quality of fruits and vegetables. In: Postharvest handling. Elsevier-Academic Press, Amsterdam

Bartel J, Krumsiek J, Theis FJ (2013) Statistical methods for the analysis of high-throughput metabolomics data. Comput Struct Biotechnol J 4:e201301009. [Green Version]

Bartłomiej S, Justyna RK, Ewa N (2012) Bioactive compounds in cereal grains—occurrence, structure, technological significance and nutritional benefits—a review. Food Sci Technol Int 18:559–568

Blaney D, Diehl H (2009) The optimal diet: the official CHIP cookbook. Autumn House Publishing, Hagerstown

Botella-Pavía P, Besumbes Ó, Phillips MA, Carretero-Paulet L, Boronat A et al (2010) Regulation of carotenoid biosynthesis in plants: evidence for a key role of hydroxymethylbutenyl diphosphate reductase in controlling the supply of plastidial isoprenoid precursors. Plant J 40:188–199

Bravo L (1998) Polyphenols: chemistry, dietary sources, metabolism, and nutritional significance. Nutr Rev 56:317–333

Castada HZ, Hanas K, Barringer SA (2019) Swiss cheese flavor variability based on correlations of volatile flavor compounds, descriptive sensory attributes, and consumer preference. Food Secur 8:78

Cedillo-Jimenez CA, Feregrino-Perez AA, Guevara-González RG, Cruz-Hernández A (2020) MicroRNA regulation during the tomato fruit development and ripening: a review. Sci Hortic 270:109435

Chong J, Soufan O, Li C, Caraus I, Li S, Bourque G, Wishart DS, Xia J (2018) MetaboAnalyst 4.0: towards more transparent and integrative metabolomics analysis. Nucleic Acids Res 46:W486–W494. [Green Version]

Coyago-Cruz E, Corell M, Moriana A, Mapelli-Brahm P, Hernanz D et al (2019) Study of commercial quality parameters, sugars, phenolics, carotenoids and plastids in different tomato varieties. Food Chem 277:480–489

Dorais M, Ehret D, Papadopoulos A (2008) Tomato (Solanum lycopersicum) health components: from the seed to the consumer. Phytochem Rev 7:231–250

Eichelmann F, Schwingshackl L, Fedirko V, Aleksandrova K (2016) Effect of plant-based diets on obesity-related inflammatory profiles: a systematic review and meta-analysis of intervention trials. Obes Rev 17:1067–1079

Fantini E, Falcone G, Frusciante S, Giliberto L, Giuliano G (2013) Dissection of tomato lycopene biosynthesis through virus-induced gene silencing. Plant Physiol 163:986–998

Fenwick GR, Griffiths NM, Heaney RK (1983) Bitterness in Brussels sprouts (Brassica oleracea L var gemnifera): the role of glucosinolates and their breakdown products. J Sci Food Agric 34:73–80

Florkowski WJ, Shewfelt RL, Prussia SE (2014) A systems approach. In: Postharvest handling, 3rd edn. Academic Press, Cambridge

Gerbens-Leenes PW, Nonhebel S, Krol MS (2010) Food consumption patterns and economic growth. Increasing affluence and the use of natural resources. Appetite 55:597–608

Glanz K, Basil M, Maibach E, Goldberg J, Snyder D (1998) Why Americans eat what they do: taste, nutrition, cost, convenience, and weight control concerns as influences on food consumption. J Am Diet Assoc 98:1118–1126

Goff SA, Klee HJ (2006) Review plant volatile compounds: sensory cues for health and nutritional value? Science 311:815–819

Harker FR, Gunson FA, Jaeger SR (2003) The case for fruit quality: an interpretative review of consumer attitudes, and preferences for apples. Postharvest Biol Technol 28:333–347

Hopfer H, Nelson J, Ebeler SE, Heymann H (2015) Correlating wine quality indicators to chemical and sensory measurements. Molecules 20:8453–8483. [Green Version]
Kader AA (2002) Postharvest technology of horticultural crops. University of California Agriculture and Natural Resources, Berkeley
Kearney J (2010) Food consumption trends and drivers. Philos Trans R Soc B Biol Sci 365:2793–2807
Klee HJ, Tieman DM (2013) Genetic challenges of flavor improvement in tomato. Trends Genet 29:257–262
Krauss S, Schnitzler WH, Grassmann J, Woitke M (2006) The influence of different electrical conductivity values in a simplified recirculating soilless system on inner and outer fruit quality characteristics of tomato. J Agric Food Chem 54:441–448
Kushad MM, Brown AF, Kurilich AC, Juvik JA, Klein BP, Wallig MA, Jeffery EH (1999) Variation of glucosinolates in vegetable crops of Brassica oleracea. J Agric Food Chem 47:1541–1548
Li C, Hou X, Qi N, Liu H, Li Y et al (2021) Insight into ripening-associated transcription factors in tomato: a review. Sci Hortic 288:110363
Liu W, Yin D, Li N, Hou X, Wang D, Li D, Liu J (2016) Influence of environmental factors on the active substance production and antioxidant activity in potentilla fruticosa L. and its quality assessment. Sci Rep 6:28591
Liu PP, Yin JF, Chen GS, Wang F, Xu YQ (2018) Flavor characteristics and chemical compositions of oolong tea processed using different semi-fermentation times. J Food Sci Technol 55:1185–1195
Manach C, Scalbert A, Morand C, Rémésy C, Jimenez L (2004) Polyphenols: food sources and bioavailability. Am J Clin Nutr 79:727–747
Mata CI, Magpantay J, Hertog MLATM, Van de Poel B, Nicolaï BM (2021) Expression and protein levels of ethylene receptors, CTRs and EIN2 during tomato fruit ripening as affected by 1-MCP. Postharvest Biol Technol 179:111573
Mathieu S, Cin VD, Fei Z, Li H, Bliss P, Taylor MG, Klee HJ, Tieman DM (2009) Flavour compounds in tomato fruits: identification of loci and potential pathways affecting volatile composition. J Exp Bot 60:325–337. [Green Version]
McEvoy CT, Temple N, Woodside JV (2012) Vegetarian diets, low-meat diets and health: a review. Public Health Nutr 15:2287–2294
McMacken M, Shah S (2017) A plant-based diet for the prevention and treatment of type 2 diabetes. J Geriatr Cardiol 14:342
Paulionis L (2008) The changing face of food and nutrition in Canada and the United States: opportunities and challenges for older adults. J Nutr Elder 27:277–295
Popkin BM, Caballero IB, Popkin BM (2002) The dynamics of the dietary transition in the developing world, The nutrition transition diet and disease in the developing world food science and technology international series. Academic Press, London, pp 111–129
Tilahun S, Park D, Seo M, Jeong C (2017) Review on factors affecting the quality and antioxidant properties of tomatoes. Afr J Biotechnol 16:1678–1687
Wang F, Huang Y, Wu W, Zhu C, Zhang R, Chen J, Zeng J (2020) Metabolomics analysis of the peels of different colored citrus fruits (Citrus reticulata cv. 'Shatangju') during the maturation period based on UHPLC-QQQ-MS. Molecules 25:396. [Green Version]
Wang S, Qiang Q, Xiang L, Fernie AR, Yang J (2023) Targeted approaches to improve tomato fruit taste. Hortic Res 10(1):uhac229
Waseem M, Li N, Su D, Chen J, Li Z (2019) Overexpression of a basic helix–loop–helix transcription factor gene, SlbHLH22, promotes early flowering and accelerates fruit ripening in tomato (Solanum lycopersicum L.). Planta 250:173–185
Wishart DS, Feunang YD, Marcu A, Guo AC, Liang K, Vázquez-Fresno R, Sajed T, Johnson D, Li C, Karu N et al (2018) HMDB 4.0: the human metabolome database for 2018. Nucleic Acids Res 46:D608–D617

Disease Resistance at the Molecular Level: Omics-Based Approaches for Tomato Defense

15

Asma Khalil, Kinza Fatima, Muhammad Sadaqat, and Muhammad Tahir ul Qamar

Abstract

Tomato (*Solanum lycopersicum*) is a globally significant crop due to its nutritional value and extensive cultivation. However, various pathogens severely impact tomato production, including bacteria, fungi, viruses, and nematodes, resulting in significant yield losses. Developing disease-resistant tomato cultivars is one of the most sustainable methods for successful production. The recent advancements in omics techniques allow for enhanced coverage of plant transcriptomes, proteomes, and metabolomes during pathogen attacks and regulation of the response after infection. Genomics-assisted breeding and gene editing techniques such as quantitative trait loci (QTL) mapping, marker-assisted selection, and CRISPR-Cas9 offer precise methods within the high-throughput method of developing disease-resistant tomato varieties. Transcriptomics provides insights into differentially expressed genes (DEGs) during pathogen

A. Khalil · M. Tahir ul Qamar (✉)
Integrative Omics and Molecular Modeling Laboratory, Department of Bioinformatics and Biotechnology, Government College University Faisalabad (GCUF), Faisalabad, Pakistan
e-mail: asmakhalil.201705701@gcuf.edu.pk; tahirulqamar@gcuf.edu.pk

K. Fatima
Integrative Omics and Molecular Modeling Laboratory, Department of Bioinformatics and Biotechnology, Government College University Faisalabad (GCUF), Faisalabad, Pakistan

College of Natural & Agricultural Sciences, University of California, Riverside, USA
e-mail: kfati002@ucr.edu

M. Sadaqat
Integrative Omics and Molecular Modeling Laboratory, Department of Bioinformatics and Biotechnology, Government College University Faisalabad (GCUF), Faisalabad, Pakistan

UMR CNRS 6553 Ecosystèmes, Biodiversité, Evolution (ECOBIO), Université de Rennes 1, Rennes, France
e-mail: muhammad.sadaqat@univ-rennes.fr

M. Waseem et al. (eds.), *Omics Approaches for Tomato Yield and Quality Trait Improvement*, https://doi.org/10.1007/978-981-96-3890-1_15

attacks. Noncoding RNAs (ncRNAs) regulate gene expression and signaling pathways crucial for plant immunity. On the other hand, proteomics and metabolomics reveal critical proteins and metabolites involved in tomato defense responses. This comprehensive cataloging of omics resources has raised attention to the necessity of integrating omics methods to ensure effective resource use and an enhanced understanding of the molecular process. The information provided in this chapter will help us understand tomato defense responses against diseases, the molecular mechanisms involved in plant disease resistance, and the effective utilization of omics resources for tomato crop improvement.

Keywords

Solanum lycopersicum · Tomato · OMICS · Quantitative trait loci (QTL) · Disease resistance · Protein–protein interactions (PPI) · Pathogens · Noncoding RNAs (ncRNAs) · High-throughput methods · TFs · R-genes

15.1 Introduction

15.1.1 Background of Tomato (*Solanum lycopersicum*)

Tomato is one of the most significant vegetable crops globally, often considered a fruit but treated as a vegetable. It is the most popular non-starchy vegetable consumed globally, with production reaching 180 million metric tons in 2019, equating to an average consumption of 63 grams per person every day (Kitinoja et al. 2019). China is the world's top producer of tomatoes, closely followed by India and Turkey. According to FAOSTAT 2005 (http://faostat.fao.org), tomatoes rank as the second-most consumed vegetable, after potatoes. Furthermore, it is an invaluable model organism for understanding the genetic features associated with disease and pest resistance and the molecular mechanism involved in plant-pathogen interactions (Hoshikawa et al. 2021).

Solanum is the most substantial genus within the Solanaceae family, containing between 1250 and 1700 species. The physical and ecological variety of *Solanum* species is noteworthy. These species can be found on temperate and tropical continents, and are known for their remarkable morphological diversity (Wakil et al. 2018). *Solanum* is of great commercial significance because it includes agricultural and other species that produce toxic or therapeutic chemicals. The tomato processing industry originated in 1869 with the establishment of Henry John Heinz's ketchup company in Pittsburgh, USA (Hoenig 2014).

Tomatoes and their products are popular nutritious food and a good source of minerals, vitamins, and antioxidants. Macro and microelements along with vitamins A and C, are present in tomato fruit, which strongly support the health of humans. Hence Its consumption is increasing globally. Flavonoids are mostly produced in the peel of tomato fruit (Oliveira et al. 2013). Additionally, tomatoes contain modest amounts of ascorbic acid, providing approximately 40% of the recommended dietary vital vitamins (Hanson et al. 2004). In recent years, OMICS-based approaches have revolutionized our understanding of plant defense mechanisms at the molecular level, offering new strategies for enhancing disease resistance in crops.

15.1.2 Pathogens Affecting Tomato

In addition to being edible, tomatoes contain several secondary metabolites that can be used as antioxidants and may also have anti-cancer properties. However, tomato plants are susceptible to numerous pathogens, such as bacteria, fungi, viruses, and nematodes, which can severely impact yield and quality (Panno et al. 2021). These pathogens can attack in several different ways. Some colonize the tissue in the plant, some invade the plant's tissue, and others may go for specific areas of the plant, such as the roots, stems, and leaves (Fig. 15.1, Table 15.1) (Lucas 2009). Late blight (*Phytophthora infestans*) is one of the most common tomato fungal diseases, responsible for 20–70% of the crop's total economic loss (Goutam et al. 2018). Nematodes particularly infect tomato roots, forming galls that impair root function and nutrient uptake. Root-knot disease by the nematode *Meloidogyne* sp. is the other most devastating and widespread disease in tomatoes (Shilpa et al. 2022). These pathogens frequently induce symptoms such as necrosis, tissue senescence, reduced yield, and, in extreme cases, the ability to kill the tomato plant (Tsitsigiannis et al. 2008).

The genome of tomato plant viruses can consist of both DNA and RNA. Geminiviruses can affect both monocot and dicot plants, while begomoviruses primarily infect arrays of dicot host species, such as *Capsicum annuum* and

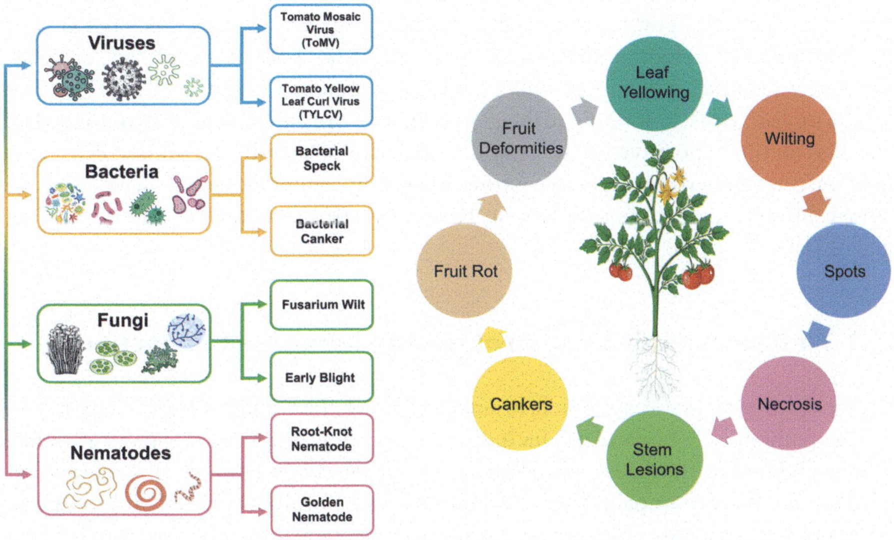

Fig. 15.1 Common pathogens affecting tomato and their disease symptoms

Table 15.1 Pathogens and the species that cause diseases in tomato plants

Pathogen	Species
DNA virus	Beet curly top virus (BCTV), Ageratum yellow vein virus (AYVV), Geminiviruses, Begomoviruses, Tomato yellow leaf curl Sardinia virus (TYLCSV)
RNA virus	Tomato mottle mosaic virus (ToMMV), Tobacco etch virus (TEV), Tomato chlorosis virus (ToCV), Alfalfa mosaic virus (AMV), Tomato spotted wilt virus (TSWV), Cucumber mosaic virus (CMV), Pepino mosaic virus (PepMV), Chickpea chlorotic dwarf virus (CpCDV), Potato virus Y (PVY), Tomato brown rugose fruit virus (ToBRFV),
Viroids	Tomato apical stunt viroid (TASVd), Potato spindle tuber viroid (PSTVd)
Bacteria	*Xanthomonas axonopodis* pv. *Vesicatoria*, *Pseudomonas syringae* pv. tomato, *Ralstonia solanacearum*, *Pseudomonas corrugate*, *Erwinia carotovora* subsp. *Carotovora*,
Fungi	*Verticillium dahlia*, *Rhizoctonia solani*, *Alternaria solani*, *Pyrenochaeta lycopersici*, *Septoria lycopersici*, *Leveillula Taurica*, *Fusarium oxysporum*, *Phytophthora infestans*
Nematod	*Meloidogyne* sp.

Carica papaya. They also damage both monocot and dicot plants (Sawangjit et al. 2005). Begomoviruses replicate within the nucleus of the host and are identified by a twinned capsid morphology. They possess either monopartite or bipartite circular single-stranded DNA. According to Rojas et al., these viruses employ double-stranded DNA (dsDNA) and a rolling circle replication (RCR) process to replicate. Among begomoviruses, the tomato yellow leaf curl virus (TYLCV) has the most substantial impact on tomato production, particularly in subtropical and tropical areas worldwide (Rojas 2004).

Tomato RNA viruses are all single-stranded, with genomes that can be monopartite, bipartite, or tripartite. A new RNA virus known as the tomato brown rugose fruit virus (ToBRFV) affects tomatoes, spreading rapidly through mechanical contact with objects including plant waste, growth media, contaminated soil, circulating water, farming operations, and cultural equipment (Salem et al. 2023). The Alfalfa mosaic virus, belonging to the genus Alfamovirus in the family Bromoviridae, consists of three positive-sense single-stranded RNA (ssRNA) molecules. This virus infects numerous vegetable crops, including eggplant, peas, beans, celery, pepper, alfalfa, tomato, borage, mung bean, chickpea, and tumble pigweed (Ong et al. 2020).

15.1.3 Phenotypic Evaluation of Tomato Response to Infections

It refers to a systematic approach to understanding the tomato plant's response to various pathogens. This process includes the selection of different tomato varieties and the pathogens of interest, ensuring a comprehensive assessment of genetic variability and pathogen impact. Visual symptoms are carefully observed and recorded, including leaf yellowing, wilting, spots, necrosis, stem lesions, cankers, and fruit rot or deformities. These symptoms provide immediate and visible indicators of infection severity and progression (Kiirika et al. 2013).

Tomato leaf curl virus (TLCV) is the most destructive virus, spread by whiteflies (Bemisia tabaci) in a circulative manner. Twenty-seven tomato cultivars were examined in a field trial for the source of resistance against (TLCVD). Some genotypes showed strong resistance, which can be utilized in breeding strategies to create ToLCV-resistant varieties (Zeshan et al. 2016). Caruso et al. assessed 37 local Sicilian tomato ecotypes for their response to tomato brown rugose fruit virus (ToBRFV) infection. The severity of the ToBRFV disease varied within the ecotypes. Although none of the ecotypes demonstrated high resistance, On the other hand, the two Pop27 and Pop35 showed intermediate resistance to the virus and good tolerance, indicating potential for selective breeding (Caruso et al. 2024). The effectiveness of plant growth-promoting rhizobacteria (PGPR), specifically *Pseudomonas fluorescens* and *Bacillus subtilis*, in enhancing tomato resistance to tomato mosaic virus (ToMV) has been evaluated. It was demonstrated that these PGPRs drive tomato plants to mount defensive mechanisms and develop ToMV resistance (Sofy et al. 2019).

15.1.4 Molecular Mechanisms Involved in Tomato Plant Disease Resistance

A complex network of molecular processes, such as pathogen detection, singling transmission, and defense response stimulation, interact to produce plant disease resistance. Throughout their evolutionary history, plants have developed multiple active and passive resistance mechanisms that they may employ in coordination to fend off pathogen invasion (Andersen et al. 2018). Active mechanisms are quickly triggered following pathogen invasion and encompass several key responses, such as reactive oxygen species (ROS) synthesis, hypersensitive response (HR) initiation, and the fortification or repair of the cell wall. However, passive resistance in plants involves pre-existing physical barriers on the cell surface and intracellular substances toxic to pathogens, like phenols, unsaturated lactones, and antibacterial compounds (Hutcheson 1998). In plants, pattern-triggered immunity is thought to be the first line of inducible protection. Pathogen-associated molecular patterns, or PAMPs, are characteristics of non-self-microbial signatures that may be detected and serve as canonical triggers. Plants utilize pattern recognition receptors (PRRs) that activate the PAMP immunity (PTI), leading to ROS production and pathogenesis-related (PR) protein expression (Hutcheson 1998). Furthermore, certain pathogen effectors are detected by resistance (R) proteins, such as nucleotide-binding site leucine-rich repeat (NBS-LRR) proteins. This leads to the initiation of effector-triggered immunity (ETI), which is often characterized by localized 'hypersensitive response' (HR), a programmed cell death (PCD) (Wu et al. 2014). Moreover, transcription factors (TFs) such as WRKY, MYB, and NAC are engaged in regulating the expression of defense-related genes, and defense response coordination is done by phytohormones like ethylene (ET), salicylic acid (SA), and jasmonic acid (JA)

(Erpen et al. 2018). Noncoding RNAs (ncRNAs), such as microRNAs (miRNAs) and long noncoding RNAs (lncRNAs), further regulate gene expression and signaling pathways important for plant defense (Zaynab et al. 2018).

15.2 Objectives

The chapter aims to explore the molecular mechanisms underlying tomato disease resistance and provide a comprehensive understanding of omics technologies such as genomics, transcriptomics, proteomics, and metabolomics contributing to the identification and functional analysis of genes and pathways involved in tomato defense responses. By utilizing contemporary sequencing technologies and sophisticated bioinformatics tools, this chapter will highlight recent advancements in quantitative trait loci (QTL) mapping and genome-wide association studies (GWAS), which support the development of disease-resistant tomato cultivars.

15.3 Integrated Omics Approaches to Study Defense Mechanism in Tomato

The yield of tomato crops is negatively impacted by the attack of many pathogens, including fungi, viruses, and bacteria. Plants have devised various tactics to fend off these attacks, like altering gene expression and activating several metabolic pathways. Recent advancement in omics approaches allows the enhanced protection of plant transcriptomes, proteomes, and metabolomes during pathogen attacks and the alterations of the response to subsequent infection. Optimizing crop protection through breeding strategies or ad hoc transgenic procedures demands awareness of plant response mechanisms by omics resources (Fig. 15.2) (Das et al. 2021).

15.3.1 Genomics-Based Approaches

The study of genes and genomes is known as genomics, and it includes an examination of their composition, dynamics, mapping, epi-genomics, metagenomics, and genome editing (Chaitanya 2019). Plant biologists can create physical and genetic maps to discover qualities of interest by using structural genomics, which studies chromosomal organization and sequence polymorphisms (Dixon et al. 2007). Conversely, functional genomics investigates how genes control particular features. The term epi-genomics describes the study of epigenetic modifications that take place at the genomic level, such as histone modifications, DNA methylation, or small RNA methylation (Przybyla and Gilbert 2022). Multiple genomics can facilitate identifying and characterizing genes responsible for plant resistance against pathogens. Additionally, the molecular process of plant-pathogen interactions can be understood by genomic approaches, which can be useful for developing specific pathogen management strategies (Peyraud et al. 2017). Metagenomics deals with

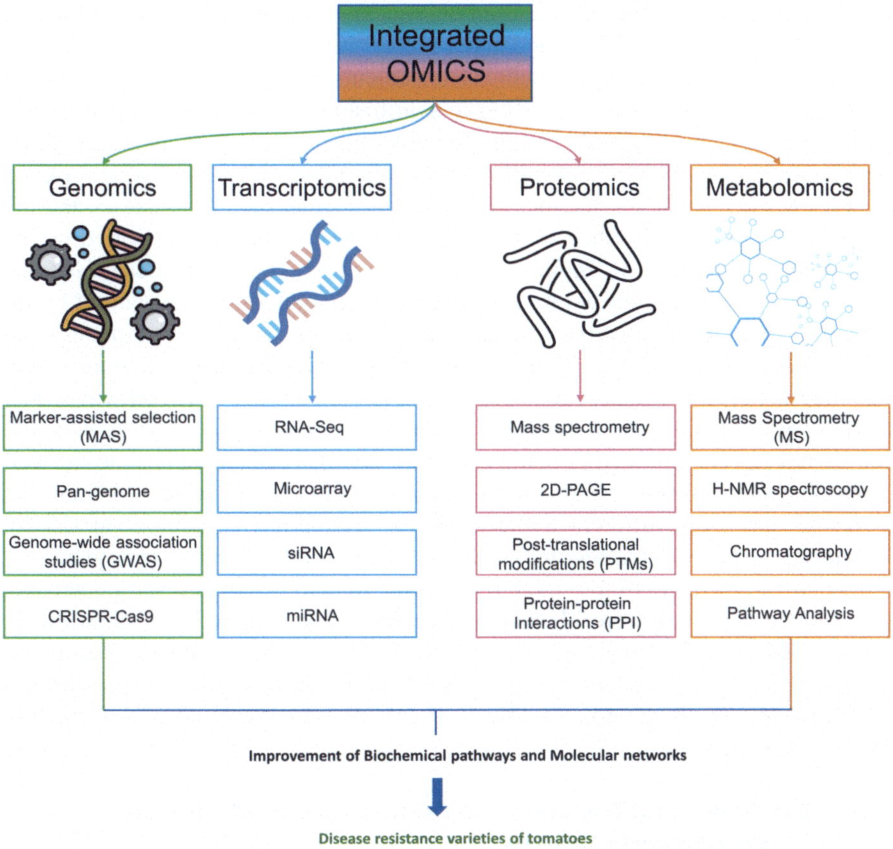

Fig. 15.2 Integrative multi-omics techniques to confer disease defense in tomato plants

mutational occurrences directing genetic alterations in mutant characteristics. Integrating these genomic approaches with systems biology and advanced technologies like CRISPR/Cas9 genome editing enhances our ability to manipulate specific genes for improved crop resilience (Anton and Roberts 2021). Moreover, pangenomics has emerged as a recent approach to study intraspecies diversity providing deeper insights into the molecular evolution mechanisms contributing to plant resistance against environmental stresses.

15.3.1.1 Genomic Resources of Tomato

The number of plant genomes that have been fully sequenced has been continuously rising since the 2000 complete sequencing of the first plant genome, Arabidopsis thaliana. Currently, multiple plant genomes, including those of rice, poplar, and grapevine, have been fully sequenced (Flagel and Blackman 2012). Under the International Solanaceae Genome Project (SOL), the genome of the tomato was sequenced through the bacterial artificial chromosome BAC approach, which started at the end of 2004. A substantial amount of sequence information, such as >200,000 expressed sequence tags (ESTs) and about 90,000 BAC-ends, has been made public for tomato species, coinciding with advancements in genomic investigations in plants (Shirasawa and Hirakawa 2013). The utilization of DNA markers has facilitated genetics, genomics, and breeding in numerous plant species, including tomatoes. These markers help to create detailed genetic map, which are crucial for QTL analysis, marker-assisted selection, and association analysis. For tomatoes and the larger Solanaceae family, structural and functional genomics are supported by the expanding availability of bioinformatics resources. The Solanaceae Genome Network (SGN), accessible at http://www.sgn.cornell.edu, offers a wealth of bioinformatics resources focused on structural and functional genomics for tomatoes and the broader Solanaceae family. This platform provides extensive genomic data, including genome sequences, gene expression patterns, metabolite pathways, phenotypes, and quantitative trait loci (QTLs). These resources support various genomic studies and research endeavors (Chiusano et al. 2008). Yano and colleagues as well as Barone and colleagues have analyzed more publicly available websites that aggregate data from different tomato "omics" techniques (Barone et al. 2009). Nonetheless, much work is required to convert unprocessed data into information of biological significance. Before the release of the tomato genome, several databases dedicated to tomatoes were already in existence. These included KaTomicsDB, TOMATOMA, miSolRNA, MoTo DB, SolEST, and the Tomato Functional Genomics Database. For the Solanaceae study community, these databases offer a wealth of biological material, as well as genetic and biochemical information (Suresh et al. 2014).

15.3.1.2 R-Genes and Transcription Factors-Mediated Disease Resistance in Tomato

The advances in molecular techniques, genomic tools, and high-throughput sequencing of the early twenty-first century drove the identification of R-genes from various crops. Through genetic engineering and marker-assisted selection (MAS), these genes have been introduced in agricultural cultivars to create disease-resistant crops (Delseny et al. 2010).

R-gene-Loci have high Polymorphism, particularly in sequences that encode leucine-rich-repeat-motif-amino acids. In tomatoes, Resistance (R) genes perform important functions in providing defense against various pathogens, such as viruses, bacteria, fungi, and nematodes (Table 15.2) (Chaudhary and Atamian 2017).

The DNA sequence analysis of the *Arabidopsis* RPP1, RPP5, and RPP8 families, and the tomato Hcr9s and Hcr2s, has been well documented in recent research. Like the homologs of RPP5 and RPP1, Hcr2s have different numbers of leucine-rich repeats (LRRs). This variability suggests that this process is crucial for recognizing the pathogen effectors and activating immune responses to combat infection (Dixon et al. 2000).

TFs are proteins, important for the modulation of stress-responsive genes during bacterial and fungal infections in the plant. These factors contain multiple domains such as a conserved DNA-binding domain, a transcription regulation domain (which

Table 15.2 List of R-genes conferring resistance against disease in tomatoes

Gene	Resistance against the disease	Pathogen	Reference
Cf-16	Leaf mold	*Cladosporium fulvum*	Zhang et al. (2023)
Cf-2, 4, 5, 6, 9, 11	Leaf mold	*Cladosporium fulvum*	Zhao et al. (2022)
I3	Fusarium wilt	*Fusarium oxysporum f.* sp. *Lycopersici*	Swett et al. (2023)
Ve1, Ve2	Verticillium wilt	*Verticillium* Spp.	Kawchuk et al. (2001)
Pto	Bacterial speck	*Pseudomonas floridensis*	Eckshtain-Levi et al. (2019)
Mi-1	Soil-borne diseases	*Meloidogyne* spp.	Padilla-Hurtado et al. (2022)
Tm1, Tm2, and *Tm2[a]*	Tomato mosaic virus (ToMV)	Tobamovirus	Panthee et al. (2013)
Sw-5	Tomato spotted wilt virus disease	Tomato spotted wilt virus	Rodríguez-Negrete et al. (2023)
Ty-1	Tomato yellow leaf curl virus (TYLCV)	*Geminiviridae*, genus *Begomovirus*	Koeda and Kitawaki (2024)
Ph-1, Ph-2, Ph-3, Ph-4, Ph-5-1, and *Ph-5-2*	Late blight	*Phytophthora infestans*	Arafa et al. (2022)

can be an activator or suppressor region), an oligomerization site, and a nuclear targeting signal (Meraj et al. 2020). TFs play crucial roles in transcriptional reprogramming, involving infinitely highly coordinated molecular, biochemical, and physiological modifications. The families such as bZIP, NAC, MYB, WRKY, Apetala2/Ethylene Responsive Factors composed TFs are implicated in multiple defensive pathways with crucial roles in immune responses to infections (Ng et al. 2018). Recent omics technologies have greatly improved our ability to identify the TFs throughout the genome. The accessibility of the whole tomato genome arrangement has made it conceivable to portray quality families, including the TFs, at an extensive level. Transcriptome analyses enable the distinguishing of genes coding for TFs (Naik et al. 2023).

SlERF2 is a member of the Ethylene Response Factor (ERF), a family of TFs in tomatoes, and plays a significant role in maintaining the plant's defense mechanisms against a range of pathogens. For example, *ERF2* may enhance the overexpression of tomato fruit resistance to *Botrytis cinerea* enhanced by the production of methyl jasmonate (MeJA) in response to various disease conditions, leading to increased ethylene synthesis, PR proteins, and phenolic content. MeJA has been shown to induce disease resistance in tomato fruits that overexpress *ERF2* (Yu et al. 2018). Campos et al. reported, that various transcription factors are engaged in tomatoes resistance to biotic stresses, such as bacterial, viral, and fungal infections. They can regulate the expression of crucial resistance genes and other defense-related pathways, thereby effectively improving the tomato plant's capability to detect and respond to biotic stresses (Campos et al. 2022).

15.3.1.3 Tomato Pan-Genome

The swift advancements in sequencing technologies, with their lower sample costs, high-throughput SNP genotyping systems, and bioinformatics tools, have transformed crop improvement programs. The availability of the tomato reference genome has enabled a comprehensive analysis of genetic variation via SNPs and small insertions/deletions (indels) by resequencing numerous accessions. This has revealed the domestication history and wild introgression events in tomatoes. However, a growing body of research has shown that large structural variants (SVs), including present absent variations (PAVs) and copy number variations (CNVs), also contribute significantly to plant adaptive diversity and functional variations (Van Deynze et al. 2007). Gao et al. constructed a pan-genome a pan-genome by processing Illumina short reads and using a 'map-to-pan' approach, based on resequencing data from 725 accessions of the Lycopersicon clade, which includes *S. lycopersicum* and its closely related wild species, *Solanum pimpinellifolium*, *Solanum cheesmaniae*, and *Solanum galapagense* (SCG) (Van Deynze et al. 2007). A pan-SV map constructed from long-read sequencing of 100 varied tomato lines and a graph pan-genome incorporating variant data from 838 tomato genomes were used by Oxford Nanopore and PacBio to detect structural variants more effectively (Alonge 2021). These studies indicate that structural variants (SVs) significantly participate in phenotypic variation and can be highly effective in GWAS (Jin et al. 2023). Super-pan-genome was constructed by 11 distinct tomato species and revealed potential genes or alleles that enhance yields in modern cultivated tomatoes (Li et al. 2023a).

15.3.1.4 Genomics-Assisted Breeding Approaches for Tomato Defense

QTL Mapping and Marker-Assisted Breeding

Genomics-assisted breeding has significantly advanced the development of tomato varieties with enhanced disease resistance. By leveraging genomic tools, researchers can identify and incorporate resistance genes with greater precision and efficiency (Ercolano et al. 2012).

Tomatoes are abundant in quantity and variety of molecular markers, positioned on various molecular maps. The real application of MAS in tomato breeding dates back to 30 years, when the acid phosphatase (Aps-11 locus) isozyme marker was used as an indirect selection tool for breeding nematode resistance (Osei et al. 2018). Recently, more than 2500 markers, such as RFLP, EST, SSR, and COS, have been mapped to the 12 tomato chromosomes, forming a high-density genetic map. Genes that provide resistance against more than 35 diseases have been found and mapped in tomatoes (Merk 2010). Particularly, the disease resistance controlled by major R-genes has been intensively investigated. QTL mapping techniques have been utilized to identify genomic regions associated with quality traits in processing tomatoes, and a total of 130 QTLs representing 38 traits have been identified so far. Breeding strategies that utilized molecular markers have enabled the introgression resistance alleles from wild tomato species into domesticated varieties, as demonstrated in studies focusing on black-mold resistance and Verticillium wilt resistance (Xu et al. 2013). Strategies involving the transfer of pathogen resistance genes in tomatoes facilitated by molecular markers have shown promising results, emphasizing the importance of marker-assisted approaches in enhancing breeding efforts. These advancements in this approach could be used for all traits that are controlled by many genes with different influences on the phenotype, such as yield, disease resistance, and fruit quality (Mores et al. 2021).

Genome-Wide Association Studies (GWAS)

Modern tomato cultivars exhibit extensive genetic diversity and have not made it easier to study complex features due to extended artificial selection. On the other hand, wild cousins of tomatoes provide a significant allelic variation reservoir. GWAS have identified significant associations between complex traits and phenotypic diversity within natural populations across various tomato species (Uffelmann et al. 2021). Next-generation sequencing methods in conjunction with GWAS provide improved mapping resolution for pinpointing the genetic areas that confer disease resistance in tomatoes (Saidi et al. 2020).

Through genetic variation and recombination events within germplasm collections, the GWAS technique provides higher mapping resolution than linkage mapping, which is a clear advantage. As a result, GWAS is now a common technique used in tomato germplasm collections around the world to identify SNPs linked to agronomic traits. As an illustration, a GWAS was carried out in tomatoes utilizing 182 SSR markers to identify chromosomal areas linked to 28 distinct volatile compounds that characterize tomato flavor (Zhang et al. 2016). GWAS identified SNPs associated with disease resistance in tomatoes across multiple tests. Candidate genes linked to Fusarium wilt resistance were identified, including those encoding leucine-rich repeat-like and disease-resistance proteins (Kawicha et al. 2023). GWAS investigated QTL for bacterial wilt resistance in a core collection of 191 tomato varieties, revealed the significant phenotypic variation, and was validated in commercial cultivars, indicating their potential for marker-assisted selection in tomato breeding programs (Nguyen et al. 2021).

15.3.1.5 CRISPR-Cas9 and Gene Editing for Targeted Mutagenesis

In recent years, CRISPR-Cas9 technology has been employed to knock out key tomato genes implicated in resistance pathways to study their function in granting immunity against infections or to create disease-resistant tomato plants (Tyagi et al. 2021).

The CRISPR-Cas9 method was applied to tomatoes for the first time in 2014. At the moment, CRISPR/Cas genome editing has been used to decrease pathogen virulence in cases of *F. oxysporum*, *B. cinerea*, *Oidium lycopersici*, *Phytophthora capsici*, *P. syringae pv.* tomato, TYLCV, and TYLCSV, or to generate resistant tomato plants. This can be achieved by targeting either pathogen (e.g., PKS4, CP, IR, and Rep sequences) or tomato (e.g., *SlMlo1*, *Solyc08g075770*, *SlMAPK3*) genes (Talakayala et al. 2022).

The removal of Argonaute 7 produced wiry morphologies; the mutants first produced leaves with petiole-free leaflets, then leaves with lamina-free leaflets. By focusing on the coat protein and replicase loci of the virus genome, Tashkandi et al. developed tomato plants resistant to (TYLCV) using CRISPR-Cas9 technology (Tashkandi 2018). Key genes implicated in resistance pathways have also been knocked out using CRISPR-Cas9 technology to examine their capacity to provide immunity against viruses. Nekrasov et al. eliminated the mildew resistance locus (Mlo1), which encodes a membrane-associated protein, to completely protect tomato plants against the powdery mildew fungus *Oidium neolycopersici* (Nekrasov et al. 2017). Moreover, Koseoglou blocks the expression of the callose synthase-encoding Powdery Mildew Resistance 4 (*PMR4*) gene (Koseoglou 2017).

15.3.2 Transcriptome-Mediated Tomato Defense Mechanism

Tomato plants are infected by various pathogens that cause symptoms such as wilts, leaf spots or blights, fruit spots, and rots. These symptoms lower yield and reduce product quality. As a result, tomato becomes a crucial culture for high-throughput techniques that allow for differentially expressed genes (DEGs) identification in response to infections and broad transcriptome profiling. Transcriptomics measures the differential expression levels of transcripts spatially and temporally across distinct developmental stages and under different physiological circumstances. It also determines the amount of RNA and transcriptional structure (Dong and Chen 2013).

The transcriptome can be studied through various strategies. The introduction of hybridization-based techniques (microarray) and sequencing-based techniques such as serial analysis of gene expression (SAGE), massively parallel signature sequencing (MPSS), and cap analysis of gene expression (CAGE) has started to elucidate the involvement of multiple genes (Murphy 2002). High-throughput sequencing of RNA (RNA-seq) is the most recent technique. The unbiased detection of all RNA molecules in a sample, including uncommon and novel transcripts, alternative splicing processes, and noncoding RNAs, is made possible by RNA-seq instead of microarrays (Fig. 15.3) (Han et al. 2015).

Saidi et al. utilized the microarray technique for transcriptome analysis of tomato plants infected by *P. infestans* and *Colletotrichum coccodes*. A total of 51 DEGs were identified and revealed significant upregulation of Solyc06g007510.3 and downregulation of Solyc05g008850.3 genes, indicating their functions in regulating auxin metabolism during infections and potential targets for the development of disease-resistant tomato cultivars (Saidi et al. 2020).

Chenet et al. use RNA-seq analysis to compare transcriptional changes in a TYLCV-resistant tomato breeding line CLN2777A (R) and a TYLCV-susceptible tomato breeding line TMXA48-4-0 (S) in response to TYLCV infection. This analysis revealed a higher proportion of upregulated DEGs in the tomato R line compared with the S line. The upregulated genes can be beneficial for identifying the important defense-related genes in tomatoes for TYLCV disease management (Chen et al. 2013).

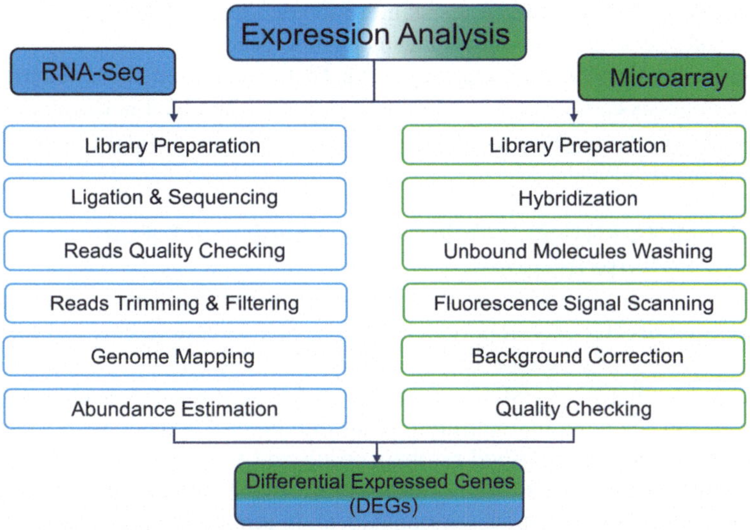

Fig. 15.3 Workflow for identifying the DEGs through microarrays and RNA-seq analysis

Pombo and colleagues (2014) utilized RNA-seq technology to identify genes involved in unique immune responses to *P. syringae* in tomato plants to understand the molecular basis of this response (Gomez-Casati et al. 2016). Tomato Fusarium crown and root rot (FCRR) is a highly damaging soil-borne disease. Beijing Genomics Institute (BGI) (Shenzhen, China) prepared the RNA-seq libraries and performed the sequencing through the DNBseq platform. RNA-seq analysis identified 189 DEGs in the 'plant-pathogen interaction' pathway and revealed that MAPK and WRKY gene families play important roles in Frl-mediated resistance to *F. oxysporum* in tomatoes. This emphasizes the significance of these immune response pathways for inhibiting pathogen development and boosting disease resistance. These pathways include hormone regulation involving JA and SA (Sun et al. 2022).

Barad et al. (2017) used transcriptomic analysis to investigate the response of wild-type (WT) and RNAi-SlPH tomatoes during colonization to *Colletotrichum gloeosporioides* infection. The Illumina HiSeq 2500 V3 platform was used for RNA-seq analysis, revealing that WT tomatoes upregulated 1190 genes and downregulated 631 genes, inducing pathways related to jasmonic acid and glutamate biosynthesis. On the other hand, RNAi-SlPH tomatoes increased sensitivity to the disease by upregulating phenylpropanoid production and downregulating sugar metabolism. These findings revealed that a lower fruit acidity and a higher pH may affect the interaction between the pathogen and the host.

Milc et al. (2019) conducted comparative RNA-seq-based transcriptomic analysis on two tomato cultivars: resistant Mogeor and susceptible Moneymaker, to investigate the differences in their transcriptomic responses to *P. lycopersici*. RNA analysis revealed that both cultivars shared the same reactions to fungal infection, suggesting overlapping signaling pathways. However, the susceptible genotype exhibited more DEGs than the resistant genotype. The pyl gene in the resistant genotype may be a recessive form of a susceptible gene.

These multiple studies demonstrated that DEGs identified by Mircoarrays and RNA-seq analysis can reveal key genes and pathways involved in the plant defense response under varied pathogen stress and can be valuable for developing disease-resistant tomato cultivars.

15.3.2.1 ncRNA-Mediated Molecular Mechanism in Tomato Defense Response

ncRNAs are a cluster of RNAs that are transcribed from the genome but not translated into proteins and play important roles in gene expression, thus controlling the mechanism underlying defense responses (Altayli 2020).

Small noncoding RNAs (sRNAs) are essential regulators of gene expression that are specifically involved in controlling plant immunity. Plants contain a variety of sRNA classes that are identified by their origin and biogenesis. Small interfering RNAs (siRNAs) and microRNAs (miRNAs) are two of them that primarily block gene expression at the post-transcriptional level. In tomatoes, miR482 and miR2118 specifically function as important regulators in triggering immune responses to pathogen invasion. Among these, miR482 turns its focus toward genes associated with disease resistance and initiates RNA polymerase VI to produce secondary siRNAs. miR482-mediated RNA silencing is suppressed by pathogen invasion, which permits defense protein genes to gradually express in response to viral and bacterial assaults. MiRNAs that are upregulated after infection include miR162, miR168, miR396, miR398, miR408, and miR447, whereas miR160, miR169, miR170, and miR391 are downregulated. These findings suggest that the regulatory network governing tomato resistance to *Botrytis cinerea* may include miR319, miR394, and miRn1. These miRNAs have the potential to be used in the development of antiviral tactics as well as biomarkers for tracking the spread of viruses (Jin and Wu 2015).

lncRNAs are vital to a plant's defense system during viral infection. TYLCV, transmitted by the whitefly *Bemisia tabaci*, is one of the most devastating viruses affecting tomatoes globally. Infected plants exhibit yellowing of young tissues; the leaves curled upward, delayed development, and significantly reduced yield. Many differentially expressed lncRNAs have been identified in tomatoes. Specific lncRNAs, such as lncRNA0850, lncRNA0372, lncRNA1514, lncRNA0957, lncRNA0519, and lncRNA0494, show significant responses after silencing lncRNA0957 through virus-induced gene silencing, it was found that the signs of genetically modified tomato plants were considerably diminished, suggesting its involvement in the defense network against TYLCV (Wang et al. 2018).

Pseudomonas syringae pv. tomato (Pst) is a bacterial pathogen that injects effector proteins into tomato cells. These effector proteins disturb the PTI response and enhance ETI. Significant changes in lncRNA expression have been observed at different time points after Pst infection. The expression of genes encoding glutathione S-transferases (GST) during ETI and PTI is regulated by these lncRNAs, which also play a role in photosynthesis. Identification of lncRNA regulatory modules associated with Pst resistance was made easier by transcriptome data, which supports the functions of protein-coding genes (Li et al. 2023b).

15.3.3 Proteomics-Based Approaches to Explore Tomato Defense Mechanism

Proteomics-based methods have been utilized to investigate the complex aspects of plant defenses against various pathogens (El Hadrami et al. 2012).

Proteomics is an investigative field that uses various methods, such as mass spectrometry and two-dimensional polyacrylamide gel electrophoresis (2D-PAGE), to identify and quantify the proteomes in the cell under specific conditions. On the other hand, a yeast-two hybrid system is utilized to detect cellular proteins that interact with particular proteins under investigation. Numerous studies have employed reversed-phase liquid chromatography coupled with mass spectrometry (LC-MS) (Quirino et al. 2010).

In their initial study on xylem sap proteins in tomatoes infected with *F. oxysporum f.*sp. *lycopersici* (Fol), Houterman et al. identified 21 proteins specific to the tomato-pathogen interaction and 13 proteins present in infected plants. They used mass spectrometry analysis following 2D-PAGE fractionation. Although the identification capacity of the method is limited, three small proteins and enzymes were predicted during tomato colonization (Sant'Ana and Lefsrud 2018).

Additionally, Balmant et al. used the cysteine labeling tandem mass tags proteomics technology to investigate the protein redox activities of *P. syringae* tomato resistant (PtoR) and susceptible (prf3) genotypes. Their findings revealed 90 putative redox-regulated proteins associated with carbohydrate metabolism, cell wall biogenesis, brassinosteroid production, and cysteine. These proteins may have the potential for redox regulation and signaling in the defense mechanisms of tomato plants (Sant'Ana 2019).

15.3.3.1 Post-translational Modifications (PTMs) in Tomatoes' Immune Response

PTMs are adaptable regulatory modifications critical for plant immune systems. Crucially, PTMs perform important functions in cross-signaling that act as fine-tuning processes to regulate cellular responses to pathogen infection (De Vega et al. 2018).

Effectors have the potential to break the stability of the host proteins. *Pseudomonas syringae* pv. tomato DC3000 produces the effector AvrPtoB, which has a plant U-box (PUB) type E3 ubiquitin ligase domain at its carboxyl terminus. In addition to myristoylation, the effector AvrPto, which directly interacts with the host protein Pto in resistant tomato plants, undergoes phosphorylation as a second type of PTM (Yin et al. 2019).

Respiratory Burst Oxidase Homolog D (RBOHD) is involved in the production of reactive oxygen species (ROS) during the hypersensitive response. S-Nitrosylation of RBOHD modulates its activity, ensuring a controlled ROS production that can effectively limit pathogen spread without causing excessive damage to plant tissues (Liu and Zhang 2021). Overall, the PTMs are important for the dynamic regulation of protein functions in tomato plants, enabling them to respond swiftly and effectively to pathogen attacks.

15.3.3.2 Tomato Protein–Protein Interactions (PPIs)

PPIs are crucial for comprehending the molecular processes underlying metabolic pathways and signal transduction at both cellular and systemic levels. PPI has become a crucial determinant of whether plant interactions with pathogens lead to disease or effective plant defense. Plant resistance genes and pathogen avirulence genes interact genetically to allow plants to detect pathogens and initiate defense mechanisms (Bogdanove 2002).

Tomato plants have developed intricate defense mechanisms to combat pathogens, heavily relying on protein–protein interactions. These interactions are essential for recognizing threats, signaling, and executing defense responses. Yeast-two hybrid screens were used to find tomato proteins that interact with either the Tm-22-LRR or the tomato mosaic virus (ToMV) movement protein. The ToMV movement protein and Tm-22-LRR were used as the appropriate baits in a screening of a tomato cDNA library. During these screenings, it was discovered that SGT1 interacted with Tm-22 and that the Rubisco small subunit (RbCS) interacted with the movement protein of the tomato mosaic virus (ToMV). It was demonstrated that type I J-domain proteins MP-Interacting Protein 1 (MIP1) interacted with the ToMV movement protein and Tm-22. Nicotiana benthamiana MIP1s (NbMIP1s) are a class of J-domain proteins that may function as molecular chaperones to aid in protein folding and preserve protein structure, and they are necessary for Tm-22-mediated resistance against TMV and ToMV (Chaudhary and Atamian 2017).

Researchers used a novel method combining lectin affinity chromatography with two-dimensional liquid chromatography and MALDI-MS to enrich and identify glycoproteins in ripe tomato fruit (Ruiz-May and Rose 2013). In silico tools and databases play a crucial role in studying PPI interactions. Predicted Tomato Interactome Resource (PTIR) utilizes computational methods to predict interactions among tomato proteins based on orthologous interactions from model organisms (Sant'Ana and Lefsrud 2018).

15.3.4 Metabolomics in Defense Priming of Tomatoes

The metabolome refers to the complete array of metabolites produced by metabolic pathways inside the plant system. Representing the biochemical phenotype of the cell, metabolites are the end products of gene expression (Weckwerth 2003). Metabolites have distinct chemical and physical characteristics; analytical methods and separation are needed to create metabolic profiles for plant samples. Various analytical methods have been utilized in plant systems to measure the levels of metabolites, such as thin-layer chromatography (TLC), gas or liquid chromatography within mass spectrometry (GC/LC-MS), liquid chromatography-electrochemistry-mass spectrometry (LCEC-MS), NMR, direct infusion mass spectrometry (DIMS), and capillary electrophoresis liquid-chromatography-mass spectrometry (Dikobe et al. 2023). In cereal and fruit crops such as rice, maize, barley, tomato, and pear, metabolites play a critical role in responding to different biotic stressors (Allwood et al. 2021).

Recent work on the interactions between plants and fungi has shown that tomato phenylalanine ammonia-lyase (PAL) plays a critical role in providing resistance against *V. dahlia*, implies that the phenylpropanoid pathway's enzymes interact with those involved in secondary metabolism related to defense, fortifying cell walls (Ali et al. 2018).

Metabolic engineering was used to enhance tomato crops' production of the secondary metabolite anthocyanin. H-NMR spectroscopy was used to identify metabolites involved in tomato-nematode interactions, with a focus on substances related to root defense against nematodes (Pentimone et al. 2020). Consequently, metabolomics provides a thorough understanding of numerous functions ranging from the genome to the metabolome and related phenotypic features when paired with other omics technologies like transcriptomics, proteomics, and genomics (Vailati-Riboni et al. 2017).

15.4 Challenges and Future Directions

S. lycopersicum is a globally significant crop due to its excessive cultivation and nutritional value. As the world population increases, the need for food also rises. Thus, it is important to generate tomato varieties that are very high-yielding, resistant to pathogens, and of the finest quality. Even though tomatoes are in great demand worldwide, diseases remain a major constraint responsible for the low yield. The tomato plant possesses an inducible defense to inhibit attack by pathogens such as bacteria, fungi, viruses, and nematodes through metabolic, biochemical, and molecular mechanisms. However, understanding the complex interactions between tomato plants and diverse pathogens remains challenging due to the intricate nature of these relationships.

Advanced Omics approaches have emerged as a promising alternative to classical breeding strategies. Powered by mass sequencing and advanced data processing through omics technologies, it promises to revolutionize agricultural practices. They significantly enhance breeding programs' accuracy, facilitating the development of disease-resistant cultivars that require fewer chemical controls and supporting more environmentally friendly, sustainable agriculture. However, several challenges need to be addressed to fully harness the potential of omics techniques in developing robust, disease-resistant tomato varieties (Fig. 15.4). Tomatoes exhibit extensive genetic diversity, with numerous cultivars and wild relatives possessing unique defense traits.

Tomatoes exhibit extensive genetic diversity and are subjected to various environmental stresses, such as temperature fluctuations, drought, salinity, and pathogen pressure. These factors can alter gene expression and metabolic pathways, affecting

Fig. 15.4 Common challenges associated with tomato research

the tomato plant's defense mechanisms. Advanced computational methods and machine learning algorithms are needed to integrate and analyze massive amounts of omics data effectively. This integration can reveal novel regulatory networks and key defense-related genes, providing deeper insights into plant-pathogen interactions. Regardless, it is plausible to believe that many of the technological advancements mentioned here will lead to more sustainable, disease-resistant varieties that can meet global food security and societal demands now and in the future.

15.5 Conclusion

Tomato is a major food crop worldwide; however, it is greatly affected by pathogens which affects its yield. Recent multi-omics approaches and their advancements have enabled researchers to study the underlying mechanisms that control plant responses to various environmental stresses and further help in developing resilient varieties. Tomato plants are also susceptible to various pathogens and virus attacks, which can be indicated through alteration in their phenotypes. Major genetic mechanisms to help tomato plants survive against these stresses include R-genes, TFs, and other response regulators. Further, the tomato pan-genome is a major resource for exploring genes that provide resistance against pathogens. These genes can further be manipulated to produce more sustainable varieties by integrating the genomic transcriptomic, protein interaction, and metabolomic information.

References

Ali M, Cheng Z, Ahmad H, Hayat S (2018) Reactive oxygen species (ROS) as defenses against a broad range of plant fungal infections and case study on ROS employed by crops against Verticillium dahliae wilts. J Plant Interact 13(1):353–363

Allwood JW, Williams A, Uthe H, van Dam NM, Mur LA, Grant MR, Pétriacq P (2021) Unravelling plant responses to stress—The importance of targeted and untargeted metabolomics. Metabolites 11(8):558

Alonge MC (2021) Pan-genomics and the structural diversity of plant genomes. Johns Hopkins University

Altayli E (2020) Regulator non-coding RNAs: miRNA, siRNA, piRNA, lncRNA, circRNA. J Clin Med Kazakhstan 6(60):29–39

Andersen EJ, Ali S, Byamukama E, Yen Y, Nepal MP (2018) Disease resistance mechanisms in plants. Genes (Basel) 9(7):339

Anton BP, Roberts RJ (2021) Beyond restriction modification: epigenomic roles of DNA methylation in prokaryotes. Annu Rev Microbiol 75(1):129–149

Arafa RA, Kamel SM, Taher DI, Solberg SØ, Rakha MT (2022) Leaf extracts from resistant wild tomato can be used to control late blight (Phytophthora infestans) in the cultivated tomato. Plants (Basel) 11(14):1824

Barad S, Sela N, Dubey AK, Kumar D, Luria N, Ment D, Cohen S, Schaffer AA, Prusky D (2017) Differential gene expression in tomato fruit and Colletotrichum gloeosporioides during colonization of the RNAi–SlPH tomato line with reduced fruit acidity and higher pH. BMC Genomics 18:1–21

Barone A, Matteo A, Carputo D, Frusciante L (2009) High-throughput genomics enhances tomato breeding efficiency. Curr Genomics 10(1):1–9

Bogdanove AJ (2002) Protein-protein interactions in pathogen recognition by plants. Plant Mol Biol 50:981–989

Campos MD, Felix MDR, Patanita M, Materatski P, Albuquerque A, Ribeiro JA, Varanda CJB (2022) Defense strategies: the role of transcription factors in tomato–pathogen interaction. Biology (Basel) 11(2):235

Caruso AG, Panno S, Ragona A, Peiró R, Vetrano F, Moncada A, Miceli A, La Marra CM, Galipienso L, Rubio L (2024) Screening local sicilian tomato ecotypes to evaluate the response of tomato brown rugose fruit virus infection. Agronomy 14(3):412

Chaitanya K (2019) Genome and genomics. Springer

Chaudhary R, Atamian HJ (2017) Resistance-gene-mediated defense responses against biotic stresses in the crop model plant tomato. J Plant Pathol Microbiol 8(2):1000404

Chen T, Lv Y, Zhao T, Li N, Yang Y, Yu W, He X, Liu T, Zhang B (2013) Comparative transcriptome profiling of a resistant vs susceptible tomato (Solanum lycopersicum) cultivar in response to infection by tomato yellow leaf curl virus. PLoS One 8(11):e80816

Chiusano ML, D'Agostino N, Traini A, Licciardello C, Raimondo E, Aversano M, Frusciante L, Monti L (2008) ISOL@: an Italian SOLAnaceae genomics resource. BMC Bioinf 9:1–11

Das PP, Rana S, Muthamilarasan M, Kannan M, Ghazi IA (2021) Omics approaches for understanding plant defense response. Omics Technol Sustain Agric Global Food Secur 1:41–83

De Vega D, Newton AC, Sadanandom A (2018) Post-translational modifications in priming the plant immune system: ripe for exploitation? FEBS Lett 592(12):1929–1936

Delseny M, Han B, Hsing YI (2010) High throughput DNA sequencing: the new sequencing revolution. Plant Sci 179(5):407–422

Dikobe T, Masenya K, Manganyi MC (2023) Molecular technologies ending with 'omics': the driving force toward sustainable plant production and protection. F1000Research 12:480

Dixon MS, Golstein C, Thomas CM, Van der Biezen EA, Jones JD (2000) Genetic complexity of pathogen perception by plants: the example of Rcr3, a tomato gene required specifically by Cf-2. Proc Natl Acad Sci USA 97(16):8807–8814

Dixon RA, Bouton JH, Narasimhamoorthy B, Saha M, Wang ZY, May GD (2007) Beyond structural genomics for plant science. Adv Agron 95:77–161

Dong Z, Chen Y (2013) Transcriptomics: advances and approaches. Sci China Life Sci 56:960–967

Eckshtain-Levi N, Lindeberg M, Vallad G, Martin G (2019) The tomato Pto gene confers resistance to Pseudomonas floridensis, an emergent plant pathogen with just nine type III effectors. Plant Pathol 68(5):977–984

El Hadrami A, El-Bebany AF, Yao Z, Adam LR, El Hadrami I, Daayf F (2012) Plants versus fungi and oomycetes: pathogenesis, defense and counter-defense in the proteomics era. Int J Mol Sci 13(6):7237–7259

Ercolano MR, Sanseverino W, Carli P, Ferriello F, Frusciante L (2012) Genetic and genomic approaches for R-gene mediated disease resistance in tomato: retrospects and prospects. Plant Cell Rep 31:973–985

Erpen L, Devi HS, Grosser JW, Dutt M (2018) Potential use of the DREB/ERF, MYB, NAC and WRKY transcription factors to improve abiotic and biotic stress in transgenic plants. Plant Cell Tissue Organ Cult (PCTOC) 132:1–25

Flagel LE, Blackman BK (2012) The first ten years of plant genome sequencing and prospects for the next decade. In: Plant genome diversity volume 1: plant genomes, their residents, and their evolutionary dynamics. Springer, pp 1–15

Gomez-Casati DF, Pagani MA, Busi MV, Bhadauria V (2016) Omics approaches for the engineering of pathogen resistant plants. Curr Issues Mol Biol 19(1):89–98

Goutam U, Thakur K, Salaria N, Kukreja S (2018) Recent approaches for late blight disease management of potato caused by Phytophthora infestans. In: Fungi and their role in sustainable development: current perspectives. Springer Singapore, Singapore, pp 311–325

Han Y, Gao S, Muegge K, Zhang W, Zhou B (2015) Advanced applications of RNA sequencing and challenges. Bioinform Biol Insights 9:BBI.S28991

Hanson PM, Yang R-Y, Wu J, Chen J-T, Ledesma D, Tsou SC, Lee T-C (2004) Variation for antioxidant activity and antioxidants in tomato. J Am Soc Hort Sci 129(5):704–711

Hoenig JM (2014) The canned and the fresh: the making and remaking of American food culture. The Pennsylvania State University
Hoshikawa K, Pham D, Ezura H, Schafleitner R, Nakashima K (2021) Genetic and molecular mechanisms conferring heat stress tolerance in tomato plants. Front Plant Sci 12:786688
Hutcheson SW (1998) Current concepts of active defense in plants. Annu Rev Phytopathol 36(1):59–90
Jin W, Wu F (2015) Characterization of miRNAs associated with Botrytis cinerea infection of tomato leaves. BMC Plant Biol 15:1–14
Jin S, Han Z, Hu Y, Si Z, Dai F, He L, Cheng Y, Li Y, Zhao T, Fang L (2023) Structural variation (SV)-based pan-genome and GWAS reveal the impacts of SVs on the speciation and diversification of allotetraploid cottons. Mol Plant 16(4):678–693
Kawchuk LM, Hachey J, Lynch DR, Kulcsar F, Van Rooijen G, Waterer DR, Robertson A, Kokko E, Byers R, Howard RJ (2001) Tomato Ve disease resistance genes encode cell surface-like receptors. Proc Natl Acad Sci USA 98(11):6511–6515
Kawicha P, Tongyoo P, Wongpakdee S, Rattanapolsan L, Duangjit J, Chunwongse J, Suwor P, Sangdee A, Thanyasiriwat T (2023) Genome-wide association study revealed genetic loci for resistance to fusarium wilt in tomato germplasm. Crop Breed Appl Biotechnol 23(1):e43532311
Kiirika LM, Stahl F, Wydra K (2013) Phenotypic and molecular characterization of resistance induction by single and combined application of chitosan and silicon in tomato against Ralstonia solanacearum. Physiol Mol Plant Pathol 81:1–12
Kitinoja L, Odeyemi O, Dubey N, Musanase S, Gill GS (2019) Commodity system assessment studies on the postharvest handling and marketing of tomatoes in Nigeria, Rwanda and Maharashtra, India. J Hortic Postharvest Res 2(Special Issue-Postharvest Losses):1–14
Koeda S, Kitawaki A (2024) Breakdown of Ty-1-based resistance to tomato yellow leaf curl virus in tomato plants at high temperatures. Phytopathology 114(1):294–303
Koseoglou E (2017) The study of SlPMR4 CRISPR/Cas9-mediated tomato allelic series for resistance against powdery mildew. Wageningen University, Wageningen
Li N, He Q, Wang J, Wang B, Zhao J, Huang S, Yang T, Tang Y, Yang S, Aisimutuola P (2023a) Super-pangenome analyses highlight genomic diversity and structural variation across wild and cultivated tomato species. Nat Genet 55(5):852–860
Li Q, Shen H, Yuan S, Dai X, Yang C (2023b) miRNAs and lncRNAs in tomato: roles in biotic and abiotic stress responses. Front Plant Sci 13:1094459
Liu Y, Zhang H (2021) Reactive oxygen species and nitric oxide as mediators in plant hypersensitive response and stomatal closure. Plant Signal Behav 16(12):1985860
Lucas JA (2009) Plant pathology and plant pathogens. Wiley
Meraj TA, Fu J, Raza MA, Zhu C, Shen Q, Xu D, Wang Q (2020) Transcriptional factors regulate plant stress responses through mediating secondary metabolism. Genes (Basel) 11(4):346
Merk HL (2010) Identification and mapping of new tomato late blight resistance genes in an accession of the wild tomato species, Solanum pimpinellifolium
Milc J, Bagnaresi P, Aragona M, Valente M, Biselli C, Infantino A, Francia E, Pecchioni N (2019) Comparative transcriptome profiling of the response to Pyrenochaeta lycopersici in resistant tomato cultivar Mogeor and its background genotype—susceptible Moneymaker. Funct Integr Genomics 19:811–826
Mores A, Borrelli GM, Laidò G, Petruzzino G, Pecchioni N, Amoroso LGM, Desiderio F, Mazzucotelli E, Mastrangelo AM, Marone D (2021) Genomic approaches to identify molecular bases of crop resistance to diseases and to develop future breeding strategies. Int J Mol Sci 22(11):5423
Murphy D (2002) Gene expression studies using microarrays: principles, problems, and prospects. Adv Physiol Educ 26(4):256–270
Naik B, Kumar V, Rizwanuddin S, Chauhan M, Choudhary M, Gupta AK, Kumar P, Kumar V, Saris PEJ, Rather MA (2023) Genomics, proteomics, and metabolomics approaches to improve abiotic stress tolerance in tomato plant. Int J Mol Sci 24(3):3025
Nekrasov V, Wang C, Win J, Lanz C, Weigel D, Kamoun S (2017) Rapid generation of a transgene-free powdery mildew resistant tomato by genome deletion. Sci Rep 7(1):482

Ng DW, Abeysinghe JK, Kamali M (2018) Regulating the regulators: the control of transcription factors in plant defense signaling. Int J Mol Sci 19(12):3737

Nguyen TT, Le NT, Sim S-C (2021) Genome-wide association study and marker development for bacterial wilt resistance in tomato (Solanum lycopersicum L). Sci Hortic 289:110418

Oliveira AB, Moura CF, Gomes-Filho E, Marco CA, Urban L, Miranda MRA (2013) The impact of organic farming on quality of tomatoes is associated to increased oxidative stress during fruit development. PLoS One 8(2):e56354

Ong SN, Taheri S, Othman RY, Teo CH (2020) Viral disease of tomato crops (Solanum lycopesicum L): an overview. J Plant Dis Prot 127:725–739

Osei MK, Prempeh R, Adjebeng-Danquah J, Opoku JA, Danquah A, Danquah E, Blay E, Adu-Dapaah H (2018) Marker-assisted selection (MAS): a fast-track tool in tomato breeding. In: Recent advances in tomato breeding and production. IntechOpen, pp 93–113

Padilla-Hurtado B, Morillo-Coronado Y, Tarapues S, Burbano S, Soto-Suárez M, Urrea R, Ceballos-Aguirre N (2022) Evaluation of root-knot nematodes (Meloidogyne spp) population density for disease resistance screening of tomato germplasm carrying the gene Mi-1. Chil J Agric Res 82(1):157–166

Panno S, Davino S, Caruso AG, Bertacca S, Crnogorac A, Mandić A, Noris E, Matić S (2021) A review of the most common and economically important diseases that undermine the cultivation of tomato crop in the mediterranean basin. Agronomy 11(11):2188

Panthee DR, Brown AF, Yousef GG, Ibrahem R, Anderson C (2013) Novel molecular marker associated with Tm2a gene conferring resistance to tomato mosaic virus in tomato. Plant Breed 132(4):413–416

Pentimone I, Colagiero M, Rosso LC, Ciancio A (2020) Omics applications: towards a sustainable protection of tomato. Appl Microbiol Biotechnol 104:4185–4195

Peyraud R, Dubiella U, Barbacci A, Genin S, Raffaele S, Roby D (2017) Advances on plant–pathogen interactions from molecular toward systems biology perspectives. Plant J 90(4):720–737

Pombo MA, Zheng Y, Fernandez-Pozo N, Dunham DM, Fei Z, Martin GB (2014) Transcriptomic analysis reveals tomato genes whose expression is induced specifically during effector-triggered immunity and identifies the Epk1 protein kinase which is required for the host response to three bacterial effector proteins. Genome Biol. 15:1–16

Przybyla L, Gilbert LA (2022) A new era in functional genomics screens. Nat Rev Genet 23(2):89–103

Quirino B, Candido E, Campos P, Franco O, Krüger R (2010) Proteomic approaches to study plant–pathogen interactions. Phytochemistry 71(4):351–362

Rodríguez-Negrete EA, Guevara-Rivera EA, Arce-Leal ÁP, Leyva-López NE, Méndez-Lozano J (2023) A novel tomato spotted wilt virus isolate encoding a noncanonical NSm C118F substitution associated with Sw-5 tomato gene resistance breaking. Mol Plant Pathol 24(10):1300–1311

Rojas A (2004) A complex of begomoviruses affecting tomato crops in Nicaragua

Ruiz-May E, Rose JK (2013) Progress toward the tomato fruit cell wall proteome. Front Plant Sci 4:159

Saidi A, Hajibarat Z, Hajibarat Z (2020) Transcriptome analysis of Phytophthora infestans and Colletotrichum coccodes in tomato to reveal resistance mechanisms. Asia-Pacific J Mol Biol Biotechnol 28:39–51

Salem NM, Jewehan A, Aranda MA, Fox A (2023) Tomato brown rugose fruit virus pandemic. Annu Rev Phytopathol 61(1):137–164

Sant'Ana DVP (2019) The impact of wavelength in plant's response to extreme light-induced stress. McGill University (Canada)

Sant'Ana DVP, Lefsrud MH (2018) Tomato proteomics: tomato as a model for crop proteomics. Sci Hortic 239:224–233

Sawangjit S, Chatchawankanphanich O, Chiemsombat P, Attathom T, Dale J, Attathom S (2005) Molecular characterization of tomato-infecting begomoviruses in Thailand. Virus Res 109(1):1–8

Shilpa N, Sharma P, Thakur V, Sharma A, Rana R, Kumar P (2022) A status-quo review on management of root knot nematode in tomato. J Hortic Sci Biotechnol 97(4):403–416

Shirasawa K, Hirakawa H (2013) DNA marker applications to molecular genetics and genomics in tomato. Breed Sci 63(1):21–30
Sofy AR, Sofy MR, Hmed AA, El-Dougdoug NK (2019) Potential effect of plant growth-promoting rhizobacteria (PGPR) on enhancing protection against viral diseases. Field crops: sustainable management by PGPR. Springer, pp 411–445
Sun Y, Yang H, Li J (2022) Transcriptome analysis reveals the response mechanism of frl-mediated resistance to Fusarium oxysporum f sp. radicis-lycopersici (FORL) infection in tomato. Int J Mol Sci 23(13):7078
Suresh BV, Roy R, Sahu K, Misra G, Chattopadhyay D (2014) Tomato genomic resources database: an integrated repository of useful tomato genomic information for basic and applied research. PLoS One 9(1):e86387
Swett CL, Del Castillo Múnera J, Hellman E, Helpio E, Gastelum M, Lopez Raymundo E, Johnson H, Oguchi R, Hopkins A, Beaulieu J (2023) Monitoring for a new I3 resistance gene-breaking race of F. oxysporum f. sp. lycopersici (Fusarium wilt) in California processing tomatoes following recent widespread adoption of resistant (F3) cultivars: challenges with race 3 and 4 differentiation methods. Front. Plant Sci 14:1088044
Talakayala A, Ankanagari S, Garladinne M (2022) CRISPR-Cas genome editing system: A versatile tool for developing disease resistant crops. Plant Stress 3:100056
Tashkandi M (2018) CRISPR/Cas9-mediated viral interference in plants. Genome Biol 16:1–11
Tsitsigiannis DI, Antoniou PP, Tjamos SE, Paplomatas EJ (2008) Major diseases of tomato, pepper and egg plant in green houses. Eur J Plant Sci Biotechnol 2(1):106–124
Tyagi S, Kumar R, Kumar V, Won SY, Shukla P (2021) Engineering disease resistant plants through CRISPR-Cas9 technology. GM Crops Food 12(1):125–144
Uffelmann E, Huang QQ, Munung NS, De Vries J, Okada Y, Martin AR, Martin HC, Lappalainen T, Posthuma D (2021) Genome-wide association studies. Nat Rev Methods Primers 1(1):59
Vailati-Riboni M, Palombo V, Loor JJ (2017) What are omics sciences? Periparturient diseases of dairy cows: a systems biology approach. Springer International Publishing, Cham, pp 1–7
Van Deynze A, Stoffel K, Buell CR, Kozik A, Liu J, van der Knaap E, Francis D (2007) Diversity in conserved genes in tomato. BMC Genomics 8:1–9
Wakil W, Brust GE, Perring TM (2018) Tomato and management of associated arthropod pests: past, present, and future. In: Sustainable management of arthropod pests of tomato. Academic Press, pp 3–12
Wang Y, Zou W, Xiao Y, Cheng L, Liu Y, Gao S, Shi Z, Jiang Y, Qi M, Xu T (2018) MicroRNA1917 targets CTR4 splice variants to regulate ethylene responses in tomato. J Exp Bot 69(5):1011–1025
Weckwerth W (2003) Metabolomics in systems biology. Annu Rev Plant Biol 54(1):669–689
Wu L, Chen H, Curtis C, Fu ZQ (2014) Go in for the kill: how plants deploy effector-triggered immunity to combat pathogens. Virulence 5(7):710–721
Xu J, Ranc N, Muños S, Rolland S, Bouchet J-P, Desplat N, Le Paslier M-C, Liang Y, Brunel D, Causse M (2013) Phenotypic diversity and association mapping for fruit quality traits in cultivated tomato and related species. Theor Appl Genet 126:567–581
Yin J, Yi H, Chen X, Wang J (2019) Post-translational modifications of proteins have versatile roles in regulating plant immune responses. Int J Mol Sci 20(11):2807
Yu W, Zhao R, Sheng J, Shen L (2018) SlERF2 is associated with methyl jasmonate-mediated defense response against Botrytis cinerea in tomato fruit. J Agric Food Chem 66(38):9923–9932
Zaynab M, Fatima M, Abbas S, Umair M, Sharif Y, Raza MA (2018) Long non-coding RNAs as molecular players in plant defense against pathogens. Microb Pathog 121:277–282
Zeshan M, Khan M, Ali S, Arshad M (2016) Phenotypic evaluation of tomato germplasm for the source of resistance against tomato leaf curl virus disease. JAPS: J Anim Plant Sci 26(1)
Zhang J, Zhao J, Liang Y, Zou ZJE (2016) Genome-wide association-mapping for fruit quality traits in tomato. Euphytica 207:439–451
Zhang D, Li H, Liu G, Xie L, Feng G, Xu X (2023) Mapping of the Cladosporium fulvum resistance gene Cf-16, a major gene involved in leaf mold disease in tomato. Front Genet 14:1219898
Zhao T, Pei T, Jiang J, Yang H, Zhang H, Li J, Xu XJHPJ (2022) Understanding the mechanisms of resistance to tomato leaf mold: a review Hortic. Plant J 8(6):667–675

Future Horizons: Emerging "Omics" Technologies and Challenges in Tomato

16

Zakir Ullah, Javed Iqbal, Banzeer Ahsan Abbasi, Shumaila Ijaz, Shumaila Ahmad, Tabassum Yaseen, Rooma Waqar, Amin Fathi, Rashid Iqbal, Ghulam Murtaza, Ammara Saleem, Sobia Kanwal, and Tariq Mahmood

Abstract

Tomato (*Solanum lycopersicum*) breeding has changed significantly with the advent of "omics" technologies, which provide unheard-of opportunities to increase yield, quality, and resistance. The chapter explores the transformative impact of emerging "omics" technologies on tomato breeding, highlighting their potential to enhance yield quality, stress resistance, and health-promoting attributes. With advancements like transcriptomics, metabolomics, high-throughput phenotyping, and gene editing, breeders can now decipher complex metabolic pathways and genetic networks more effectively. The sequencing of the high-quality tomato genome in 2012 has been pivotal, accelerating the integration of these technologies. Despite challenges in data integration and the need for sophisticated bioinformatics tools, the chapter underscores the promising future

Zakir Ullah and Javed Iqbal contributed equally with all other contributors.

Z. Ullah · S. Ahmad · R. Waqar · T. Mahmood
Department of Plant Sciences, Faculty of Biological Sciences, Quaid-i-Azam University, Islamabad, Pakistan
e-mail: zakirullah@bs.qau.edu.pk; tmahmood@qau.edu.pk

J. Iqbal (✉) · T. Yaseen
Department of Botany, Bacha Khan University, Khyber Pakhtunkhwa, Pakistan
e-mail: tabassumyaseen@bkuc.edu.pk

B. A. Abbasi (✉)
Department of Botany, Rawalpindi Women University, Rawalpindi, Pakistan

S. Ijaz
Department of Burn and Plastic Surgery, Shenzhen Institute of Translational Medicine, The First Affiliated Hospital of Shenzhen University, School of Biomedical Engineering, Shenzhen University Medical School, Shenzhen University, Shenzhen, China

A. Fathi
Department of Agronomy, Ayatollah Amoli Branch, Islamic Azad University, Amol, Iran

M. Waseem et al. (eds.), *Omics Approaches for Tomato Yield and Quality Trait Improvement*, https://doi.org/10.1007/978-981-96-3890-1_16

of "omics" in developing superior tomato varieties. It provides a comprehensive review of current trends and future directions in the field, emphasizing the role of innovative technologies in achieving global food security and sustainability goals.

Keywords

Genomics · Transcriptomics · Proteomics · Metabolomics · Tomato breeding · High-throughput phenotyping · Bioinformatics tools

16.1 Introduction

16.1.1 Overview of "Omics" in Plant Breeding

Numerous challenges encountered by field crops negatively impact agricultural productivity. These unfavorable conditions cause plants to adjust morphologically, physiologically, biochemically, molecularly, and cellularly (Alkan et al. 2015a). Temperature-induced stress can result in oxidative damage in plants, whether occurring in isolation or in conjunction with other stressors (Chen et al. 2018). Investigating the mechanisms by which plants have evolved stress tolerance provides crucial insights for developing innovative breeding strategies to produce stress-tolerant crops (Chen et al. 2017). Traditionally, efforts to enhance agricultural yield have relied on conventional breeding techniques, primarily based on phenotypic selection (Das et al. 2015). However, over the past two decades, advancements in genomics have significantly enhanced the precision and efficiency of breeding programs for various temperate crop species (Huang et al. 2017). In recent decades, numerous biotechnological tools have emerged to elucidate the pathways and mechanisms underlying plant responses to environmental stress (Jiang et al. 2016). Advances in tissue or single-cell level omics techniques, including transcriptomics, proteomics, metabolomics, miRNA omics, and phenomics, have shown enormous potential in advancing crop improvement and

R. Iqbal
Department of Agronomy, Faculty of Agriculture and Environment, The Islamia University of Bahawalpur, Bahawalpur, Pakistan

Department of Life Sciences, Western Caspian University, Baku, Azerbaijan
e-mail: rashid.iqbal@iub.edu.pk

G. Murtaza
School of Agriculture, Yunnan University, Kunming, Yunnan, China

A. Saleem
Institute of Botany, University of the Punjab, Lahore, Pakistan

S. Kanwal
Department of Biology and Environmental Sciences, Allama Iqbal Open University, Islamabad, Pakistan
e-mail: sobia.kanwal@aiou.edu.pk

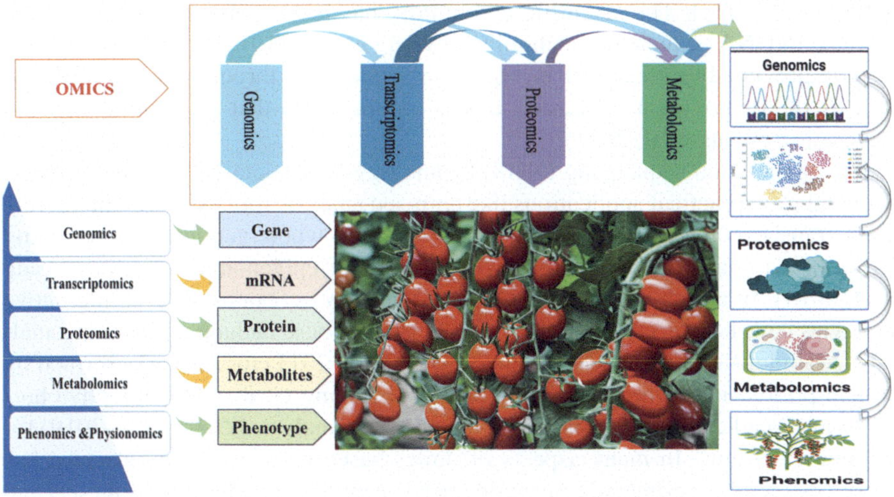

Fig. 16.1 Omics' technologies (genomics, transcriptomics, proteomics, metabolomics) in plant breeding

ensuring global food security (Fig. 16.1) (Li et al. 2015, 2018). Additionally, the scientific community has been dedicated to developing intelligent systems and genomic resources leveraging artificial intelligence (AI) to bridge the gap between genotype-phenotype and provide practical methods for cultivating climate-resilient crop species (Li et al. 2015).

Machine learning models, notably, can predict complex traits and analyze large datasets, thereby enhancing cultivar production and enabling informed decisions regarding the development of specific traits, including temperature resilience (Li et al. 2019). Concurrently, intelligent breeding strategies, such as speed breeding, shorten breeding cycles and expedite plant growth, facilitating the creation of new cultivars and accelerating the development of high-yielding, stress-tolerant plant varieties (Liu et al. 2015a, 2017). The genetic foundations of the gene networks involved in crop development and stress responses must be found using integrated omics approaches (Liu et al. 2016). To pinpoint the mechanisms behind plant growth, yield, stress responses, and other economically significant characteristics of essential food and fiber crops including wheat, rice, maize, tomato, and cotton, multi-omics methods have been widely used (Mittler et al. 2017).

16.1.2 Systematic Development of "Omics" in Plant Science

Gene discovery was aided, and a broad understanding of gene function was supplied by genomic research conducted in model plants in the first ten years of the twenty-first century (Niu et al. 2019). In the second decade, omics-based techniques have enabled us to address the intricate global biological systems that underlie many plant functions, thanks to recent technological advancements

(Ullah et al. 2024). The scientific community has been transformed by a recent advance in DNA sequencing technology (Wang et al. 2019). These developments include previously unheard-of increases in sequencing throughput and scale, as well as the deployment of several unique applications that go beyond genome sequencing (Wang et al. 2021).

Next-generation sequencing (NGS) technology is accelerating genome projects while offering practical applications like chip-seq analysis for DNA-protein interactions, transcriptome and non-coding RNA analysis, whole-genome re-sequencing for variation analysis, and quantitative detection of epigenomic dynamics (Yang et al. 2014). In addition to transcriptional regulatory networks, several alternative strategies have been established (Yang et al. 2017). These include interactome analysis (Arabidopsis Interactome Mapping Consortium 2011) for networks formed by protein-protein interactions, hormone analysis (Zhang et al. 2020) for phytohormone-mediated cellular signaling, and metabolome analysis (Zhang et al. 2019) for metabolic systems. In many aspects of omics-based research, bioinformatics has proven essential for efficiently managing heterogeneous genome-scale data sets and extracting valuable insights (Yang et al. 2021). Thus, databases including omics data from more recent innovative analytical platforms make up the core of systems analysis (Priyadarshi et al. 2021).

Here are some current developments in plant omics and bioinformatics along with an overview of popular platforms such as the Genome Analyzer/Hiseq (Illumina Solexa), the SOLiD (Life Technologies), and the 454 FLX (Roche). The single-molecule sequencing devices PacBio RS (Pacific Biosciences) and Heliscope (Helicos) (Priyadarshi et al. 2021; Cao et al. 2022) are examples of more modern systems, as is the semiconductor chip-based Ion Torrent (Life Technologies) (Priyadarshi et al. 2021). The current Illumina Hiseq2000 short reader, for example, can generate 600 Gb of sequence data in 100 bp 2 paired-end reads in a 400 bp run, whereas the Roche 454 platform, an NGS long reader, can generate one million reads in a 400 bp run (Ijaz et al. 2024).

16.1.3 Importance of "Omics" in Modern Agriculture

Before we take a drink, omics can tell us if the steak we are about to savor will be juicy and tender or if the wine in your glass will be sweet or dry (Varshney et al. 2018). They can also give you a genetic map that will help you produce the darkest red tomato imaginable. The panomic arsenal of omic instruments is contributing greatly to crop protection, raising agricultural output for food, feed, and energy, improving the quality, flavor, and nutritional makeup of food crops, and having a big impact on agricultural economics (Mahmood et al. 2022). For more effective pest control, omics have shed light on the molecular processes behind plant tolerance to herbicides and insect resistance to insecticides (Fig. 16.2). Improved cultivars and a better knowledge of the mechanisms behind weed and insect resistance can result from the linking of genes to characteristics, which increases scientific confidence (Raza et al. 2023). Omics enables a systems biology perspective to comprehend the intricate interplay of genes, proteins, and metabolites inside the ultimate phenotype.

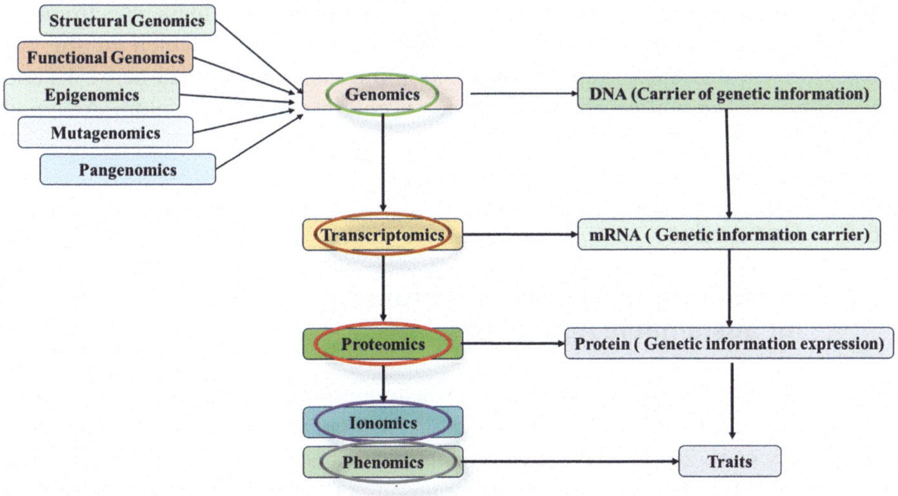

Fig. 16.2 Systematic development of "Omics" in plant science

To grow and maintain crops, this integrated strategy mainly uses computer analysis, bioinformatics, chemical analytical techniques, and other biological disciplines (Bibi et al. 2023).

16.1.4 Advanced Omics Strategies in Tomato Research and Breeding

According to (Alotaibi et al. 2021), tomato is the Solanaceae family member that has been studied the most. Inbreeding tolerance, short generation time, fundamental diploid genetics, well-established genetic transformation technology, and large, well-characterized genetic resources are the main reasons for this (Singh et al. 2019). Globally, the tomato is a key fruit and an essential component of processed foods like ketchup and tomato puree. The increasing needs of consumers for a wide range of varietals, improved flavor and quality, year-round availability, and extra health advantages need scientists to explore the genetic and regulatory systems behind biochemical properties (Yoosefzadeh Najafabadi et al. 2023). In the postgenomic age, omics techniques such as transcriptomics, proteomics, and metabolomics are essential instruments for clarifying important biological functions. (Chen et al. 2023). To improve quality and yield, tomato metabolic networks have been analyzed using transcriptomics and metabolomics, either alone or in tandem. Gaining an understanding of these networks is essential to enhancing tomato plant biochemical composition by enhancing flavor, quality, stress resistance, and health-promoting metabolites (such as antioxidants). (Watson et al. 2022) Sequencing of the tomato genome has made it possible to use this genetic information to annotate gene expression arrays more precisely (Ye et al. 2022). Tomato's study has made extensive use of metabolic approaches, which combine nuclear magnetic resonance

(NMR) spectroscopy and chromatographic separation with mass spectrometry (MS). These techniques are essential for studying metabolites, which are the last phenotypic expressions of homeostasis within intricate biochemical networks and affect desirable qualities including taste, scent, texture, and color (Alahmad et al. 2024).

16.2 Genomics

16.2.1 Breeding with Genomic Assistance for Sustainable Agriculture

Molecular markers have brought about a revolution in crop breeding by improving the speed and precision of trait selection without the need for phenotyping (Großkinsky et al. 2018). The use of sequencing technology has been strengthened by recent advancements. Breeding programs use a variety of genomics-assisted breeding (GAB) techniques, such as genomic editing (PAGE/RAGE), advanced backcross quantitative trait loci (AB-QTL), marker-assisted backcrossing (MABC), marker-assisted selection (MAS), marker-assisted recurrent selection (MARS), and marker-assisted backcrossing (MABC) (Muthamilarasan et al. 2019). Using MAS, crops including wheat, rice, and maize have greatly increased their ability to withstand drought. A greater collection of markers is used in genomic selection (GS), which improves prediction accuracy and leads to achievements in disease resistance and abiotic stress tolerance. Adding GWAS results has improved GS models (Ijaz et al. 2023a). The understanding of genetic variety has improved thanks to selective sweeps and haplotype-based GWAS, which has helped national breeding programs improve their germplasm (Mochida and Shinozaki 2011; Jamil et al. 2020).

Pangenomes, which comprise both the essential and redundant genomes, reflect the whole genetic repertoire of a species and incorporate all its variants. Crops such as wheat, rice, soybean, sesame, and tomato have shown structural variations in their pangenomes, which reduces the impact of single-sample bias. A pangenome of tomatoes derived from 725 different accessions revealed 351 Mbp of sequences absent from the reference genome and a 4-bp alteration in the regulatory region of the TomLoxC gene that impacts the flavor of the fruit. Pangenomics is being improved by cutting-edge long-read sequencing and bioinformatics technologies, which are helping identify critical genes for crop development (Yu et al. 2023; Ullah et al. 2023a).

16.2.2 Transforming Agriculture: Genomic-Assisted Plant Breeding

Plant breeding likely began as societies transitioned from hunter-gatherers to agrarian communities (Arabidopsis Interactome Mapping Consortium 2011). Crop domestication, the modification of wild plants for sustainable farming, started between 9000 and 11,000 years ago. Early farmers used selection techniques, and over centuries, crop trading diversified agriculture and promoted breeding. Scientific breeding emerged in the 1800s with Gregor Mendel's experiments on garden peas, which established the rules of inheritance through hybridization. Initially overlooked, Mendel's work was rediscovered in 1900 by William Bateson, leading to the integration of Mendel's findings with the Boveri-Sutton chromosome theory, forming the basis of classical genetics (Lin et al. 2011; Ullah et al. 2023b). This advancement propelled plant breeding into a new era, utilizing transmission genetics to develop new breeding populations. Early breeding relied on phenotypic assessments, but the introduction of the Maxam-Gilbert sequencing technique in the late 1970s revolutionized the field by providing nucleotide sequence data for molecular marker creation (Mochida and Shinozaki 2011).

Then came three generations of sequencing technologies: the first, known as Sanger sequencing; the second, known as next-generation sequencing, mostly used by Illumina; and the third, known as Pacific BioScience and Oxford Nanopore (Ullah et al. 2023c; Aksoy et al. 2022). These technologies made use of single nucleotide polymorphisms (SNPs) and other molecular markers to extend the use of genomic selection (GS) and marker-assisted selection (MAS). Full genome sequence information became crucial for the introgression of desirable traits and understanding gene expression under stress conditions through transcriptome sequencing (RNAseq) (Tahir ul Qamar et al. 2020). Breeders can now identify marker-trait associations (MTA) by genotyping populations and analyzing phenotypic-genotypic relationships. This integration of big data with traditional MAS approaches enhances genomics-assisted selection (GAS), improving breeding programs by reducing linkage drag and false positives (Paulpriyaa et al. 2022). Genomic selection (GS) involves predicting breeding values using genome-wide markers. This approach, rooted in bioinformatics, statistics, and machine learning, represents a new paradigm in plant breeding (Esposito et al. 2019).

16.2.3 Transforming Plant Breeding: The Role of Whole Genome References

Although completeness varies, the generation of whole genome references for plant species has changed due to the decreasing cost of nucleotide sequencing. Some references are published as drafts, while others have advanced to pseudomolecules attached to chromosomes (Roychowdhury et al. 2023). Pseudomolecule references are particularly valuable for plant breeders, providing comprehensive locus information for accurate molecular marker creation and gene identification (Tripodi et al. 2018). Draft genomes, aiding in precise genotyping, are beneficial for crop breeding. As of October 2020, 69 Viridiplantae genomes have been sequenced, with data available in databases such as NCBI, Phytozome, and Plant GDB (Bokszczanin 2019). Early challenges in plant genome assembly included large sizes, repetitive DNA, and polyploidy. Initial efforts focused on first-generation sequencing technologies, but model species like Arabidopsis, rice, and maize were sequenced.

Next-generation sequencing (NGS) technologies have reduced costs and resolved assembly issues, leading to the sequencing of over 250 angiosperm species. The genome database for angiosperms (GDA) provides comprehensive data on these genomes. Whole genome resources facilitate the creation of accurate genotyping schemes, enhancing genomics-assisted breeding (GAB). This method, expanding on MAS and GAS concepts, allows breeders to apply precise genotyping in field trials (Naqvi et al. 2022; Razzaq et al. 2021).

16.2.4 Genomics in Tomatoes

The tomato is the main member of the Solanaceae family, sometimes referred to as the nightshades, due to its economic importance to agriculture (Bayram and Gökırmaklı 2018). Additionally, it serves as a model organism for studies on fruit ripening, disease resistance, and plant development. As a result, during the past few decades, a vast array of genetic and molecular resources has been developed for the species (Singha et al. 2022). These resources include molecular markers, mapping populations, mapping populations, introgression lines, mutant collections, microarray chips, high-density genetic maps, BAC and EST libraries, and a high-quality draft genome sequence that was made public in 2012 (Bayram and Gökırmaklı 2018; Singha et al. 2022; Rasheed et al. 2017).

16.2.5 The Tomato Genome Project: An Understanding of a Model Plant

Tomatoes have a high degree of genome conservation at the macro and micro synteny levels with another Solanaceae, are diploids with a small genome size (about 900 Mb) and are accessible as homozygous inbreds (Ullah 2017). As a guide for wild tomatoes and other species in the larger Solanaceae family, the sequenced genome can be used. The wild cousins of tomatoes, the majority of which are crossable with tomatoes, have likewise maintained the allelic diversity needed for environmental and ecological adaptations (Fraser et al. 2020). The nuclear genome of tomatoes is made up of 12 chromosomes and around 950 Mbp of DNA, of which 28% is coding sequence, 11% is transposon, 59% is noncoding sequence, and 2% is organellar sequence (Li et al. 2020). Centromeric heterochromatic areas, which lack genes, make up around 77% of the chromosomal DNA (Devanna et al. 2023; Sampath et al. 2023). The tomato genome is thought to include 35,000 genes, the majority of which are in the distal euchromatic regions of the chromosomes (Narad et al. 2022). The tomato genome has a gene density of around 6.7 kb/gene, which is like that of Arabidopsis and rice (Vidal et al. 2023).

The subsequent study (Rosa-Martínez et al. 2023) found that the intergenic spacer regions contain a significant proportion of methylated tomato euchromatin. A part of the International Solanaceae Genome Project (SOL) is this endeavor (Ijaz et al. 2023b). Over the next 10 years, this large-scale research project (http://www.

sgn.cornell.edu/solanaceae-project/index.pl) will investigate the physical, evolutionary, and functional genomics of the Solanaceae family. It was initiated in 2003 (Muthuramalingam et al. 2019). However, the first important objective of the study is to determine high-quality sequencing for the euchromatic regions of tomato chromosomes as a reference for the Solanaceae (Ahmed et al. 2024). As of March 2007, around 17% of the target locations had been sequenced. Simultaneously, efforts are underway to enhance the tomato EST database and carry out more genome organization investigations on tomatoes (Muthuramalingam et al. 2019). According to this source around 214,000 ESTs have been produced so far. The tomato genome is predicted to encode about 35,000 genes, mostly located in euchromatic areas of the 12 tomato chromosomes, based on a study of the tomato EST database and many sequenced BAC libraries. Even though the tomato species' full gene repertoire is not represented by the EST-derived unigene sets, this quantity of DNA still makes up less than 25% of the total nuclear DNA in tomatoes (Ahmed et al. 2024; Singh et al. 2022). Data from 17,000 BACs that Syngenta recently gave to the *L. esculentum* × *L. pennellii* ILs were accessible to the tomato sequencing project (Mitra et al. 2021).

16.2.6 Tomato Sequencing Project

With participation from 14 different nations, an international team started the tomato sequencing project in 2004 (Ullah et al. 2023d). The strategy started with sequencing a BAC tiling route of the euchromatin, which makes up less than 25% of the 900 Mb genome but includes over 90% of the genes (Ullah et al. 2023d; Gatica-Arias 2020). The International Tomato Annotation Group (ITAG) annotated the genome by providing predicted gene models, proteins, cDNA alignments, repeats, and CDS. Versioned and available on the Sol Genomics Network website are the genome assembly and the annotations (Hasanuzzaman et al. 2013; Bineau 2022). A draft genome for *S. pimpinellifolium* was also made public by the consortium. It estimated a 0.6% divergence from the reference genome, "Heinz 1706," and identified multiple introgressed regions from *S. pimpinellifolium* that were identified in the "Heinz 1706" genome. This work provides the basis for breeding efforts that incorporate advantageous genes from wild tomato species into commercial germplasm using introgression analysis (Sato and Tabata 2016).

16.2.6.1 Published Research Enabled by the Tomato Genome

The high-quality *S. lycopersicum* "Heinz 1706" genome is now publicly available, which has greatly increased our understanding of tomatoes, especially in terms of bioinformatics studies, candidate gene identification linked to fruit development and ripening, and genomic reference for other Solanaceae species (Sato and Tabata 2016). These works demonstrate how the tomato genome may be used as a reference for closely related Solanaceae, which makes it possible to study genome evolution, orthologous genes, gene families, and gene mapping in tomato and other divergent plants (Pramanik et al. 2024). Breeding is moving away from marker-assisted selection and toward genome-wide methods thanks to modern sequencing technologies. Additionally, systems biology investigations of gene networks, transcriptomics, proteomics, and epigenomics are furthering fundamental research. Short reads from next-generation sequencing are challenging to put together in the absence of a high-quality reference genome. SNP discovery and divergence analysis have been made possible by using the tomato genome for reference-guided assembly using *S. pimpinellifolium* Illumina data (Tomato Genome Consortium, 2012). In the development of bioinformatics programs and other genome projects, such as the assembly of the Centaurea solstitial transcriptome (Sato and Tabata 2016) and the prediction of gene structures using Genome Threader (Muthuramalingam et al. 2019), it is also utilized as a benchmark. Analyzing the location and structure of genes as well as estimating their function based on sequence similarity constitute gene annotation. The chromosomal location of SlscADH1, an alcohol dehydrogenase involved in fruit ripening, and the identification of a single copy of the sulfite reductase gene (SiR) that protects against sulfite toxicity were both made possible by the tomato reference assembly (Sato and Tabata 2016; Pramanik et al. 2024). To ensure there were no phenotypic-related flaws, the PSY1 sequence of a yellow-fruited mutant was compared to the "Heinz 1706" sequence (Hajyzadeh and Egi 2023). Other research has mapped the target locus for RIPENING INHIBITOR (RIN) and identified proteins linked to reproductive barriers (Bineau 2022) using data from the tomato genome to predict gene function within the tomato and similar species (Sato and Tabata 2016).

16.2.6.2 Epigenetics and Gene Expression

Research on gene expression and epigenetics has been made easier by the sequencing of the tomato genome. To study changes in genomic methylation during fruit ripening, genome resequencing in conjunction with bisulfite sequencing has been used. This has shown differentially methylated areas with RIN binding sites (Hajyzadeh and Egi 2023). Thanks to the annotation of the tomato genome, it is now possible to identify miRNA-mRNA combinations in gene silencing experiments (Foolad and Panthee 2012) and make sure that there are no off-target effects (Bineau 2022). High-quality reference genomes, such as the tomato genome, are necessary for read mapping in RNA-seq technology, which is often employed in gene expression research (Sato and Tabata 2016). This has been used in research on the expression of the SUN, OFP, GABBY transcription factors (Ahmed et al. 2024), auxin or cytokinin-induced differential expression (Singh et al. 2022), and arbuscular mycorrhizal fungi (Hajyzadeh and Egi 2023).

16.2.6.3 Phenotype to Genotype Studies

Identifying DNA sequences' chromosomal locations is essential for creating markers for trait mapping. The tomato physical map has been utilized to fine-map bacterial spot resistance loci from *S. pennellii* introgressions (Foolad and Panthee 2012) and to find markers associated with yellow leaf curl resistance (Tiwari et al. 2022). To map TERMINATING FLOWER (TMF) and track its gene expression, markers based on the *S. pimpinellifolium* assembly have also been created (Hasan et al. 2021). Using annotations of the tomato genome, the Marker2sequence QTL tool was shown to be useful in identifying genes associated with QTLs (Sato and Tabata 2016). Sim et al. (Singh et al. 2022) examined population structure in farmed tomatoes and *S. pimpinellifolium* by mapping SNPs from the SolCAP tomato array to the genome (Pramanik et al. 2024).

16.2.7 Advancements in Tomato Breeding: From Molecular Markers to Genomic Selection

S. lycopersicum is one of the most significant vegetable crops in the world. It has special qualities, including an abundance of antioxidant chemicals and minerals (potassium, magnesium, and phosphorus) that boost our immune system and prevent cancer and cardiovascular disease (Vu et al. 2020). The tomato is a diploid autogenous plant with a short life cycle and a small genome (around 900 Mb). The tomato species has been created as a model plant for the development of several genetic and molecular tools, such as a high-quality draft genome sequence, high-density genetic maps, high-throughput molecular markers, introgression lines, and mutant collections (Tomato Genome Consortium) (Vu et al. 2020; Osei et al. 2018). Re-sequencing hundreds of genomes from landraces, cultivars, and wild relatives has also revealed a relatively low molecular diversity, but a significant occurrence of chromosomal rearrangements arising from remnants of wild introgressions (Lee et al. 2015).

Furthermore, the preservation of genome segments from wild relatives contributes significantly to the genetic diversity observed in the tomato gene pool under culture. These segments are used to introrse agronomically useful qualities like disease resistance and quality attributes (Sato and Tabata 2016). The creation of many molecular marker systems in the early 1980s fundamentally altered the course of plant breeding. Based on the interspecific populations of *L. pennellii* LA716 and *L. esculentum* LA925 (E6203), a highly saturated tomato reference linkage map is available, and expanding tomato sequencing studies are yielding more data to help construct more resolving genetic markers (Sato and Tabata 2016; Pramanik et al. 2024). Nonetheless, a significant problem with tomato crops continues to be the scarcity of DNA markers that may identify variation within the farmed species (Hajyzadeh and Egi 2023). Most recent efforts have been on using intraspecific high-resolution genetic markers like single nucleotide polymorphisms (SNPs) and insertion and deletion (InDels) to find variation among closely related individuals (Sharma et al. 2019; Ashraf and Foolad 2013).

By utilizing marker-assisted selection (MAS) to incorporate chromosomal segments carrying single genes or quantitative trait loci (QTLs) into classical phenotypic selection (PS), molecular markers were primarily incorporated into the process to enhance plant selection (Simko et al. 2021; Tiwari et al. 2022) contains reviews of some research on the identification of tomato QTLs and important genes that confer resistance to biotic and environmental stresses (Tiwari et al. 2022). Genomic selection (GS), which uses genome-wide markers to forecast breeding values, may be very helpful to breeding programs for complicated traits governed by several minor genes. The previous year saw the publication of groundbreaking research on the application of GS to generate tomato lines resistant to bacterial spots (Foolad and Sharma 2004), improve the number of soluble solids in fruit weight and content in F1 tomato varieties (Ibitoye and Akin-Idowu 2011), and transfer yield-related characteristics in tomato varieties from related wild species. Better and more reliable GS techniques are still needed for tomatoes. It is still necessary to optimize several GS programs in this species for tomato breeding, thus the impact of variables affecting the accuracy and performance of the model has not yet been evaluated (Prasanna et al. 2023). Among them, further research should be done on the genetic makeup of species, phenotyping techniques, population size, genetic relationships among individuals in the population, genotyping platforms, metrics for measuring marker quality, and GS scheme design (Gupta et al. 2010). Here, we highlight the key developments in plant GS and go over the primary obstacles to its use in tomato breeding. Also discussed are tomato GS schemes both within and across breeding generations and the possibility of choosing parents according to their estimated GEBV (He et al. 2014). The NGS-based approach does not require prior knowledge of the nucleotide sequence information, in contrast to array- or PCR-based genotyping systems. For species without a reference genome, NGS-based genotyping can therefore be broadly used; nevertheless, it is not as accurate as array- or PCR-based systems, nor is it repeatable across several trials. We've already talked about those platforms' specifics (Wai et al. 2020). We believe that the present review article should include a check on those platforms' popularity over the previous ten years. Now, Taqman (Applied Biosystems), SNPlex (Applied Biosystems), BioMark HD (Fluidigm), Axiom Biobank (Affymetrix), Infinium II (Illumina), GoldenGate (Illumina), and iPlex (Sequenome) are the commercially available SNP genotyping methods that rely on arrays or PCRs (Rashid et al. 2014).

16.2.8 Marker-Assisted Selection (MAS) and Genomic Selection (GS)

Molecular markers were used in plant breeding programs to create the original concept of molecular association selection (MAS) due to its objectivity and repeatability, which were enabled by tighter genome segmentation (Orellana et al. 2010). Therefore, the identification of genetic differences in various populations associated with agriculturally relevant traits that underpin potential genes is an essential field of research in crop genetics and breeding science. Scientists have been working on a precise MTA for a productive and successful MAS (Wai et al. 2020). At the beginning of the molecular marker era, some experimented with single-marker analysis (SMA) for MTA. This process involves creating a regression model and performing an F-test analysis of variance (ANOVA) for each marker locus using phenotypic data based on the pertinent statistical models (Prasanna et al. 2023). Currently, SMA is used more often than MAS for a quick assessment of genotype and phenotypic associations. The two main streams of MTA using genomic approaches now are genome-wide association studies and QTL analysis (GWAS). Using bi-parental populations and linkage mapping, QTLs are essentially detected. Newly developed genomics technologies improve this process' accuracy by allowing for higher marker mapping counts, which results in high-density genetic maps (He et al. 2014).

To overcome these limitations, the mapping resolution may be greatly enhanced by employing newly generated bi-parental populations, such as a multi-parent advanced generation intercross (MAGIC) population (He et al. 2014; Wai et al. 2020). High throughput genotyping, which enables researchers to examine a huge number of mapped individuals, somewhat counteracts these advantages. To improve MTA resolution, GWAS, on the other hand, may consider a far greater number of historical recombination events for the population being studied (Tiwari et al. 2022). With this innovative and potent technique, high-throughput genotyping enables the evaluation of diversity panels, core collections, and even nested-association mapping (NAM) populations (Gupta et al. 2010). Although GWAS has a better resolution than QTL mapping, because of the subpopulation structures of the populations under analysis, it may produce false positives or true negatives. Even though GWAS uses a sizable sample of the population if the proper selection of population members is overlooked, the entire procedure may still be jeopardized because of an increase in genetic variation (Wai et al. 2020). Genotyping mistakes can arise from polyploidy (or paleopolyploid) when allelic variants and sub-genomic (or paralogous) variations are confused. As said, the biallelic nature of SNPs makes it impossible to fix this issue until reference genomes are firmly established. Another issue with GWAS is the exclusion of rare alleles when filtering genotyping data based on minor allele frequency (MAF) (Rashid et al. 2014).

Target accessions are chosen with the use of markers as the last stage in genomics-assisted breeding. Traditionally, breeders used their judgment to make selections based on phenotypic characteristics. Nonetheless, many contemporary selection tactics depend on the creation of MTA in combination with genetic markers, which are widely distributed and highly objective in plant genomes. Since the creation of molecular markers, MAS has been utilized in the breeding of many crops, including rice (Rashid et al. 2014), wheat (Orellana et al. 2010), sorghum (Gupta et al. 2009), and soybean (Liu et al. 2015b). Dissecting genomic areas becomes hard if favor alleles are closely related to harmful alleles, even if a breeder can cultivate many accessions. Alternatively, geneticists began examining haplotypes according to linkage disequilibrium (LD). Plant scientists may use all available genotype and phenotypic data to determine genomic estimated breeding values (GEBVs) by combining big data from genomics with haplotype data. The following part will go into more detail about the new selection method known as GS, which was developed in response to the new idea (Kumar et al. 2023).

16.2.9 Functional Validation of Abiotic Stress Tolerant Genes in Tomato Plant

Genetic engineering is one of the many methods available for enhancing modern tomato plants. The discovery of several gene families and mechanisms that affect how plants respond to abiotic stresses and, consequently, increase yield has been made conceivable by recent technological advancements in functional genomics (Liu et al. 2015b). Genetic modification (GM) technology makes it possible to acquire tomato plants that are resistant to abiotic challenges more quickly and effectively than with traditional selective breeding, which increases food yield (Wang et al. 2013; Ishihara et al. 2012). report that the SlSnRK22.1 gene, which regulates abscisic acid signaling, is overexpressed and that the SlSnRK2.2 gene, which regulates osmotic stress signal transduction, is more sensitive to osmotic stress (Ishihara et al. 2012; Dai et al. 2017). investigated the low-temperature stress in tomatoes and discovered that the SlMPK7 gene impacted stress-associated gene transcription, improving resistance to chilling and controlling ROS homeostasis by starting cellular antioxidant mechanisms. When (Dai et al. 2017) investigated the overexpression of ShDHN/Shabrochaites genes, they observed an accumulation of protein with detergent and chaperone-like properties as well as increased tolerance to cold, drought, and salt.

According to (Cui et al. 2017), overexpression of the transcriptional factor, transcriptional regulatory stress-related gene SpWRKY1 led to increased abiotic tolerance. According to (Cui et al. 2017), plants that are resistant to cold can express the transcriptional regulator stress-related gene, which is produced when the AtDREB1A gene overexpresses under the Lip9 promoter. According to (Gupta et al. 2013), numerous structural genes involved in the anthocyanin biosynthesis pathway were stimulated when the LeAN2 gene was overexpressed. This anthocyanin accumulation improved heat resistance. Tomato plants are capable of withstanding cold stress because of the AtGRX gene, which also maintains cellular redox homeostasis (Bineau et al. 2021). Like this, (Kashyap et al. 2020) identified the MdSOS2L1 gene, which controls ion-driving transport pathways and is implicated in signal transduction proteins; overexpression of the gene led to an improvement in salt tolerance. Researchers (Liu et al. 2015b) studied the SlSAM1 gene in tomatoes, which catalyzes the conversion of ATP and L-methionine into S-adenosylmethionine, is involved in the production of ethylene and polyamines, and gives plants alkali tolerance (Kumar et al. 2023). silenced the vis-1 gene to increase tomato fruit quality under heat stress using small interfering RNA technology.

The activities of a tomato SlGRAS10 gene were characterized using abiotic stresses including salt stress and drought (Ishihara et al. 2012). RNA interference (RNAi) was used to downregulate SlGRAS10, producing dwarf plants with shorter internode lengths, smaller leaves, and a greater concentration of flavonoids (Dai et al. 2017). These plants demonstrated heat resistance. It has been reported that abscisic acid (ABA) is important for plants' response to abiotic stresses. Members of the basic leucine zipper (bZIP) transcription factor subfamily, which includes abscisic acid-responsive element binding factor (ABF) and abscisic acid-responsive element binding protein (AREB), are associated with ABA (Gupta et al. 2013). The cultivated tomato has the elements of both SlAREB1 and SlAREB2. Salinity and drought tolerance were enhanced in the tissues of the leaves and roots by the overexpression of SlAREB1 and SlAREB2 (Bineau et al. 2021). In SlAREB1-overexpressing lines, it was discovered that the transcription regulation, oxidative stress, lipid transfer proteins (LTPs), and late embryogenesis abundant protein genes were all increased (Kashyap et al. 2020). These results imply that these genes could be reactive to pathogenic microorganisms and might be implicated in abiotic stress. Sixty-six potential G2-like genes were found and categorized into five groups (I to V) by (Saidi et al. 2020) based on the gene's structure, motif composition, and phylogenetic analysis. There was variation in the distribution of G2-like genes among the 12 chromosomes. Four tandemly duplicated SlGlk genes and nine pairs of tandemly duplicated gene segments were discovered. Within the SlGlks promoter regions, a variety of cisregulatory elements (CREs) linked to stress and hormones are revealed by studies on CREs. SlGlks were shown to be expressed in response to a variety of abiotic stimuli by RNA-seq (Karlova et al. 2011).

According to (Li et al. 2023), drought shocks dramatically upregulated the expression of SlDWARF, SlCPD, and BIN2 while downregulating the expression of TCH4-related genes in SLB3-silenced seedlings. These findings showed that SLB3 gene silencing decreased tomato plant drought tolerance and had an impact on BR signal transduction, which is likely the reason tomato plant drought tolerance varies for responses to biotic and abiotic stress (Ruzicka et al. 2010). investigated the functions of tomato histone H3 lysine methyltransferases Set Domain Group 33 (SDG33) and SDG34. The H3K36 and H3K4 methylations, as well as the gene expression associated with many activities and responses to biotic and abiotic stimuli, were altered in the SDG33 and SDG34 mutants (Ruzicka et al. 2010). The double mutant displayed increased tolerance to drought, consistent with independent and additional functions, whereas the single mutants remained resistant. Mutants had better recovery and survival both during and after the drought, and they retained a higher water status (Alkan et al. 2015b). Notably, decreased trimethylation in H3K4 and H3K36 as well as the production of negative regulators in stressed plants contribute to the mutants' resilience to stress. Nineteen PDI genes were found in tomatoes following the first thorough genome-wide investigation of the tomato PDI gene family by (Basim et al. 2021). Eight of the twelve tomato chromosomes were discovered to have an uneven distribution of these genes, and three pairs of paralogous genes were identified to have segmental duplications (Zhao et al. 2019). The majority of PDI genes showed variable expression across various fruit tissues and developmental stages, according to expression profiling research. Moreover, cold and other stressors including starvation and water scarcity only marginally stimulated a smaller fraction of the PDI genes, but high temperatures, salt, and abscisic acid treatments strongly activated most of the genes (Sant'Ana and Lefsrud 2018). In response to ABA therapy and abiotic challenges, the PDI gene family—which includes SlPDI1-1, SlPDI1-3, SlPDI1-4, SlPDI2-1, SlPDI4-1, and SlPDI5-1—exhibited dominant expression, suggesting that they may be involved in the regulation of abiotic pressures in tomatoes (Fig. 16.3).

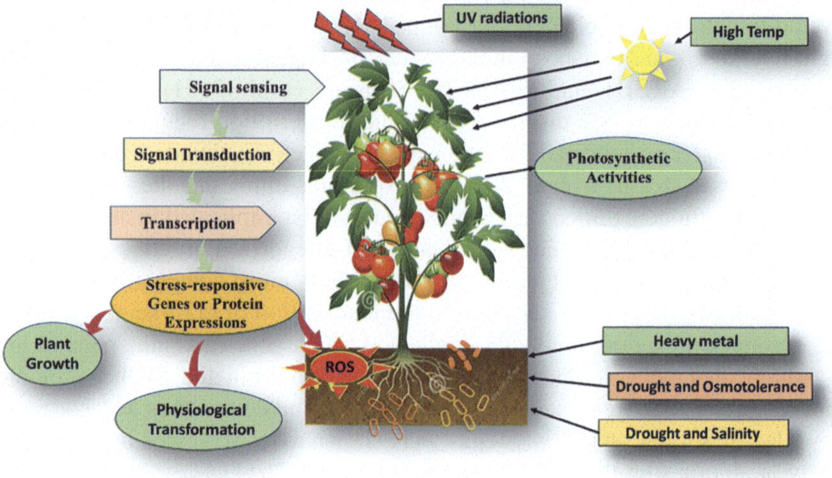

Fig. 16.3 Sketch showing tomato genes associated with abiotic stress tolerance. (Adopted from Naik et al. (2023))

16.2.10 Genomic Methods to Combat Abiotic Stress

Through a variety of mechanisms, such as transcription, translation, calcium regulation, phytohormone, sugar, and lipid signaling, as well as primary and secondary metabolism, plants can withstand high temperatures and survive. Many abiotic factors, such as heat stress during bud development, which results in stigma (style) exertion in tomato flowers, affect the placement and maturity of the male (anther cone) and female (style) organs (Naqvi et al. 2022). The ability to withstand thermal stress is hereditary. Plants exhibit a range of physiological responses in response to heat stress, such as fruit destruction, shoot and root development retardation, and abscission and senescence of leaves (Rasheed et al. 2017). Plant yield is thus greatly decreased. Unfavorable environmental factors including salt stress, drought, and extreme temperatures impair plant growth and agricultural output. GRAS genes are members of plant-specific transcription factors (TFs), which are known to have a few functions in the growth and development of plants. According to many studies, the GRAS protein family is essential for plant growth and development in response to abiotic stressors (Bokszczanin 2019; Narad et al. 2022). The functional characterization of the tomato SlGRAS10 gene under abiotic stresses such as salt stress and drought was shown by (Vidal et al. 2023). Smaller leaves and shorter internodes were produced in dwarf plants when RNA interference (RNAi) was used to downregulate SlGRAS10 (Muthuramalingam et al. 2019).

Additionally, compared to wild-type plants, SlGRAS10-RNAi plants showed increased resilience to abiotic stimuli such as salt, dehydration, and abscisic acid (Mitra et al. 2021). The researchers enhanced the tomato plant's osmotic potential, flavonoid biosynthesis, and ROS scavenging mechanism to demonstrate the significant role of SlGRAS10 as a stress-tolerant transcription factor in a particular form of abiotic stress tolerance. Figure 16.4 provides a schematic representation of how genetic techniques are used to create tomato plants that can withstand abiotic stress (Hasanuzzaman et al. 2013). Therefore, by decreasing long-term shoot Na + retention, overexpression of the yeast gene HAL5 in tomatoes increases their resilience to salt stress. Reduced Na + transport from roots to shoots was the cause of this, regardless of the degree of salt stress. Tomato plants that overexpressed AtNHX1 showed greater salt tolerance, according to (Sato and Tabata 2016). The researchers found that AtNHX1 increased K+ absorption at the tonoplast as well as K+ dispersion throughout cells. Most of the bZIP family's members' activities in tomatoes were examined by (Pramanik et al. 2024); basic region/leucine zipper (bZIP) transcription factors are important modulators of the ABA-mediated stress response in plants. According to the researchers' hypothesis, SlbZIP1 is essential for salt and drought stress tolerance (Hajyzadeh and Egi 2023).

Another study revealed that the expression of SlMYB102 was greater in tomato plant roots and mature fruits than in vegetative organs. Researchers found that the SlMYB102 promoter region is sensitive to hormones, abiotic stress, and light. The proline biosynthesis pathway and C-repeat binding transcription factor (CBF), enhance tomato cold stress resilience and might be linked to SlMYB102 (Wang et al. 2013).

16.3 Transcriptomes

Charles Auffray coined the word "transcriptomics" in 1996, referring to the study of RNA patterns inside of cells. The most recent progress in transcriptomic technology has made it possible to describe a cell's transcriptome and investigate the molecular basis for crop growth (Dai et al. 2017). According to (Kashyap et al. 2020), RNA is the primary mediator of gene expression in the fundamental theories of molecular biology. In a study of mRNA expression levels in a particular cell population, transcriptomics, also referred to as expression profiling provides information on expressed sequence tags (EST) in a particular tissue (Karlova et al. 2011). According to (Basim et al. 2021), transcriptomics studies alterations in gene expression patterns brought on by different biotic and abiotic stresses. Transcriptome libraries for gene annotation may now be obtained at a reasonable cost because of advancements in NGS technology (Zhao et al. 2019). Analyzing transcriptome data sheds light on many functional markers associated with stress resistance, including SSR and SNPs (Ruzicka et al. 2010). Normalizing the data using different statistical modeling techniques allows for the analysis of qualitative counts of transcript data and differential gene expression (Basim et al. 2021; Zhao et al. 2019).

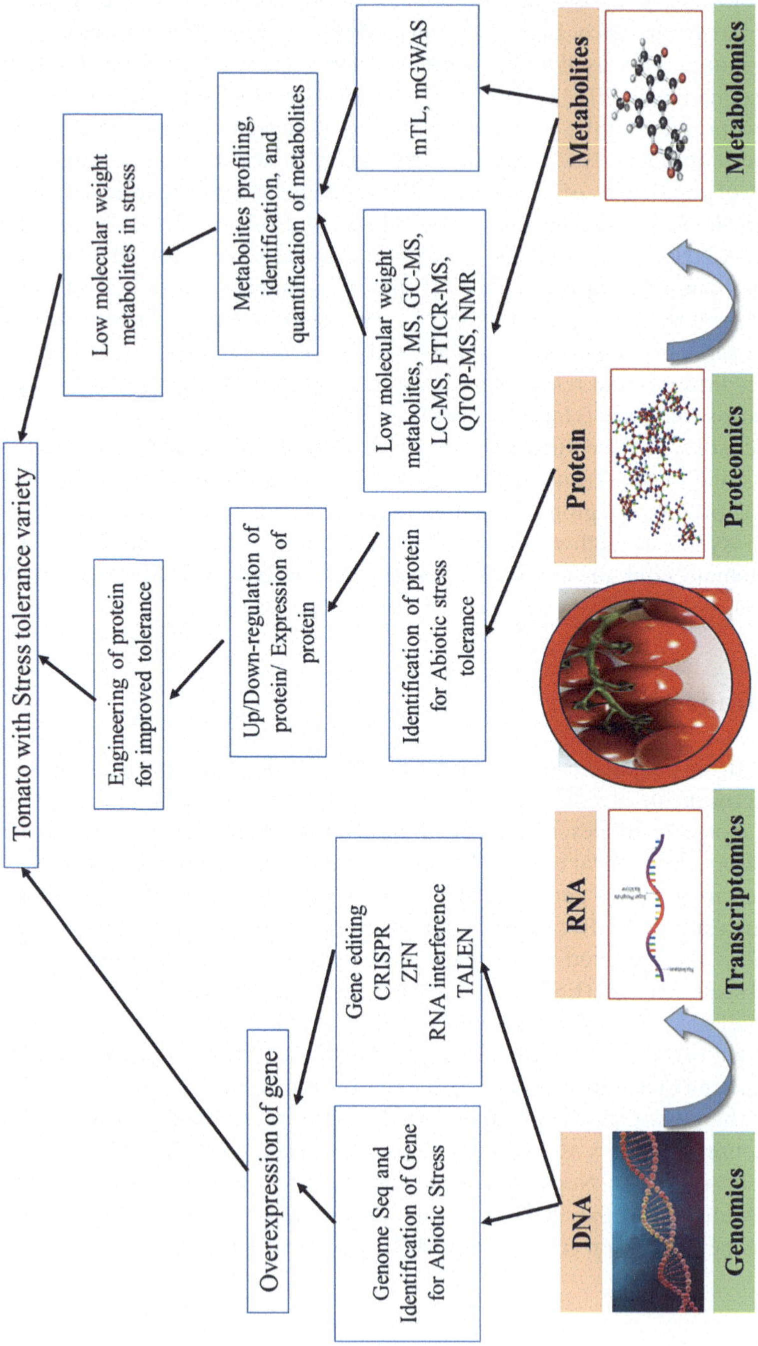

Fig. 16.4 Genomic approaches to combat abiotic stress in tomatoes. (Adopted from Naik et al. (2023))

16.3.1 Insights from Tomato Transcriptome Studies

Tomato transcriptome research provides comprehensive insights into gene expression gradients during fruit development, the spatial distribution of regulatory and structural genes, transcriptional programs and hormonal pathways, novel transcripts and regulatory mechanisms, and the impact of viral infections (Sant'Ana and Lefsrud 2018). Several transcriptome changes involve the process of fruit development and ripening of tomatoes (Wu et al. 2017). Findings indicated differential expressions of 7734 genes during fruit development and ripening were highlighted. This study revealed that photosynthesis-related genes were downregulated as the fruit ripened, while those involved in ethylene signaling and biosynthesis were upregulated, indicating a shift in the fruit ripening during the metabolic process (Razzaq et al. 2021). These studies collectively demonstrate the role of transcriptome analysis in understanding various aspects of tomato biology. From fruit development and ripening to stress responses and quality traits, the insights gained from these studies are invaluable for crop improvement (Wu et al. 2017).

16.3.2 Advances in Tomato Breeding Through Transcriptomic Profiling

Over the past two decades, transcriptomic profiling has significantly advanced tomato breeding by identifying key genes and regulatory networks involved in various stress responses, fruit development, and quality traits (Camargo et al. 2015). Studies utilizing RNA sequencing (RNA-Seq) have provided an understanding of genetic and molecular mechanisms underlying drought tolerance, disease resistance, and fruit ripening. For instance, RNA-Seq analyses have identified differentially expressed genes (DEGs) involved in the response to biotic and abiotic stresses, which are crucial for developing stress-resistant tomato varieties (Dong et al. 2022; Ramesh et al. 2017). Furthermore, combining transcriptomic data with genome-wide association studies (GWAS) has augmented the genetic architecture of important traits such as flavor, nutritional content, and yield. These integrative approaches have facilitated the identification of candidate genes and metabolic pathways that can be targeted for improving tomato crop quality and resilience to environmental challenges. This comprehensive approach is pivotal in addressing the challenges posed by climate change and ensuring sustainable tomato production (Su et al. 2021; Zhang and Shang 2024).

16.3.3 Key Findings in Transcriptomic Profiling of Tomato Plant

The transcriptomes of tomato fruit during seven developmental periods (7, 14, 21, 28, 35, 42, and 49 days after flowering) were examined using two tomato cultivars (Ullah 2017; Muthuramalingam et al. 2019). Twenty-six thousand three hundred and ninety-seven genes were identified to be expressed at least once during the developmental phase in the two tomato cultivars. A co-expression study identified several transcriptional factors whose expression patterns are associated with genes related to the production of ascorbic acid, carotenoid, and flavonoid in tomato plants (Ahmed et al. 2024; Hajyzadeh and Egi 2023).

- To provide a useful catalog of RNA-Seq differences caused by salt, a comparative transcriptome analysis of *S. chilense* (wild) and cultivated tomato in response to NaCl was conducted again to discover the numerous genes and pathways related to salinity stress. The relationships between the genes involved in the Ca2+, auxin, and ethylene-mediated signal were discovered by this study (Tiwari et al. 2022).
- The root of the susceptible (Mi–) and resistant (Mi+) cultivars "Moneymaker" and "Motelle," respectively, showed evidence of effective resistance in tomato (*Solanum lycopersicum*) due to infection by the RKN (Root-knot nematode, or *Meloidogyne incognita* spp.). Broad gene suites are typically impacted by RKN infection; at some point throughout the infection process, more than half of the probes on the array detected altered gene regulation in infected and uninfected root tissue. Transcriptome analysis revealed that 217 genes were regulated throughout the RKN infection period, which corresponded to the development of feeding sites. Additionally, 58 genes, including glycosyltransferase, showed differential regulation in resistant roots as compared to uninfected roots. (Gupta et al. 2009; Karlova et al. 2011).
- The Affymetrix tomato genome array analysis revealed that PI365967 had fewer responsive genes to salt stress compared to Moneymaker. The SOS pathway was more active in PI365967, resulting in lower Na + accumulation in its shoots. Salicylic acid signaling and a more effective detoxification system, indicated by specific gene inductions, were unique to PI365967. Additionally, better root growth during salt stress in PI365967 could be due to the downregulation of a high-affinity nitrate transporter (Dong et al. 2022).
- The whole transcriptome sequencing revealed that the TYLCV-resistant tomato line CLN2777A (R) had a higher proportion of upregulated DEGs (58.37%) compared to the susceptible line TMXA48-4-0 (S) (9.17%). In the R line, key defense-related genes, including WRKY transcription factors and protein kinases, were upregulated or unchanged, unlike in the S line where they were downregulated (Ramesh et al. 2017). This study highlights the gene expression patterns and defense mechanisms contributing to TYLCV resistance in the R tomato line (Su et al. 2021).
- For the first time, a study combining morphological, molecular, and biochemical approaches discovered changes in tomato fruit from mycorrhizal plants. Mycorrhizal symbiosis in roots not only altered fruit gene expression but also affected flowering phenology, fruit ripening, and amino acid profiles, indicating significant metabolic modifications. Hormonal signaling, particularly involving ethylene and methyl jasmonate, plays a key role in these mycorrhiza-induced changes (Zhang and Shang 2024).

- When tomato plants were treated with a commercial formulation of *A. nodosum* extract (Stimplex), RNA-seq differential gene expression analysis showed that 635 genes were upregulated, and 456 genes were downregulated. It was hypothesized that favorable plant responses that allowed for improved tomato plant development and stress tolerance were caused by gene regulation induced by seaweed extract (Wang et al. 2016).
- A transcriptome atlas of tomato root regions revealed the transcriptional responses to cytokinin and auxin, identifying orthologues of hormone-regulated genes from other species and confirming recent findings in tomatoes. The study also mapped the spatial distribution of genes related to plant defense, secondary metabolite production, and general metabolism across different root regions (Tanigaki et al. 2015).
- In the tomato/potato heterografting study, transcriptome profiling of the tomato scion revealed 209 upregulated genes, some of which are involved in starch and sucrose biosynthesis. Phenotypic analysis showed minimal changes in the tomato scion, suggesting that the tomato scion encourages stolon development into aerial stems and sprouting but is less successful at generating chemicals or signals that cause tuberization (Herraiz et al. 2016).
- A comparative RNA-sequencing analysis of tomatoes exposed to various abiotic and biotic stresses identified genes involved in defense responses. Sixty-seven genes were associated with responses to at least four different stresses, including genes involved in the biosynthesis of cell walls, signaling, hormone pathways, and transcription factors. This study provides a valuable list of genes for potential use in genetic engineering to enhance tomato resistance to multiple stresses (Penin et al. 2019).
- Another transcriptomic study examines the whole genome transcriptome dynamics in various tissue types of *S. lycopersicum*, particularly root tissue. It finds that 80–85% of genes are regulated during root growth, while 15–16% of genes, involved in JA, ethylene, and auxin pathways, govern root-specific growth aspects affecting root architecture. The findings specify a foundation for further studies to link hormonal pathways with regulatory genes to enhance root growth and agronomic traits in the Solanaceae family (Yeats et al. 2010).
- Another study addressed the challenge of drought in tomatoes by comparing transcriptomic profiles of tomato leaves under controlled and drought conditions. Drought impacted reduced leaf water content and increased malondialdehyde, ion leakage, oxidative stress, and antioxidant enzyme activities. RNA sequencing identified thousands of differentially expressed genes (DEGs), including 742 transcription factors (TFs) of AP2/ERF, NAC, MYB, bHLH, bZIP, WRKY, and HB families, after two hours and 499 TFs after four hours. The study validated the expression of 31 TFs and identified six drought-responsive marker genes, providing a foundation for enhancing drought tolerance in tomatoes (Zhou et al. 2009).

16.3.4 Challenges

RNA-seq is superior to microarray hybridization but faces challenges such as low sequencing coverage in single-cell data, difficulties in evaluating computational results due to plant genome complexity, and issues with read alignment to reference genomes (Parker et al. 2013). The process of creating cDNA libraries involves multiple manipulations, complicating transcript profiling. Computational challenges include data storage, retrieval, and processing, as well as minimizing errors and eliminating low-quality reads (Balmant et al. 2015). High-throughput machine learning algorithms are needed to analyze the vast amounts of RNA-seq data effectively. MiRNAs play roles in gene silencing and are involved in various plant growth developmental stages, signal transmission, disease resilience, and genetic engineering, though many of their biological functions remain unknown (Zhou et al. 2009).

16.4 Proteomics

One of the main areas of study for functional genomics is now protein mapping, profiling, and proteomics. Humans have been mostly identified by proteomics research on malignant samples. Despite this, the expansion of agriculture is a crucial area for ensuring food security (Parker et al. 2013). To achieve crop improvement, the identification of proteins and their alterations under stress circumstances is crucial for understanding the significance of complex plant proteomes (Fig. 16.5).

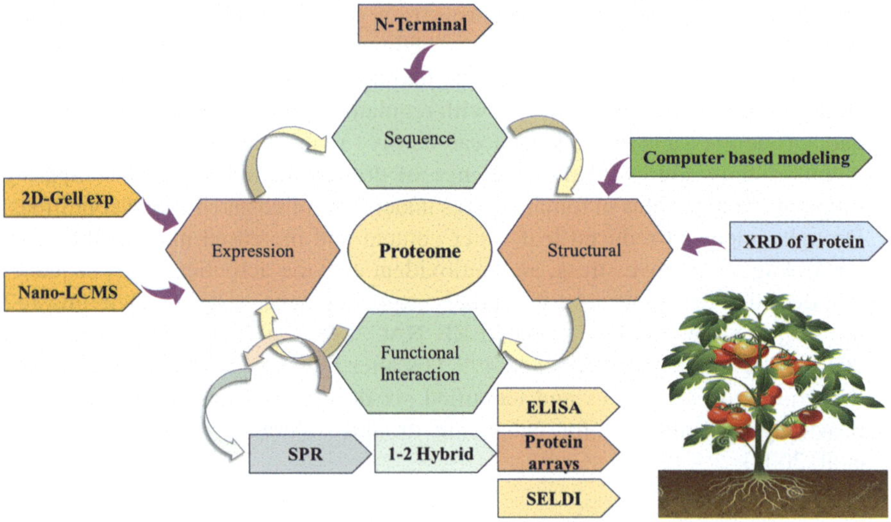

Fig. 16.5 Understanding tomato molecular physiology with a proteomic approach

16.4.1 The Tomato Genome and Proteome Databases and Software

Proteome databases are constructed from sequences of amino acids derived from anticipated genomic gene models and unigene transcripts. Unigene transcripts are identified using methods such as Hidden Markov models for coding region detection, EstScan for identifying potential coding regions (Balmant et al. 2015), Netstat for predicting initiation and translational sites (Zhou et al. 2011), or by determining the longest open reading frame (Zhang and Shang 2024). The International Tomato Annotation Group (iTAG) maintains the tomato database, version 3.20 containing 30,868 annotated genes. Databases like plant genome and systems biology (PGSB) and the tomato genomic resources database (TGRD) offer additional proteomics and genomics data. TGRD, centered on the Heinz 1706 cultivar, integrates genes, microRNAs, simple sequence repeats (SSRs), and quantitative trait loci (QTLs) across the twelve tomato chromosomes. It also provides high-quality single nucleotide polymorphisms (SNPs) and analysis tools for SSRs and SNPs (Wang et al. 2016). The Sol Genomics Network (SGN) offers the latest tomato genome assembly (SL4.0) incorporating improved repeat region annotations and an extensive protein-coding gene set. These comprehensive resources empower researchers and breeders to investigate diverse parameters of tomato biology, such as fruit development, disease resistance, and nutrient metabolism (Wang et al. 2016).

Sol Genomics Network (SGN) has developed "tomato secretome." This secretome's goal is to investigate the proteins involved in the secretory pathway, those found in the cell wall, and proteins released outside of the plasma membrane. Secretome proteins are crucial for plant growth, communication, and stress response (Tanigaki et al. 2015). The secretome is useful for examining the mechanical and chemical properties of the cell wall as well as the texture of the fruit, particularly for fleshy fruits like tomatoes (Herraiz et al. 2016).

FASTA sequence databases, compatible with search tools like Mascot, BLAST, and Protein Pilot, were traditionally accessed via FTP. Today, a variety of search algorithms, including Mascot, SEQUEST, Comet, X! Tandem-GPMDB, MS Amanda, and OMSSA, underpin protein identification in proteomics workflows. These algorithms are integrated into diverse software packages such as Bionic, Proteome Discoverer, Scaffold, MassLynx MS, ProteinPilot, MaxQuant, and Searchgui, each offering unique capabilities in terms of input file formats, search engine compatibility, data analysis pipelines, visualization options, and additional parameters (Penin et al. 2019).

16.4.2 Plant Proteomics from Stress Conditions

The creation of plants that are resilient to stress is significant in the context of food security. A thorough examination of gene networks and plant response regulators is necessary to accurately balance energy, adaptability, and plant growth (Yeats et al. 2010). The tomato has become the Solanaceae family's model plant; this table illustrates its use as a model for crop proteomics (Table 16.1).

Table 16.1 Key outcomes on tomato proteomics under stress

Plant organ	Stress/condition	Proteins	References
Root	Biotic, abiotic (salt, Cd)	1214+	Liu et al. (2015b)
Leaf	Biotic (Phytophthora, bacteria, virus), abiotic (drought)	2600+ spots	Kumar et al. (2023)
Apoplastic fluid	Biotic (Ralstonia)	335	Wang et al. (2013)
Xylem sap	Nutrient deficiency (Fe, Mn)	643	Ishihara et al. (2012)
Fruit	Development, ripening, biotic (fungus)	8588+	Dai et al. (2017)
Pericarp	Ripe	8588	Cui et al. (2017)
Graft union	Heat	700–900	Gupta et al. (2013)
Pollen	Biotic (heat)	365+	Kashyap et al. (2020)
Floral pedicel	Ethylene	1429	Saidi et al. (2020)
Stem/petiole	Secondary chemicals	1552	Karlova et al. (2011)
Cotyledon	Biotic (fungus)	48 phosphoproteins	Li et al. (2023)

16.4.3 Proteins in Plant Stress: A Spatial Perspective

Proteins are pivotal in plant stress responses, driving phenotypic adaptations through altered physiological functions. Their subcellular localization, crucial for their roles, is often compartmentalized to organelles like the nucleus, mitochondria, and chloroplasts. Traditional differential centrifugation isolates major cell fractions (nuclear, microsomal, mitochondrial, and plastidial) for proteomic analysis (Zhou et al. 2009). These proteins participate in diverse processes, including disease resistance, redox regulation, photosynthesis, defense, metabolism, and protein synthesis (Parker et al. 2013).

Spatial proteomics, particularly imaging-based approaches, offers direct visualization of protein localization. While demanding in terms of resources and technology, it provides unparalleled insights into protein distribution. Complementary to this, co-immunoprecipitation or proximity labeling methods like APEX or BioID, coupled with mass spectrometry, identify protein interaction networks, aiding in spatial proteomics (Tiwari et al. 2022; Foolad and Sharma 2004; Ibitoye and Akin-Idowu 2011). Organellar mapping through proteome profiling is a more accessible approach, fractionating cell lysates to isolate organelles and identifying their protein constituents. This method enables tracking protein movement between compartments, offering a systems-level perspective on cellular processes (Prasanna et al. 2023).

16.4.4 Post-Translational Modifications (PTMs) and Plant Stresses

Post-translational modifications (PTMs) profoundly influence protein function, stability, and localization, playing a critical role in plant stress responses. Diverse PTMs, including phosphorylation, glycosylation, and lipidation, shape protein structure and function, varying across cell types, tissues, and developmental stages. Protein isoforms, generated through PTMs, splicing, and degradation, contribute to proteome complexity. Studies on tomato cultivars revealed shared stress-related proteins, but also cultivar-specific isoforms (Hasan et al. 2021; Tiwari et al. 2022). While transcriptional regulation is often emphasized in plant stress responses, PTMs offer a more nuanced perspective (Table 16.2). Large-scale PTM analysis is essential for understanding plant stress adaptation and developing improved crop varieties (Gupta et al. 2010).

Table 16.2 Proteins to contend the abiotic stress resilience in tomato

Protein	Function	Result	References
Systemin peptide	Sodium exclusion, antioxidant potential protease inhibitor	High salt tolerance	Zhang and Shang (2024)
COR15 protein	Folds and attached to the chloroplast membranes support leaf cells	Cold tolerance	Wang et al. (2016)
P5CS protein	Proline biogenesis	Salt and drought tolerance	Tanigaki et al. (2015)
ERF family	Lowers cell injury and improves tolerance against freezing	Heat, cold, and flood tolerance	Herraiz et al. (2016)
SpPKE1	Intermingle with an F-box protein	Drought tolerance	Penin et al. (2019)
LEA proteins	Avoids membrane seepage, and protein maintenance	Water balance and ion impounding	Yeats et al. (2010)
Dehydrins (DHNs)	Improves tolerance to drought and saline stress	Saline and drought stress	Zhou et al. (2009)
2,3-butanediol	Drought and chilling response	Reduced effects of abiotic stresses	Parker et al. (2013)
E42 and LA3120	Promotes plant growth and lowers stresses	Better plant growth under water stress	Balmant et al. (2015)
RING zinc finger	Plant growth and reduction of abiotic stresses	Reduced stresses on plant	Zhou et al. (2011)
Histidine kinase	Minimize abiotic stresses	Providing cellular Na + homeostasis	Mazzeo et al. (2014)
SlbZIP1	Decreases ABA-mediated stress	Diminished salt and drought stress	Bovy et al. (2007)
MdVHA-B	Enhances resilience to drought stress	Enhances tolerance	Karlova et al. (2011)
LeHSP21.5	Improved tolerance to tunicamycin-ER stress inducer	Increased tolerance	Li et al. (2023)

Adopted from Naik et al. (2023)

16.4.5 The Fruit Proteome of Tomato

Ripening plays a major role in determining fruit quality. It is further examined for proteins linked to ripening, proteins that are regulated, and proteins that are specific to a cultivar. The assessment procedure is dependent on both high- and low-throughput procedures that are started using a gel-based technique (2DE). To identify genes and proteins that are essential to fruit traits, such as an ethylene-responsive transcriptional coactivator, glutathione S-transferase (GST)/peroxidase, oxalyl-CoA decarboxylase, etc., quantitative reverse transcription-PCR (qRT-PCR) analyses and microarray-based transcriptomics were integrated with proteomic data (Mazzeo et al. 2014; Bovy et al. 2007).

16.5 Metabolomics

16.5.1 Introduction to Metabolomics

Metabolomics is the inclusive assessment of endogenic metabolites, aiming to steadily identify and compute metabolites from a biological sample. This field is distinct from metabolic profiling, which targets specific compounds and their biochemical transformations. Metabolites, which are small-molecule structures such as lipids, amino acids, and carbohydrates, play critical roles in biological systems and disease phenotypes (Zhu et al. 2018). The approach has proven superior for pattern recognition investigations of biological samples and has demonstrated the ability to measure rapid changes in metabolite concentrations caused by biochemical processes like signaling cascades (Luna et al. 2020).

16.5.2 Techniques for Metabolite Profiling

Metabolites are identified primarily using mass spectrometry (MS) and nuclear magnetic resonance (NMR). These systems, usually coupled with chromatography, enable metabolite separation, detection, and quantification (Iula et al. 2022). Due to the metabolite diversity, integrated analytical approaches are essential. MS, especially with HILIC and high-resolution capabilities, enhances metabolite identification and reduces errors. Tools like MyCompoundID aid in MS data interpretation (Khapte et al. 2022). NMR, while less sensitive, provides structural information crucial for metabolite identification (Hurtado et al. 2013).

16.5.3 Impact of Metabolomics on Tomato Breeding

Metabolomics has significant implications for tomato breeding, offering insights into the metabolic changes associated with fruit development, ripening, and stress responses (Zhu et al. 2018). By profiling the metabolome of different tomato varieties, researchers can identify metabolites linked to desirable traits such as flavor, nutritional quality, and stress tolerance. This information can inform breeding programs aimed at enhancing these traits. For instance, metabolomic studies have identified key metabolites and metabolic pathways involved in the response of tomatoes to abiotic stresses, such as cold and drought (Luna et al. 2020). The identification of these metabolites helps in understanding the mechanisms of stress tolerance and in acquiring tomato varieties that can withstand harmful environmental conditions (Khapte et al. 2022). Additionally, metabolomics has been used to explore the differences between wild and cultivated tomatoes, identifying metabolites associated with fruit size, shape, and color, which are critical for breeding programs focused on improving fruit quality (Hurtado et al. 2013). Moreover, metabolomics can identify biomarkers for disease resistance in tomatoes. For example, the metabolomic analysis of tomato plants infected with pathogens can reveal metabolites involved in defense responses, which can be targeted in breeding programs to develop disease-resistant varieties (Laxman et al. 2018).

16.5.4 Tomato Metabolomics for Stress Reduction

Evaluation of biological processes in complicated contexts is aided by metabolomics. It is estimated that there are more than 200,000 metabolites in the kingdom of plants, ranging from polar to nonpolar and volatile to nonvolatile. A metabolite's ultimate variety necessitates the assembly of many analytical systems to investigate all its chemical features (Ruzicka et al. 2010). Table 16.3 lists the several metabolites that tomatoes use to tolerate stress.

16.5.5 Overview of Analytical Approaches

Metabolomics leverages separation techniques such as liquid/gas chromatography (LC/GC) combined with mass spectrometry (MS), nuclear magnetic resonance (NMR), or Fourier transform ion resonance (FTIR) for metabolite profiling and quantification (Janni et al. 2019). Targeted metabolomics focuses on specific compounds using reference standards, ensuring accurate identification (Hurtado et al. 2013). Untargeted metabolomics aims to comprehensively profile metabolites but suffers from limited identification and data complexity (Laxman et al. 2018). To bridge this gap, semi-targeted and pseudo-targeted approaches have emerged. Semi-targeted analysis screens for metabolites founded on compound-specific data without reference standards, while pseudo-targeted metabolomics uses MRM for ion pair detection without reference values (Janni et al. 2019).

Single-cell metabolomics has emerged as a powerful tool for investigating cellular heterogeneity. Initial studies relied on protoplasts, but recent advancements focus on minimally invasive techniques. Methods such as pipette-based aspiration and electrospray ionization-coupled probes enable direct sampling of individual

Table 16.3 Metabolites of tomato plants are involved in abiotic stress tolerance

Metabolites	Function	Result	References
Melatonin	Inhibits protein/membrane damage	Abiotic stress tolerance	Bovy et al. (2007)
ABP19a protein	Involved in development and drought response	Drought tolerance	Zhu et al. (2018)
SpUSP	Reduces oxidative stress	Drought tolerance	Luna et al. (2020)
Solyc04g014600	Protein profiling of phloem and exudations	Drought tolerance	Iula et al. (2022)
SlSRN1	Disease and drought resistance	Abiotic stress tolerance	Khapte et al. (2022)
SlNAC4	Fruit ripening, carotenoid accumulation	Oxidative stress response	Luna et al. (2020)
SlNAM2	Flower-edge morphogenesis	Prevents water loss	Iula et al. (2022)
SlNAC1	Viral replication, chill tolerance	Tolerance to freezing and pathogens	Khapte et al. (2022)
GOBLET	Identifies leaflet borders	Inhibits transpiration	Hurtado et al. (2013)
SlMAPK3	Germination, chlorophyll content, root biomass	Cadmium and drought tolerance	Laxman et al. (2018)
C2H2-type zinc finger protein	Growth and development	Abiotic stress responses	Khapte et al. (2022)

Adopted from Naik et al. (2023)

cells, preserving cellular integrity and microenvironment. These approaches offer unprecedented insights into intracellular metabolic processes (Hurtado et al. 2013).

Mass spectrometry imaging (MSI) offers a label-free approach to visualize and compute the distribution of molecules within a sample. For MSI, methods such as matrix-assisted laser desorption ionization (MALDI), laser-ablation electrospray ionization (LAESI), secondary ion mass spectrometry (SIMS), and laser desorption/ionization (LDI) are often employed. The ease of matrix-free approaches (LDI, DESI, LAESI) is making them more popular, even if SIMS is still the best in terms of spatial resolution (Laxman et al. 2018). Metabolomics, the study of metabolites, is valuable for understanding plant stress responses as core metabolite structures are conserved across species. Advances in mass spectrometry and spectroscopy generate vast and complex datasets, increasing the risk of false discoveries (Janni et al. 2019). To address this, researchers are developing new software tools, databases, and data analysis methods. Recent breakthroughs in deep learning (DL) and machine learning (ML) enable the construction of detailed metabolic models for various plant species under different conditions. These models are crucial for developing crop varieties resistant to environmental challenges (Cao et al. 2024).

16.5.6 Metabolomics of Plant–Herbivore Interactions

By boosting putrescine hydroxycinnamoyl transferases and generating phenol amides, tomato plants protect themselves against the Tuta absoluta pest (Bovy et al. 2007). Tolerant tomato varieties accumulate organic acids in response to infestation. Comparisons between tomato and eggplant revealed that tomatoes have lower levels of phenols, jasmonic acid, sugars, and amino acids, making them more susceptible to insect attack (Tanigaki et al. 2015). However, tomato plants can benefit from symbiotic relationships with microbes like *Trichoderma harzianum* and *Rhizophagus irregularis*, which enhance the production of defensive compounds like steroidal glycoalkaloids, fatty acid amides, and carnitine-derived metabolites, deterring pests such as *Manduca sexta*. Additionally, secondary metabolites like tannins and lignins provide physical and chemical barriers against herbivores by binding to proteins and strengthening plant cell walls. Furanocoumarins are toxic compounds that disrupt insect DNA, leading to cell death (Zhu et al. 2018).

Tomato plants employ various defense mechanisms against herbivores. The phytohormone jasmonic acid, triggered by the signaling peptide systemin, activates defenses against insects like *Spodoptera littoralis*, *Tetranychus urticae*, and *Manduca sexta* larvae (Iula et al. 2022). Volatile terpenoids stored in glandular trichomes are released upon herbivore attack, directly or indirectly triggering plant defenses (Khapte et al. 2022). Additionally, volatile compounds like methyl salicylate and green-leaf volatiles are produced, further enhancing defense responses. Salicylic acid, a precursor to methyl salicylate, can be applied exogenously to protect against pests like *Bemisia tabaci*, which also transmits the Tomato yellow leaf curl virus (Hurtado et al. 2013). High flavonoid levels in tomato plants deter *B. tabaci*. Beyond aerial attacks, tomato plants also face threats from root-feeding nematodes. Mycorrhizal associations with *Glomus mosseae* and intercropping with crown daisy can reduce root-knot nematode infestations (Laxman et al. 2018).

16.5.7 Effects of Chemicals, Pesticides, and Epigenetic Alterations on Tomato Metabolomics

The use of plant hormones and pesticides to bolster plant defenses, alleviate stress, and improve condition and yield is a widespread practice in traditional agriculture. However, there is limited research on the impact of these treatments on plant metabolites (Janni et al. 2019). For instance, in tomatoes, the administration of Ethephon (2-chloroethyl phosphonic acid, or CEPA) changes the composition of the cell wall and decreases photosynthesis, stomatal conductance, and glucose turnover. Supplementing with jasmonate and nitric oxide reduces the effects of salt stress by boosting antioxidant enzymes and raising levels of proline, glycine betaine, and flavonoids (Cao et al. 2024). Spermidine foliar sprays help tomatoes cope with low-iron stress by modifying osmoregulatory systems, metabolism, and photosynthetic efficiency, and by upregulating certain transcription factors (Janni et al. 2019). Excessive pesticide use can induce oxidative stress in tomato seedlings. For example, pesticides like emamectin benzoate, alpha-cypermethrin, and imidacloprid have been shown to elevate ROS levels and reduce germination rates, root biomass, chlorophyll content, and carotenoid levels. Alternative strategies to enhance plant defense have emerged, such as using β-aminobutyric acid to induce resistance against multiple pathogens, combining fluopimomide with *Bacillus methylotrophic* for disease management, and using indole-3-acetic acid (IAA) to induce fighting against *Fusarium oxysporum lycopersici*. Natural pesticides like azadirachtin have also gained popularity for inducing systemic resistance (Wang et al. 2024).

Tomato fruit ripening is regulated by epigenetic modifications that control gene expression. DNA methylation, for example, is essential for normal tomato fruit development (Di Gioia et al. 2013). The DNA methyltransferase SlMET1 regulates development-related genes to ensure proper flower and fruit production. Epimutations in genes like the SQUAMOSA promoter binding protein-like (SBP-box) gene at the colorless non-ripening (cnr) locus can prevent normal ripening (Di Gioia et al. 2013). DNA demethylation activates genes necessary for ripening and flavor development. Cold storage at 4 °C depletes flavor volatiles exactly for transient cytosine methylations in ripening-related genes. Histone modifications, such as methylation and acetylation, also play roles in tomato fruit ripening (Wang et al. 2011). Histone H3K4 demethylase SlJMJ7 and histone demethylase SlJMJ6 regulate ripening by modifying methylation levels. The polycomb repressive complex 2 (PRC2) inhibits ripening through H3K27 trimethylation. Histone deacetylases (HDACs) such as SlHDA3, SlHDA1, and SlHDA4 interact with MADS-box proteins to regulate reproductive development and stress tolerance. The HD2 family genes SlHDT3 and SlHDT1 affect carotenoid accumulation and ethylene biosynthesis, while the histone acetyltransferase gene SlHAF1 is crucial for ripening (Kamaluldeen et al. 2014). N6-methyladenosine (m6A) modification of mRNA affects several physiological functions, such as RNA metabolism and stress, m6A regulates mRNA stability, splicing, and translation efficiency. During tomato fruit expansion, m6A modifies genes involved in hormone responses and endoreduplication. m6A also mediates fruit ripening, with mutations in the RNA demethylase gene SlALKBH2 delaying the process (Lopez and Satti 1996).

16.6 Phenomics

Phenomics is the organized study of phenotypes on a large amount, focusing on the comprehensive analysis of physical and biochemical attributes of organisms in response to genetic and ecological factors. This field aims to link the genotype-to-phenotype gap by encompassing a wide range of traits from the molecular to the whole-plant level. (Wang et al. 2011). It integrates high-throughput phenotyping with advanced data analysis, allowing for the study of complex trait interactions and their genetic basis (Kamaluldeen et al. 2014).

16.6.1 Technologies in Phenomics

- *Imaging Technologies*: Various imaging technologies are used in plant phenomics to capture different traits. These technologies include visible-light imaging for assessing morphological traits, fluorescence imaging for evaluating photosynthetic efficiency, thermal imaging for monitoring water status and stomatal conductance, and hyperspectral imaging for analyzing biochemical composition (Janni et al. 2019).
- *Sensor Technologies*: These include spectral reflectance sensors for estimating biomass and nutrient status, LiDAR (Light Detection and Ranging) for analyzing 3D plant architecture, and chlorophyll fluorescence sensors for measuring photosynthetic performance (Balmant et al. 2015).
- *Robotics and Automation*: Automated phenotyping platforms enhance the efficiency and accuracy of data collection. Various systems such as conveyor belt systems for high-throughput imaging, robotic arms for sample handling and data collection, and automated irrigation and environmental control systems (Zhu et al. 2018).
- *Data Management and Analysis*: Effective data handling is essential in phenomics, highlighting the significance of machine learning algorithms for image analysis and trait extraction, big data management systems for storing and accessing phenotypic data, and statistical tools for analyzing complex trait relationships (Tanigaki et al. 2015).

16.6.2 Applications of Phenomics in Tomato Improvement

There are many applications of phenomics in tomato breeding enlisted here:

1. *Yield Improvement*: Phenomics has been applied to enhance tomato yield (Wang et al. 2016) in Frontiers in Plant Science used phenomics to study yield-related traits by analyzing 3D plant architecture and its relation to yield. They identified quantitative trait loci (QTLs) linked with plant height and branching, leading to the development of high-yielding tomato lines through marker-assisted selection.
2. *Stress Tolerance*: Phenomics also aids in improving stress tolerance in tomatoes (Yeats et al. 2010) in Plant Physiology applied thermal imaging to assess plant water status and identified genotypes with improved water use efficiency. They discovered genes associated with drought tolerance, which are valuable for breeding programs aimed at developing drought-resistant tomato varieties.
3. *Fruit Quality*: Assessing tomato fruit quality has been enhanced through phenomics (Parker et al. 2013) in Plant Methods developed a phenomics approach using visible and near-infrared spectroscopy for non-destructive quality evaluation. They analyzed traits such as fruit color, firmness, and soluble solids content, identifying genetic markers associated with these quality traits.
4. *Disease Resistance*: In plant pathology, phenomics is used for early disease detection (Balmant et al. 2015) in Plant Disease reviewed the application of

hyperspectral imaging to identify spectral signatures associated with various tomato diseases. This approach has facilitated the development of rapid screening methods for breeding disease-resistant tomato varieties.

5. *Nutrient Use Efficiency*: Improving nutrient use efficiency in tomatoes is another application of phenomics (Zhou et al. 2011) in Frontiers in Plant Science studied nitrogen use efficiency by analyzing plant growth and chlorophyll content under varying nitrogen levels. They identified tomato genotypes with improved nitrogen use efficiency and discovered genetic markers for breeding nitrogen-efficient varieties.
6. *Photosynthetic Efficiency*: Phenomics has also been applied to study photosynthetic traits in tomatoes (Zhu et al. 2018) in the Journal of Experimental Botany used chlorophyll fluorescence imaging to assess photosynthetic performance, identifying genetic variation in photosynthetic efficiency among tomato genotypes. These findings have informed strategies to improve photosynthetic efficiency in tomato breeding programs.

Therefore, phenomics integrates advanced technologies and comprehensive data analysis to enhance the understanding and improvement of various traits in tomatoes, including yield, stress tolerance, fruit quality, disease resistance, nutrient use efficiency, and photosynthetic efficiency. These advancements in phenomics hold significant potential for optimizing tomato breeding programs and achieving sustainable agricultural practices.

16.7 Integration of "Omics" Data

Such powerful technologies provide knowledge of the functioning system of genes, which is explained by bioinformatics. Through the integration of several omics technologies, bioinformatics also specifies available resources for computational modeling and simulation analysis. A variety of software packages and bioinformatics tools have been utilized to assess the multi-omics properties in agricultural research. Computational resources have not only provided a simple way to retrieve user-friendly datasets, but have also enabled the storing, cataloging, and investigation of the accessible data. For crop sciences, many multi-omics databases have been constructed.

16.8 Key Findings in Multi-Omic Analysis of Tomato Plant

The model plant Arabidopsis has made progress in identifying the genes involved in abiotic stress tolerance, whereas crop species like tomatoes have made less progress in this regard (Janni et al. 2019). Because of its physio-anatomical characteristics, the tomato is an excellent model plant for researching long-distance transport of saline ions (Wang et al. 2024). According to (Di Gioia et al. 2013), the SlSOS1 gene plays a significant role in the partitioning of Na + in the tomato plant's shoot, which is not the case in Arabidopsis. Since too much Na + is particularly harmful to tomato

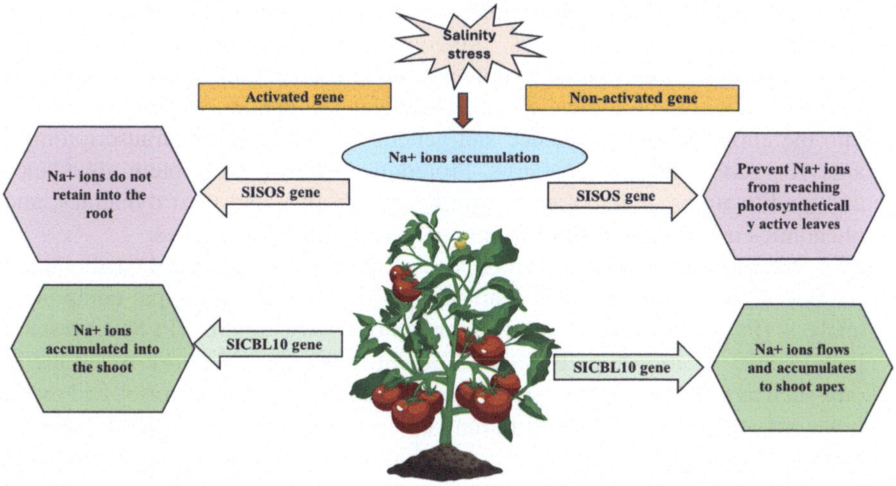

Fig. 16.6 Partitioning of Na^+Ions under saline stress in tomato

young leaves, there is a buildup of Na + in the tomato's shoot that differs in young vs mature/old leaves. While SlSOS1's primary function in Arabidopsis is to extrude Na + out of the root, in tomatoes, SlSOS1 is essential for Na + sectionalization in tomato shoot tissues, keeping Na + ions contained in the mature portions while preventing them from leaking to the photosynthetically active portions of the plant, or the leaves (Wang et al. 2011).

The function of calcineurin B-like protein 10 (CBL10) in tomato and Arabidopsis was recently described as another example. A shared characteristic of cbl10 loss-of-function mutants in Arabidopsis and tomatoes is their extreme susceptibility to ion toxicity in the first stages of the shoot, coupled with a markedly reduced buildup of Na + under salt stress (Kamaluldeen et al. 2014). A comprehensive study on Na + separation was conducted in the shoot to understand the salt-hypersensitive phenotype of the tomato plant and the cbl10 mutant accumulates less Na + (Lopez and Satti 1996). The cbl10 tomato mutant exhibited a completely different Na + accumulation pattern, with Na + ions only traveling to the apex of the shoot rather than accumulating in the root or shoot (Fig. 16.6). The SlCBL10 gene may oversee shielding the shoot apex from saline, based on the damage shown in the apex of the mutant plants. Therefore, a different distribution of Na + throughout the shoot accounts for the fact that a modest concentration of Na + induces a larger salt sensitivity in cbl10 mutants (Kusano and Fukushima 2013).

16.8.1 Modern Integrative Approaches in "Omics"

Integrative "omics" approaches combine data from multiple high-throughput technologies to provide a comprehensive view of biological systems. In tomato breeding, these approaches help identify key genes, pathways, and networks associated with important traits.

1. *Multi-omics Data Integration*: Integrative "omics" involves combining data from genomics, transcriptomics, proteomics, metabolomics, and phenomics. This comprehensive approach enables a detailed recognition of the complex interactions within biological systems. For instance, genomics provides insights into the entire genome sequence and genetic variations, while transcriptomics analyzes gene expression patterns. Proteomics investigates protein abundance and modifications, metabolomics profiles metabolites and their dynamics, and phenomics involves high-throughput analysis of plant phenotypes.
2. *Data Integration Methods*: Various statistical and computational methods are employed for data integration. Statistical approaches like principal component analysis (PCA), correlation analysis, and partial least squares (PLS) regression help identify relationships between different data types. Network-based methods, such as gene co-expression, protein-protein interaction networks, and metabolic networks, reveal how different components interact. Several machine learning techniques, including random forests, support vector machines, and deep learning algorithms, are increasingly used to analyze complex data sets and predict biological outcomes.
3. *Data Integration Platforms and Tools*: Several platforms and tools facilitate multi-omics data integration. Pathway databases like KEGG, MapMan, and PlantCyc provide valuable information on metabolic and regulatory pathways. Network visualization tools such as Cytoscape and Gephi help in visualizing complex interaction networks. Multi-omics data integration platforms like mixOmics, MOFA, and DIABLO enable the integration and analysis of diverse data types.
4. *Functional Annotation and Ontologies*: Functional annotation tools and ontologies standardize the description of gene functions and traits. Gene ontology (GO) provides a standardized vocabulary for gene function description, while plant ontology (PO) offers terms specific to plant anatomy and development. Trait ontology (TO) describes plant traits, facilitating consistent data annotation and interpretation.
5. *Systems-Level Modeling Approaches*: Various modeling approaches are used to study biological systems at different scales. Constraint-based modeling, such as flux balance analysis (FBA), is used for metabolic network analysis. Kinetic modeling, based on ordinary differential equations (ODEs), simulates dynamic processes. Multi-scale modeling integrates models across different biological scales, providing a comprehensive understanding of complex systems.

16.8.2 Case Study of Integrated "Omics" in Tomato Breeding

- *Fruit Quality Improvement*:
  1. A study was conducted in tomato plant cells regarding integrated transcriptomics and metabolomics to improve tomato fruit quality (Kumar et al. 2021). In this study key regulatory genes and metabolic pathways involved in fruit

ripening were analyzed through gene expression and metabolite profiles during tomato fruit development. Under stressful conditions, improving fruit quality involves key genes associated with the isoprenoid pathway, which is crucial for the biosynthesis of carotenoids and tocopherols. Carotenoids, such as lycopene, and tocopherols (vitamin E) are vital for human health and are targeted for biofortification in crops. These compounds are derived from geranylgeranyl pyrophosphate (GGPP), a common precursor in the isoprenoid pathway. Phytoene Synthase 1 (PSY1) is a critical rate-limiting enzyme in carotenoid biosynthesis, converting GGPP into phytoene. The regulation of PSY1 is essential for carotenoid accumulation and is influenced by various transcription factors (TFs) like RIN and FUL1, as well as post-translational regulators such as SGR1 and OR. SGR1 negatively regulates PSY1 activity, reducing carotenoid levels, while OR positively regulates PSY1, enhancing fruit color and quality. Tocopherol (vitamin E) biosynthesis is dependent on Phytol Kinase 5 (VTE5), which also plays a role in stress tolerance (Faryad et al. 2021).

Downregulation of VTE5 leads to reduced tocopherol content, affecting carbohydrate and lipid metabolism and consequently impacting fruit yield and quality. Tocopherols help stabilize membranes and combat oxidative stress, which is especially important under high temperature and light stress (Kumar et al. 2021). Additionally, chloroplasts contribute to photosynthesis and carotenoid accumulation, while abiotic stress can induce ROS accumulation, affecting chloroplast function and triggering the transition to chromoplasts, which influences fruit ripening and quality (Kusano and Fukushima 2013). Key genes such as ORANGE (OR), which promotes chromoplast differentiation, MADS-box AGAMOUS-like transcription factors (TAGLs), which are involved in fruit development and ripening, and GOLDEN 2-LIKE 2 (SlGLK2), which regulates chloroplast development and fruit photosynthesis, are central to this process. Understanding these genes and their regulatory networks provides insight into strategies for improving tomato fruit quality under severe environmental conditions (Wang et al. 2024).

- In tomato fruit, PHYTOCHROME INTERACTING FACTORS (PIFs) play a crucial role in light signaling that affects fruit ripening and nutritional quality. Upon light exposure, PHYTOCHROMES (PHYs) move into the nucleus, where they interact with PIFs, leading to the degradation of these transcription factors and thereby lifting their repression on other genes. This process is vital for regulating fruit ripening and quality. Specifically, PIF3 has been identified as a mediator of light-dependent tocopherol (vitamin E) biosynthesis during fruit ripening. Key PIF genes such as SlPIF1a, SlPIF3, and SlPIF4 are involved in controlling carotenoid and tocopherol levels (Lopez and Satti 1996).

16.9 Challenges of Multi-Omics Technologies

Despite their strengths, genomics and transcriptomics face several challenges. Managing and analyzing the large volume of data generated requires robust tools for data cleaning, filtering, assembly, alignment, variant calling, and annotation. Variability in data normalization and scaling, especially between different types of datasets like RNA-Seq and small RNA-Seq, complicates data integration. The advanced development of single-cell sequencing technologies and longer-read sequencing introduces additional complexities, such as ensuring adequate sequence coverage and developing appropriate statistical analyses (Di Gioia et al. 2013). Addressing these challenges requires continuous improvements in computational tools and methods, as well as effective strategies for data integration and interpretation.

- Genomics and transcriptomics rely on accurate nucleic acid amplification and sequencing from biological samples to generate comprehensive data on genomes and transcriptomes. However, a primary challenge is interpreting these datasets in the context of biological function. Integrating proteomics and metabolomics data with genomics and transcriptomics can address this limitation by bridging the gap between genetic information and observable traits (Faryad et al. 2021).
- Proteomics faces challenges related to data standardization and analysis. The lack of consensus on data formatting, cleaning, and normalization methods complicates comparisons and integrations of proteomic datasets. Additionally, sample preparation and processing, including protein extraction and digestion, can introduce variability that affects the reliability of results. Efforts to address these challenges include ongoing development of standardized protocols and improvements in data analysis methodologies. Proteomics, which involves quantifying proteins and their modifications, faces several unique challenges (Lozada et al. 2022). Moreover, isolation and detection of proteins, particularly those embedded in membranes or present in low abundances, can be problematic. Techniques like mass spectrometry (MS) and liquid chromatography (LC) often require extensive sample preparation and may suffer from variability due to the diverse chemical properties of peptides and proteins. This variability can affect peptide detection and quantification, resulting in issues with data consistency and reliability. Advances in MS sensitivity and isotopic labeling have improved quantification accuracy, but challenges related to normalization, data imputation, and the handling of absent data persist (Faryad et al. 2021).
- Metabolomics presents challenges related to data analysis and sample processing. The complexity of metabolite data requires careful experimental design, sample collection, and data analysis, including alignment, filtering, and imputation. The variability in data structure, normalization, and scaling across different platforms adds to the complexity. Additionally, identifying unknown metabolites and ensuring consistency between targeted and untargeted analyses can be challenging (Kamaluldeen et al. 2014). Addressing these challenges involves optimizing experimental protocols and enhancing data analysis techniques to

improve the reliability and interpretability of metabolomic data. However, metabolites, like proteins, cannot be amplified, and only a fraction of metabolites in a sample can be reliably identified and quantified. The chemical diversity of metabolites and the lack of standardized data formats and analysis pipelines further complicate the interpretation of metabolomics data. Efforts are underway to address these challenges, including standardization of methodologies and improved annotation of metabolites, to enhance the reliability and utility of metabolomics (Kusano and Fukushima 2013).

16.9.1 Shared Issues Among Omics Database Systems

There are several shared issues related to the omics database system.

1. *Data Handling*: All omics platforms must address the challenges of managing large datasets, including data filtering, cleaning, imputation, normalization, and scaling. There is no universally accepted standard for these processes, which means that different analysis pipelines can yield varying results. The continuous evolution of computational tools and methodologies necessitates meticulous documentation of each analysis pipeline, including software versions and parameter choices, to ensure reproducibility and transparency.
2. *Annotation*: Annotating biomolecules presents a significant challenge, particularly for non-standard model organisms with less well-characterized genomes and proteomes. While standard model organisms benefit from extensive annotation tools, non-standard species often require comparative approaches to enhance transcript and peptide annotation. Developing comprehensive peptide reference libraries and iterative annotation strategies can improve the accuracy of annotations. In metabolomics, the challenge is even greater due to the unknown chemical repertoire of many organisms, which complicates the annotation of metabolites without detailed transcriptomic and proteomic data.
3. *Study Design and Analytical Assumptions*: Robust study design and appropriate statistical or machine learning approaches are crucial for reliable omics analysis. Study design must account for potential biases and ensure proper sample selection and experimental unit independence. Statistical models and machine learning algorithms need to be chosen based on the nature of the omics data and the specific research questions. Assumptions underlying these methods, such as the normality of residuals in linear models, must be validated to avoid false inferences. The selection of analytical approaches should align with whether omics variables are predictors or outcomes, and transformations or alternative models may be required to meet analytical assumptions.
4. *Statistical Power*: Omics studies often grapple with the issue of balancing the sample number with the number of quantified molecules, which can lead to false positives and false negatives. Different omics technologies have varying capabilities for multiplexing samples and detecting differences in abundance, which affects the statistical power of the analysis. Integrating data from multiple omics

platforms requires careful consideration of these factors to ensure robust and reproducible results.

5. *Data Archiving and Sharing*: Standardization of nomenclature, data formatting, and public access to multi-omics datasets is crucial for effective data sharing and comparison. While significant progress has been made in standardizing genomic, transcriptomic, and proteomic data, metabolomics and other omics fields are still developing standardized practices. Following established minimum information standards, such as MIAME for microarray data and MIAPE for proteomics, can facilitate the creation of interoperable protocols and scalable datasets for global research efforts.

16.10 Application of Advanced "Multi-Omics" Technologies in Tomato Breeding

1. *Metabolite Quantitative Trait Loci (mQTL) Analysis*: mQTL analysis links genetic variation with metabolite levels, aiding in tomato trait improvement (Kamaluldeen et al. 2014) and uses introgression lines (ILs) from *Solanum lycopersicum* and *S. pennellii* to identify mQTLs, enhancing biochemical traits and identifying key genetic hubs.
2. *Parthenocarpy*: Omics approaches reveal molecular mechanisms underlying parthenocarpy, the ability to produce fruit without seed (Lopez and Satti 1996). Scientists highlighted the roles of auxin, ethylene, gibberellins, and cell division in parthenocarpic fruit development (Pandey et al. 2021). Downregulation of CHS in tomatoes, involved in the flavonoid pathway, led to parthenocarpy suggesting flavonoids play crucial roles in fruit development (Kumar et al. 2021).
3. *Transcript Profiling of Pat3/Pat4 Mutant*: Pascual 2009 performed transcript profiling of the pat3/pat4 mutant, highlighting the involvement of cell-related features like cell division, cell cycle events, and phytohormones such as gibberellins (GAs) and ethylene in regulating parthenocarpy fruit development.
4. *Improvement of Other Traits Using Omics Approaches*: Various multi-omics approaches improve traits such as fruit architecture, and biochemical properties. Gene expression and metabolite composition in tomato peel and flesh (Kumar et al. 2021), and UV-B light's effect on flavonoid content, come up with enhancing fruit quality in tomatoes (Faryad et al. 2021).

16.11 Multi-dimensional Goals in Tomato Breeding

(a) *Precision breeding*: The integration of multi-omics data with high-resolution phenotyping enables the development of predictive models for trait performance. This approach allows for the targeted selection of optimal genetic combinations to achieve desired traits, enhancing the precision and efficiency of breeding programs (Cao et al. 2024).

(b) *Speed breeding*: Combining "omics" data with accelerated breeding techniques, such as genomic selection and rapid generation cycling, shortens breeding cycles. This strategy facilitates the rapid identification and introgression of beneficial alleles, accelerating the development of new tomato varieties (Wang et al. 2024).

(c) *Genome editing optimization*: Using "omics" data to identify optimal targets for CRISPR-Cas9 editing allows for precise modifications of genes controlling key traits. This leads to the development of non-transgenic, improved tomato varieties with enhanced disease resistance, yield, and quality (Di Gioia et al. 2013).

(d) *Climate-resilient tomato varieties*: Integrating "omics" data with climate models helps identify genetic factors contributing to stress tolerance. This knowledge is used to develop tomato varieties adapted to future climate scenarios, ensuring sustainable production under changing environmental conditions (Wang et al. 2011).

(e) *Biofortification*: Utilizing metabolomics and genomics to enhance the nutritional content of tomatoes addresses global nutritional deficiencies. Breeding tomatoes with improved vitamin, mineral, and antioxidant profiles contributes to better human health and nutrition (Kusano and Fukushima 2013).

(f) *Personalized tomato breeding*: Tailoring tomato varieties to specific consumer preferences or health needs involves utilizing metabolomics and sensory analysis to optimize flavor profiles. This approach leads to the development of tomatoes with enhanced health-promoting compounds, catering to personalized dietary requirements (Kumar et al. 2021).

(g) *Microbiome engineering*: Integrating plant and microbial "omics" data optimizes plant-microbiome interactions for improved growth and stress tolerance. Developing tomato varieties with an enhanced ability to recruit beneficial microbes improves overall plant health and productivity (Faryad et al. 2021).

(h) *Increased crop productivity*: The development of high-yielding tomato varieties adapted to diverse environments and improved resource use efficiency through precision breeding can significantly increase global tomato production and food security. These advancements help meet the growing demand for tomatoes while ensuring sustainable agricultural practices (Lozada et al. 2022).

(i) *Enhanced sustainability*: Breeding tomatoes with improved pest and disease resistance reduces reliance on chemical inputs, contributing to more sustainable farming practices. This approach aligns with global efforts to reduce environmental impact and promote eco-friendly agricultural systems (Kamaluldeen et al. 2014). These emerging trends and technologies highlight the transformative potential of "omics" approaches in tomato breeding, offering innovative solutions to enhance crop performance, nutritional value, and sustainability.

16.12 Emerging Trends and Future Perspectives in "Omics" Technologies in Tomato Breeding

(a) *Single-cell omics*: Such technologies, including single-cell RNA sequencing, single-cell proteomics, and single-cell metabolomics, allow for the analysis of cellular heterogeneity within plant tissues. These technologies provide detailed insights into cell-specific responses to environmental stresses and developmental processes. In tomato breeding, single-cell "omics" can help identify specific cellular pathways that contribute to stress resistance and optimal growth, leading to more targeted breeding strategies (Pandey et al. 2021).
(b) *Spatial "omics"*: Spatial "omics" technologies, including spatial transcriptomics and imaging mass spectrometry, enable the mapping of gene expression and metabolite distribution across plant tissues. These technologies offer detailed insights into tissue-specific regulation of physiological processes in tomatoes. By understanding the spatial organization of gene expression and metabolites, researchers can better manipulate these processes to improve fruit quality and yield of tomato plants (Faryad et al. 2021).
(c) *Long-read sequencing*: Technologies like nanopore sequencing and PacBio SMRT sequencing provide improved genome assembly and detection of structural variants. These long-read sequencing technologies allow for more accurate characterization of tomato genomes, including complex regions and repetitive elements. This level of detail is crucial for identifying and utilizing genetic diversity to improve various traits in tomato plants (Kusano and Fukushima 2013).
(d) *CRISPR-based Omics*: *CRISPR*-based "omics," including CRISPR activation/interference (CRISPRa/CRISPRi) and CRISPR-Cas13 RNA editing, are used for genome-wide functional genomics studies. These technologies facilitate rapid identification and validation of genes controlling a few important traits in tomatoes, like disease resistance and fruit quality. The precision and efficiency of CRISPR-based techniques accelerate the breeding process by enabling targeted genetic modifications (Faryad et al. 2021).
(e) *Multi-omics data integration*: Machine learning and AI algorithms are employed to integrate diverse "omics" datasets, building predictive models for complex traits. Integrating data from multi-omics databases provides a comprehensive understanding of trait regulation, leading to more efficient breeding strategies (Kusano and Fukushima 2013).
(f) *Epigenomics*: Epigenomic technologies like whole-genome bisulfite sequencing, ChIP-seq, and ATAC-seq are used to study, histone modifications, chromatin accessibility, and DNA methylation These studies reveal epigenetic regulation of tomato development and stress responses, offering new targets for breeding programs aimed at improving crop resilience and performance (Janni et al. 2019).
(g) *Phenomics advancements*: High-resolution imaging, hyperspectral analysis, and 3D modeling are noninvasive, high-throughput phenotyping technologies. They enable the rapid assessment of complex traits and their correlation with genetic and environmental factors. These advancements in phenomics facilitate

the identification of superior genotypes and enhance breeding efficiency (Faryad et al. 2021; Lozada et al. 2022).

16.13 Machine Learning and Artificial Intelligence in Breeding

The utilization of machine learning (ML) and artificial intelligence (AI) in tomato breeding has revolutionized crop improvement, offering tools for yield estimation, disease detection, quality grading, and predictive modeling. Studies have demonstrated the use of unmanned aircraft systems (UAS) with ML for yield prediction, deep learning for generating fruit shapes from DNA, and AI for automated tomato sorting and disease detection. Techniques like deep transfer learning and machine vision have enhanced the accuracy of disease classification and quality grading. Additionally, integrating phenotypic data with ML for yield prediction and using computer vision for growth stage monitoring have further improved the efficiency and precision of tomato breeding programs. These advancements support smarter farming practices and better decision-making in tomato breeding goals.

16.14 Conclusion

In conclusion, the integration of advanced "omics" technologies into tomato breeding heralds a transformative era with immense potential to enhance tomato yield, quality, and resilience significantly. Genomics, transcriptomics, proteomics, and metabolomics offer a comprehensive understanding of key traits' genetic and molecular foundations, enabling the development of superior tomato varieties. The advancements in sequencing technologies, gene editing tools, and high-throughput phenotyping are critical in accelerating the breeding process and addressing the complex challenges faced by breeders. However, successfully applying these technologies requires overcoming hurdles such as data integration, developing robust bioinformatics tools, and fostering interdisciplinary collaboration. As we navigate these challenges, the future of tomato breeding looks promising, with "omics" technologies providing innovative strategies to meet global food security and sustainability goals. This chapter underscores the pivotal role of "omics" in shaping the future of tomato cultivars, emphasizing both the potential and the challenges that lie ahead in this rapidly evolving field.

References

Ahmed HGMD, Zeng Y, Yang X, Fatima N, Faisal A (2024) Crop genome sequencing and their application for crop improvement. In: Omics-based techniques for global food security. John Wiley & Sons, pp 1–26

Aksoy E, Yılmaz H, Kayıhan C (2022) The revolution of omics Technology in Plant Science. In: Principles and practices of Omics and genome editing for crop improvement. Springer International Publishing, Cham, pp 23–56

Alahmad A, Edelman L, Bouteiller M, Castel L, Riah-Anglet W, Bennegadi-Laurent N, Trinsoutrot-Gattin I (2024) Unveiling the impact of soil prebiotics on Rhizospheric microbial functionality in Zea mays L. Agriculture 14(7):1115

Alkan N, Davydov O, Sagi M, Fluhr R, Prusky D (2015a) Botrytis cinerea counteracts salicylic acid-mediated basal defense of tomato by up-regulating its salicylic acid-like effector export machinery. Mol Plant-Microbe Interact 28(3):307–319

Alkan N, Friedlander G, Ment D, Prusky D, Fluhr R (2015b) Simultaneous transcriptome analysis of C olletotrichum gloeosporioides and tomato fruit pathosystem reveals novel fungal pathogenicity and fruit defense strategies. New Phytol 205(2):801–815

Alotaibi F, Alharbi S, Alotaibi M, Al Mosallam M, Motawei M, Alrajhi A (2021) Wheat omics: classical breeding to new breeding technologies. Saudi J Biol Sci 28(2):1433–1444

Arabidopsis Interactome Mapping Consortium (2011) Evidence for network evolution in an Arabidopsis interactome map. Science (New York, NY) 333(6042):601

Ashraf M, Foolad MR (2013) Crop breeding for salt tolerance in the era of molecular markers and marker-assisted selection. Plant Breed 132(1):10–20

Balmant KM, Parker J, Yoo MJ, Zhu N, Dufresne C, Chen S (2015) Redox proteomics of tomato in response to Pseudomonas syringae infection. Hortic Res 2:15043

Basim H, Basim E, Tombuloglu H, Unver T (2021) Comparative transcriptome analysis of resistant and cultivated tomato lines in response to Clavibacter michiganensis subsp. michiganensis. Genomics 113(4):2455–2467

Bayram M, Gökırmaklı Ç (2018) Horizon scanning: how will metabolomics applications transform food science, bioengineering, and medical innovation in the current era of foodomics? Omics 22(3):177–183

Bibi S, Ullah R, Burni T, Ullah Z, Kazi M (2023) Impact of resorcinol and biochar application on the growth attributes, metabolite contents, and antioxidant systems of tomato (Lycopersicon esculentum Mill.). ACS Omega 8(48):45750–45762

Bineau E (2022) Combining genetic association and transcriptome analyses to bring back flavor into tomato hybrid varieties (Doctoral dissertation, Université d'Avignon)

Bineau E, Diouf I, Carretero Y, Duboscq R, Bitton F, Djari A et al (2021) Genetic diversity of tomato response to heat stress at the QTL and transcriptome levels. Plant J 107(4):1213–1227

Bokszczanin KŁ (2019) Future breeding strategies. In: The pear genome. Springer, pp 301–315

Bovy A, Schijlen E, Hall RD (2007) Metabolic engineering of flavonoids in tomato (Solanum lycopersicum): the potential for metabolomics. Metabolomics 3:399–412

Camargo RDA, Herai RH, Santos LN, Bento FM, Lima JE, Marques-Souza H, Figueira A (2015) De novo transcriptome assembly and analysis to identify potential gene targets for RNAi-mediated control of the tomato leafminer (Tuta absoluta). BMC Genomics 16:1–17

Cao P, Zhao Y, Wu F, Xin D, Liu C, Wu X, Qi Z (2022) Multi-omics techniques for soybean molecular breeding. Int J Mol Sci 23(9):4994

Cao Y, She J, Li Z, Liu Y, Tian T, You Q et al (2024) TomAP: a multi-omics data analysis platform for advancing functional genomics research in tomatoes. New Crops 1:100002

Chen L, Li X, Gao J, Wang Z, Zhao Y (2017) Silencing of a proteinase inhibitor gene in tomato confers enhanced thermotolerance and resistance to fungal pathogens. Plant Physiol Biochem 118:45–54

Chen H, Liu R, Shen C, Ma Q, Li Z, Luo M (2018) INVINH1, a novel proteinase inhibitor, enhances cold stress tolerance by suppressing the expression of cold-responsive genes in transgenic tomato plants. Plant Physiol Biochem 132:186–194

Chen C, Wang J, Pan D, Wang X, Xu Y, Yan J et al (2023) Applications of multi-omics analysis in human diseases. MedComm 4(4):e315

Cui J, Luan Y, Jiang N, Bao H, Meng J (2017) Comparative transcriptome analysis between resistant and susceptible tomato allows the identification of lnc RNA 16397 conferring resistance to Phytophthora infestans by co-expressing glutaredoxin. Plant J 89(3):577–589

Dai Q, Geng L, Lu M, Jin W, Nan X, He PA, Yao Y (2017) Comparative transcriptome analysis of the different tissues between the cultivated and wild tomato. PLoS One 12(3):e0172411

Das A, Kamal A, Reddy M, Baranwal V (2015) SDG34: a key gene in response to stress in tomato plants. J Plant Biochem Biotechnol 24(3):313–318
Devanna BN, Pramesh D, Prashant SH, Rajashekara H, Kumari M, Raghu S, Sharma TR (2023) Multi-omics approaches and their applications for disease management in cereal crops. In: Advances in plant disease management. CRC Press, pp 175–208
Di Gioia F, Signore A, Serio F, Santamaria P (2013) Grafting improves tomato salinity tolerance through sodium partitioning within the shoot. HortScience 48(7):855–862
Dong X, Wang Y, Yan Y, Wang H (2022) Transcriptome analysis of the development of Pedicel Abscission Zone in Tomato. Horticulturae 8(10):865
Esposito S, Carputo D, Cardi T, Tripodi P (2019) Applications and trends of machine learning in genomics and phenomics for next-generation breeding. Plan Theory 9(1):34
Faryad A, Aziz F, Tahir J, Kousar M, Qasim M, Shamim A (2021) Integration of Omics technologies for crop improvement. Protein Pept Lett 28(8):896–908
Foolad MR, Panthee DR (2012) Marker-assisted selection in tomato breeding. Crit Rev Plant Sci 31(2):93–123
Foolad MR, Sharma A (2004, June) Molecular markers as selection tools in tomato breeding. In I International Symposium on Tomato Diseases 695, pp 225–240
Fraser PD, Aharoni A, Hall RD, Huang S, Giovannoni JJ, Sonnewald U, Fernie AR (2020) Metabolomics should be deployed in the identification and characterization of gene-edited crops. Plant J 102(5):897–902
Gatica-Arias A (2020) The regulatory current status of plant breeding technologies in some Latin American and the Caribbean countries. Plant Cell Tissue Organ Cult 141(2):229–242
Großkinsky DK, Syaifullah SJ, Roitsch T (2018) Integration of multi-omics techniques and physiological phenotyping within a holistic phenomics approach to study senescence in model and crop plants. J Exp Bot 69(4):825–844
Gupta V, Mathur S, Solanke AU, Sharma MK, Kumar R, Vyas S et al (2009) Genome analysis and genetic enhancement of tomato. Crit Rev Biotechnol 29(2):152–181
Gupta PK, Kumar J, Mir RR, Kumar A (2010) 4 marker-assisted selection as a component of conventional plant breeding. Plant Breed Rev 33(4):145–217
Gupta S, Shi X, Lindquist IE, Devitt N, Mudge J, Rashotte AM (2013) Transcriptome profiling of cytokinin and auxin regulation in tomato root. J Exp Bot 64(2):695–704
Hajyzadeh M, Egi Y (2023) Innovative and ground breaking technologies for sustainable agriculture. in sürdürülebilir tarim ve gida güvenliği (pp. 247–266). iksad yayınevi. https://doi.org/10.5281/zenodo.8292204
Hasan N, Choudhary S, Naaz N, Sharma N, Laskar RA (2021) Recent advancements in molecular marker-assisted selection and applications in plant breeding programmes. J Genet Eng Biotechnol 19(1):128
Hasanuzzaman M, Nahar K, Fujita M, Ahmad P, Chandna R, Prasad MNV, Ozturk M (2013) Enhancing plant productivity under salt stress: relevance of poly-omics. In: Salt stress in plants: signalling, omics and adaptations. Springer, New York, pp 113–156
He J, Zhao X, Laroche A, Lu ZX, Liu H, Li Z (2014) Genotyping-by-sequencing (GBS), an ultimate marker-assisted selection (MAS) tool to accelerate plant breeding. Front Plant Sci 5:484
Herraiz FJ, Blanca J, Ziarsolo P, Gramazio P, Plazas M, Anderson GJ et al (2016) The first de novo transcriptome of pepino (Solanum muricatum): assembly, comprehensive analysis and comparison with the closely related species S. caripense, potato and tomato. BMC Genomics 17:1–17
Huang J, Liu J, Han X, Sun T, Li Y, Zhou Q (2017) SlGRAS10, a GRAS transcription factor, is involved in the regulation of osmotic potential, flavonoid production, and ROS scavenging in tomato under abiotic stress. Plant Sci 254:1–10
Hurtado M, Vilanova S, Plazas M, Gramazio P, Herraiz FJ, Andújar I, Prohens J (2013) Phenomics of fruit shape in eggplant (Solanum melongena L.) using Tomato Analyzer software. Sci Hortic 164:625–632
Ibitoye DO, Akin-Idowu PE (2011) Marker-assisted-selection (MAS): a fast track to increase genetic gain in horticultural crop breeding. Afr J Biotechnol 10(55):11333–11339

Ijaz S, Iqbal J, Abbasi BA, Tufail A, Ullah Z, Sharifi-Rad J, Uddin S (2023a) Biofortification: lessons from the past and strategies for future food security. In: Legumes biofortification. Springer International Publishing, Cham, pp 521–545

Ijaz S, Iqbal J, Abbasi BA, Ullah Z, Yaseen T, Kanwal S et al (2023b) Rosmarinic acid and its derivatives: current insights on anticancer potential and other biomedical applications. Biomed Pharmacother 162:114687

Ijaz S, Iqbal J, Abbasi BA, Tufail A, Ullah Z, Yaseen T, Iqbal R (2024) Heavy metal–polluted arable land and its consequences: a global scenario. In: Biochar-assisted remediation of contaminated soils under changing climate. Elsevier, pp 71–99

Ishihara T, Mitsuhara I, Takahashi H, Nakaho K (2012) Transcriptome analysis of quantitative resistance-specific response upon Ralstonia solanacearum infection in tomato. PLoS One 7(10):e46763

Iula G, Miras-Moreno B, Rouphael Y, Lucini L, Trevisan M (2022) The complex metabolomics crosstalk triggered by four molecular elicitors in tomato. Plants 11(5):678

Jamil IN, Remali J, Azizan KA, Nor Muhammad NA, Arita M, Goh HH, Aizat WM (2020) Systematic multi-omics integration (MOI) approach in plant systems biology. Front Plant Sci 11:944

Janni M, Coppede N, Bettelli M, Briglia N, Petrozza A, Summerer S et al (2019) In vivo phenotyping for the early detection of drought stress in tomato. Plant Phenomics 2019:6168209

Jiang Y, Liang G, Yu D (2016) The involvement of bHLH transcription factor SlBZR1D in tomato's salt tolerance. J Exp Bot 67(17):5037–5049

Kamaluldeen J, Yunusa IA, Zerihun A, Bruhl JJ, Kristiansen P (2014) Uptake and distribution of ions reveal contrasting tolerance mechanisms for soil and water salinity in okra (Abelmoschus esculentus) and tomato (Solanum esculentum). Agric Water Manag 146:95–104

Karlova R, Rosin FM, Busscher-Lange J, Parapunova V, Do PT, Fernie AR et al (2011) Transcriptome and metabolite profiling show that APETALA2a is a major regulator of tomato fruit ripening. Plant Cell 23(3):923–941

Kashyap SP, Prasanna HC, Kumari N, Mishra P, Singh B (2020) Understanding salt tolerance mechanism using transcriptome profiling and de novo assembly of wild tomato Solanum chilense. Sci Rep 10(1):15835

Khapte PS, Kumar P, Wakchaure GC, Jangid KK, Colla G, Cardarelli M, Rane J (2022) Application of phenomics to elucidate the influence of rootstocks on drought response of tomato. Agronomy 12(7):1529

Kumar R, Sharma V, Suresh S, Ramrao DP, Veershetty A, Kumar S et al (2021) Understanding omics driven plant improvement and de novo crop domestication: some examples. Front Genet 12:637141

Kumar M, Prusty MR, Pandey MK, Singh PK, Bohra A, Guo B, Varshney RK (2023) Application of CRISPR/Cas9-mediated gene editing for abiotic stress management in crop plants. Front Plant Sci 14:1157678

Kusano M, Fukushima A (2013) Current challenges and future potential of tomato breeding using omics approaches. Breed Sci 63(1):31–41

Laxman RH, Hemamalini P, Bhatt RM, Sadashiva AT (2018) Non-invasive quantification of tomato (Solanum lycopersicum L.) plant biomass through digital imaging using phenomics platform. Indian J Plant Physiol 23:369–375

Lee JM, Oh CS, Yeam I (2015) Molecular markers for selecting diverse disease resistances in tomato breeding programs. Plant Breed Biotechnol 3(4):308–322

Li X, Wu J, Shen C (2015) NAC transcription factor JUNGBRUNNEN1 improves tomato's abiotic stress tolerance. Plant Physiol Biochem 95:1–11

Li J, Zhang H, Wang W, Qi Z (2018) DEAD-box RNA helicase SlDEAD23 and SlDEAD35 are involved in tomato's abiotic and biotic stress responses. Plant Mol Biol 96(1–2):165–180

Li Y, Li S, Zhao M (2019) Suppression of SlNL33 expression improves ascorbate accumulation and stress tolerance in tomato. J Exp Bot 70(18):4855–4865. https://doi.org/10.1093/jxb/erz250

Li Q, Sapkota M, van der Knaap E (2020) Perspectives of CRISPR/Cas-mediated cis-engineering in horticulture: unlocking the neglected potential for crop improvement. Hortic Res 7:36

Li J, Cao L, Xie Q, Chen G, Hu Z (2023) Transcriptome sequencing and analysis during seed growth and development in tomato. Sci Hortic 310:111763

Lin M, Zhou X, Shen X, Mao C, Chen X (2011) The predicted Arabidopsis interactome resource and network topology-based systems biology analyses. Plant Cell 23(3):911–922

Liu F, Wang Z, Ren H, Shen C, Liu J (2015a) MdSWEET17, a drought stress-responsive gene, regulates fructose content in tomato. Plant Sci 236:45–56

Liu H, Yu C, Li H, Ouyang B, Wang T, Zhang J et al (2015b) Overexpression of ShDHN, a dehydrin gene from Solanum habrochaites enhances tolerance to multiple abiotic stresses in tomato. Plant Sci 231:198–211

Liu R, Shen C, Yang C, Ma Q, Chen H (2016) BEL1-like genes regulate numerous biological processes in plants via TALE proteins and are involved in the heat, cold, and drought stress response. J Plant Physiol 200:81–92

Liu J, Chen L, Zhang Y, Fan X, Li R (2017) Overexpression of SlMAPK3 and SlMPK7 enhances resistance to chilling, cadmium, and drought stresses in tomato. Plant Physiol Biochem 118:99–108

Lopez MV, Satti SME (1996) Calcium and potassium-enhanced growth and yield of tomato under sodium chloride stress. Plant Sci 114(1):19–27

Lozada DN, Bosland PW, Barchenger DW, Haghshenas-Jaryani M, Sanogo S, Walker S (2022) Chile pepper (Capsicum) breeding and improvement in the "multi-omics" era. Front Plant Sci 13:879182

Luna E, Flandin A, Cassan C, Prigent S, Chevanne C, Kadiri CF et al (2020) Metabolomics to exploit the primed immune system of tomato fruit. Metabolites 10(3):96

Mahmood U, Li X, Fan Y, Chang W, Niu Y, Li J et al (2022) Multi-omics revolution to promote plant breeding efficiency. Front Plant Sci 13:1062952

Mazzeo MF, Cacace G, Ferriello F, Puopolo G, Zoina A, Ercolano MR, Siciliano RA (2014) Proteomic investigation of response to FORL infection in tomato roots. Plant Physiol Biochem 74:42–49

Mitra D, Chaudhary P, Verma D, Khoshru B, Senapati A, Mahakur B et al (2021) Bioinformatics' role in studying microbe-mediated biotic and abiotic stress tolerance. In: Microbial management of plant stresses. Woodhead Publishing, pp 203–219

Mittler R, Finka A, Goloubinoff P (2017) How do plants feel the heat? Trends Biochem Sci 42(6):444–451

Mochida K, Shinozaki K (2011) Advances in omics and bioinformatics tools for systems analyses of plant functions. Plant Cell Physiol 52(12):2017–2038

Muthamilarasan M, Singh NK, Prasad M (2019) Multi-omics approaches for strategic improvement of stress tolerance in underutilized crop species: a climate change perspective. Adv Genet 103:1–38

Muthuramalingam P, Jeyasri R, Krishnan SR, Pandian STK, Sathishkumar R, Ramesh M (2019) Integrating the bioinformatics and omics tools for systems analysis of abiotic stress tolerance in Oryza sativa (L.). In: Advances in plant transgenics: methods and applications. Springer, pp 59–77

Naik B, Kumar V, Rizwanuddin S, Chauhan M, Choudhary M, Gupta AK et al (2023) Genomics, proteomics, and metabolomics approaches to improve abiotic stress tolerance in tomato plant. Int J Mol Sci 24(3):3025

Naqvi RZ, Siddiqui HA, Mahmood MA, Najeebullah S, Ehsan A, Azhar M, Asif M (2022) Smart breeding approaches in post-genomics era for developing climate-resilient food crops. Front Plant Sci 13:972164

Narad P, Gupta R, Sengupta A (2022) Plant metabolomics: a new era in the advancement of agricultural research. In: Bioinformatics in agriculture. Academic Press, pp 139–160

Niu Q, He Z, Li J (2019) Functional characterization of tomato transcription factors SlMBP8, SlHB2, and SlAGO4A in response to salt and drought stress. Plant Mol Biol 99(4–5):389–399

Orellana S, Yanez M, Espinoza A, Verdugo I, Gonzalez E, Ruiz-Lara S, Casaretto JA (2010) The transcription factor SlAREB1 confers drought, salt stress tolerance and regulates biotic and abiotic stress-related genes in tomato. Plant Cell Environ 33(12):2191–2208

Osei MK, Prempeh R, Adjebeng-Danquah J, Opoku JA, Danquah A, Danquah E et al (2018) Marker-assisted selection (MAS): a fast-track tool in tomato breeding. In: Recent advances in tomato breeding and production. InTech, pp 93–113

Pandey AK, Savani AK, Singh P (2021) The early blight of tomato: omics interventions toward controlling disease spread and development. In: Omics technologies for sustainable agriculture and global food security, vol 1. Springer, pp 85–108

Parker J, Koh J, Yoo MJ, Zhu N, Feole M, Yi S, Chen S (2013) Quantitative proteomics of tomato defense against Pseudomonas syringae infection. Proteomics 13(12–13):1934–1946

Paulpriyaa K, Kalpanadevib V, Rajalakshmic K, Jeevithad M, Shalinie CB, Eluvakkala T (2022) A critical review on omics technology in plant genetics and breeding. J Pharm Negat Results:3682–3699

Penin AA, Klepikova AV, Kasianov AS, Gerasimov ES, Logacheva MD (2019) Comparative analysis of developmental transcriptome maps of Arabidopsis thaliana and Solanum lycopersicum. Genes 10(1):50

Pramanik B, Das A, Debnath S (2024) 12 advancement in gene editing. In: CRISPR and plant functional genomics. CRC Press, p 206

Prasanna HC, Rai N, Hussain Z, Yerasu SR, Tiwari JK (2023) Tomato: breeding and genomics. Vegetable Sci 50:146–155

Priyadarshi, R., Sinha, P., Dasari, A., & Sundaram, R. M. (2021). Application of "omics" technologies in crop breeding. Omics technologies for sustainable agriculture and global food security (II), 25–45. Springer

Ramesh SV, Williams S, Kappagantu M, Mitter N, Pappu HR (2017) Transcriptome-wide identification of host genes targeted by tomato spotted wilt virus-derived small interfering RNAs. Virus Res 238:13–23

Rasheed A, Hao Y, Xia X, Khan A, Xu Y, Varshney RK, He Z (2017) Crop breeding chips and genotyping platforms: progress, challenges, and perspectives. Mol Plant 10(8):1047–1064

Rashid B, Husnain T, Riazuddin S (2014) Genomic approaches and abiotic stress tolerance in plants. In: Emerging technologies and management of crop stress tolerance. Academic Press, San Diego, pp 1–37

Raza A, Mubarik MS, Sharif R, Habib M, Jabeen W, Zhang C, Varshney RK (2023) Developing drought-smart, ready-to-grow future crops. Plant Genome 16(1):e20279

Razzaq A, Kaur P, Akhter N, Wani SH, Saleem F (2021) Next-generation breeding strategies for climate-ready crops. Front Plant Sci 12:620420

Rosa-Martínez E, Bovy A, Plazas M, Tikunov Y, Prohens J, Pereira-Dias L (2023) Genetics and breeding of phenolic content in tomato, eggplant and pepper fruits. Front Plant Sci 14:1135237

Roychowdhury R, Das SP, Gupta A, Parihar P, Chandrasekhar K, Sarker U et al (2023) Multi-omics pipeline and omics-integration approach to decipher plant's abiotic stress tolerance responses. Genes 14(6):1281

Ruzicka DR, Barrios-Masias FH, Hausmann NT, Jackson LE, Schachtman DP (2010) Tomato root transcriptome response to a nitrogen-enriched soil patch. BMC Plant Biol 10:1–19

Saidi A, Hajibarat Z, Hajibarat Z (2020) Transcriptome analysis of Phytophthora infestans and Colletotrichum coccodes in tomato to reveal resistance mechanisms. Asia-Pacific J Mol Biol Biotechnol 28:39–51

Sampath V, Rangarajan N, Sharanappa CH, Deori M, Veeraragavan M, Ghodake BD, Kaushal K (2023) Advancing crop improvement through CRISPR technology in precision agriculture trends-a review. Int J Environ Clim Change 13(11):4683–4694

Sant'Ana DVP, Lefsrud M (2018) Tomato proteomics: tomato as a model for crop proteomics. Sci Hortic 239:224–233

Sato S, Tabata S (2016) Tomato genome sequence. In: Functional genomics and biotechnology in Solanaceae and Cucurbitaceae crops. Springer, pp 1–13

Sharma P, Thakur S, Negi R (2019) Recent advances in breeding of tomato-a review. Int J Curr Microbiol App Sci 8(3):1275–1283

Simko I, Jia M, Venkatesh J, Kang BC, Weng Y, Barcaccia G et al (2021) Genomics and marker-assisted improvement of vegetable crops. Crit Rev Plant Sci 40(4):303–365

Singh A, Shannon CP, Gautier B, Rohart F, Vacher M, Tebbutt SJ, Lê Cao KA (2019) DIABLO: an integrative approach for identifying key molecular drivers from multi-omics assays. Bioinformatics 35(17):3055–3062

Singh AK, Mishra P, Kashyap SP, Karkute SG, Singh PM, Rai N et al (2022) Molecular insights into mechanisms underlying thermo-tolerance in tomato. Front Plant Sci 13:1040532

Singha DL, Das D, Paswan RR, Chikkaputtaiah C, Kumar S (2022) Novel approaches and advanced molecular techniques for crop improvement. In: Plant-microbe interactions. CRC Press, pp 1–27

Su L, Xie Y, He Z, Zhang J, Tang Y, Zhou X (2021) Network response of two cherry tomato (Lycopersicon esculentum) cultivars to Cadmium stress as revealed by transcriptome analysis. Ecotoxicol Environ Saf 222:112473

Tahir ul Qamar M, Faryad A, Bari A, Zahid B, Zhu X, Chen LL (2020) Effectiveness of conventional crop improvement strategies vs. omics. In: Environment, climate, plant and vegetation growth. Springer, pp 253–284

Tanigaki Y, Higashi T, Takayama K, Nagano AJ, Honjo MN, Fukuda H (2015) Transcriptome analysis of plant hormone-related tomato (Solanum lycopersicum) genes in a sunlight-type plant factory. PLoS One 10(12):e0143412

Tiwari JK, Yerasu SR, Rai N, Singh DP, Singh AK, Karkute SG et al (2022) Progress in marker-assisted selection to genomics-assisted breeding in tomato. Crit Rev Plant Sci 41(5):321–350

Tripodi P, Massa D, Venezia A, Cardi T (2018) Sensing technologies for precision phenotyping in vegetable crops: current status and future challenges. Agronomy 8(4):57

Ullah Z (2017) Estimation of genetic diversity in the germplasm of Brassica Campestris (L.) using morphological and biochemical markers. J Bot 1(1):45–59

Ullah Z, Iqbal J, Abbasi BA, Ijaz S, Sharifi-Rad J, Yaseen T, Mahmood T (2023a) How genome editing can be helpful in the biofortification of legumes. In: Legumes biofortification. Springer International Publishing, Cham, pp 207–232

Ullah R, Ullah Z, Iqbal J, Chalgham W, Ahmad A (2023b) Aspartic acid-based Nano-copper induces resilience in Zea mays to applied Lead stress via conserving photosynthetic pigments and triggering the antioxidant biosystem. Sustain For 15(16):12186

Ullah Z, Iqbal J, Abbasi BA, Akhtar W, Kanwal S, Ali I et al (2023c) Assessment of Gus expression induced by anti-sense Os PPO gene promoter and antioxidant enzymatic assays in response to drought and heavy metal stress in transgenic Arabidopsis thaliana. Sustain For 15(17):12783

Ullah Z, Gul F, Iqbal J, Abbasi BA, Kanwal S, Chalgham W et al (2023d) Biogenic synthesis of multifunctional silver oxide nanoparticles (Ag2ONPs) using Parieteria alsinaefolia delile aqueous extract and assessment of their diverse biological applications. Microorganisms 11(4):1069

Ullah Z, Iqbal J, Gul F, Abbasi BA, Kanwal S, Elsadek MF, Mahmood T (2024) Biogenic synthesis, characterization, and in vitro biological investigation of silver oxide nanoparticles (AgONPs) using Rhynchosia capitata. Sci Rep 14(1):10484

Varshney RK, Singh VK, Kumar A, Powell W, Sorrells ME (2018) Can genomics deliver climate-change ready crops? Curr Opin Plant Biol 45:205–211

Vidal A, Matus T, Gainza F, Malnoy M (2023). Leveraging system biology and new breeding technologies for water stress tolerance in grapevines. In PlantEd: genome editing in plants: 4th PlantEd Conference, Porto, Portugal, 18–20 September 2023 (p. 51). PT.

Vu TV, Das S, Tran MT, Hong JC, Kim JY (2020) Precision genome engineering for the breeding of tomatoes: recent progress and future perspectives. Front Genome Editing 2:612137

Wai AH, Naing AH, Lee DJ, Kim CK, Chung MY (2020) Molecular genetic approaches for enhancing stress tolerance and fruit quality of tomato. Plant Biotechnol Rep 14:515–537

Wang X, Geng S, Ri YJ, Cao D, Liu J, Shi D, Yang C (2011) Physiological responses and adaptive strategies of tomato plants to salt and alkali stresses. Sci Hortic 130(1):248–255

Wang Y, Tao X, Tang XM, Xiao L, Sun JL, Yan XF et al (2013) Comparative transcriptome analysis of tomato (Solanum lycopersicum) in response to exogenous abscisic acid. BMC Genomics 14:1–14

Wang RH, Yuan XY, Meng LH, Zhu BZ, Zhu HL, Luo YB, Fu DQ (2016) Transcriptome analysis provides a preliminary regulation route of the ethylene signal transduction component, SlEIN2, during tomato ripening. PLoS One 11(12):e0168287

Wang Z, Li H, Zhao Y, Yang J (2019) Overexpression of SlAIM1 enhances salt and oxidative stress tolerance in tomato. Plant Physiol Biochem 135:80–89

Wang Z, Liu Y, Li H (2021) CRISPR/Cas9 targeted mutagenesis of SlLBD40 improves drought tolerance in tomatoes. Plant Biotechnol J 19(3):381–391

Wang R, Yang Y, Xu K, Wang T, Elsadek M, Yuan L et al (2024) Multi-omics analysis reveals improvement of tomato quality by grafting on goji rootstock. Food Qual Saf 8:fyae023

Watson ER, Taherian Fard A, Mar JC (2022) Computational methods for single-cell imaging and omics data integration. Front Mol Biosci 8:768106

Wu Q, Park S, Kirkham MB, Williams KA (2017) Transcriptome analysis reveals potential mechanisms for inhibition of intumescence development by UV radiation in tomato. Environ Exp Bot 134:130–140

Yang F, Zhang H, Shen C (2014) SlbZIP1 transcription factor enhances salt and drought stress tolerance in tomato by regulating stress-responsive genes. Plant Mol Biol 86(6):665–676

Yang L, Wang Y, He Q (2017) Overexpression of Saussurea involucrata dehydrin gene enhances cold and drought tolerance in tomato. J Plant Physiol 210:24–34

Yang Y, Saand MA, Huang L, Abdelaal WB, Zhang J, Wu Y et al (2021) Applications of multi-omics technologies for crop improvement. Front Plant Sci 12:563953

Ye Q, Hickey J, Summers K, Falatovich B, Gencheva M, Eubank TD, Guo NL (2022) Multi-omics immune interaction networks in lung cancer tumorigenesis, proliferation, and survival. Int J Mol Sci 23(23):14978

Yeats TH, Howe KJ, Matas AJ, Buda GJ, Thannhauser TW, Rose JK (2010) Mining the surface proteome of tomato (Solanum lycopersicum) fruit for proteins associated with cuticle biogenesis. J Exp Bot 61(13):3759–3771

Yoosefzadeh Najafabadi M, Hesami M, Eskandari M (2023) Machine learning-assisted approaches in modernized plant breeding programs. Genes 14(4):777

Yu X, Liu Z, Sun X (2023) Single-cell and spatial multi-omics in the plant sciences: technical advances, applications, and perspectives. Plant Commun 4(3):100508

Zhang A, Shang Q (2024) Transcriptome analysis of early lateral root formation in tomato. Plants 13(12):1620

Zhang J, Liu H, Chen R (2019) SlABIG1, a salt stress negative regulator gene, enhances salt tolerance in tomatoes by knockout. Plant Cell Rep 38(8):1025–1035

Zhang H, Zhao M, Li J (2020) Overexpression of SlHY5 enhances cold tolerance in tomato. Plant Sci 294:110435

Zhao T, Liu W, Zhao Z, Yang H, Bao Y, Zhang D et al (2019) Transcriptome profiling reveals the response process of tomato carrying Cf-19 and Cladosporium fulvum interaction. BMC Plant Biol 19:1–12

Zhou S, Sauvé R, Thannhauser TW (2009) Proteome changes induced by aluminium stress in tomato roots. J Exp Bot 60(6):1849–1857

Zhou S, Sauvé RJ, Liu Z, Reddy S, Bhatti S, Hucko SD et al (2011) Heat-induced proteome changes in tomato leaves. J Am Soc Hortic Sci 136(3):219–226

Zhu G, Wang S, Huang Z, Zhang S, Liao Q, Zhang C et al (2018) Rewiring of the fruit metabolome in tomato breeding. Cell 172(1):249–261

17 Genomic Strategies for Enhancing Quantitative and Qualitative Traits in Tomato

Sarita Pandey, Abhishek Anand, Monika Jain,
Shuvadeep Halder, and Md Kamaruzzaman

Abstract

Marker-assisted selection (MAS) uses markers inter-connected with desired qualities to choose lines with superior alleles and develop improved lines. Advances in genomics technologies have significantly reduced the genotyping cost, by utilizing genome-wide marker data to select lines having higher breeding value. Genomic selection (GS) is a modern breeding technique that leverages genome-wide marker data to estimate breeding values, offering a powerful approach to tackling complex traits. By integrating phenotypic and genotypic data, GS trains predictive models to calculate genomic estimated breeding values (GEBVs). This method not only enhances selection accuracy, efficiency, and intensity but also accelerates genetic gains by reducing the period of the selection cycle, ultimately increasing the overall rate of genetic improvement. The availability of low-value genotyping systems has made large-scale genotyping data collections reasonable, allowing for the deployment of GS in different varieties of crop species. This chapter delves into the core principles and groundbreaking concepts of GS breeding, exploring its significant impact on advancing crop improvement in tomatoes.

Keywords

Tomato · Genomics · Genomic selection · Abiotic and biotic stress

S. Pandey (✉) · A. Anand · M. Jain · S. Halder · M. Kamaruzzaman
The Neotia University, Sarisha, West Bengal, India

M. Waseem et al. (eds.), *Omics Approaches for Tomato Yield and Quality Trait Improvement*, https://doi.org/10.1007/978-981-96-3890-1_17

17.1 Introduction

Solanum lycopersicum of the family *Solanaceae*, often referred to as nightshades, comprises of 3000 species and plays an effective role in global agriculture due to the economic and nutritional value of its members (TanujaBuckseth 2023). Within the Solanaceae, tomatoes stand out with over 7000 varieties cultivated worldwide (Letstalk Science 2021). This vegetable crop is the most extensively grown garden produce and the second most consumed crop after potatoes. Tomato is indigenous to Mexico, Peru, and regions of Central, South, and Southern America (Bionity 2023). Its journey from these native lands to becoming a global agricultural staple is marked by its adaptability and the relentless efforts of breeders to enhance its yield and quality traits. Contemporary tomato breeding programs are increasingly harnessing the power of genomics to accelerate the development of better-quality tomato varieties. Genomics has opened up new possibilities for understanding and manipulating the genetic foundations of economically important traits of tomato, and adaptability to different environments. Through genomics, researchers can identify specific genes and genetic variations linked to these desirable traits, allowing for more targeted and effective breeding approaches.

The taxonomic journey of the tomato has been marked by a series of reclassifications and debates among botanists. However, in the mid-eighteenth century, botanist Philip Miller contested Linnaeus' taxonomy. Miller argued that tomatoes warranted their own distinct genus, Lycopersicon, due to differences he observed between tomatoes and other Solanum species. Recent advances in genetics and molecular biology research have forced taxonomists to reevaluate how tomatoes are classified. Detailed genetic studies revealed that tomatoes share closer evolutionary relationships with other Solanum species than previously understood. As a result, modern taxonomists have reclassified tomatoes back into the genus Solanum, reinforcing Linnaeus' original taxonomy.

17.2 Taxonomy and Botanical Description

While most tomatoes are cultivated as annual crops, there are also biennial and perennial varieties, as they are perennial herbaceous in nature. The plant has an indeterminate to determinate growth pattern and can reach a height of 3 meters. The tomato fruit is a straightforward fruit with fleshy seeds that has all the characteristics of berries. It has a diploid chromosome number of $2n = 2x = 24$. This chromosomal structure was first identified by Barton in 1950. A few popular varieties of tomato and their properties are presented in Table 17.1. The tomato genome is relatively small, comprising approximately 950 megabases (Mb), with about 30% consisting of repetitive sequences primarily present in thehetero-chromatin regions (Van der Hoeven et al. 2002). These repetitive elements significantly influence genome structure and function, affecting gene regulation and stability. High-resolution mapping and sequencing have identified numerous genes linked to important traits for example fruit size, flavor, color, disease resistance, and stress

Table 17.1 Different popular varieties of tomato in India and their properties

Varities	Potential yield (q or t /ha)	Properties
Arkavishesh	750–800 q/ha	Best variety for tomato processing.
ArkaApeksha	800–900 q/ha	Best variety for tomato processing.
ArkaRakshak	75–80 t/ha	High-yielding F1 hybrid, Suitable for fresh market and processing.
ArkaMeghali	18 t/ha	Open-pollinated variety bred for fresh market, Suitable for rainfed cultivation.
ArkaAbha	43 t/ha	Open-pollinated tomato variety for fresh market, resistant to bacterial wilt.

tolerance. This genetic information is crucial for marker-assisted breeding (MAB) and other modern breeding techniques.

17.3 Origin, Distribution, and Production Ratio of Tomato

Wild varieties of tomatoes were originally mostly yellow and small, resembling today's cherry tomatoes. These early forms originated in Mesoamerica around 500 years ago and are the ancestors of most modern cultivated varieties. The Spanish first brought tomatoes to Europe in 1523. From there, the tomato's popularity spread to Italy by 1544 and England by 1597. By 1650, tomatoes had reached the Philippines and Malaysia, and they finally made their way to North America in the eighteenth century (Liedl et al. 2013; Peralta and Spooner 2006). Today, tomatoes are cultivated globally, supported by extensive genetic resources. The global gene bank collection includes over 36,431 tomato accessions, ensuring the preservation and availability of diverse genetic material for breeding and research (Genesys-PGR 2023). This extensive repository plays a significant role in maintaining the wide genetic diversity essential for the ongoing advancement of tomato varieties. However, human intervention and selective breeding process, while enhancing specific traits, have also caused reduction in genetic diversity. If this trend continues unchecked, we risk losing the genetic diversity inherent in the original domesticated fruit (OECD 2017). Wild tomato species are a treasure trove of traits that can be utilized to enhance modern tomato varieties. Essential economic traits of crops have been improved by drawing on the genetic pool of wild relatives (Sanchez-Pena et al. 2004). The integration of genes from wild tomato species has led to significant progress in tomato breeding, such as heightened pathogen resistance, better tolerance to abiotic stresses like drought and salinity, and enhanced nutritional value. These genetic enhancements are vital for developing tomato varieties capable of thriving in a wide range of diverse environmental conditions.

The state-wise production (tones) for India has been presented according to the National Horticulture Board for the year 2021–2022 (Table 17.2). Worldwide production of tomato is highest in China which is around 60 M tons and around 20 M in tons in India according to FAOSTAT world's top 10 producers of tomato, as

Table 17.2 Status of tomato production by different states of India

State	Production (tons)	Share (%)
Madhya Pradesh	2970	14.63
Andhra Pradesh	2217	10.92
Karnataka	2077	10.23
Tamil Nadu	1489	7.34
Orissa	1432	7.06
Gujarat	1395	6.87
West Bengal	1284	6.33
Chhattisgarh	1149	5.66
Maharashtra	1125	5.54
Bihar	951	4.68

Source: National Horticulture Board
Indian Production of TOMATO (2021–22)

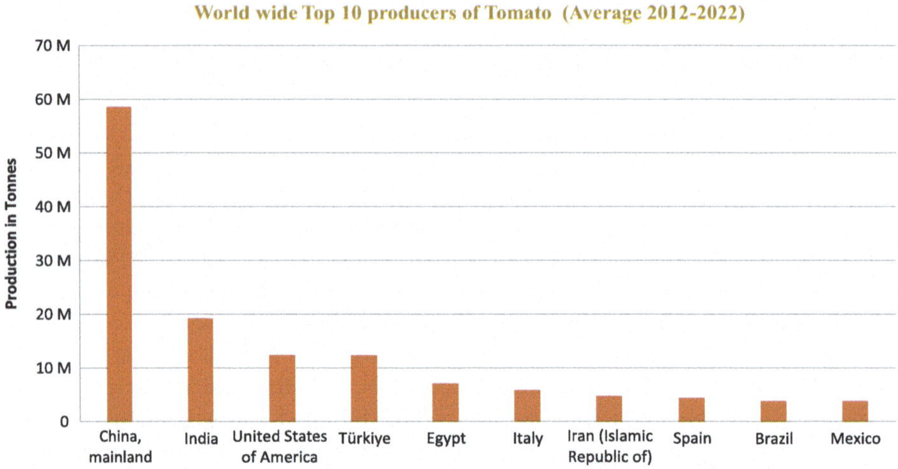

Fig. 17.1 Worldwide top 10 producers of tomato. (Source: FAOSTAT (27April, 2024))

shown in Fig. 17.1. Similarly, the world's maximum production of tomato was recorded in the year 2021 and lowest in 2012, and world's maximum area harvested of tomato was recorded in the year 2012 and the lowest in the year 2015 (Fig. 17.2). Similarly, the maximum production of tomato in India is recorded in the year 2021 and lowest in 2012, and the maximum area harvested in case of tomato in India is recorded in the year 2022 and lowest in the year 2012 (Fig. 17.3). Production share of tomato by region is shown in Fig. 17.4.

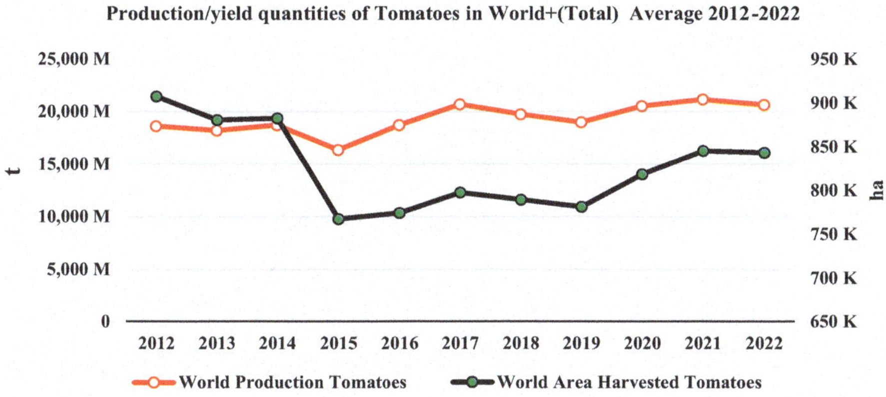

Fig. 17.2 Production/yield quantities of tomatoes in World+ (Total). (Source: FAOSTAT (27April, 2024))

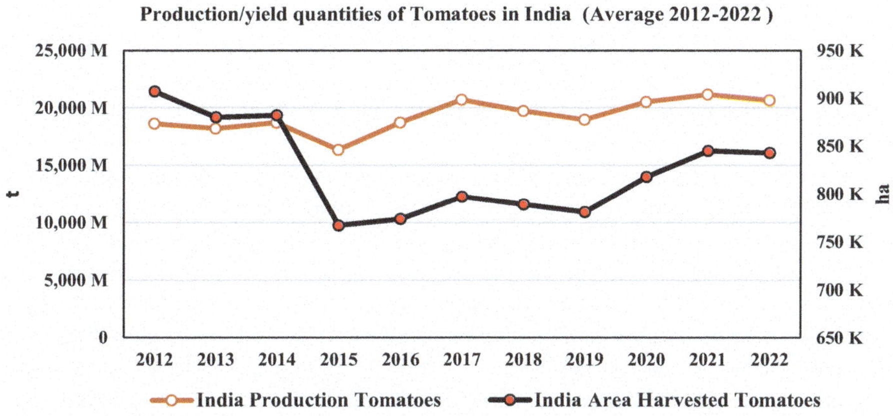

Fig. 17.3 Production/yield quantities of tomatoes in India. (Source: FAOSTAT (27 April, 2024))

17.4 Breeding Objectives and Efforts in Tomato

The primary goals of tomato breeding are to increase fruit quality for the fresh market as well as processing categories, along with yield (Tiwari et al. 2022). Through heterosis breeding yield increase is the key objective of almost all the breeding programs (Kumar et al. 2021). The best approach for high production, consistent fruit, and varietal protection is hybrid creation. A second breeding goal is resistance to pests and insects. The third one is stress tolerance such as salinity, heat, drought, and chilling injury is an additional target under climate change scenarios, and to achieve it, mutation breeding plays a major role. In conclusion, it is critical to enhance fruit quality for both processed and fresh tomatoes (Liedl et al. 2013). The main environmental factors that hinder plant growth and drastically reduce agricultural yield are salinity and drought. Important disease-resistant genes, particularly

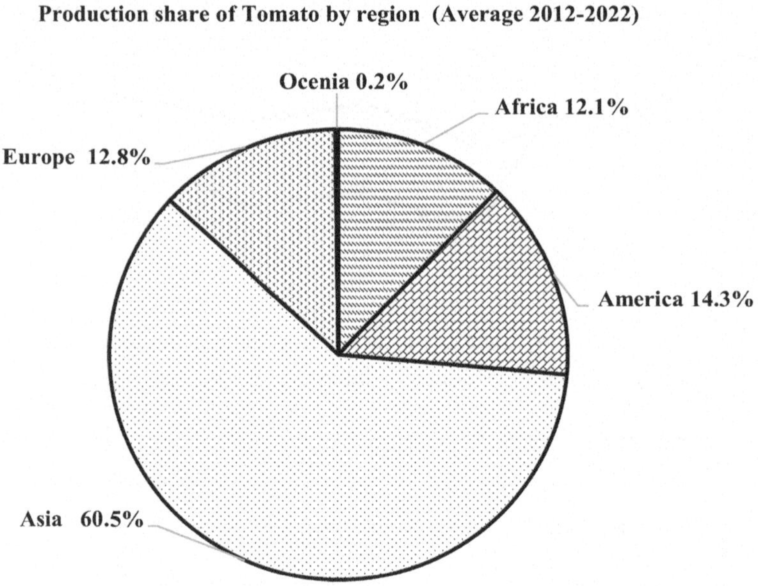

Fig. 17.4 Production share of tomato by region. (Source: FAOSTAT (27 April, 2024))

against bacterial wilt, leaf curl, and late blight, have been identified and introduced from wild species (Sanchez-Pena et al. 2004; Foolad 2006). Developing tomato varieties that can withstand abiotic stresses viz., heat, salinity, and drought is crucial for maintaining productivity in varying climatic conditions. Improving fruit quality traits, for example flavor, color, nutritional content, and shelf life, is a significant breeding objective. Additionally, enhancing productivity and yield, including fruit size and number, remains a key focus to fulfill the demands of global horticulture (Grandillo et al. 1999). Modern breeding objectives also align with sustainable agricultural practices. This includes developing tomato varieties that require fewer inputs, such as water and fertilizers, and are resistant to biotic stress, thereby reducing the need for chemical interventions (Foolad et al. 2006). During the 1950s, tomato breeding became more specialized to provide varieties for both the processing and fresh markets. Till the 1980s, most of the tomato breeding programs focused on morphology-based selection, which faced challenges like phenotype screening environments, need for large space and large population, environment interactions, genotype, and reduced trait response with low heritability and linkage drag. The entire genetic makeup of germplasm or breeding lines may not always be able to be fully characterized by the phenotypic selection process (Tanksley and McCouch 1997). With the introduction of molecular marker technology, there has been a significant increase in interest in genetic mapping and the identification of molecular markers for tomato breeding, particularly in the development of resistance to diseases and pests.

17.5 Tomato Breeding Through Marker-Assisted Selection

Most commercial tomato cultivars were created by conventional breeding techniques in combination with phenotypic selection (PS). Molecular markers and marker-assisted selection (MAS) technology have introduced an innovative approach to tomato genetics and breeding research. These technologies have been extensively utilized in the characterization and identification of genes and quantitative trait loci (QTLs) associated with various agriculturally significant traits in tomatoes, such as disease and insect resistance, as well as tolerance to abiotic stresses. Additionally, molecular markers have played a crucial role in genetic mapping. Notably, the tomato was among the first agricultural species for which breeding genetic maps and markers were developed (Tanksley et al. 1992). The use of molecular markers as an indirect selection criterion for tomato breeding was initially suggested by (Rick and Fobes 1974). Around 45 years ago, the first use of MAS in tomato breeding was initiated when the acid phosphatase (Aps-1) locus isozyme marker was employed as an indirect method to select for resistance toward nematode (Medina-Filho and Stevens 1980). Unlike PS, which is constrained by seasonal growth patterns, molecular markers are detectable at any growth stage, providing breeders with the flexibility to select plants based on their convenience. The advent of molecular markers dispersed throughout the tomato genome has led to the creation of numerous genetic maps, including a high-density linkage map derived from a hybrid of *S. lycopersicum and S. pennellii* (Pillen et al. 1996; Fulton et al. 2002). PCR-based markers, such as cleaved amplified polymorphic sequence, random amplified polymorphic DNA, simple sequence repeat, amplified fragment length polymorphism, sequence characterized amplified region, single nucleotide polymorphism, and insertion-deletion markers, are now extensively used in tomato genetic mapping and breeding (Table 17.3). However, a review of the literature

Table 17.3 Application of MAS in tomato breeding programs

Molecular marker	Trait	No. of makers	Reference
SNPs	Salt stress tolerance and powdery mildew resistance	5528	Viquez-Zamora et al. (2013)
	Heat tolerance	93	Cappetta et al. (2021)
	Agronomic trait	2,154,571	Yang et al. (2023)
AFLP	Drought tolerance	2	Orellana et al. (2010)
	Salt stress tolerance and powdery mildew resistance	31	Viquez-Zamora et al. (2013)
	Genotype	337	Yamamoto et al. (2016)
	Morphological trait	15	Ruggieri et al. (2018)
	Genotype	9200	Prasanna et al. (2023)
SSR	Genotype	31	Hernandez-Bautista et al. (2016)
	Genotype	61	Hernandez-Bautista et al. (2020)

reveals that many identified markers are neither polymorphic within tomato breeding populations nor consistently validated across different genotypes, limiting their utility in crop improvement (Foolad et al., 2006). MAS offers a quicker, more cost-effective, and more efficient alternative to PS for many basic disease-resistance traits in tomatoes. Over 35 disease-resistance genes have been identified and mapped in tomatoes (Van Ooijen et al. 2007). Nonetheless, varietal screening may still be necessary to confirm MAS results and validate the link between the markers and the traits of interest. Despite its benefits, MAS faces challenges in tomato breeding, particularly with complex traits governed by quantitative trait loci (QTLs) that are often inconsistent, population-specific, or weakly linked, making them less useful for breeding. Issues such as linkage drag and the recovery of desired horticultural traits have been problematic when transferring QTLs from wild species using MAS (Collard et al. 2005). Additionally, MAS has seen limited application in improving abiotic stress resistance in tomatoes despite considerable efforts to identify and map QTLs related to these characteristics. To maximize the potential of MAS, it is crucial to address these challenges by enhancing the validation and application of molecular markers across various genotypes and breeding populations. Furthermore, integrating MAS with other advanced genomic techniques could significantly expedite the development of superior tomato genotypes.

17.6 MAS for Resistance to Fungal Diseases in Tomato

Fusarium oxysporum f. sp. *lycopersici* (FOL) is the causal organism for fusarium wilt which is a significant threat to tomato crops. The gene I-2 provides resistance to FOL race 2, has been extensively studied, and is commonly used in tomato breeding programs (Sim et al. 2012). Gray leaf spot, although a relatively minor disease in most tomato-growing regions, is caused by four different species of *Stemphylium*. Resistance to gray leaf spot is mainly conferred by an incomplete-dominant gene known as Sm, which has been successfully transferred into cultivated tomato varieties, enhancing resistance to this disease (Bashi et al. 1973). Fusarium crown and root rot (FCRR), caused by *F. oxysporum f.* sp. *radicis-lycopersici*, is a significant soil-borne disease. The Frl gene, which confers resistance to FCRR, has been mapped to tomato chromosome 9, near the Tm-22 gene. This resistance was identified in an induced mutant of *Solanum peruvianum* (Vakalounakis et al. 1997). Integrating the Frl gene into cultivated tomato varieties is a valuable strategy for improving resistance to FCRR, thereby enhancing yield and quality traits in tomato production. Resistance genes such as Cf-2, Cf-4, Cf-5, and Cf-9 have been identified and introgressed into cultivated tomato varieties, providing effective disease management options against tomato leaf mold (causal organism: *Cladosporium fulvum*) (Huang et al. 2000). Early blight of tomato mostly caused by the fungi both *Alternaria tomatophila* and *Alternaria solani* is among the most widespread tomato diseases. Researchers have identified key QTLs on chromosomes 5 and 9 that confer resistance to early blight, offering valuable targets for breeding programs (Zhang et al. 2003). The genetic basis of resistance toward bacterial wilt in tomatoes has been extensively studied, and numerous sources of resistance have been identified (Scott et al. 2005). Several QTLs linked to bacterial wilt resistance have been mapped, facilitating the development of resistant cultivars through MAS. For example, QTLs on chromosomes 6 and 12 have been shown to provide significant resistance to bacterial wilt (Carmeille et al. 2006). Tomato mosaic virus (ToMV), a highly stable plant virus that can be mechanically transmitted, poses a serious threat to tomato production, particularly in greenhouse settings (Sobir et al. 2000). Molecular markers linked to the genes (Tm-1, Tm-2, and Tm-22) enable breeders to quickly and accurately select for ToMV resistance, thereby accelerating the development of resistant cultivars (Ling et al. 2013). Tomato yellow leaf curl virus (TYLCV), a monopartite geminivirus spread by whiteflies, is a major threat to tomato crops in tropical and subtropical regions. MAS, using molecular markers linked to the resistance genes (Ty-1, Ty-2, and Ty-3), facilitates the rapid and precise selection of TYLCV-resistant plants, expediting the breeding of resistant cultivars while maintaining other desirable agronomic traits.

17.7 Genomic Selection (GS)

In order to advance genetically complicated features in a variety of important agricultural crops, including tomatoes, genomic selection, an advanced breeding strategy, has shown tremendous promise (Cappetta et al. 2020). According to Meuwissen et al. (2001), this method makes use of genome-wide markers to forecast a line's or population's genomic estimated breeding value (GEBV), which enables more precise selection and may greatly increase genetic gain. It has been shown that in tomato breeding, GS accelerates the proportion of genetic gain in elite line selection and other breeding methods, such as backcross. Predicting a candidate genotype's breeding value for selection is made possible by GS, allowing for the collection of small genetic influences through genome-wide markers. Despite its potential, GS in tomato breeding also faces challenges, such as genomic constraints and the need for precise phenotyping. However, with the integration of computer science and the development of novel selection tactics, GS schemes are becoming more defined, paving the way for a promising breeding framework to optimize tomato improvement methods in comparison to traditional breeding methods. GS represents a novel and auspicious approach in tomato breeding, offering significant advantages in respect to accuracy, efficiency, and genetic gain. As GS schemes continue to evolve, they hold the prospective to revolutionize tomato breeding and add to the development of superior tomato varieties. Plant breeding strategies for characters with polygenic inheritance can become more effective because of genomic selection (GS). The potential breeding value of an individual is ascertained using extensive genetic data, such as SNPs. It is now possible to scan complete genomes due to the availability of thousands of markers produced by recent high-throughput genotyping (HTG) techniques. A candidate for selection's phenotypic value can be predicted using the allelic-associated marker loci with phenotypes. A genomic estimated breeding value, or GEBV, is calculated for each breeding candidate and expressed as a unique function of marker effects. To generate the GEBVs of a testing set (TST), which is genotyped but not phenotyped, GS combines both genotypic and phenotypic data from a training population (TRN) in a training set (TRS). The GS model will be utilized to forecast the breeding values of non-phenotyped individuals in the subsequent selection step. The first work on GS simulation in plants was published by Meuwissen and colleagues (Meuwissen et al. 2001). The scientists compared the best linear unbiased prediction, Bayesian prediction, and linear regression methodologies for figuring out the association between estimated and true breeding values so as to develop suitable GS models in crops.

17.8 Importance and Scope of GS in Tomato

GS is a great tool in plant breeding, including for horticultural crops like tomatoes. It involves using genomic information for the prediction of the breeding value of crops and making selections early in the breeding cycle, thereby speeding up the development of new cultivars with desirable traits. The first step in genomic selection is to collect high-quality genomic data. This involves sequencing or genotyping a population of tomato plants to identify single nucleotide polymorphisms (SNPs) or other genetic markers across the genome. A training population is phenotyped for desirable attributes (e.g., disease resistance, yield, fruit quality) based on a representative sample of the breeding population. After this, a prediction model is developed using the genotypes and phenotypes. GS employs statistical models to predict an individual's genetic value based on their genetics. Bayesian techniques and genomic best linear unbiased prediction (GBLUP) are examples of common models. These models relate the traits that are seen to the genetic markers. Once the prediction model is created, it can be used to forecast, only from genotypes, the breeding values of newly acquired or untested plants. This allows breeders to select the best candidates for crossing without the need for extensive phenotyping, which is laborious and costly. By predicting the performance of tomato plants early in their development, breeders can focus their efforts on the most promising lines, reducing the time and resources needed for field trials. The breeding cycle can be accelerated by genomic selection, leading to the quicker production of tomato varieties with enhanced qualities like greater yield, better flavor, resistance to disease, and stress tolerance. Breeders can increase the genetic gains made during each cycle of selection by utilizing genomic information, which enhances the overall efficacy of breeding programs. Many genes frequently regulate traits like yield and disease resistance. Multiple features can be selected at the same time through genomic selection, which produces superior cultivars with a well-balanced set of desired attributes.

17.9 Implementation of GS Schema in Tomato

According to Contaldi et al. (2021), a meticulous assessment of critical elements is necessary for the determination of the ideal settings for a GS experiment in a tomato. According to Robertsen et al. (2019), the accuracy of genomic estimated breeding values (GEBVs) in the plant selection process depends on the precision of phenotyping and genotyping methods. Key factors influencing this accuracy include the size of the training population (TRN size), marker density, marker technology, genome structure, and the extent of marker linkage disequilibrium. The accuracy of measures pertaining to phenotyped traits is a critical component of modern breeding strategies that rely on genomic technologies (Esposito et al. 2020).

Araus and Cairns (2014) highlight that scalable digital technologies can accelerate the selection process, reduce the reliance on manual data annotation, and enhance the precision of phenotyping. To improve prediction accuracy, it is essential to develop and continually update a tomato trait reference network (TRN) for each selection cycle, as both genetic diversity and gene frequencies may change within the segregating population (Liabeuf et al. 2018). To capture a broad range of informative loci, a sufficient number of markers must be employed (Zhang et al. 2019). High-density marker panels can be efficiently created through genotyping-by-sequencing (GBS). Alternatively, conserved SNPs associated with candidate mutations can be identified directly using the cDNA-based GBS approach, such as restriction site-associated RNA sequencing (RAR-seq) (Alabady et al. 2015). The recently developed single primer enrichment technology (SPET) offers a tailored approach for tomato genotyping, enhancing panel design and increasing multiplexing capabilities (Barchi et al. 2019). Previous genotyping data can be leveraged to refine marker sets for subsequent stages of breeding. For instance, Liabeuf et al. (2018) improved upon the original "SolCAP array" of 7700 SNPs (Sim et al. 2012) to better assess populations with low recombination. Additionally, Zhang et al. (2019) suggest that the minor allele frequency (MAF) threshold can influence prediction accuracy. It is also crucial to develop methods for integrating validated genomic signatures into tomato breeding programs effectively. Selection signatures, resulting from human-driven changes in genomic regions, impact agronomic traits of interest (Smykal et al. 2018). Identifying these signatures is key to understanding population history and the genetic mechanisms driving phenotypic differentiation among cultivars (Gao et al. 2019).

17.10 Factors Affecting Genomic Prediction

Generally, higher marker densities enhance prediction accuracy across most genomic selection (GS) models, including BLUP, LASSO, machine learning (ML), and deep learning approaches. However, for methods requiring Markov chain Monte Carlo (MCMC) iteration, such as Bayesian models, convergence can sometimes be slow, impacting performance (Norman et al. 2018; Zhang et al. 2017; Zhang et al. 2019). In cases where there is strong linkage disequilibrium (LD) among markers, a lower density of markers—ranging from a few hundred to a few thousand—can still achieve high prediction accuracy in breeding populations. However, this effectiveness can depend on the trait architecture and heritability, making it trait-specific (Werner et al. 2018). Additionally, maintaining an extremely high marker density can be financially burdensome. Effective genomic selection should balance marker density with cost considerations, suggesting a reasonable marker density of at least 2000 SNPs to ensure robust prediction accuracy without excessive costs (Abed et al. 2018). Even with a reduced density, increasing multiplexing levels can significantly reduce genotyping costs while maintaining effective prediction accuracy.

17.11 Challenges of GS Application in Tomato Breeding Program

The effective application of genomic selection (GS) in tomato breeding necessitates careful optimization of various agricultural practices, including field trial management, sample collection, seed production, sequencing, and phenotyping (Bassi et al. 2016). Effective management of a GS-assisted breeding program involves several considerations, such as the level of population inbreeding, the number of plants to be analyzed, and marker metrics. Genotyping for genomic estimated breeding values (GEBVs) typically requires about three months, a significant improvement over traditional breeding methods, which often take considerably longer (Rahim et al. 2020). The main advantage of GS is its ability to accelerate breeding cycles, facilitating faster selection of improved varieties. In GS-assisted breeding for tomato F1 hybrid improvement, the selection of parental lines is based on their estimated breeding value—the projected average performance of their offspring—which requires accurate estimation. Yamamoto et al. (2017) verified the effectiveness of GS models developed with 96 large-fruited F1 tomato varieties, validating these models by segregating populations from crosses. By leveraging progeny phenotypic and genotypic data for traits like total fruit weight and soluble solids content, these models successfully predicted effective parental pairings for creating superior hybrids. However, prediction accuracy varied with characteristics and parental pairings. In small populations with limited genetic variability, the benefits of GS are reduced, as the need to fix advantageous alleles can lead to increased inbreeding (Cappetta et al. 2020).

17.12 Artificial Intelligence (AI) and ML in Tomato Genomics

In this section, we have discussed about the pivotal role of AI and ML in the genomics approach for the advancement of quantitative and qualitative characters in tomato (Jubair and Domaratzki 2023). AI and ML are instrumental in advancing the field by leveraging computational power to analyze and interpret complex genomic data for trait improvement. AI and ML algorithms enhance data integration from diverse omics layers, allowing for a holistic understanding of the genetic basis of characters. Predictive modeling techniques, such as Random Forests and Gradient Boosting, are employed to forecast quantitative traits like yield and disease resistance, while deep learning methods process high-resolution phenotypic images to assess qualitative traits such as fruit color and shape. Furthermore, AI-driven tools streamline Genome wide association studies (GWAS) and GS by identifying significant genetic markers more accurately and efficiently, while machine learning models refine these findings by evaluating gene-environment interactions. AI and ML also facilitate the design of optimized breeding programs and precision breeding strategies, accelerating the development of tomato varieties with improved quality traits. The integration of AI and ML into genomics offers a transformative approach to identifying genetic variations, predicting trait outcomes, and improving tomato breeding processes, ultimately leading to more robust and high-quality tomato varieties

17.13 Conclusion

The integration of genomics into tomato breeding represents a revolutionary advancement in agricultural science, fundamentally transforming how researchers and breeders approach the enhancement of yield, quality, and resistance traits. The utilization of advanced genomic tools and techniques, including MAS and GS, has enabled substantial progress in developing tomato varieties with superior attributes. Traditional tomato breeding methods, which relied heavily on phenotypic selection, have now evolved into a more sophisticated paradigm that utilizes molecular markers and genomic insights. This shift has led to significant breakthroughs in overcoming major production challenges such as developing resistance genes against stresses such as gray leaf spot, fusarium wilt and tomato yellow leaf curl virus (TYLCV) demonstrates the powerful impact of genomics in addressing critical issues that affect tomato production. The advent of genomic selection further enhances breeding efficiency by accelerating genetic gains and allowing for more precise improvements in complex traits. This approach has the latent to expedite the production of tomato cultivars with desirable traits, such as increased enhanced fruit quality, yield, and improved resilience to environmental stresses. As technology continues to advance, the future of tomato breeding is poised for even greater innovations. The integration of genomics with emerging technologies, such as genome editing and multi-omics approaches, promises to unlock new opportunities for creating tomato varieties that meet the evolving demands of global agriculture. AI and ML significantly advance the genomics approach for enhancing tomato traits by providing powerful tools for precise data analysis, predictive modeling, and trait optimization. These technologies streamline the identification of genetic variations and improve breeding strategies, leading to the development of tomato varieties with superior quantitative and qualitative attributes. These advancements will not only improve the productivity and sustainability of tomato cultivation but also ensure the continued relevance and contribution of tomato crops to food security worldwide. In summary, the genomic approach has significantly advanced the field of tomato breeding, providing powerful tools for the enhancement of yield, quality, and resilience. The ongoing evolution of genomic technologies will drive further progress in developing superior tomato varieties, reaffirming their crucial role in global agriculture and food systems.

References

Abed A, Perez-Rodriguez P, Crossa J, Belzile F (2018) When less can be better: how can we make genomic selection more cost-effective and accurate in barley? Theor Appl Genet 131:1873–1890. https://doi.org/10.1007/S00122-018-3120-8

Alabady MS, Rogers WL, Malmberg RL (2015) Development of transcriptomic markers for population analysis using restriction site associated RNA sequencing (RARseq). PLoS One 10:48–55

Araus JL, Cairns JE (2014) Field high-throughput phenotyping the new crop breeding frontier. Trends Plant Sci pp 19:52–61

Barchi L, Acquadro A, Alonso D, Aprea G, Bassolino L, Demurtas O et al (2019) Single Primer Enrichment Technology (SPET) for high-throughput genotyping in tomato and eggplant germplasm. Front Plant Sci 10:1005

Bashi E, Pilowsky M, Rotem J (1973) Resistance in tomatoes to stem phyliumfloridanum and S. botryosum f. sp. Lycopersici. Phytopathology 63:1542–1544

Bassi FM, Bentley AR, Charmet G, Ortiz R, Crossa J (2016) Breeding schemes for the implementation of genomic selection in wheat (*Triticum Spp.*). Plant Sci 242:23–36

Bionity (2023) Tomato. Bionitycom; https://www.bionity.com/en/encyclopedia/Tomato.html

Budhlakoti N, Mishra DC, Rai A (2019) Package 'MTGS'. pp 1–6

Cappetta E, Andolfo G, Di Matteo A, Barone A, Frusciante L, Ercolano MR (2020) Accelerating Tomato breeding by exploiting genomic selection approaches. Plan Theory 9:1236

Cappetta E, Andolfo G, Guadagno A, Di Matteo A, Ercolano MR et al (2021) Tomato genomic prediction for good performance under high-temperature and identification of loci involved in thermotolerance response. Hortic Res 8:212

Carmeille A, Caranta C, Dintinger J, Prior P, Luisetti J, Besse P (2006) Identification of QTLs for Ralstoniasolanacearum race 3-phylotype II resistance in tomato. Theor Appl Genet 113:110–121

Collard BCY, Jahufer MZZ, Brouwer JB, Pang ECK (2005) An introduction to markers, quantitative trait loci (QTL) mapping and marker assisted selection for crop improvement: the basic concepts. Euphytica 142:169–196

Contaldi F, Cappetta E, Esposito S (2021) Practical workflow from high throughput Genotyping to Genomic Estimated Breeding Values (GEBVs). In: Crop breeding methods and protocols. Springer, Berlin/Heidelberg

Esposito S, Carputo D, Cardi T, Tripodi P (2020) Applications and trends of machine learning in genomics and phenomics for next-generation breeding. Plants 9:34

Foolad MR, Merk HL, Ashrafi H, Kinkade MP (2006) Identification of new sources of late blight resistance in tomato and mapping of a new resistance gene. In 22 nd Annual Tomato Disease Workshop. p 4

Fulton TM, Van der Hoeven R, Eannetta NT, Tanksley SD (2002) Identification, analysis and utilization of conserved ortholog set markers for comparative genomics in higher plants. Plant Cell 14:1457–1467

Gao L, Gonda I, Sun H, Ma Q, Bao K, Tieman DM, Burzynski-Chang EA et al (2019) The tomato pan-genome uncovers new genes and a rare allele regulating fruit flavor. Nat Genet 51:1044–1051

Genesys, P. G. R. (2023). Gateway to genetic resources. 2023

Grandillo S, Ku HM, Tanksley SD (1999) Identifying the loci responsible for natural variation in fruit size and shape in tomato. Theor Appl Genet 99:978–987

Hernandez-Bautista A, Lobato-Ortiz R, Garcia-Zavala JJ, Parra-Gomez MA et al (2016) Implications of genomic selection for obtaining F2: 3 families of tomato. Sci Hortic 207:7–13

Hernandez-Bautista A, Lobato-Ortiz R, García-Zavala JJ, Cruz-Izquierdo S, Chavez-Servia JL et al (2020) Prediction accuracy of genomic selection models for earliness in tomato. Chil J Agric Res 80:505–514

Huang CC, Cui YY, Weng CR, Zabel P, Lindhout P (2000) Development of diagnostic PCR markers closely linked to the tomato powdery mildew resistance gene Ol-1 on chromosome 6 of tomato. Theor Appl Genet 101:918–924

Jubair S, Domaratzki M (2023) Crop genomic selection with deep learning and environmental data: a survey. Front Artif Intell 5:1040295. https://doi.org/10.3389/frai.2022.1040295

Kumar K, Sharma D, Singh J, Chandel N (2021) Heterosis for yield and yield attributing traits in tomato (*Solanumlycopersicum* L.). Vegetable Sci 48:41–48

Letstalkscience (2021) Tomato taxonomy. https://letstalkscience.ca/educational-resources/backgrounders/tomato-taxonomy

Liabeuf D, Sim SC, Francis DM (2018) Comparison of marker-based genomic estimated breeding values and phenotypic evaluation for selection of bacterial spot resistance in tomato. Phytopathology 108:392–401

Liedl BE, Labate JA, Stommel JR, Slade A, Kole C (2013) Genetics, genomics and breeding of tomato. CRC Press, New York, p 480

Ling KS, Li R, Panthee DR, Gardner RG (2013) First report of Potato spindle tuber viroid naturally infecting greenhouse tomatoes in North Carolina. Plant Dis 97:148–148

Medina-Filho H, Stevens M (1980) Tomato breeding for nematode resistance: survey of resistant varieties for horticultural characterisitcs and genotype of acid phosphatase. Acta Hort 100:383–391

Meuwissen TH, Hayes BJ, Goddard ME (2001) Prediction of total genetic value using genome-wide dense marker maps. Genetics 157:1819–1829

Norman A, Taylor J, Edwards J, Kuchel H (2018) Optimizing genomic selection in wheat: effect of marker density, population size and population structure on prediction accuracy. G3 Genes|Genomes|Genetics 8:2889–2899. https://doi.org/10.1534/G3.118.200311

OECD (2017) Tomato (*Solanumlycopersicum*). In: Safety assessment of transgenic organisms in the environment, volume 7: OECD consensus documents, harmonisation of regulatory oversight in biotechnology. OECD Publishing, Paris. https://doi.org/10.1787/9789264279728-en

Orellana S, Yanez M, Espinoza A, Verdugo I, Gonzalez E et al (2010) The transcription factor SlAREB1 confers drought, salt stress tolerance and regulates biotic and abiotic stress-related genes in tomato. Plant Cell Environ 33(12):2191–2208

Peralta IE, Spooner DM (2006) Morphological characterization and relationships of wild tomatoes (Solanum L. Section Lycopersicon). A Festschrift for William G. D'Arcy. T. B. Croat, V. C. Hollowell and R. C. Keating. Missouri Bot Gard Press 104:227–257

Pillen K, Pineda O, Lewis C, Tanksley SD (1996) Status of genome mapping tools in the taxon Solanaceae. In: Paterson A (ed) Genome mapping in plants. R.G. Landes, Austin, pp 281–308

Prasanna HC, Rai N, Hussain Z, Yerasu SR, Tiwari JK (2023) Tomato: breeding and genomics. Vegetable Sci 50:146–155

Rahim MS, Bhandawat A, Rana N, Sharma H, Parveen A, Kumar P, Madhawan A et al (2020) Genomic selection in cereal crops: methods and applications. In: Gosal SS, Wani SH (eds) Accelerated plant breeding: cereal crops. Springer, Berlin/Heidelberg

Rick C, Fobes J (1974) Association of an allozyme with nematode resistance. Tomato Genet Coop Rep 24:25

Robertsen CD, Hjotrtshøj RL, Janss LL (2019) Genomic selection in cereal breeding. Agronomy 9:95

Ruggieri V et al (2018) An improved assembly and annotation of the melon (*Cucumismelo* L.) reference genome. Sci Rep 8:8088. https://doi.org/10.1038/s41598-018-26416-2

Sanchez-Pena P et al (2004) Sources of resistance to whitefly (Bemisia ssp.) in wild populations of (*S. lycopersicumvar. cerasiformeDunal*) Spooner GJ, Anderson RK, and Jansen in Northwestern Mexico. Genet Resour Crop Evol 53:711–719

Scott JW, Wang JF, Hanson PM (2005) Breeding tomatoes for resistance to bacterial wilt, a global view. Proc 1st Intl Symp Tomato Dis 695:161–172

Sim SC et al (2012) High-density SNP genotyping of tomato (*Solanumlycopersicum* L.) Reveals patterns of genetic variation due to breeding. PLoS One 7(9):e45520

Smykal P, Nelson MN, Berger JD, von Wettberg EJB (2018) The impact of genetic changes during crop domestication. Agronomy 8:119

Sobir OT, Murata M, Motoyoshi F (2000) Molecular characterization of the SCAR markers tightly linked to the Tm-2 locus of the genus Lycopersicon. Theor Appl Genet 101:64–69

Tanksley SD, McCouch SR (1997) Seed banks and molecular maps: unlocking genetic potential from the wild. Science 277:1063–1066

Tanksley SD, Ganal MW, Prince JP, Devicente MC, Bonierbale MW, Broun P et al (1992) High-density molecular linkage maps of the tomato and potato genomes. Genetics 132:1141–1160

TanujaBuckseth HM (2023) Improvement in tomato (*Solanumlycopersicum* L.) Presentation transcript

Tiwari JK, Yerasu SR, Rai N et al (2022) Progress in marker-assisted selection to genomics-assisted breeding in tomato. Crit Rev Plant Sci 41:321–350

Vakalounakis DJ, Laterrot H, Moretti A, Ligoxigakis EK, Smardas K (1997) Linkage between Frl (Fusariumoxysporum f. sp.radicis-lycopersici resistance) and Tm-2 (tobacco mosaic virus resistance-2) loci in tomato (Lycopersiconesculentum). Ann Appl Biol 1:319

Van der Hoeven R et al (2002) Deductions about the number, organization, and evolution of genes in the tomato genome based on analysis of a large expressed sequence tag collection and selective genomic sequencing. Plant Cell 14(7):1441–1456

Van Ooijen G, van den Burg HA, Cornelissen BJ, Takken FL (2007) Structure and function of resistance proteins in solanaceous plants. Annu Rev Phytopathol pp 45:43–72

Viquez-Zamora M, Vosman B, van de Geest H, Bovy A, Visser RG et al (2013) Tomato breeding in the genomics era: insights from a SNP array. BMC Genomics 14:1–13

Werner CR, Voss-Fels KP, Miller CN, Qian W, Hua W, Guan CY et al (2018) Effective genomic selection in a narrow-gene pool crop with low density markers: Asian rapeseed as an example. Plant Genome 11:170084. https://doi.org/10.3835/plantgenome2017.09.0084

Yamamoto E, Matsunaga A, Onogi A, Kajiya-Kanegae H, Minamikawa M et al (2016) A simulation-based breeding design that uses whole-genome prediction in tomato. Sci Rep 6:19454

Yamamoto E, Matsunaga H, Onogi A, Ohyama A, Miyatake K et al (2017) Efficiency of genomic selection for breeding population design and phenotype prediction in tomato. Heredity 118:202–209

Yang J, Liu Y, Liang B, Yang Q, Li X et al (2023) Genomic basis of selective breeding from the closest wild relative of large-fruited tomato. Hortic Res 10:142

Zhang LP, Lin GY, Nino-Liu D, Foolad MR (2003) Mapping QTLs conferring early blight (Alternariasolani) resistance in a Lycopersiconesculentum x L. hirsutum cross by selective genotyping. Mol Breed 12:3–19

Zhang X, Perez-Rodríguez P, Burgueno J, Olsen M, Buckler E et al (2017) Rapid cycling genomic selection in a multiparental tropical maize population. G3: Genes Genomes Genet 7:2315–2326. https://doi.org/10.1534/G3.117.043141

Zhang H, Yin L, Wang M, Yuan X, Liu X (2019) Factors affecting the accuracy of genomic selection for agricultural economic traits in maize, cattle, and pig populations. Front Genet 10:189. https://doi.org/10.3389/FGENE.2019.00189/BIBTEX

The manufacturer's authorised representative in the EU is Springer Nature Customer Service Centre GmbH, Europaplatz 3, 69115 Heidelberg, Germany. If you have any concerns regarding our products, please contact ProductSafety@springernature.com

Printed and bound by CPI Group (UK) Ltd, Croydon, CR0 4YY
01/10/2026
02225928-0015